Catalytic Asymmetric Reactions of Conjugated Nitroalkenes

Catalytic Asymmetric Reactions of Conjugated Nitroalkenes

Irishi N. N. Namboothiri, Meeta Bhati, Madhu Ganesh,
Basavaprabhu Hosamani, Thekke V. Baiju, Shimi Manchery,
and Kalisankar Bera

CRC Press
Taylor & Francis Group
Boca Raton London New York

CRC Press is an imprint of the
Taylor & Francis Group, an **informa** business

First edition published 2020
by CRC Press
6000 Broken Sound Parkway NW, Suite 300, Boca Raton, FL 33487-2742

and by CRC Press
2 Park Square, Milton Park, Abingdon, Oxon, OX14 4RN

ISBN: 978-0-367-43382-6 (hbk)
ISBN: 978-0-367-53562-9 (pbk)
ISBN: 978-1-003-00304-5 (ebk)

Typeset in Times
by codeMantra

Dedicated to Professor Alfred Hassner on the occasion of his 90th birthday

Contents

List of Abbreviations xi
Foreword xiii
Preface xv
Authors xvii

Introduction 1

Chapter 1 Catalytic Asymmetric Michael Addition of 1,3-Dicarbonyls to Nitroalkenes 7

1.1 Introduction 7
1.2 Addition of 1,3-Diketones 7
1.3 Addition of Malonates 14
1.4 Addition of β-Ketoesters 26
1.5 Addition of 2-Hydroxynaphthoquinones 33
1.6 Addition of Hydroxycoumarin 35
1.7 Addition of Meldrum's Acid 37
1.8 Conclusions 39
References 39

Chapter 2 Catalytic Asymmetric Michael Addition of Aldehydes to Nitroalkenes 43

2.1 Introduction 43
2.2 Addition of Aldehyde Enolates 43
2.3 Addition of Umpolung of Aldehydes: The Stetter Reaction 71
2.4 Conclusions 73
References 73

Chapter 3 Catalytic Asymmetric Michael Addition of Ketones to Nitroalkenes 77

3.1 Introduction 77
3.2 Addition of Aliphatic and Aromatic Ketones 77
3.3 Addition of α-Ketoesters, α-Ketoamides and α-Ketophosphonates 107
3.4 Addition of β-Ketosulfones 111
3.5 Conclusions 112
References 112

Chapter 4 Catalytic Asymmetric Michael Addition of Miscellaneous Carbonyl Compounds to Nitroalkenes 115

4.1 Introduction 115
4.2 Addition of Esters/Lactones: Iminoesters, Furanones, Oxazolones and Isoxazolones 115
4.3 Addition of Amides/Lactams: Oxindoles, Pyrazolones and Pyrrolidones ... 122
4.4 Conclusions 134
References 134

Chapter 5 Catalytic Asymmetric Friedel–Crafts Reactions of Nitroalkenes 137

5.1 Introduction 137
5.2 Friedel–Crafts Reaction 137
5.3 Conclusions 158
References 158

Chapter 6 Catalytic Asymmetric Michael Addition of Miscellaneous Carbon-centered Nucleophiles to Nitroalkenes 161

6.1 Introduction 161
6.2 Addition of Nitroalkanes 161
6.3 Addition of Cyanides 166
6.4 Addition of Boronic Acids 168
6.5 Addition of Malononitrile 172
6.6 Addition of Organo-Zinc Reagents 174
6.7 Conclusions 180
References 180

Chapter 7 Catalytic Asymmetric Michael Addition of Heteroatom-centered Nucleophiles to Nitroalkenes 183

7.1 Introduction 183
7.2 Addition of Phosphorus-centered Nucleophiles 183
7.3 Addition of Sulphur-centered Nucleophiles 188
7.4 Addition of Nitrogen-centered Nucleophiles 192
7.5 Addition of Oxygen-centered Nucleophiles 195
7.6 Addition of Arsenic-centered Nucleophiles 197
7.7 Conclusions 198
References 198

Chapter 8 Catalytic Asymmetric Cycloadditions of Nitroalkenes 201

8.1 Introduction 201
8.2 [3+2] Cycloadditions 201
8.3 [4+2] Cycloadditions 215
8.4 Other Cycloadditions 221
8.5 Conclusions 223
References 223

Chapter 9 Catalytic Asymmetric Reduction of Nitroalkenes 227

9.1 Introduction 227
9.2 Metal Catalyzed Reduction 227
9.3 Organocatalyzed Reduction 232
9.4 Bioreduction 233
9.5 Conclusions 235
References 235

Chapter 10 Catalytic Asymmetric Synthesis of Cycloalkanes via Cascade Reactions of Nitroalkenes 237

10.1 Introduction 237
10.2 Synthesis of Cycloalkanes 237
10.3 Conclusions 253
References 254

Chapter 11 Catalytic Asymmetric Synthesis of Aryl-Fused Heterocycles via Cascade Reactions of Nitroalkenes 257

11.1 Introduction 257
11.2 Synthesis of Aryl-Fused Heterocycles 257
11.3 Conclusions 273
References 273

Chapter 12 Catalytic Asymmetric Synthesis of Five- and Six-Membered Heterocycles via Cascade Reactions of Nitroalkenes 275

12.1 Introduction 275
12.2 Synthesis of Five- and Six-Membered Nitrogen Heterocycles 275
12.3 Synthesis of Five- and Six-Membered Oxygen Heterocycles 278
12.4 Synthesis of Spiro-Oxindoles 287
12.5 Conclusions 293
References 293

Index 295

List of Abbreviations

BINAP	2,2′-Bis(diphenylphosphino)-1,1′-binaphthyl
BINOL	1,1′-Binaphthalene-2,2′-diol
BOPA	Bis(oxazolinylphenyl)amide
BOX	Bis-oxazoline
Cp/Rh(R)-prophos	Pentamethylcyclopentadienyl rhodium (II)-1,2-bis(diphenylphosphino)propane
$Cu(CH_3CN)_4PF_6$	Tetrakis(acetonitrile)copper(I)hexafluorophosphate
$Cu(OTf)_2.C_6H_6$	Benzene complex of copper(II) trifluoromethanesulfonate
Cu/Eu/Cu	Trimetallic complex of copper/europium/copper
Cu/Mn	Copper-manganese heterobimetals
CuOtBu	Copper(I)*tert*-butoxide
CPME	Cyclopentyl methyl ether
DABCO	1,4-Diazabicyclo[2,2,2]octane
DBAD	Di-*tert*-butyl azodicarboxylate
DBU	1,8-Diazabicyclo[5,4,0]undec-7-ene
DDQ	2,3-Dichloro-5,6-dicyano-1,4-benzoquinone
DFT	Density functional theory
DHQA	Dihydroquinine
[DHQD]2PYR	Hydroquinidine-2,5-diphenyl-4,6-pyrimidinediyl diether
DMA	*N,N*-Dimethyl acetamide
DMAP	4-*N,N*-Dimethylaminopyridine
DMF	Dimethylformamide
DMM	Diaminomethylenemalononitrile
DMSO	Dimethylsulfoxide
Duanphos	(1R,1′R,2S,2′S)-(+)-2,2′-Di-t-butyl-2,3,2′,3′-tetrahydro-1,1′-bi-1H-isophosphindole
DIPEA	Diisopropylethylamine
ESI-MS	Electrospray ionization mass spectroscopy
Fe_3O_4/PVP@SiO_2/ProTMS	(*S*)-Diphenylprolinol trimethylsilyl ether embedded polyvinyl-pyrrolidone (PVP)-modified ferric oxide magnetic nanoparticle
Fe_3O_4@SiO_2	Ferric oxide adsorbed on silica support
FOXAP	2,2″-Bis(diphenylphosphino)-1,1″-biferrocene
GABA	γ-aminobutyric acid
GC	Gas chromatography
HOMO	Highest occupied molecular orbital
HPLC	High performance liquid chromatography
HRMS	High resolution mass spectrum
H-bonding	Hydrogen bonding
HMPA	Hexamethylphosphoramide
ILs	Ionic liquids
IBX	Iodoxybenzoic acid
JH-CPP	Jorgensen-Hayashi chiral porous polymer
Josiphos	(R)-1-[(S)-2-(Diphenylphosphino)ferrocenyl] ethyldicyclohexylphosphine
KYE1	Enoate reductase 1 from *kluyveromyces lactis*
LDA	Lithium diisopropylamide
LUMO	Lowest unoccupied molecular orbital

MAO	monoamine oxidase inhibitor
MBH	Morita-Baylis-Hillmann
MDO	Modularly designed organocatalyst
MNP	Magnetic nanoparticles
MS	Molecular sieves
MW	Microwave
Ns	Nosyl
NADH	Nicotinamide adenine dinucleotide
NADPH	Nicotinamide adenine dinucleotide phosphate
N-Boc	*tert*-butyloxycarbonyl protecting group
NFSI	*N*-Fluorobenzenesulfonimide
NHC	*N*-Heterocyclic carbene
NHTf	Trifluoromethane sulfonylamide
NMM	*N*-Methylmorpholine
NMR	Nuclear magnetic resonance
nbd	2,5-norbornadiene
OPR	12-Oxophytodienoate reductase isoenzyme
OTf	trifluoromethanesulfonate
OYE	Old yellow enzyme
Pd/Mn	Palladium-manganese heterobimetals
Pd-Sm-Pd	Trimetallic complex of palladium/samarium/palladium
PEG	Polyethylene glycol
Phebim	bis-(imidazolinyl)phenyl
PMHS	polymethylhydrosiloxane
PMO	Palladium comprising dicationic bipyridinium supported periodic mesoporous organosilica
p-tol-BINAP	2,2′-Bis(di-p-tolylphosphino)-1,1′-binaphthyl
PyBidine	Diphenyldiamine-derived bis(imidazolidine)pyridine
PCC	Pyridinium chlorochromate
PPY	Polypyrrole
RT	Room temperature
[Rh(nbd)$_2$]SbF$_6$	Bis(norbornadiene)rhodium(I)hexafluoroantimonate
(R,S)-Josiphos	(R)-1-[(SP)-2-(Diphenylphosphino)ferrocenyl]ethyldicyclohexylphosphine
(S)-Binapine	(3S,3′S,4S,4′S,11bS,11′bS)-(+)-4,4′-Di-t-butyl-4,4′,5,5′-tetrahydro-3,3′-bi-3H-dinaphtho[2,1-c:1′,2′-e]phosphepin
selectFluor	1-Fluoro-4-methyl-1,4-diazoniabicyclo[2.2. 2]octanebis(tetrafluoroborate)
S$_N$2'	Nucleophilic substitution unimolecular - allylic
TBDPS	*tert*-Butyldiphenylsilyl
TEA	triethylamine
TFA	Trifluoroacetic acid
TMM	trimethylenemethane
TMS	Trimethylsilyl
TS	Transition state
TsDPEN	*N*-Tosyl-1,2-diphenyl-1,2-ethylenediamine
TBME	*Tert*-butyl methyl ether

Foreword

As nitroalkenes are versatile compounds in terms of availability and reactivity, nitroalkenes have been used as benchmark substrates to demonstrate the synthetic utility of catalytic reactions. In addition, the obtained products are readily converted into important biologically active natural products and pharmaceuticals by taking advantage of the good ability of nitro groups to be transformed into various functional groups, such as hydroxylamines, amines and masked carbonyls.

Over the past few decades, the unique properties of nitroalkenes have led to a wide range of asymmetric reactions using chiral catalysts. In 2003, we demonstrated that bifunctional aminothioureas efficiently promote the asymmetric Michael addition of malonates and β-ketoesters to β-nitrostyrenes with high enantioselectivity. In the course of these studies, we recognized the synthetic utility of asymmetric reactions of nitroalkenes. Later, we developed a double Michael reaction between γ,δ-unsaturated β-ketoester and β-nitrostyrene to achieve an asymmetric total synthesis of natural products.

To date, various types of chiral catalysts including organocatalysts and metal catalysts have been discovered for catalytic asymmetric Michael addition and cycloaddition with nitroalkenes. However, because of the large number of such catalysts, no one can grasp the details of all the research results. It is, therefore, important and timely to have a comprehensive book that summarizes major developments and key concepts in this booming area of asymmetric catalysis with nitroalkenes. In this regard, Irishi Namboothiri is an appropriate author for this book as he has published a series of papers and reviews on the application of nitroalkenes in the synthesis of carbocycles and heterocycles, as well as on the Morita–Baylis–Hillman reaction of nitroalkenes.

The book contains many examples of important catalytic reactions using newly designed catalysts and new insights on the mechanistic aspects and product profiles. These features will help readers find and retrieve information easily and straightforwardly and get a rational and historical understanding of the state-of-the-art technology and knowledge. Overall, this book is an invaluable resource for students and researchers in all laboratories working on catalysis, chemical synthesis and medicinal chemistry, and it is likely to be the best monograph in the field for a long time.

Yoshiji Takemoto
Kyoto, February 2020

Preface

Conjugated nitroalkenes are versatile substrates in organic synthesis because of their ability to participate in a wide variety of reactions, including multicomponent and cascade reactions, primarily due to the strong electron-withdrawing and coordinating ability of the nitro group. The easy accessibility of nitroalkenes and the amenability of the nitro group to undergo diverse synthetic transformations have further strengthened the profile of nitroalkenes as promising substrates in both regio and stereoselective synthesis of complex designed molecules and natural products. These unique features of nitroalkenes have prompted synthetic organic chemists to employ nitroalkenes as substrates in catalytic asymmetric reactions. Over the past two decades, numerous chiral metal-based and organocatalysts have been employed in various asymmetric reactions involving nitroalkenes not only toward developing new methodologies but in targeted synthesis as well.

Over 25 reviews have appeared in the literature since 2002 on various catalytic asymmetric reactions of nitroalkenes. However, while those excellent reviews have highlighted different facets of nitroalkenes, a comprehensive coverage of the multifaceted reactivity of nitroalkenes in catalytic asymmetric reactions was probably beyond the scope of a journal review or a book chapter. Therefore, we felt that synthetic organic chemists and medicinal chemists interested in nitroalkene-derived enantioenriched molecules with key functional groups and potential bioactivity would benefit if a systematic compilation of various asymmetric synthetic methods involving nitroalkenes as substrates in the presence of various chiral catalysts developed till date is available in one place. New insights into the mechanistic aspects and stereoselectivity could be of interest to physical organic chemists as well as theoretical chemists.

This book begins with a brief introduction to the catalytic asymmetric reactions of nitroalkenes with reference to the journal reviews and the book chapters that have appeared in the literature thus far. This is followed by a chapter-wise description of the reactivity of nitroalkenes in Michael addition, cycloadditions and cascade reactions (see Introduction). We have made every effort to include all the reagents and reaction conditions in the schemes. Further, the number of examples, product yield, diastereomeric ratios and enantiomeric excess have also been included. Although the mechanism is discussed in the text in most cases, the detailed mechanistic schemes are included only in selected cases. However, all available transition state models are presented and discussed to rationalize the observed stereoselectivity. Over 650 articles, including reviews, have been covered in this book. But if there are any omissions, we apologize in advance and promise to include them in the next edition.

We are deeply grateful to Prof Takemoto who is a pioneer in the arena of asymmetric catalysis for writing a foreword to this book. We are highly indebted to Prof Viswakarma Singh for patiently and critically reading all the chapters and giving his valuable suggestions and comments. We sincerely thank our group members Pallabita Basu, Sonia Naik and Nikil Purushotham for proofreading the manuscript. We also thank the editor Dr Renu Upadhyay and her associate Ms Jyotsna Jangra for their guidance and support. Finally, INN thanks his wife Jayasree and daughters Ashcharya and Susmrithi for allowing him to spend extended hours in his office to complete this book.

Irishi N. N. Namboothiri
Meeta Bhati
Madhu Ganesh
Basavaprabhu Hosamani
Thekke V. Baiju
Shimi Manchery
Kalisankar Bera
February 2020

Authors

Irishi N. N. Namboothiri received his MSc from Mangalore University and PhD from the Indian Institute of Science, Bangalore. He did his postdoctoral research at Bar-Ilan University, Israel, University of North Texas and Columbia University. After a brief stint as a Senior Research Scientist at Sabinsa Corporation, New Jersey, he joined the Department of Chemistry, IIT Bombay where he is currently a professor. His research interests include organic synthesis, development of new synthetic methodologies, asymmetric catalysis, mechanistic studies and materials chemistry. He is an elected fellow of the Indian Academy of Sciences and National Academy of Sciences India. He is also a recipient of the Chemical Research Society of India Bronze medal. He has supervised over 20 PhDs and co-authored over 155 publications including two chapters and a book and is also a co-inventor of six patents.

Meeta Bhati was born in Jaipur, Rajasthan, India in 1990. She received her BSc degree (2011) from Maharani's College, University of Rajasthan, Jaipur. She joined the Central University of Rajasthan, Ajmer for an MSc degree in Chemistry (2013), and has also completed her PhD in Organic Chemistry (2013–2019) in the same institution under the guidance of Dr Easwar Srinivasan. A year ago, she joined as an Institute Postdoctoral Fellow with Prof I. N. N. Namboothiri, IIT Bombay (2019). Her research interest is focused on designing asymmetric organocatalysts for enantioselective reactions using sustainable catalytic protocols and "on-water" organocatalyzed reactions.

Madhu Ganesh obtained his MSc from Mangalore University in 2000 and PhD degree from IIT Bombay in 2007. His doctoral work under Prof I. N. N. Namboothiri was on probing stereoselective reactions of nitroalkenes. He undertook postdoctoral studies at Rutgers University, New Jersey and City College, New York. He worked in the industry as a Scientific Associate (2000–2002) and further as a Research Investigator (2010–2013) at Syngene International Pvt. Ltd. After working as an Assistant Professor in the Department of Chemistry, B M S College of Engineering, Bengaluru, since 2013, recently he joined as an Associate Professor in the Department of Pharmaceutical Technology/Process Chemistry, National Institute of Pharmaceutical Education & Research, Hyderabad. His research interests include the development of synthetic methods and their applications in the pharmaceutical industry.

Basavaprabhu Hosamani was born in Raichur, Karnataka, India in 1986. He received his BSc (2006) and MSc (2008) degrees in Chemistry from Bangalore University. He earned his PhD degree from Bangalore University under the supervision of Prof V. V. Sureshbabu in May 2014. After a brief stint as an Assistant Professor at an Engineering college, he worked as an Institute Postdoctoral Fellow with Prof I. N. N. Namboothiri, IIT Bombay (2015–2017) and as a DST-SERB National Postdoctoral Fellow at the Central University of Karnataka with Dr K Hanumae Gowd (2017–2018). After serving as an Assistant Professor at the HKE Society's SLN College of Engineering, Raichur, Karnataka during 2018–2019, he joined as a Scientific Officer at the Regional Forensic Science Laboratory, Kalaburagi, Karnataka. His research interests include organocatalytic asymmetric synthesis, synthesis of modified nucleosides and some biologically active molecules.

Thekke V. Baiju was born in Kasaragod, Kerala, India in 1987. He received his BSc (2007) and MSc (2009) degrees in chemistry from Kannur University, Kerala. He completed his PhD work in organic chemistry from CSIR-NIIST, Thiruvananthapuram. Until recently, he worked as an Institute Postdoctoral Fellow with Prof I. N. N. Namboothiri, IIT Bombay. Currently, he is an Assistant Professor at Government Brennen College, Thalassery, Kerala. His research interest is focused on the synthesis of carbo- and heterocyclic scaffolds by multicomponent and cascade reactions of functionalized nitroalkenes.

Shimi Manchery received her BSc degree in Chemistry from Calicut University and MSc degree from Mahatma Gandhi University, Kerala. She completed PhD from CSIR-NIIST, Thiruvananthapuram, Kerala. Until recently, she worked as a Senior Research Fellow with Prof I. N. N. Namboothiri, IIT Bombay.

Kalisankar Bera received his BSc degree in Chemistry from Haldia Government College affiliated to Vidyasagar University, West Bengal in 2006 and MSc degree in Chemistry from IIT Madras in 2008. He completed his PhD at IIT Bombay under the supervision of Prof I. N. N. Namboothiri and postdoctoral work at the University of Leipzig, Germany, and Vanderbilt University, United States. Currently, he is a Research Scientist at Syngene International Pvt. Ltd, Bengaluru. His research includes organocatalytic asymmetric reactions for enantioselective synthesis of multifunctional molecules.

Introduction

In the past 20 years, asymmetric catalysis has emerged as a powerful technique to synthesize enantiomerically enriched compounds that are required to achieve potency and selectivity in biology and medicine. This is due to the difficulties associated with accessing complex natural products of pharmaceutical relevance that contain chiral centers via the resolution of racemic mixtures or via synthesis using traditional chiron and auxiliary approaches. The increasing demand for such molecules with complex, diverse and desirable properties necessitated the development of innovative and economical methods, particularly involving catalysis, such as organocatalysis,[1–34] metal catalysis[35–48] or biocatalysis.[49–53]

Nitroalkenes are excellent Michael acceptors because of their ability to undergo facile addition of various C-, O-, N-, and S-centered nucleophiles. Michael adducts are key precursors to complex molecules, including natural products (Figure 1).[54] Nitroalkenes also participate as dienophiles and heterodienes in Diels–Alder reactions, as dipolarophiles in 1,3-dipolar cycloadditions, as well as substrates in Morita–Baylis–Hillman and Rauhut–Currier reactions. The nitro group is also amenable for transformation to various functional groups, such as hydroxylamine, amine,[55] etc., and is a masked carbonyl exhibiting umpolung reactivity.[56] Because of their versatile reactivity, nitroalkenes are often referred to as "synthetic chameleons" in organic synthesis.[57–63]

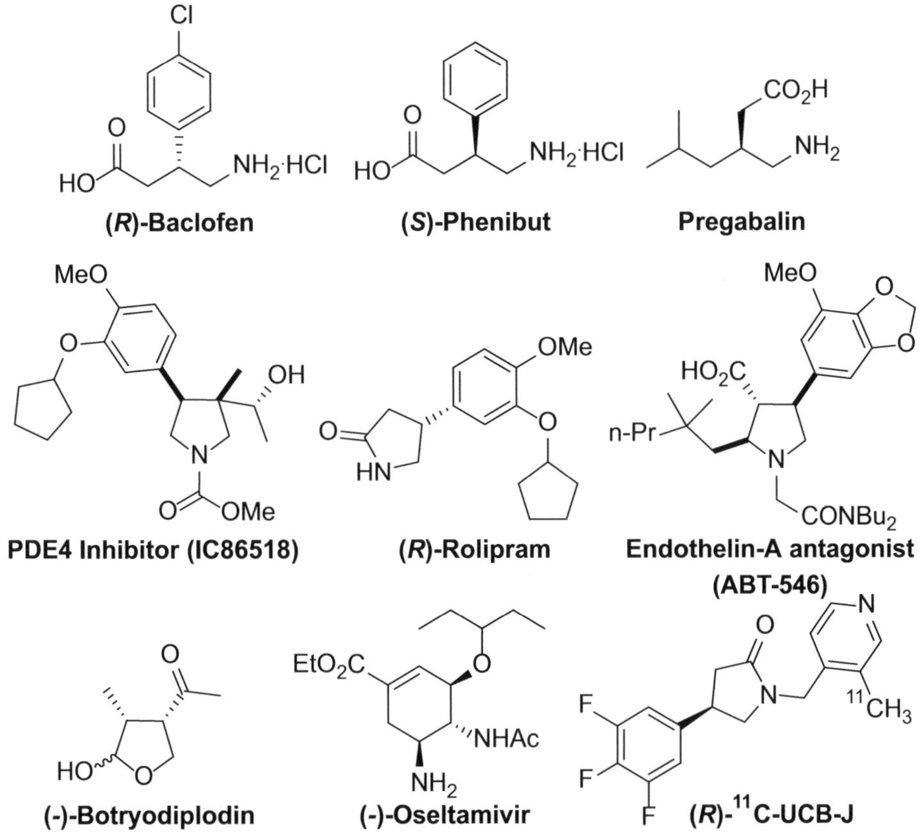

FIGURE 1 Selected biologically active molecules synthesized from nitroalkenes via asymmetric catalysis.

Nitroalkenes have become the benchmark substrates in asymmetric catalysis due to their above-mentioned reactivity profile. Catalytic asymmetric reactions of nitroalkenes have made rapid progress generating tremendous knowledge not only on synthetic strategies and tactics but also on products with diverse functionalities and biological activities. However, there has been no comprehensive literature on asymmetric reactions of nitroalkenes. A review on asymmetric Michael additions of nitroalkenes was published by Enders way back in 2002.[64] Later in 2017, Alonso and co-workers published a review on asymmetric organocatalytic conjugate additions to α,β-unsaturated nitro compounds, which covered only the reactions under organocatalytic conditions.[65] Several authors have described selected asymmetric reactions of nitroalkenes as such or as a part of their review focusing on other chemistry. For example, in a review of organocatalyzed transformations of β-ketoesters by Naicker et al., reactions with nitroalkenes were also described.[66] We also published a series of reviews on the application of nitroalkenes in the synthesis of carbocycles and heterocycles,[67–69] as well as on the Morita–Baylis–Hillman reaction of nitroalkenes[70,71] where certain asymmetric reactions were alluded. Besides, several reviews have been published on the asymmetric reactions of nitroalkenes, such as Michael addition,[39,64,72–77] cycloadditions,[78–81] Friedel–Crafts alkylation[82–84] and multicomponent cascade or domino reactions,[85–87] as well as applications of such asymmetric reactions in the synthesis of various carbocycles, heterocycles[74,82,83,88,89] and total synthesis.[90]

We felt that synthetic organic chemists and medicinal chemists interested in enantioenriched molecules with key functional groups and potential bioactivity would be of benefit if a systematic compilation of various asymmetric synthetic methods involving nitroalkenes as substrates in the presence of various chiral catalysts developed to date is available in one place. New insights into the mechanistic aspects and product profiles are also covered in this book. The book begins with catalytic asymmetric Michael additions of various carbonyl compounds to nitroalkenes. This is followed by catalytic asymmetric Friedel–Crafts reactions of nitroalkenes with indoles, pyrroles and electron-rich phenols. Subsequently, catalytic asymmetric Michael addition of other carbon- and heteroatom-centered nucleophiles to nitroalkenes and cycloadditions such as 1,3-dipolar cycloaddition, [2+2]-, [3+2]- and [4+2]-cycloadditions of nitroalkenes are discussed in detail. Catalytic asymmetric reduction and various cascade reactions are dealt with in the latter part of this book.

REFERENCES

1. Taylor, M. S.; Jacobsen, E. N., Asymmetric Catalysis by Chiral Hydrogen-Bond Donors. *Angew. Chem. Int. Ed.* **2006**, *45*, 1520.
2. Doyle, A. G.; Jacobsen, E. N., Small-Molecule H-Bond Donors in Asymmetric Catalysis. *Chem. Rev.* **2007**, *107*, 5713.
3. Enders, D.; Niemeier, O.; Henseler, A., Organocatalysis by N-Heterocyclic Carbenes. *Chem. Rev.* **2007**, *107*, 5606.
4. Bertelsen, S.; Jorgensen, K. A., Organocatalysis-After the Gold Rush. *Chem. Soc. Rev.* **2009**, *38*, 2178.
5. You, S.-L.; Cai, Q.; Zeng, M., Chiral Bronsted Acid Catalyzed Friedel-Crafts Alkylation Reactions. *Chem. Soc. Rev.* **2009**, *38*, 2190.
6. Zhang, Z.; Schreiner, P. R., (Thio)urea Organocatalysis-What can be Learnt from Anion Recognition? *Chem. Soc. Rev.* **2009**, *38*, 1187.
7. Bugaut, X.; Glorius, F., Organocatalytic Umpolung: N-Heterocyclic Carbenes and Beyond. *Chem. Soc. Rev.* **2012**, *41*, 3511.
8. Giacalone, F.; Gruttadauria, M.; Agrigento, P.; Noto, R., Low-Loading Asymmetric Organocatalysis. *Chem. Soc. Rev.* **2012**, *41*, 2406.
9. Flanigan, D. M.; Romanov-Michailidis, F.; White, N. A.; Rovis, T., Organocatalytic Reactions Enabled by N-Heterocyclic Carbenes. *Chem. Rev.* **2015**, *115*, 9307.
10. Peng, Z.; Takenaka, N., Applications of Helical-Chiral Pyridines as Organocatalysts in Asymmetric Synthesis. *Chem. Rec.* **2013**, *13*, 28.
11. Trost, B. M.; Bartlett, M. J., ProPhenol-Catalyzed Asymmetric Additions by Spontaneously Assembled Dinuclear Main Group Metal Complexes. *Acc. Chem. Res.* **2015**, *48*, 688.

12. Das, S.; Santra, S.; Mondal, P.; Majee, A.; Hajra, A., Zwitterionic Imidazolium Salt: Recent Advances in Organocatalysis. *Synthesis* **2016**, *48*, 1269.
13. Han, X.; Zhou, H.-B.; Dong, C., Applications of Chiral Squaramides: From Asymmetric Organocatalysis to Biologically Active Compounds. *Chem. Rec.* **2016**, *16*, 897.
14. Zhao, B.-L.; Li, J.-H.; Du, D.-M., Squaramide-Catalyzed Asymmetric Reactions. *Chem. Rec.* **2017**, *17*, 994.
15. Najera, C.; Miguel Sansano, J.; Gomez-Bengoa, E., Heterocycle-based Bifunctional Organocatalysts in Asymmetric Synthesis. *Pure Appl. Chem.* **2016**, *88*, 561.
16. Berkessel, A.; Groerger, H., *Asymmetric Organocatalysis*. Wiley-VCH: Weinheim, Germany: 2005.
17. Kocovsky, P.; Malkov, A. V., Ed. Organocatalysis in Organic Synthesis. *Tetrahedron* **2016**, *62*, 243.
18. Dalko, P. I., *Enantioselective Organocatalysis: Reactions and Experimental Procedures*. Wiley-VCH: Weinheim, Germany: 2007.
19. Pellissier, H., *Recent Developments in Asymmetric Organocatalysis*. Royal Society of Chemistry: Cambridge, UK: 2010.
20. Asymmetric Organocatalysis. In *Topics in Current Chemistry*; 291. List B. Ed. Springer: Berlin, Germany: 2010.
21. List, B., *Science of Synthesis: Asymmetric Organocatalysis 1, Lewis Base and Acid Catalysts*. Georg Thieme: Stuttgart, Germany: 2012.
22. Maruoka, K., *Science of Synthesis: Asymmetric Organocatalysis 2, Bronsted Base and Acid Catalysts, and Additional Topics*. Georg Thieme Verlag: Stuttgart, Germany: 2012.
23. Maruoka, K.; Shibasaki, M., *Comprehensive Chirality, Volume 6: Synthetic Methods V-Organocatalysis*. Elsevier: Amsterdam, Netherlands: 2012.
24. Torres, R. R., *Stereoselective Organocatalysis: Bond Formation Methodologies and Activation Modes*. John Wiley & Sons, Inc.: Hoboken, NJ: 2013.
25. Waser, M., *Asymmetric Organocatalysis in Natural Product Syntheses*. Springer-Verlag: Wien, Austria: 2012.
26. Dalko, P. I., *Comprehensive Enantioselective Organocatalysis: Catalysts, Reactions and Applications, Volume 1: Privileged Catalysts*. Wiley-VCH: Weinheim, Germany: 2013.
27. Dalko, P. I., *Comprehensive Enantioselective Organocatalysis: Catalysts, Reactions, and Applications, Volume 2: Activations*. Wiley-VCH: Weinheim, Germany: 2013.
28. Dalko, P. I., *Comprehensive Enantioselective Organocatalysis: Catalysts, Reactions and Applications, Volume 3: Reactions and Applications*. Wiley-VCH: Weinheim, Germany: 2013.
29. Kobayashi, Y.; Takemoto, Y., Bifunctional Guanidines as Hydrogen-Bond-Donating Catalysts. In *Guanidines as Reagents and Catalysts I*;-Selig, P., Ed. Springer International Publishing: Cham, Switzerland: 2017; pp. 71–94.
30. Takemoto, Y., Development of Chiral Thiourea Catalysts and Its Application to Asymmetric Catalytic Reactions. *Chem. Pharm. Bull.* **2010**, *58*, 593.
31. Takemoto, Y., Recognition and Activation by Ureas and Thioureas: Stereoselective Reactions Using Ureas and Thioureas as Hydrogen-Bonding Donors. *Org. Biomol. Chem.* **2005**, *3*, 4299.
32. Miyabe, H.; Takemoto, Y., Discovery and Application of Asymmetric Reaction by Multi-Functional Thioureas. *Bull. Chem. Soc. Jpn.* **2008**, *81*, 785.
33. Takemoto, Y.; Miyabe, H., The Amino Thiourea-Catalyzed Asymmetric Nucleophilic Reactions. *CHIMIA Int. J. Chem.* **2007**, *61*, 269–275.
34. Takemoto, Y.; Inokuma, T., Bifunctional Thiourea Catalysts. In *Asymmetric Synthesis II*; Christmann, M., Bräse, S., Eds., Wiley-VCH: Weinheim: 2013; pp. 233–237.
35. Ikariya, T.; Gridnev, I. D., Bifunctional Transition Metal-Based Molecular Catalysts for Asymmetric C–C and C–N Bond Formation. *Chem. Rec.* **2009**, *9*, 106.
36. Jerphagnon, T.; Pizzuti, M. G.; Minnaard, A. J.; Feringa, B. L., Recent Advances in Enantioselective Copper-Catalyzed 1,4-Addition. *Chem. Soc. Rev.* **2009**, *38*, 1039.
37. Ikariya, T.; Gridnev, I. D., Bifunctional Transition Metal-Based Molecular Catalysts for Asymmetric C–C and C–N Bond Formation. *Top. Catal.* **2010**, *53*, 894.
38. Pellissier, H., Recent Developments in Enantioselective Nickel(II)-Catalyzed Conjugate Additions. *Adv. Synth. Catal.* **2015**, *357*, 2745.
39. Heravi, M. M.; Dehghani, M.; Zadsirjan, V., Rh-Catalyzed Asymmetric 1,4-Addition Reactions to α,β-Unsaturated Carbonyl and Related Compounds: An Update. *Tetrahedron: Asymmetry* **2016**, *27*, 513.
40. Pellissier, H., Enantioselective Silver-Catalyzed Transformations. *Chem. Rev.* **2016**, *116*, 14868.
41. Hayashi, T., Rhodium-catalyzed Asymmetric Addition of Organo-boron and -Titanium Reagents to Electron-deficient Olefins. *Bull. Chem. Soc. Jpn.* **2004**, *77*, 13.

42. Hayashi, T., Rhodium-catalyzed Asymmetric 1,4-Addition of Organometallic Reagents. *Russ. Chem. Bull.* **2003**, *52*, 2595.
43. Hayashi, T., Rhodium-catalyzed Asymmetric 1,4-Addition of Organoboronic Acids and their Derivatives to Electron Deficient Olefins. *Synlett* **2001**, *2001*, 879.
44. Asymmetric Domino Reactions. In *RSC Catalysis Series*; 10. Pellissier, H., Ed. Royal Society of Chemistry: Cambridge: 2013.
45. Alexakis, A.; Krause, N.; Woodward, S., *Copper-Catalyzed Asymmetric Synthesis*. Wiley-VCH: Weinheim, Germany: 2014.
46. *Catalytic Methods in Asymmetric Synthesis: Advanced Materials, Techniques, and Applications*. Gruttadauria, M.; Giacalone, F., Ed. John Wiley & Sons, Inc.: Hoboken, N.J.: 2011.
47. Asymmetric Functionalization of C–H Bonds. In *RSC Catalysis Series*; 25. You, S.-L. Ed. Royal Society of Chemistry: Cambridge: 2015.
48. Takemoto, Y.; Miyabe, H., Asymmetric Carbon–Heteroatom Bond-Forming Reactions. In *Catalytic Asymmetric Synthesis*; Ojima, I. Ed. Wiley: Weinheim: **2013**; pp 227–267.
49. Gong, J.-S.; Shi, J.-S.; Lu, Z.-M.; Li, H.; Zhou, Z.-M.; Xu, Z.-H., Nitrile-converting Enzymes as a Tool to Improve Biocatalysis in Organic Synthesis: Recent Insights and Promises. *Crit. Rev. Biotechnol.* **2017**, *37*, 69–81.
50. Azerad, R., Biocatalysis in Organic Synthesis. *Adv. Org. Synth.* **2005**, *1*, 455–518.
51. de Souza, R. O. M. A.; Miranda, L. S. M.; Bornscheuer, U. T., A Retrosynthesis Approach for Biocatalysis in Organic Synthesis. *Chem. - Eur. J.* **2017**, *23*, 12040.
52. Kazlauskas, R., Integrating Biocatalysis into Organic Synthesis. *Tetrahedron: Asymmetry* **2004**, *15*, 2727.
53. Patel, R. N., *Stereoselective Biocatalysis*. Dekker: New York: 2000.
54. Varma, R. S.; Kabalka, G. W., Nitroalkenes in the Synthesis of Heterocyclic Compounds. *Heterocycles* **1986**, *24*, 2645.
55. Ballini, R.; Marcantoni, E.; Petrini, M. *Nitroalkenes as Amination Tools*. Ricci, A. Ed., Wiley-VCH Verlag GmbH & Co. KGaA: Weinheim, Germany: **2008**; pp. 93–148.
56. Ballini, R.; Gabrielli, S.; Palmieri, A. *Aliphatic Nitrocompounds as Versatile Building Blocks for the One-Pot Processes*. Sharma, S.K.; Mudhoo, A. Ed. CRC Press: Boca Raton, FL: 2011; pp. 53–78.
57. Barrett, A. G. M.; Graboski, G. G., Conjugated Nitroalkenes: Versatile Intermediates in Organic Synthesis. *Chem. Rev.* **1986**, *86*, 751.
58. Kabalka, G. W.; Varma, R. S., Syntheses and Selected Reductions of Conjugated Nitroalkenes. A Review. *Org. Prep. Proced. Int.* **1987**, *19*, 283.
59. Barrett, A. G. M., Heterosubstituted Nitroalkenes in Synthesis. *Chem. Soc. Rev.* **1991**, *20*, 95.
60. Perekalin, V. V.; Lipina, E. S.; Berestovitskaya, V. M.; Efremov, D. A., *Nitroalkenes: Conjugated Nitro Compounds*. John Wiley & Sons, Inc.: Chichester, UK: 1994.
61. Ferreira, A. M.; Trostchansky, A.; Ferrari, M.; Souza, J. M.; Rubbo, H., Nitroalkenes: Synthesis, Characterization, and Effects on Macrophage Activation. *Methods Enzymol.* **2008**, *441*, 33.
62. Ballini, R.; Gabrielli, S.; Palmieri, A., β-Nitroacrylates as an Emerging, Versatile Class of Functionalized Nitroalkenes for the Synthesis of a Variety of Chemicals. *Curr. Org. Chem.* **2010**, *14*, 65.
63. Turell, L.; Steglich, M.; Alvarez, B., The Chemical Foundations of Nitroalkene Fatty Acid Signaling Through Addition Reactions with Thiols. *Nitric Oxide* **2018**, *78*, 161.
64. Berner, O. M.; Tedeschi, L.; Enders, D., Asymmetric Michael Additions to Nitroalkenes. *Eur. J. Org. Chem.* **2002**, 1877.
65. Alonso, D. A.; Baeza, A.; Chinchilla, R.; Gomez, C.; Guillena, G.; Pastor, I. M.; Ramon, D. J., Recent Advances in Asymmetric Organocatalyzed Conjugate Additions to Nitroalkenes. *Molecules* **2017**, *22*, 895.
66. Govender, T.; Arvidsson, P. I.; Maguire, G. E. M.; Kruger, H. G.; Naicker, T., Enantioselective Organocatalyzed Transformations of β-Ketoesters. *Chem. Rev.* **2016**, *116*, 9375.
67. Halimehjani, A. Z.; Namboothiri, I. N. N.; Hooshmand, S. E., Nitroalkenes in the Synthesis of Carbocyclic Compounds. *RSC Adv.* **2014**, *4*, 31261.
68. Halimehjani, A. Z.; Namboothiri, I. N. N.; Hooshmand, S. E., Part I: Nitroalkenes in the Synthesis of Heterocyclic Compounds. *RSC Adv.* **2014**, *4*, 48022.
69. Halimehjani, A. Z.; Namboothiri, I. N. N.; Hooshmand, S. E., Part II: Nitroalkenes in the Synthesis of Heterocyclic Compounds. *RSC Adv.* **2014**, *4*, 51794.
70. Nair, D. K.; Kumar, T.; Namboothiri, I. N. N., α-Functionalization of Nitroalkenes and Its Applications in Organic Synthesis. *Synlett* **2016**, *27*, 2425.
71. Kaur, K.; Namboothiri, I. N. N., Morita-Baylis-Hillman and Rauhut-Currier Reactions of Conjugated Nitroalkenes. *Chimia* **2012**, *66*, 913.

72. Almaşi, D.; Alonso, D. A.; Nájera, C., Organocatalytic Asymmetric Conjugate Additions. *Tetrahedron: Asymmetry* **2007**, *18*, 299.
73. Chauhan, P.; Mahajan, S.; Enders, D., Organocatalytic Carbon–Sulfur Bond-Forming Reactions. *Chem. Rev.* **2014**, *114*, 8807.
74. Chauhan, P.; Mahajan, S.; Enders, D., Asymmetric Synthesis of Pyrazoles and Pyrazolones Employing the Reactivity of Pyrazolin-5-one Derivatives. *Chem. Commun.* **2015**, *51*, 12890.
75. Somanathan, R.; Chavez, D.; Servin, F. A.; Romero, J. A.; Navarrete, A.; Parra-Hake, M.; Aguirre, G.; Anaya de Parrodi, C.; Gonzalez, J., Bifunctional Organocatalysts in the Asymmetric Michael Additions of Carbonylic Compounds to Nitroalkenes. *Curr. Org. Chem.* **2012**, *16*, 2440.
76. Maji, B., N-Heterocyclic-carbene-Catalyzed Reactions of Nitroalkenes: Synthesizing Important Building Blocks. *Asian J. Org. Chem.* **2018**, *7*, 70.
77. Dalpozzo, R.; Mancuso, R., Enantioselective Vinylogous Reactions of 3-Alkylidene Oxindoles. *Synthesis* **2018**, *50*, 2463.
78. Moyano, A.; Rios, R., Asymmetric Organocatalytic Cyclization and Cycloaddition Reactions. *Chem. Rev.* **2011**, *111*, 4703.
79. Hashimoto, T.; Maruoka, K., Recent Advances of Catalytic Asymmetric 1,3-Dipolar Cycloadditions. *Chem. Rev.* **2015**, *115*, 5366.
80. Najera, C.; Sansano, J. M., Asymmetric 1,3-Dipolar Cycloadditons of Stabilized Azomethine Ylides with Nitroalkenes. *Curr. Top. Med. Chem.* **2014**, *14*, 1271.
81. Denmark, S. E.; Thorarensen, A., Tandem [4+2]/[3+2] Cycloadditions of Nitroalkenes. *Chem. Rev.* **1996**, *96*, 137.
82. Bartoli, G.; Bencivenni, G.; Dalpozzo, R., Organocatalytic Strategies for the Asymmetric Functionalization of Indoles. *Chem. Soc. Rev.* **2010**, *39*, 4449.
83. Dalpozzo, R., Strategies for the Asymmetric Functionalization of Indoles: An Update. *Chem. Soc. Rev.* **2015**, *44*, 742.
84. Trost, B. M.; Müller, C., Asymmetric Friedel–Crafts Alkylation of Pyrroles with Nitroalkenes Using a Dinuclear Zinc Catalyst. *J. Am. Chem. Soc.* **2008**, *130*, 2438.
85. Goudedranche, S.; Raimondi, W.; Bugaut, X.; Constantieux, T.; Bonne, D.; Rodriguez, J., Enantioselective Organocatalyzed Domino Synthesis of Six-membered Carbocycles. *Synthesis* **2013**, *45*, 1909.
86. Roca-Lopez, D.; Sadaba, D.; Delso, I.; Herrera, R. P.; Tejero, T.; Merino, P., Asymmetric Organocatalytic Synthesis of γ-Nitrocarbonyl Compounds Through Michael and Domino Reactions. *Tetrahedron: Asymmetry* **2010**, *21*, 2561.
87. Volla, C. M. R.; Atodiresei, I.; Rueping, M., Catalytic C–C Bond-Forming Multi-Component Cascade or Domino Reactions: Pushing the Boundaries of Complexity in Asymmetric Organocatalysis. *Chem. Rev.* **2014**, *114*, 2390.
88. Ley, S. V.; Gutteridge, C. E., Rapid Assembly of Polycyclic Alkaloids by Asymmetric Tandem Cycloadditions of Nitroalkenes. *Chemtracts: Org. Chem.* **1995**, *8*, 222.
89. Gabrielli, S.; Chiurchiu, E.; Palmieri, A., β-Nitroacrylates: A Versatile and Growing Class of Functionalized Nitroalkenes. *Adv. Synth. Catal.* **2019**, *361*, 630.
90. Ishikawa, H.; Hayashi, Y., Total Synthesis of Oseltamivir and ABT-341 Using One-Pot Technology. In *Asymmetric Synthesis II*; 61–66, Wiley-VCH: Weinham, Germany: 2012.

1 Catalytic Asymmetric Michael Addition of 1,3-Dicarbonyls to Nitroalkenes

1.1 INTRODUCTION

1,3-Dicarbonyls possessing an active methylene group serve as good nucleophilic partners that easily generate stable carbanions. The stability of the resulting enolate is due to the formation of an extremely stable cyclic structure with pseudoaromatic character, making it a good nucleophile. Chiral organocatalysts bearing functional groups such as thiourea, squaramide, guanidine, etc. were the catalysts of choice for the asymmetric Michael addition of various 1,3-dicarbonyls such as 1,3-diketones, malonates, β-ketoesters, Meldrum's acid and 1,3-dicarbonyl surrogates such as 2-hydroxynaphthoquinones and hydroxycoumarins. There are sporadic reports of chiral 1,2-diamine-based ligand-metal complex catalyzed Michael additions as well.

1.2 ADDITION OF 1,3-DIKETONES

Shao and co-workers reported a bifunctional amine-thiourea **C1** catalyzed asymmetric Michael addition of 1,3-dicarbonyl compound **2** to nitroalkenes **1** to afford corresponding products **3** with good enantioselectivity (Scheme 1.1).[1] The catalyst bears central and axial chiral elements that play a pivotal role in enhancing the stereochemical control of the reaction. The reaction involves an initial activation of the nitroalkene **1** by H-bonding interactions of the catalyst with the nitro group and the deprotonation of the 1,3-dicarbonyl compound **2** to form an anion which adds to the activated nitroalkene **1** to form a ternary complex (**TS-1**, Scheme 1.1). This on protonation and followed by hydrolysis yields the product **3**. In 2009, Shao and co-workers described a matched-mismatched effect of two different chiral units in an organocatalyst **C2** for an asymmetric Michael addition of acetylacetone **2** (R^1 = H, R^2 = R^3 = Me) to different nitroalkenes **1** under mild conditions to give products **3** in high yields and enantioselectivities (Scheme 1.1).[2] Both the enantiomers of the product were isolated with nearly the same enantiomeric excess in the presence of "matched" and "mismatched" organocatalysts **C2** by switching the solvent from THF to toluene.

The Wang group reported binaphthyl-derived bifunctional amine-thiourea **C3** catalyzed Michael addition of 1,3-diketone **2** to nitroalkenes **1a**.[3] As low as 1 mol% of catalyst **C3** was sufficient to promote the reaction and provided the Michael adducts **3** in high yields and enantioselectivities (Scheme 1.2). The Michael adducts **3** were further utilized for the synthesis of substituted-α-amino acids. The Bolm group synthesized ephedrine- and pseudoephedrine-derived bifunctional tertiary amine-thiourea catalysts and investigated their catalytic properties for the Michael addition of 1,3-diketone **2** to nitroalkenes **1a** (Scheme 1.2).[4] Among those catalysts, **C4** provided the products **3** in excellent yields and enantioselectivities up to 94% ee. Song and co-workers described an asymmetric Michael addition of acetylacetone **2** to nitroalkenes **1a** by employing chiral bifunctional *L*-thiazoline-thiourea derivative **C5** (Scheme 1.2).[5] In the proposed mechanism, nitroalkene **1a** gets activated by the thiourea moiety of catalyst **C5** through H-bonding interactions, whereas the tertiary *N*-atom of catalyst **C5** deprotonates the acidic proton of 1,3-diketone **2** forming a ternary complex. In addition, the synergistic steric effects of the chiral motif of the catalyst **C5** increased the stereocontrol of this reaction, offering products **3** exclusively with (S)-configuration.

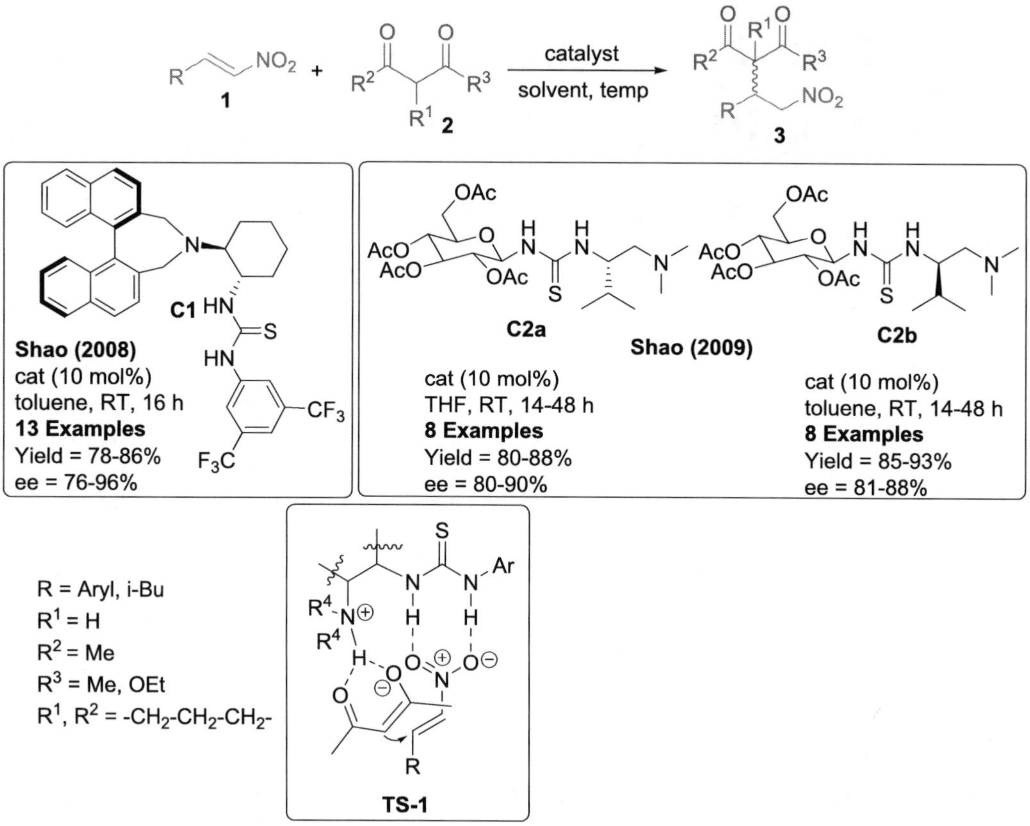

SCHEME 1.1 Michael addition of 1,3-dicarbonyl compounds to nitroalkenes using bifunctional thiourea catalysts.

In the realm of H-bonding catalysis, in 2008, the Rawal group introduced a new H-bonding catalyst based on the squaramide moiety. Squaramide exhibits better H-bonding ability than the corresponding thioureas.[6] As little as 0.5 mol% of squaramide catalyst **C6** catalyzed the reaction between acetylacetone **2** and nitroalkenes **1** to afford the Michael adducts **3** in excellent yields and enantioselectivities (Scheme 1.3). Under these reaction conditions, various aryl-substituted 1,3-diketones, acyclic and cyclic β-ketoesters also gave the addition products with nitrostyrene **1** in excellent yields and enantioselectivities, as well as with moderate-to-excellent diastereoselectivities. The Song group reported an asymmetric Michael addition of 1,3-diketones **2** to nitroalkenes **1** by employing the same squaramide-based organocatalyst **C6** in brine to give the products **3** in excellent yields and enantioselectivities (Scheme 1.3).[7] The reaction was performed with 0.5 mol% of the catalyst **C6**. Due to the hydrophobic hydration effect, the rate of reaction was observed to be faster when the reaction was performed in brine as solvent rather than in organic solvents.

Zlotin's group developed a bifunctional ionic liquid-supported recyclable organocatalyst **C7** for the asymmetric Michael reaction of 1,3-diketones **2** with nitroalkenes **1a** to give the corresponding Michael adducts **3** in excellent yields and enantioselectivities (Scheme 1.4).[8] The catalyst **C7** can be recycled and reused without any significant loss of activity and selectivity. The Miura group designed and synthesized cinchona–diaminomethylenemalononitrile (DMM) organocatalyst **C8** and employed it successfully for the asymmetric Michael addition of acetylacetone **2** to nitroalkenes **1a** under mild reaction conditions in diethyl ether to afford the respective Michael adducts **3** in high yields and enantioselectivities (Scheme 1.4).[9] A transition state model **TS-2** (Scheme 1.4) is proposed for this reaction. Trivedi and co-workers reported the design and synthesis of various

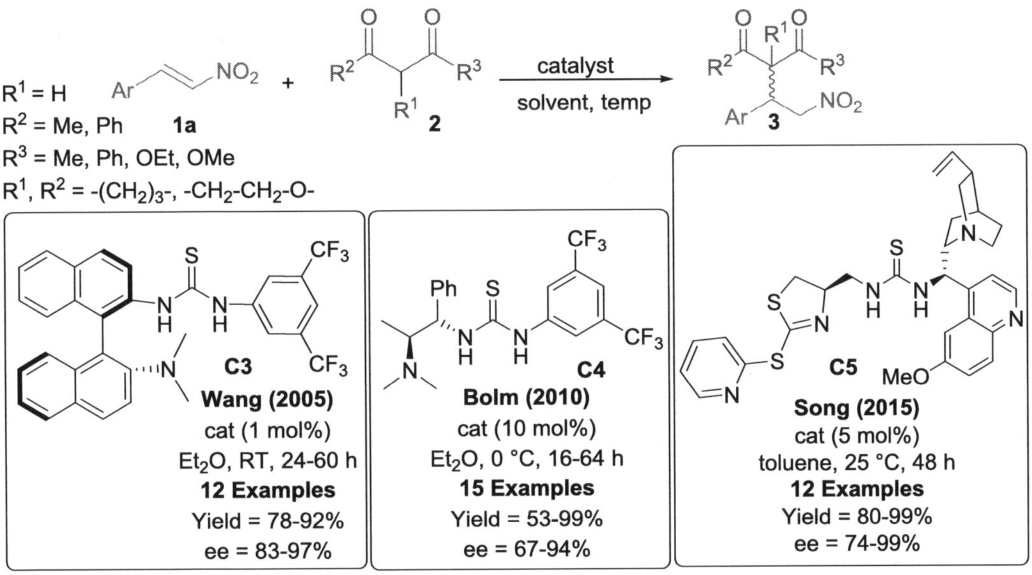

SCHEME 1.2 Various chiral thioureas as bifunctional organocatalysts in the Michael addition of 1,3-diketones to nitroalkenes.

ferrocene-based chiral bifunctional squaramides and thioureas using cinchona alkaloids.[10] Among them, the ferrocene derivative of squaramide-based bifunctional catalyst **C9** was successful in the asymmetric Michael additions of 1,3-diketone **2** to nitroalkenes **1a** (Scheme 1.4). One mol% of the

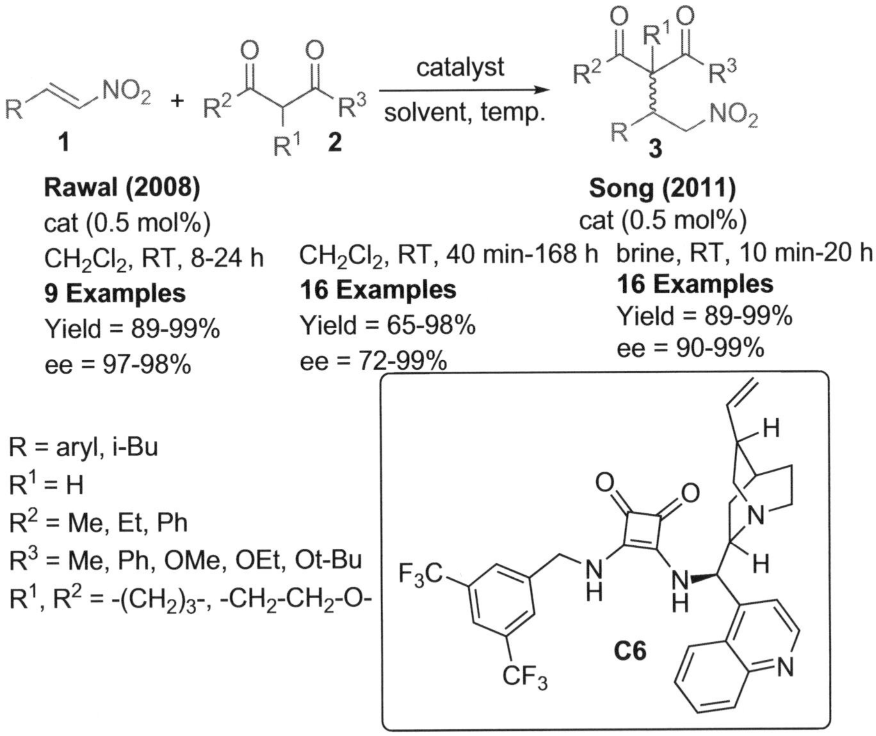

SCHEME 1.3 Asymmetric Michael addition of 1,3-diketone to nitroalkenes using squaramide-based organocatalyst.

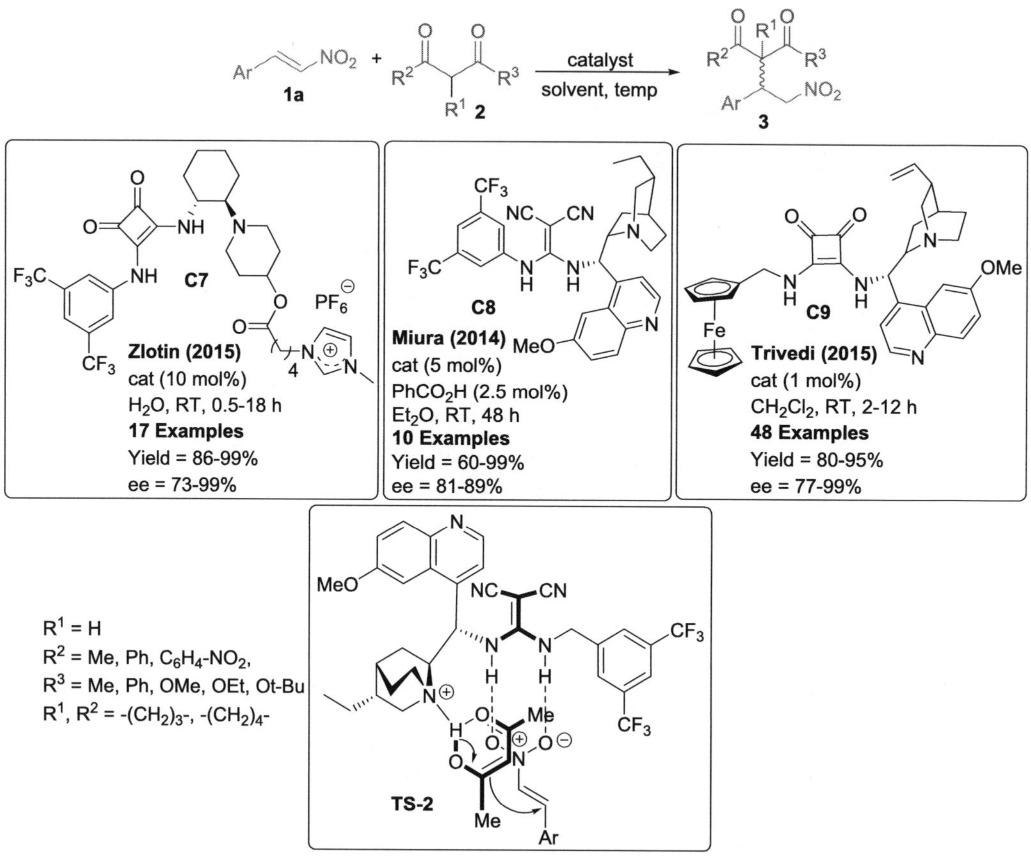

SCHEME 1.4 Ionic liquid-supported squaramide, diaminomethylenemalononitrile and ferrocene-based squaramide as organocatalysts in the Michael addition.

catalyst **C9** was used to afford the corresponding products **3** in high yields and in good-to-excellent enantio and diastereoselectivities. The proposed mechanism involves the initial activation of nitroalkene **1a** by the catalyst *via* H-bonding interactions. Similarly, the enol form of 1,3-diketone **2** also gets activated as a result of hydrogen bonding with the tertiary *N*-atom of the catalyst and deprotonating the acidic hydrogen of diketone **2** to form a carbanion to participate in the reaction and to yield the product **3**.

Dong and co-workers reported C_3-symmetric cinchonine-squaramide **C10** catalyzed enantioselective Michael addition of various 1,3-dicarbonyl compounds **2** to nitroalkenes **1a**.[11] Only 1 mol% of the catalyst **C10** was enough to obtain the corresponding adducts **3** with high enantioselectivities (Scheme 1.5). Moreover, the catalyst can be recycled and reused up to six cycles without any loss of enantioselectivity. The high selectivity has been attributed to a possible reduction in the number of diastereomeric transition states and the high steric demand of the catalyst, which reduces the flexibility of the reactants.

Tanyeli and co-workers designed and synthesized a novel bifunctional acid/base-type organocatalyst, i.e., 2-amino-DMAP-squaramide **C11** for the enantioselective Michael addition of dibenzoylmethane **2** ($R^2 = R^3 = Ph$) to nitroalkenes **1a** (Scheme 1.6).[12] The corresponding Michael adducts **3** were formed with high yields and enantioselectivities. The reaction involves a synergistic cooperation of 2-amino-DMAP and squaramide. In this reaction, the carbanion generated after the deprotonation of dibenzoylmethane **2** is stabilized by hydrogen bonding with 2-aminopyridine. Meanwhile, nitroalkene **1a** is activated by the squaramide catalyst **C11** by H-bonding interactions,

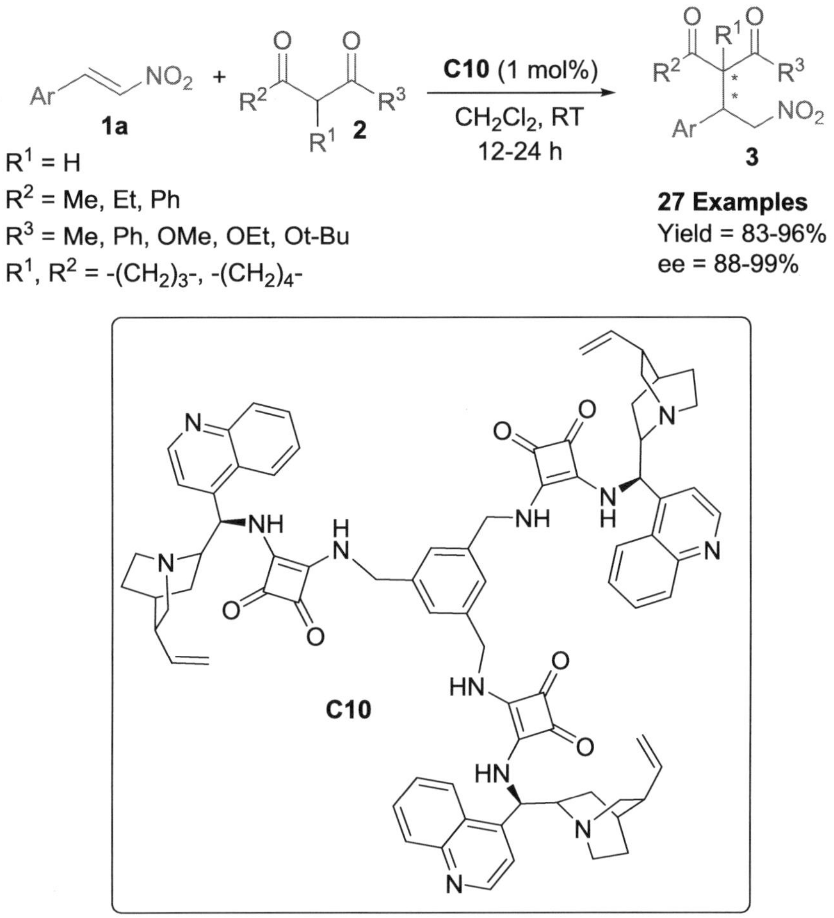

SCHEME 1.5 C_3-Symmetric cinchonine-squaramide as an organocatalyst in the Michael addition of 1,3-dicarbonyl compounds to nitroalkenes.

thus bringing both the substrates in close proximity in the chiral pocket to afford the products **3** with excellent enantioselectivity. The Pericas group, for the first time, employed a chiral squaramide catalyst **C12**, which is covalently immobilized onto a polystyrene resin, for the asymmetric Michael addition of 1,3-dicarbonyl compounds **2** to nitroalkenes **1a** in yielding products **3** with high enantioselectivities (Scheme 1.6).[13] The advantage of the catalyst **C12** is that the 1,2,3-triazole which serves as a linker provides the immobilized squaramide excellent catalytic stability and facilitates its easy recovery for further reuse.

An organocatalytic, asymmetric Michael addition of 1,3-diaryl-1,3-propanediones **2** ($R^2 = R^3 = Ar$) to nitroalkenes **1** was reported by Zhong and co-workers by employing 9-amino-9-deoxyepiquinine **C13** as an organocatalyst to afford corresponding products **3** in 86%–97% yields with 93%–99% ee (Scheme 1.7).[14] Here, the reaction proceeds via a dual activation pathway, wherein both the substrates of the reaction get activated at the same time by the catalyst **C13** to give products **3** with (*S*)-configuration (**TS-3**, Scheme 1.7).

Dong and co-workers developed a bifunctional chiral squaramide-amine organocatalyst **C14** with a pyrrolidine and a cinchona alkaloid in its core structure for the asymmetric Michael addition of 1,3-dicarbonyls **2** to various nitroalkenes **1** affording the corresponding adducts **3** in moderate-to-high yields with excellent enantio and diastereoselectivities (up to 96% yield, 96% ee, 98:2 dr, Scheme 1.7, **TS-4**).[15]

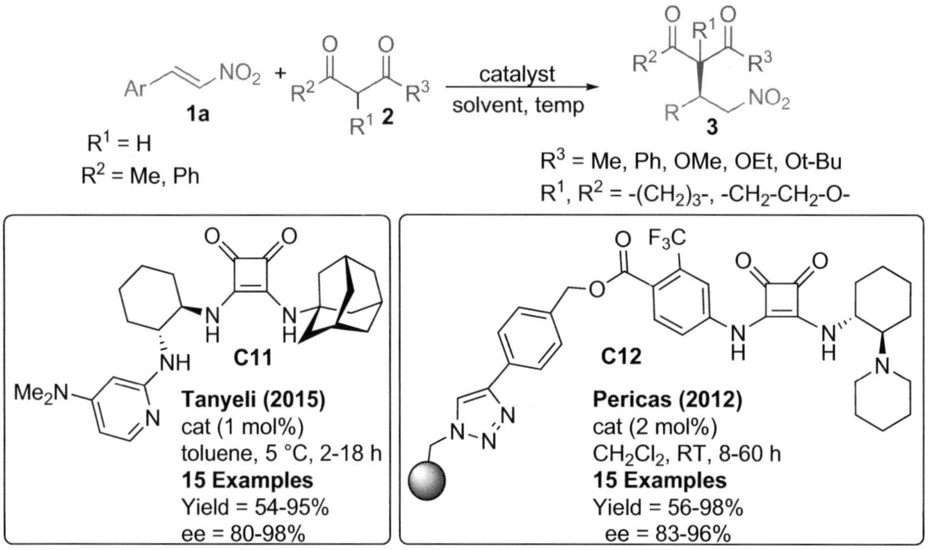

SCHEME 1.6 2-Amino-DMAP-squaramide and polystyrene-supported squaramide as organocatalysts in the Michael addition of 1,3-dicarbonyls to nitroalkenes.

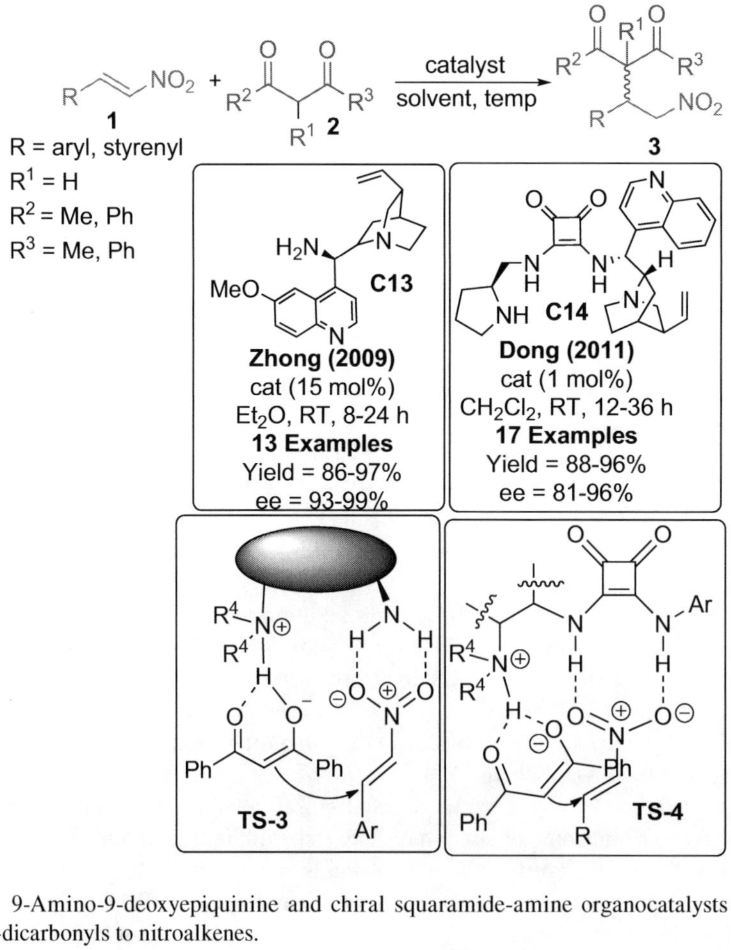

SCHEME 1.7 9-Amino-9-deoxyepiquinine and chiral squaramide-amine organocatalysts in the Michael addition of 1,3-dicarbonyls to nitroalkenes.

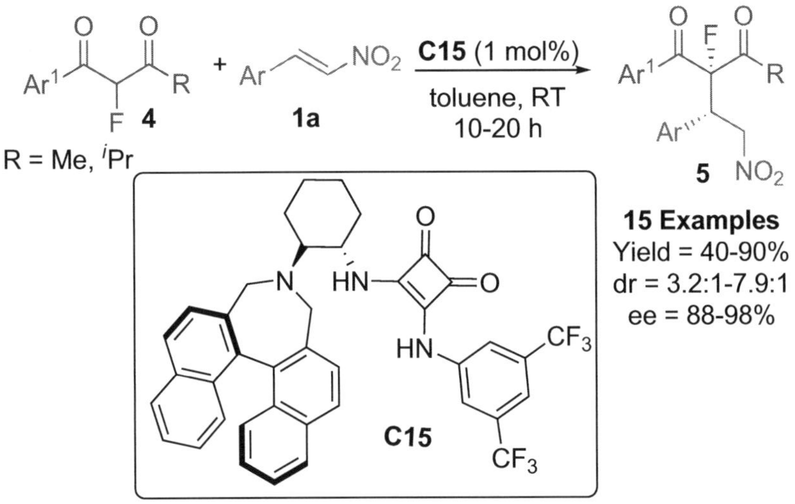

SCHEME 1.8 Asymmetric conjugate addition of 2-fluoro-1,3-diketones to nitroalkenes using binaphthyl squaramide catalyst.

Kim et al. also reported the conjugate addition of 2-fluoro-1,3-diketones **4** to nitroalkenes **1a** under mild conditions using chiral binaphthyl-modified squaramide catalyst **C15** to afford Michael adducts **5** bearing a fluorinated quaternary center (Scheme 1.8).[16]

Recently, Zlotin and co-workers designed a C_2-symmetric tertiary amine-squaramide organocatalyst **C16**, which mediated the asymmetric Michael addition of β-diketones **2** to nitrostyrenes **1a** (Scheme 1.9A).[17] The reactions proceeded efficiently, providing quantitative yields of the enantioenriched products **3a** (ee up to >99%). It was worth observing that the recycling of the unsupported catalysts was carried out up to seven cycles with no loss of catalytic efficiency.

Pan and co-workers reported a cinchona-derived bifunctional thiourea catalyst **C17a**-mediated asymmetric cascade Michael/hemiketalization/acyl transfer reaction between (*E*)-2-(2-nitrovinyl) phenols **1b** and 1,3-propanediones **2a** (Scheme 1.9b).[18] Here, it was observed that the reactions afforded the desired products **3b** in good yields (up to 78%) along with excellent enantioselectivities (up to 97%). The authors explained that the nitro functional group present in the substrate coordinates with the thiourea moiety through H-bonding interactions which will help in directing the addition of enol from the *Si*-face of the nitroalkene (**TS-5**, Scheme 1.9B). This Michael addition is then followed by intramolecular hemiketalization and retro-aldol reaction to afford the final products.

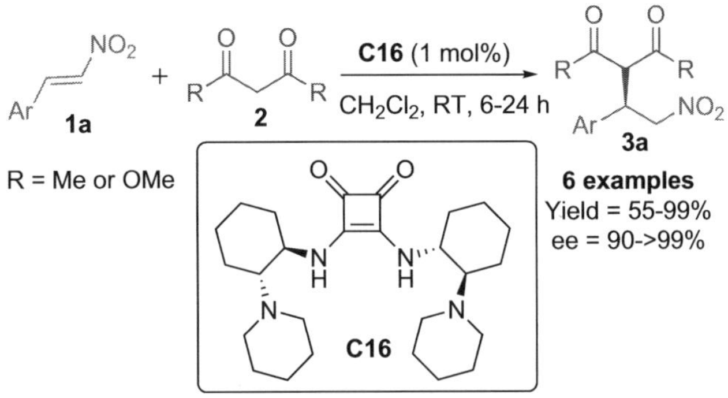

SCHEME 1.9A Asymmetric Michael addition of β-diketones to nitrostyrenes mediated by C_2-symmetric squaramide catalyst.

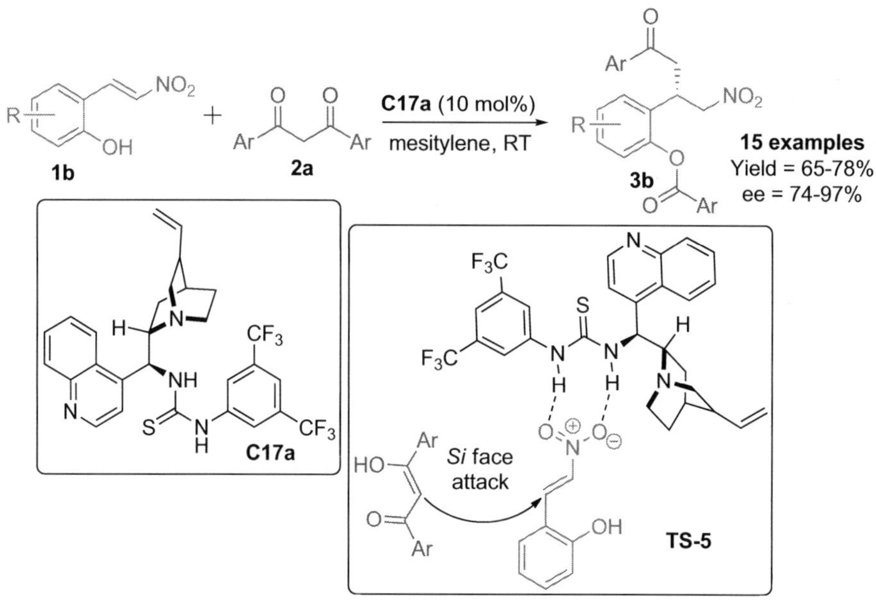

SCHEME 1.9B Cinchona-based bifunctional organocatalyst-mediated asymmetric Michael/hemiketalization/acyl transfer reaction.

1.3 ADDITION OF MALONATES

The Takemoto group reported an asymmetric Michael reaction of malonates **6** to various nitroalkenes **1** by employing a thiourea-based bifunctional organocatalyst with high enantioselectivities (Scheme 1.10).[19] The reaction was also successful without a solvent. The high yields and selectivity of the reaction were ascribed to the presence of both thiourea and a tertiary amino group within the molecule of catalyst **C17**. Li and co-workers described an asymmetric Michael addition of diethyl malonate **6** to nitroalkene **1** under solvent-free conditions using thiourea catalyst **C17**. Various aromatic- and aliphatic-substituted nitroalkenes **1** were employed to afford the corresponding adducts **7** in high yields and enantioselectivities up to 93% ee (Scheme 1.10).[20]

Zlotin and co-workers treated malonates **6** with nitroalkenes **1** in the presence of Takemoto's organocatalyst **C17** in liquid carbon dioxide at 100 bar pressure under homogeneous conditions to give the corresponding Michael adducts **7** in moderate-to-good yields and with enantioselectivities comparable with those obtained in organic solvents (Scheme 1.10).[21] Pápai et al. conducted a computational mechanistic study to confirm the mechanism of enantioselective Michael addition of 1,3-dicarbonyl compounds to nitroalkenes in the presence of thiourea-based chiral catalysts.[22] Based on the findings, a transition state model **TS-6** (Scheme 1.10) was proposed. Hydrogen bonding plays a significant role in the catalytic cycle, amine functionality activates the malonate via hydrogen bonding and the thiourea functionality orients nitroalkene for a favorable nucleophilic attack through hydrogen bonding (**TS-6**, Scheme 1.10).

Vetticat and co-workers also investigated the mechanistic aspects of the enantioselective Michael addition of malonates **6** (R^1 = Me, Et; R^2 = H) to nitroalkenes **1** (R = H, Ph) mediated through Takemoto's catalyst **C17** by performing density functional theory (DFT) calculations and C^{13} kinetic isotope effect studies (Scheme 1.10).[23] The mechanistic studies suggested that the C–C bond formation is the rate-determining step of the catalytic cycle.

The Dixon group synthesized 9-amino(9-deoxy)epicinchonine derivatives with single and double hydrogen bond donors and further utilized them for the asymmetric Michael addition of dialkyl malonate **6** (R^1 = R^2 = alkyl) to nitroalkenes **1** (Scheme 1.11).[24] However, among the various

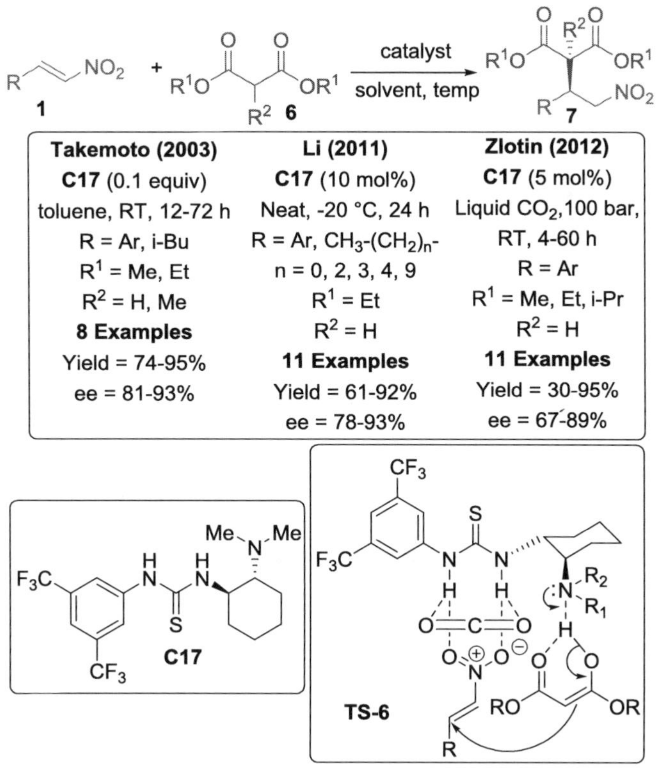

SCHEME 1.10 Michael addition of malonates to nitroalkenes using Takemoto's organocatalyst.

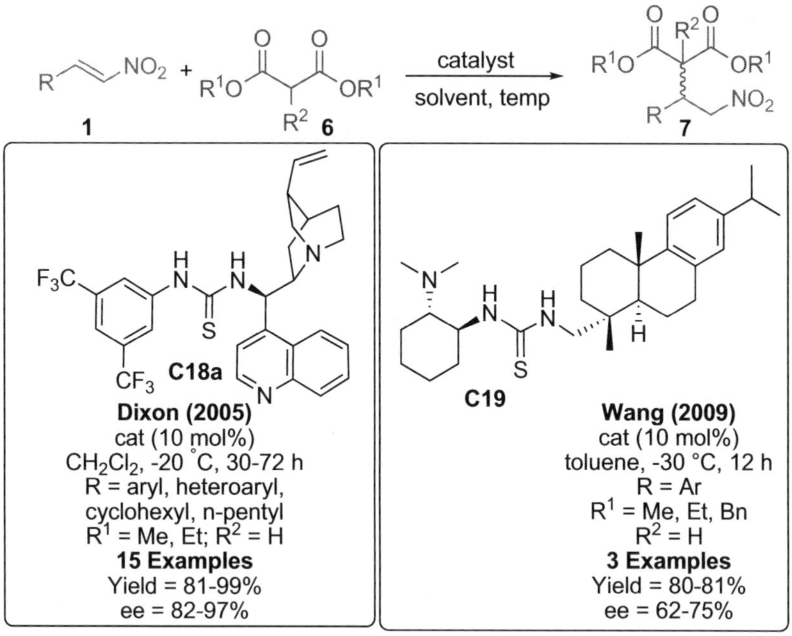

SCHEME 1.11 9-Amino(9-deoxy)epicinchonine thiourea and rosin-derived amine-thiourea as organocatalysts in the Michael addition of malonates to nitroalkenes.

catalysts employed, the thiourea derivative **C18a** provided excellent enantioselectivity. The catalyst possesses adequate positioning of Brønsted acidic and Lewis basic functional groups, which permit the required activation and organization of malonate nucleophiles **6** and nitroalkene **1** resulting in the formation of adducts **7** in good yield and enantioselectivity.

The Wang group designed and synthesized a novel class of bifunctional rosin-derived amine-thiourea catalysts **C19** for the doubly stereocontrolled enantioselective Michael addition of malonate **6** to nitroalkenes **1** (Scheme 1.11).[25] The corresponding adducts **7** containing quaternary carbon centers were formed in good yields and excellent diastereoselectivities and enantioselectivities.

The Connon group reported a thiourea-based cinchona alkaloid **C20** catalyzed asymmetric Michael addition of dimethyl malonate **6** to nitroalkenes **1**.[26] The reaction mainly operates through bifunctional catalysis wherein quinuclidine group assists in the generation of nucleophile by deprotonating the malonate **6**, and the nucleophile thus generated adds to the activated nitroalkene **1** (Scheme 1.12). However, the relative stereochemistry at C8 and C9 of the catalyst **C20** plays a key role in bifunctional catalysis. Moreover, it has been observed that C9 epimeric catalysts are more effective in terms of reaction rate and selectivity in comparison to their "natural" cinchona alkaloid counterparts. The Ma group designed and synthesized a saccharide-substituted bifunctional tertiary amine-thiourea **C21** for the enantioselective Michael addition of malonate **6** to different substituted nitroalkenes **1** (Scheme 1.12).[27] The robustness of the catalyst afforded the products **7** in good yields and enantioselectivities.

The Liu group employed chiral thiourea **C1** organocatalyst for the asymmetric Michael addition of malonates **6** to nitroalkenes **1** under mild conditions.[28] Different substituted nitroalkenes **1** reacted to afford the products **7** in excellent yields and enantioselectivity up to 97% (Scheme 1.12). Kim and co-workers described an asymmetric Michael addition of fluoromalonates **6** to aromatic nitroalkenes **1** by employing chiral amine-thiourea bifunctional organocatalyst **C1** to obtain the corresponding 2-fluoro-2-(2-nitro-1-arylethyl)malonates **7** in high yields and enantioselectivities (Scheme 1.12).[29] This reaction generates a chiral center at the carbon bearing the aromatic group and an adjacent prochiral center from the fluoromalonate.

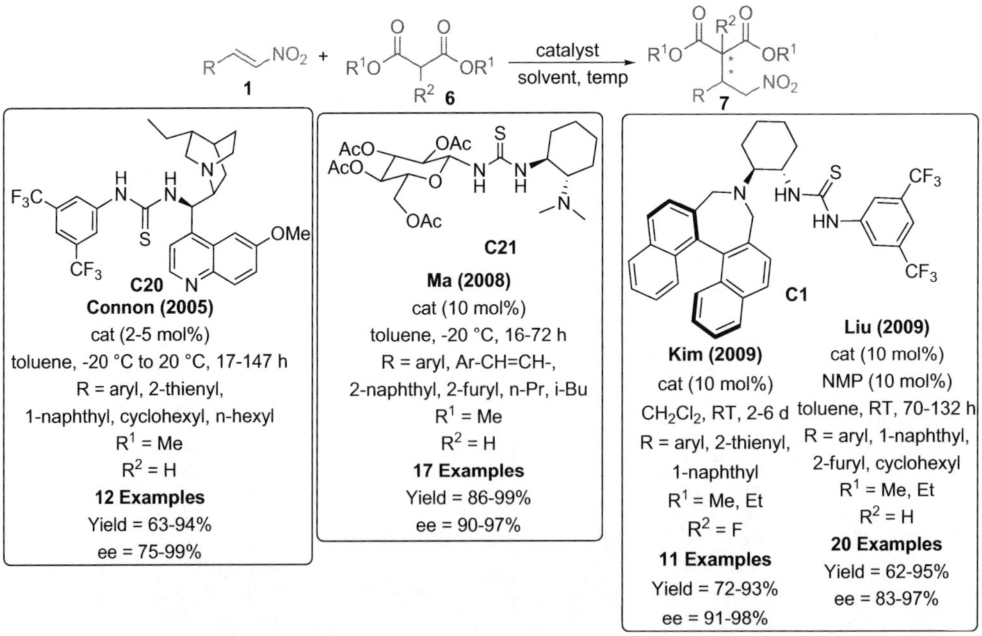

SCHEME 1.12 Cinchona, saccharide and chiral amine-derived thiourea bifunctional organocatalysts in the Michael addition of malonates to nitroalkenes.

Deng and co-workers described the use of cinchona alkaloids **C22** as organocatalysts for the enantioselective Michael addition of dimethyl malonate **6** to nitroalkenes **1** yielding the corresponding adducts **7** in high yields with good enantioselectivity (Scheme 1.13).[30] The reaction followed a first-order dependence on the catalyst **C22**, dimethyl malonate **6** and nitroalkene **1**. A comparative study of catalysts **C22a** and **C22b** revealed that significantly higher enantioselectivity was observed with **C22b**, wherein it acts as a bifunctional catalyst employing both the phenolic OH and the quinuclidine functionalities for stabilizing and organizing the transition state of the enantioselective 1,4-addition.

The Wang group utilized soluble demethylated quinine salt **C23** for the asymmetric Michael addition of malonate **6** to nitroalkenes **1** in water.[31] The environmentally friendly solvent and simple recycling procedure make this protocol attractive to yield the corresponding Michael adducts **7** with high stereoselectivity (Scheme 1.13). The Wang group developed an organocatalytic, asymmetric Michael addition of α-fluoromalonate **6** to nitroalkenes **1** catalyzed by a simple cinchona alkaloid derivative **C24**.[32] The reaction resulted in products **7** with fluorine-bearing quaternary carbon center and an adjacent chiral carbon center with good-to-excellent enantioselectivities (Scheme 1.13).

Terada et al. developed an axially chiral guanidine **C25** as an enantioselective base catalyst for the asymmetric Michael addition of 1,3-dicarbonyl compounds **6** with various nitroalkenes **1** to give a wide array of optically pure nitroalkane derivatives **7** (Scheme 1.14).[33] The Gandelman group designed and synthesized a bifunctional organocatalyst **C26** with a guanine unit and employed it for the asymmetric Michael addition of 1,3-dicarbonyl compounds **6** to different nitroalkenes **1** to obtain the products **7** in high yields and enantioselectivities (Scheme 1.14).[34] Sebesta and co-workers described an enantioselective Michael addition of dimethyl malonate **6** to nitroalkenes **1** catalyzed by squaramide catalyst **C27**. However, when the authors employed polymer-immobilized squaramides, the corresponding Michael adducts **7** were obtained with high enantiomeric purity, but with a drop in the yield (Scheme 1.14).[35]

Park, Jew and co-workers synthesized cinchona alkaloid-derived 2-aminobenzimidazole catalyst **C28** and studied the enantioselective Michael addition of dimethyl malonate **6** with various nitroalkenes **1** (Scheme 1.15).[36] The proposed transition state is shown in Scheme 1.15 (**TS-7**)

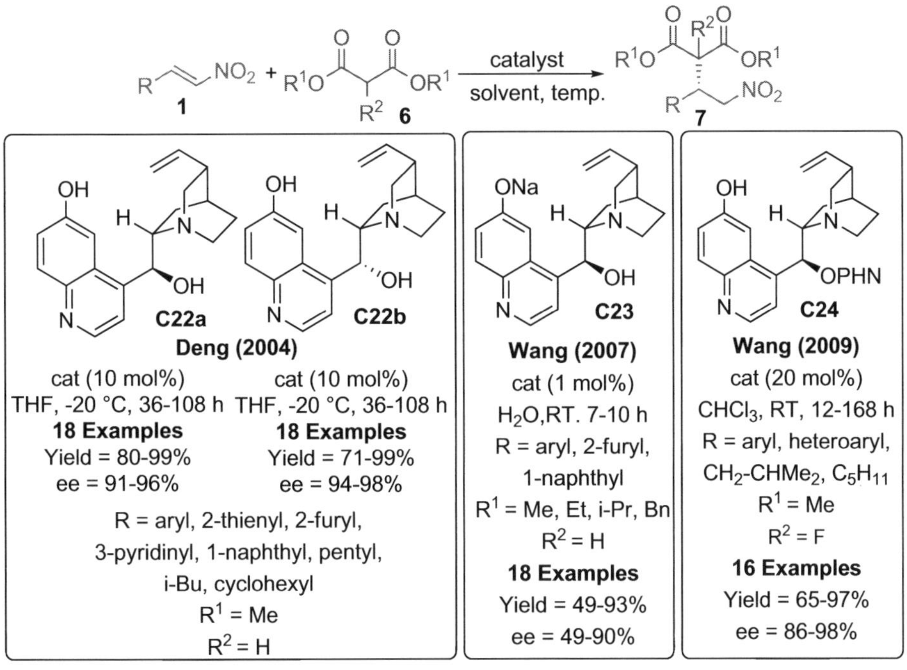

SCHEME 1.13 Cinchona alkaloids as organocatalysts in the conjugate addition of malonates to nitroalkenes.

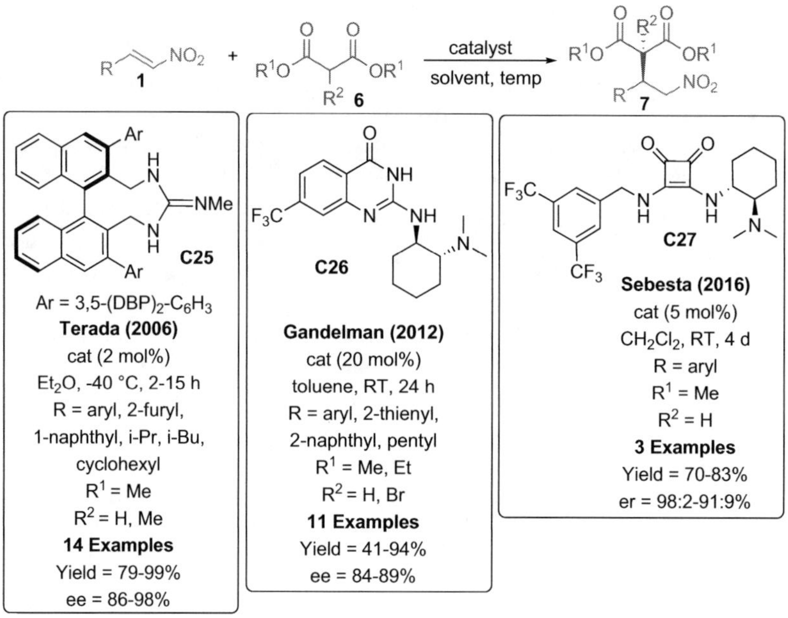

SCHEME 1.14 Guanidine and *tert*-amine-squaramide based organocatalysts in the Michael addition of malonates to nitroalkenes.

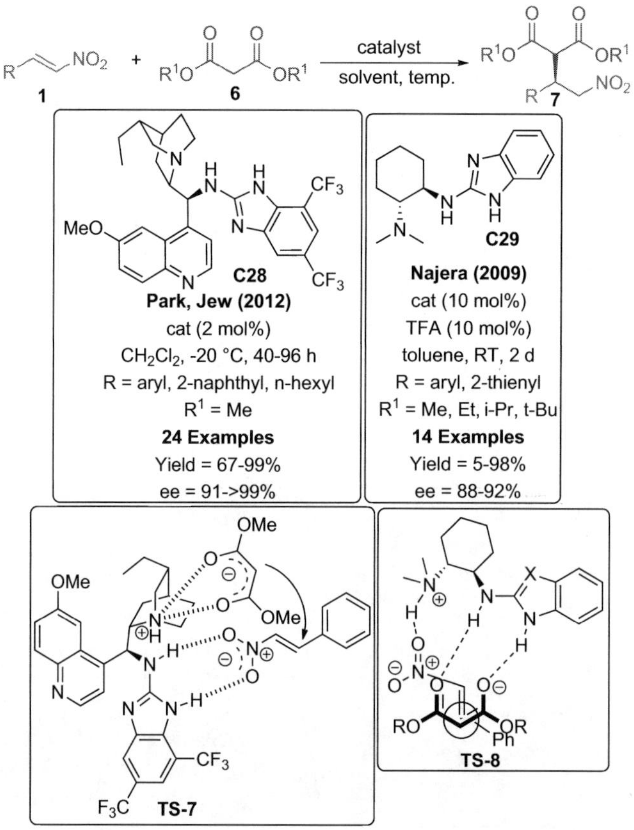

SCHEME 1.15 Benzimidazole-based organocatalysts in the Michael addition of malonates to nitroalkenes.

wherein the nitro group is activated by forming a hydrogen bond with N–H moieties of catalyst **C28**, and the acidic proton of dimethyl malonate is abstracted by the bridged nitrogen of the cinchona moiety. Najera and co-workers employed chiral *trans*-cyclohexanediamine-benzimidazole **C29** as an organocatalyst for the asymmetric conjugate addition of malonate **6** to nitroalkenes **1** in the presence of TFA as a co-catalyst in toluene.[37] The corresponding Michael adducts **7** were obtained in high yields and enantioselectivities (Scheme 1.15). Later, the catalyst **C29** was recovered by a simple acid–base workup. A mechanism was postulated based on DFT studies, wherein the protonated tertiary amine of the cyclohexanediamine activates the nitroalkene **1**, and at the same time, the benzimidazole unit of the catalyst **C29** activates the malonate **6** (**TS-8**, Scheme 1.15).

Later, the Park group reported the enantioselective Michael addition of diethyl malonate to nitroalkenes by (*S*,*S*)-*trans*-cyclohexanediamine-5,7-di-CF_3-benzimidazole **C30** as an organocatalyst (Figure 1.1).[38] A comparative study on the catalytic efficiency of **C30** with the catalyst **C29** devoid of the CF_3 group was conducted, and the results showed that **C30** provided the products **7** with higher chemical yield and enantioselectivity. They systematically investigated the effect of the CF_3 group and the influence of the co-catalyst TFA and explored the catalytic mechanism by revealing the role of guanidine and the dimethylamine moieties present in the catalyst under neutral and acidic conditions. When di-CF_3 substituted catalyst **C30** was employed as a catalyst in the presence of a TFA co-catalyst (10 mol%), the Michael adduct was not obtained. This is because of the electron-withdrawing effect of di-CF_3 which decreases the basicity of guanidine and thus could not abstract the proton of malonate. Based on this observation, the authors proposed transition state models for the reaction in the presence of catalyst **C30**, as well as in the presence of catalyst **C29** and co-catalyst TFA (Figure 1.1). In the case of **C30**, the nitro group of nitroolefin forms hydrogen bonds with two N–H moieties with the assistance of the CF_3 groups, whereas the dimethylamine of the (S,S)-*trans*-cyclohexyldiamine moiety abstracts proton from malonate (**TS-9,** Figure 1.1). Whereas in the presence of **C29** and TFA, the dimethylamine moiety of catalyst is initially protonated by TFA and the guanidine moiety acts as a base to abstract the proton of diethyl malonate to form an ion pair (**TS-10,** Figure 1.1).[39]

Lin and co-workers employed chiral bifunctional bis-alkaloid **C31** derived from pseudoenantiomeric forms dihydroquinine and dihydroquinidine (**C31a** and **C31b**, respectively) as organocatalysts for the asymmetric conjugate addition of dimethyl malonate **6** to nitroalkenes **1**.[39] The reaction

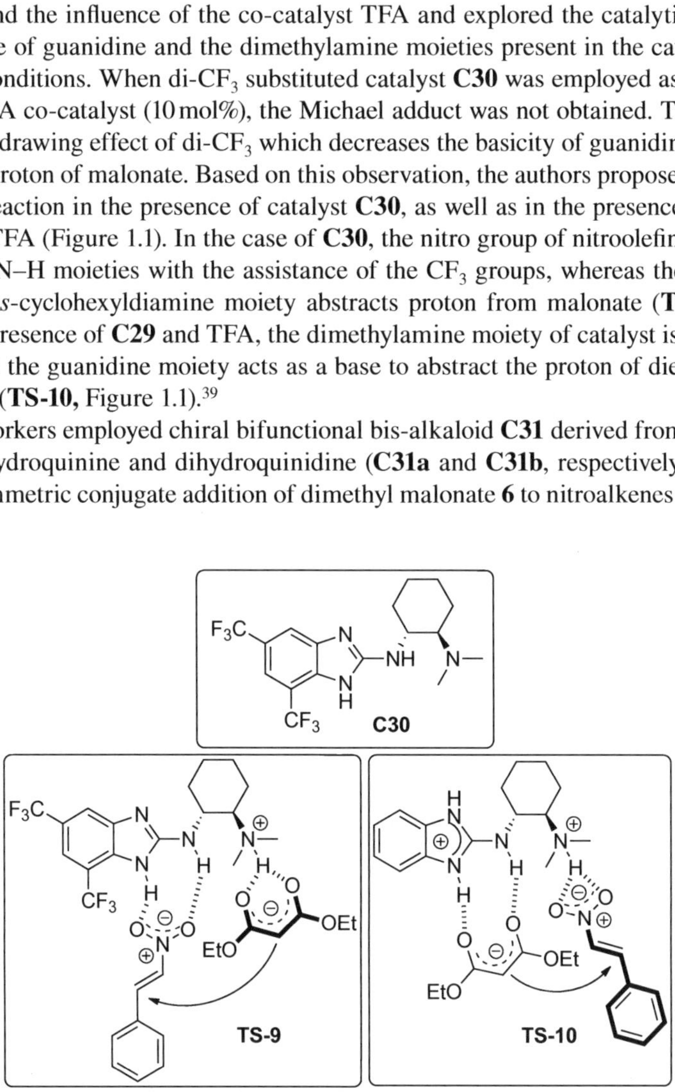

FIGURE 1.1 Proposed transition state models for the neutral conditions in the presence of **C30** and in the presence of **C29** and TFA as a co-catalyst.

offered corresponding adducts **7a** and **7b** in good yields and excellent enantioselectivities under low catalyst loading (1 mol%) (Scheme 1.16).

Evans and co-workers reported a catalytic, enantioselective Michael addition of malonate **6** to nitroalkenes **1** under ambient temperature by employing Lewis acid metal complexed by two chiral diamine ligands (Scheme 1.17).[40,41] The authors revealed by mechanistic studies that the monomeric catalyst **C32** goes through ligand exchange and subsequently activates the dicarbonyl compound **6** through enolization by the diamine ligand. The so-formed chiral enolate participates in enantioselective Michael addition to various nitroalkenes **1** (**TS-11**, Scheme 1.17). Kim and Kang developed an efficient catalytic asymmetric Michael addition of fluoromalonates **6** to nitroalkenes **1** by employing the same chiral nickel catalyst **C32** to afford the corresponding γ-nitro-α-fluoro carbonyl compounds **7** in good-to-high yields along with excellent stereoselectivities (up to 97% ee) (Scheme 1.17).[42]

The Czekelius group employed chiral C_2-symmetric 1,2-diamine **C33** based on a 1,1-bi(tetrahydroisoquinoline) as a chiral ligand in the asymmetric Ni(II)-catalyzed Michael addition of malonates **6** to nitroalkenes **1**. The reaction offered 92%–98% yields of the products **7** with 91%–99% enantioselectivity (Scheme 1.18).[43] The Kobayashi group developed a new coordinative calcium-pybox (**L1**) catalyst for the catalytic asymmetric Michael addition of malonate **6** to nitroalkenes **1** (Scheme 1.18).[44] The authors claim this reaction to be the first of its kind which uses a chiral, coordinative alkaline earth metal catalyst in the enantioselective C–C bond-forming reactions. In the proposed mechanism, the α-position of malonate **6** gets deprotonated by the calcium-pybox (**L1**) complex to form a chiral calcium enolate. This intermediate then reacts with nitroalkene **1** to form the initial 1,4-adduct. Subsequent protonation by a phenol derivative gives the Michael adduct **7** with the regeneration of the catalyst (**TS-12**, Scheme 1.18). The Xia group demonstrated the use of newly synthesized chiral diamine biisoindoline ligands **L2** by the Ni-catalyzed asymmetric Michael addition of malonates **6** to nitroalkenes **1** to obtain the corresponding Michael adducts **7** in excellent yields and enantioselectivities (Scheme 1.18).[45]

Li and co-workers immobilized chiral bis-(cyclohexyldiamine)-based Ni(II) complexes **C34a-b** within periodic mesoporous organosilica (PMO) to form a chiral, heterogeneous, PMO-supported Ni(II) catalyst.[46,47] Subsequently, this catalyst **C34a-b** was employed for the asymmetric Michael

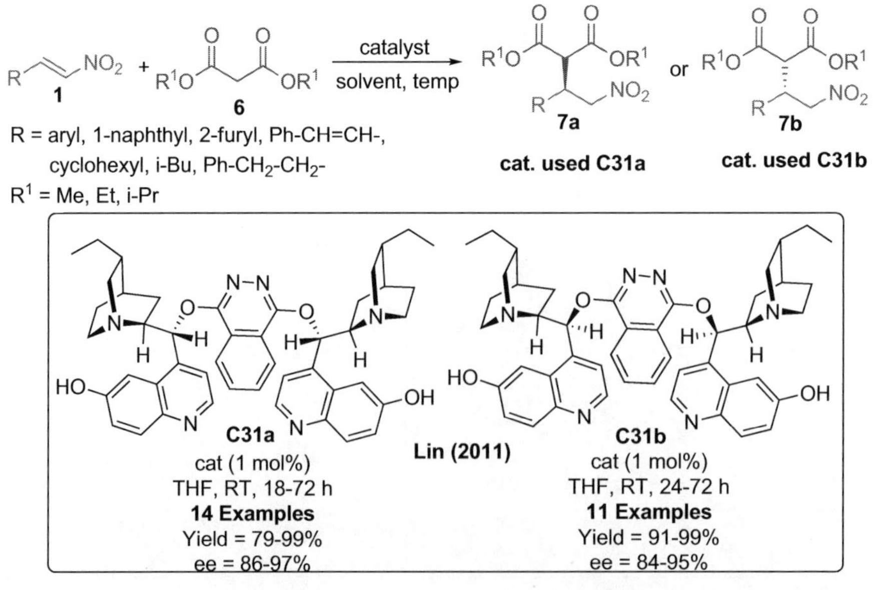

SCHEME 1.16 Bis-alkaloid derived from dihydroquinine and dihydroquinidine as organocatalysts in the Michael addition of malonates to nitroalkenes.

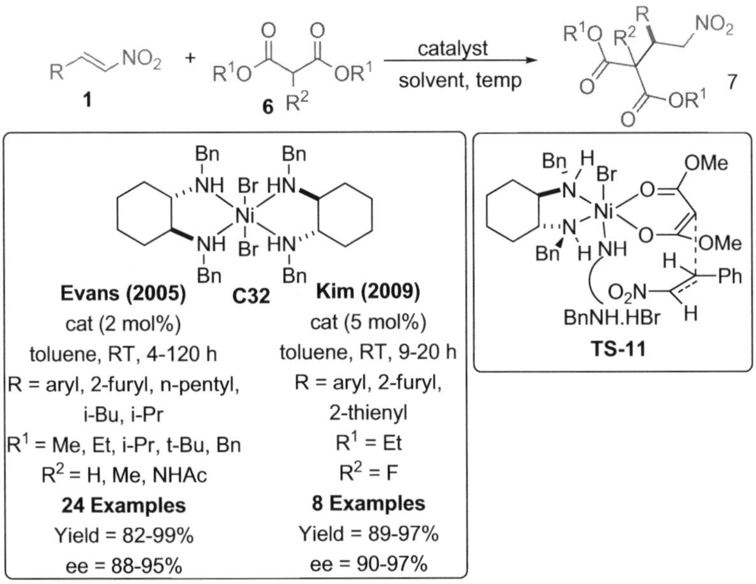

SCHEME 1.17 Chiral nickel complexes as catalysts in the Michael addition of malonates to nitroalkenes.

addition of malonates **6** to nitroalkenes **1** to give products **7** in high yields and enantioselectivities (Scheme 1.19). Interestingly, the catalyst **C34a-b** can be recovered and reused nine times without affecting its enantioselectivity.

Scherrman and co-workers reported the enantioselective 1,4-addition of dimethyl malonate **6** (R^1 = Me) to *trans*-β-nitrostyrene **1** (R = Ph) mediated by recyclable silica-supported cupreine **C35** as a catalyst using a biomass-derived solvent (Scheme 1.19).[48] This heterogeneous catalytic system was capable of producing the products **7** with good enantioselectivities and being recycled up to three to five times without losing catalytic efficiency.

Takemoto and co-workers conducted a detailed study of the structure–activity relationship of various bifunctional catalysts containing thiourea and amino group attached to the chiral center toward the asymmetric Michael addition of malonates **6** to nitroalkenes **1**.[49] Among the various catalysts screened, **C17** with 3,5-bis(trifluoromethyl)benzene and dimethyl amino group at the chiral center displayed excellent results. Using **C17**, they achieved the total synthesis of *R*-(-)-baclofen **8** following mild steps and demonstrated the formation of contiguous stereogenic centers. The Nichols group demonstrated an asymmetric Michael addition of α-substituted malonates **6** to various substituted nitroalkenes **1** in the presence of ligand **L3** to afford products **7** with a stereocenter at the carbon bearing the aromatic group and an adjacent prochiral center from the α-substituted malonate **6** (Scheme 1.20).[50] The nitro group in the adducts can be reduced to an amino group followed by intramolecular cyclization with the ester group to provide pyrrolidinones with two contiguous stereocenters, one of which is quaternary.

Bakó and co-workers employed α-D-glucopyranoside- and α-D-mannopyranoside-based crown ethers **C36** as the chiral catalysts for the asymmetric conjugate addition of diethyl malonate **6** to nitroalkenes **1** to obtain the corresponding adducts **7** in good yields and enantioselectivity of up to 99% ee (Scheme 1.20).[51] The reaction was carried out under solid–liquid phase transfer conditions. In the possible reaction pathway, it was reasoned that sodium carbonate, which is initially employed in the reaction, forms a complex with the crown ether, which in turn deprotonates the malonate **6** to form sodium enolate. Addition of this enolate to nitroalkene **1** followed by protonation yields Michael adduct **7** (**TS-13**, Scheme 1.20). The Michael adduct **7** with appropriate substitution can be elaborated to the muscle relaxant drug (*R*)-Baclofen.

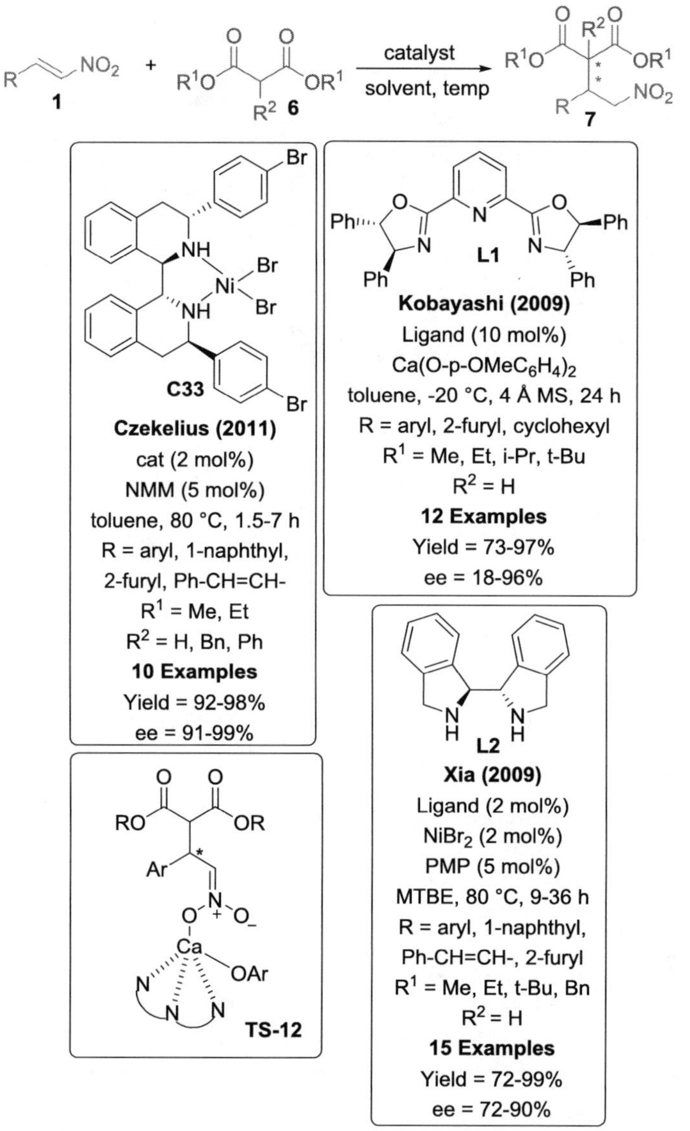

SCHEME 1.18 Ni(II)-bi(tetrahydroisoquinoline), Ni(II)-biisoindoline and calcium-pybox as catalyst systems for the enantioselective Michael addition of malonates to nitroalkenes.

Yan and co-workers described the stereoselective conjugate addition of malonates **6** to 3-nitro-2H-chromenes **9** using various chiral organocatalysts.[52] It was observed that among the catalysts employed, a bifunctional thiourea-tertiary amine-based catalyst **C37** was found to be the most effective for the reaction. Good yields and enantioselectivities were observed with different substituted 3-nitro-2H-chromenes **9** (Scheme 1.21). This reaction was assumed to proceed through a bifunctional catalytic mechanism. In the proposed mechanism, in addition to the electrostatic attraction between the enolate anion generated by the deprotonation of malonate **6** and the resulting ammonium cation, hydrogen bonding was also anticipated to assist the interaction. This was followed by the subsequent nucleophilic attack of enolate and the proton transfer, which resulted in enantioenriched chromane derivatives **10** (**TS-14**, Scheme 1.21).

The Li group described the chiral nickel complex **C32** catalyzed asymmetric Michael addition of 1,3-dicarbonyl compounds **6** to 3-nitro-2H-chromenes **9** to afford highly functionalized adducts

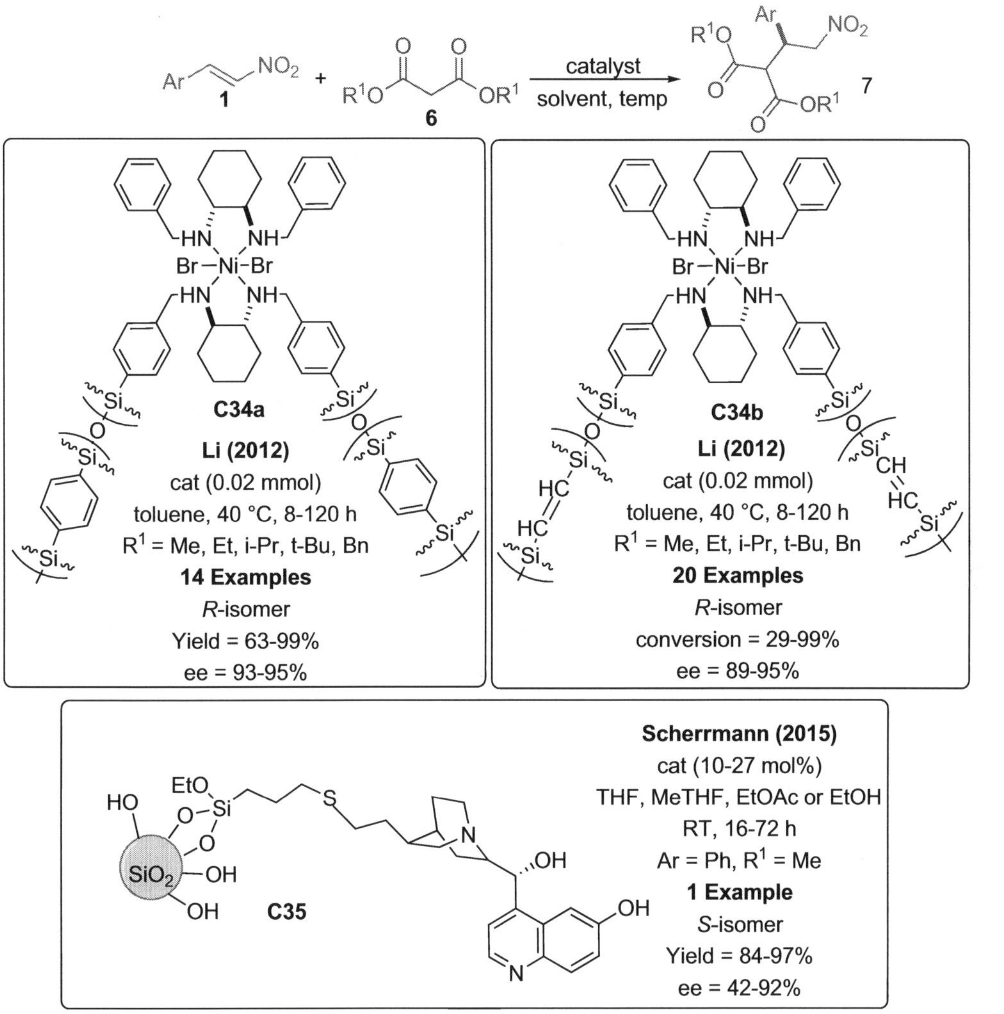

SCHEME 1.19 Organonickel-functionalized mesoporous silica and recyclable cupreine-based catalysts in the Michael addition of malonates to nitroalkenes.

10 with two contiguous carbon stereocenters in high yields and enantioselectivities (Scheme 1.21, **TS-15**).[53] The group also reported an enantioselective Michael addition of malonates **6** to 3-nitro-2H-chromenes **9** using a catalytic amount of thiourea-based cinchona alkaloids **C18b** (Scheme 1.21).[54] The corresponding adducts **10** showed excellent stereoselectivity. The reaction proceeds through a dual activation pathway. Here, both the substrates are activated by the thiourea and an amino group of the catalyst **C18b**, respectively. The enolate anion generated via the deprotonation of the malonate **6** by the amino group of the catalyst adds to the nitroalkene **9** affording the corresponding adduct **10** in high yields and good enantioselectivities.

Peng, Shao and co-workers reported an asymmetric Michael addition of 1,3-dicarbonyl compounds **6** to nitroenynes **11a** by employing cinchona alkaloid-based thiourea organocatalyst **C20**.[55] The corresponding 1,4-addition products **12a** were obtained in good yields with excellent enantioselectivities (Scheme 1.22). Interestingly, an opposite enantiomer of the previously observed product was obtained when the pseudoenantiomer derived from quinidine was used as the catalyst. The absolute configuration of the product was assumed to be '*R*' as it was depicted by the transition state. In the transition state, the deprotonated malonate approaches the *Re*-face of the nitroenyne

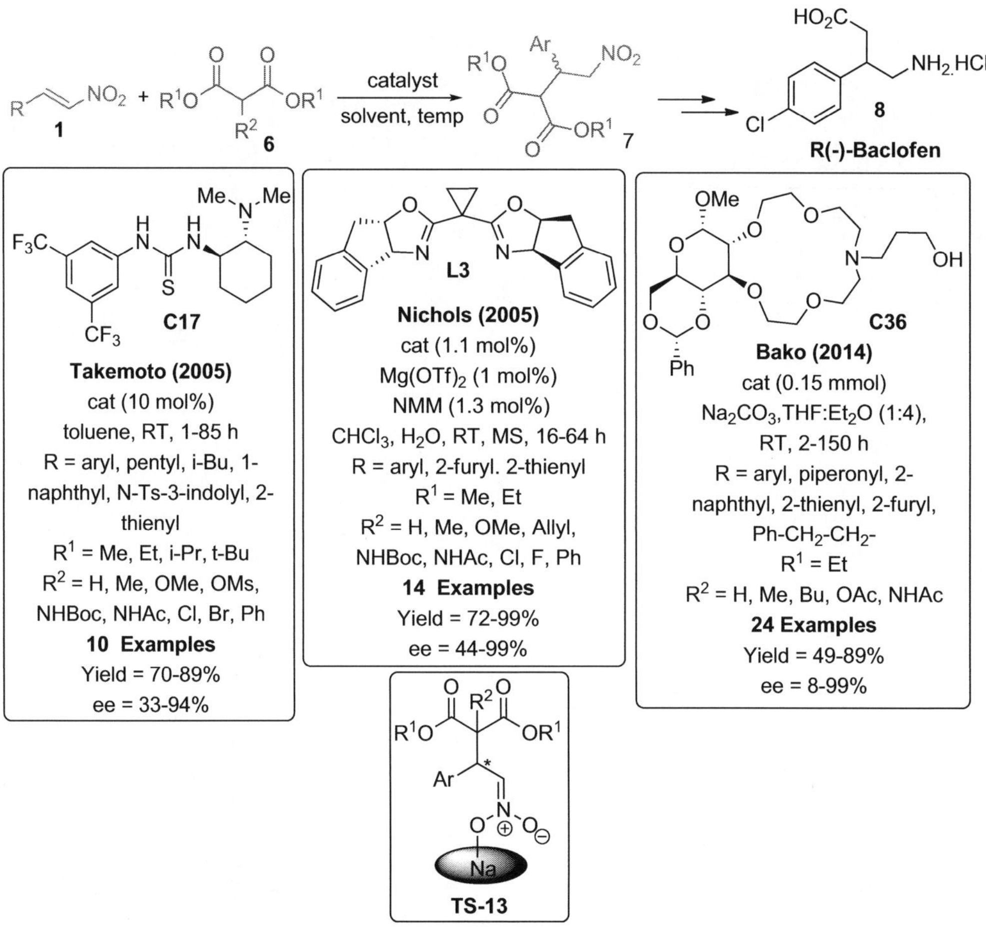

SCHEME 1.20 Bifunctional thiourea, Mg-oxazoline complex and carbohydrate-based crown ether as catalysts in the Michael addition of malonates to nitroalkenes.

11a to give the product **12a** with (*R*)-configuration (**TS-16**, Scheme 1.22). Shao and co-workers also reported an asymmetric conjugate addition of malonates **6** to nitroenynes **11a** catalyzed by a chiral nickel(II)-diamine catalyst **C38** to afford highly functionalized chiral β-alkynyl esters **12a** in good yields and enantioselectivities with a nitro group in the core structure (Scheme 1.22).[56] The Michael adducts **12** can act as potential building blocks for the stereoselective synthesis of conformationally constrained bicyclic *γ*-amino acids bearing an alkynyl side chain with an adjacent quaternary carbon chiral center. Ghosh et al. revealed the highly regio and stereoselective addition of malonates **6** to nitrodienes **11b** using a cinchona-based thiourea organocatalyst **C20b** (Scheme 1.22).[57] Similar results were obtained with 1,3-diketones and β-ketoesters. The product **12b** was amenable for further transformation by the chemoselective reduction of the nitro group followed by subsequent lactamization.

The Yan group described an asymmetric synthesis of nitrocyclopropanes **13** by an organocatalytic conjugate addition of dimethyl bromomalonate **6** to nitroalkenes **1** followed by intramolecular cyclopropanation in the presence of 6′-demethylquinine **C39** (Scheme 1.23).[58] High yields and excellent enantio and diastereoselectivities were observed in this reaction irrespective of the nature of substituents. Russo and Lattanzi reported an asymmetric Michael addition of dimethyl bromomalonate **6** to nitroalkenes **1** catalyzed by (*S*)-α,α-di-β-naphthyl-2-pyrrolidinemethanol catalyst **C40** to form adducts **7** which without isolation was treated with 1,4-diazabicyclo[2.2.2]octane (DABCO)

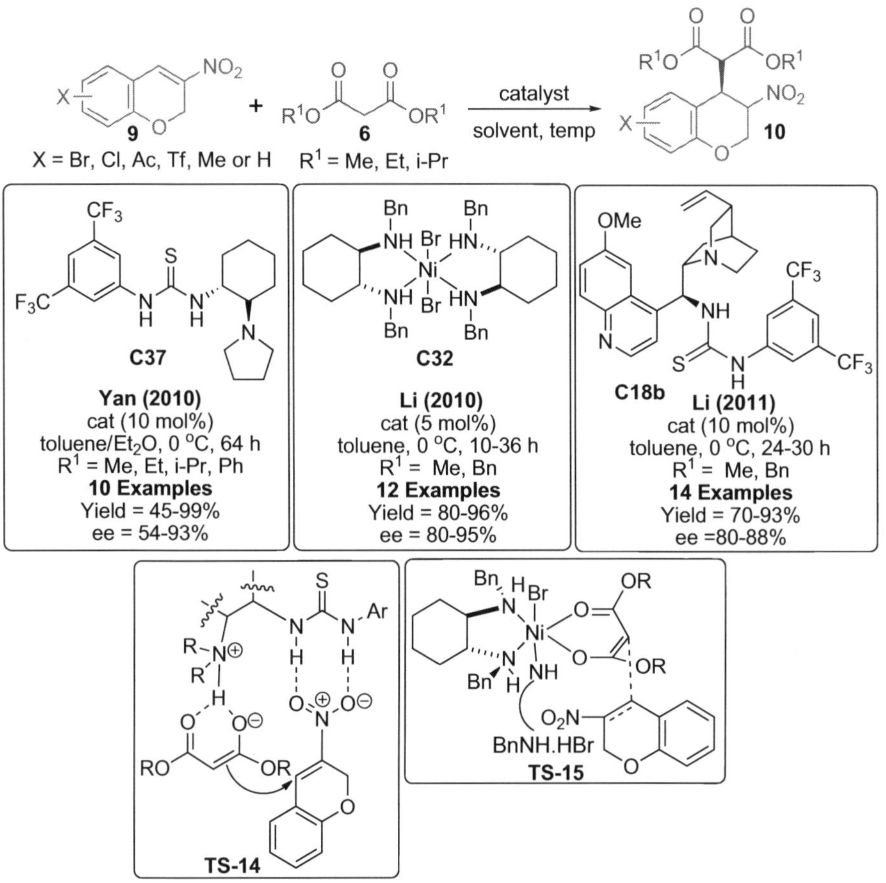

SCHEME 1.21 Asymmetric conjugate addition of malonates to nitrochromenes.

to facilitate intramolecular nucleophilic substitution which afforded functionalized nitrocyclopropanes **13** as diastereomerically pure *trans*-isomers in good yield with moderate enantioselectivity (Scheme 1.23).[59] The Kim group reported a catalytic asymmetric synthesis of functionalized nitrocyclopropanes **13** in excellent diastereoselectivities and enantioselectivities by the chiral nickel complex **C41** catalyzed Michael addition of bromomalonates **6** to nitroalkenes **1** followed by cyclization in the presence of a base (Scheme 1.23).[60] Ni catalyst activates bromomalonates **6**, and Michael addition of the bromomalonate anion occurs at the *Si*-face of the double bond of nitroalkenes **1** (**TS-17**, Scheme 1.23).

In 2006, the Connon group, for the first time, reported an organocatalytic asymmetric synthesis of functionalized nitrocyclopropanes **13** by the conjugate addition of dimethyl chloromalonate **6** to diversely substituted nitroalkenes **1** in the presence of a chiral bifunctional cinchona alkaloid-based organocatalyst **C20** (Scheme 1.24).[61] The cyclization of Michael adduct **7** to deliver the nitrocyclopropane **13** was achieved using DBU under controlled conditions with moderate enantioselectivity.

The Song group described an enantioselective Michael addition of dithiomalonates **14** to highly unreactive β,β-disubstituted nitroalkenes **15** in brine by employing a chiral squaramide catalyst **C42** (Scheme 1.25).[62] The presence of water in the reaction medium enhances hydrophobic hydration and thus increases the hydrophobic interaction between the catalyst and substrates. Pharmaceutically relevant chiral β-substituted γ-aminobutyric acid (GABA) analogs were synthesized in one pot by employing this strategy.

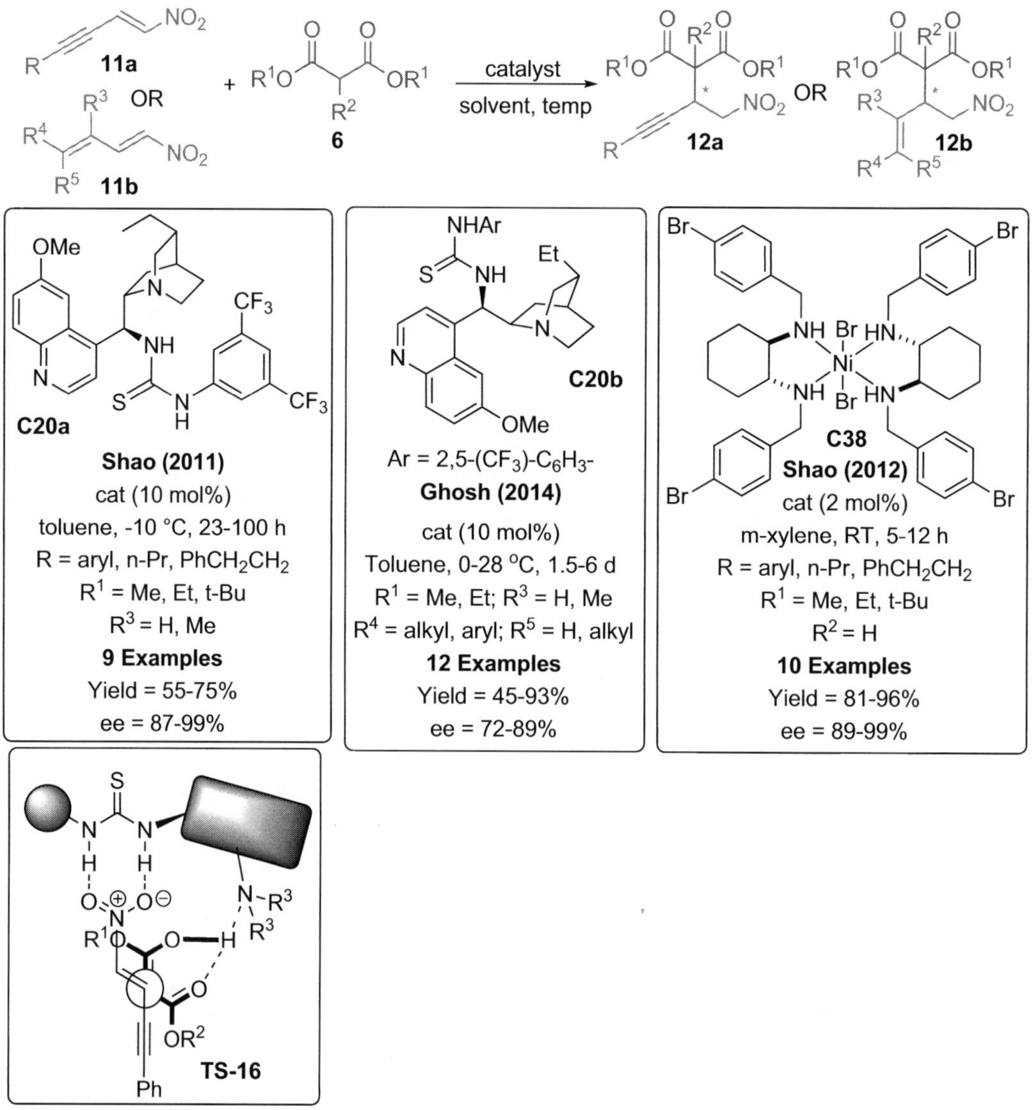

SCHEME 1.22 Asymmetric Michael addition of malonates to nitroenynes and nitrodienes.

1.4 ADDITION OF β-KETOESTERS

In 2016, Naicker and co-workers published a detailed review article on the organocatalyzed enantioselective transformations of β-ketoesters.[63] As the above review covered the literature on the asymmetric reactions of β-ketoesters with nitroalkenes, here we delineated the work of others not included in that review. However, to clarify what is discussed in that review, we provide here the references which were cited in the above review.[49,64–83]

Shibasaki and co-workers described a catalytic asymmetric conjugate addition of β-ketoesters **17** to nitroalkenes **1** using a homo-bimetallic bifunctional Co_2-Schiff base complex **C43** as a catalyst system (Scheme 1.26).[84] The reaction proceeded well and afforded the products **18** in excellent yields and enantioselectivity of up to 99%. According to the mechanistic studies, the reaction was found to operate through intramolecular cooperative interactions of the two Co-metal centers.

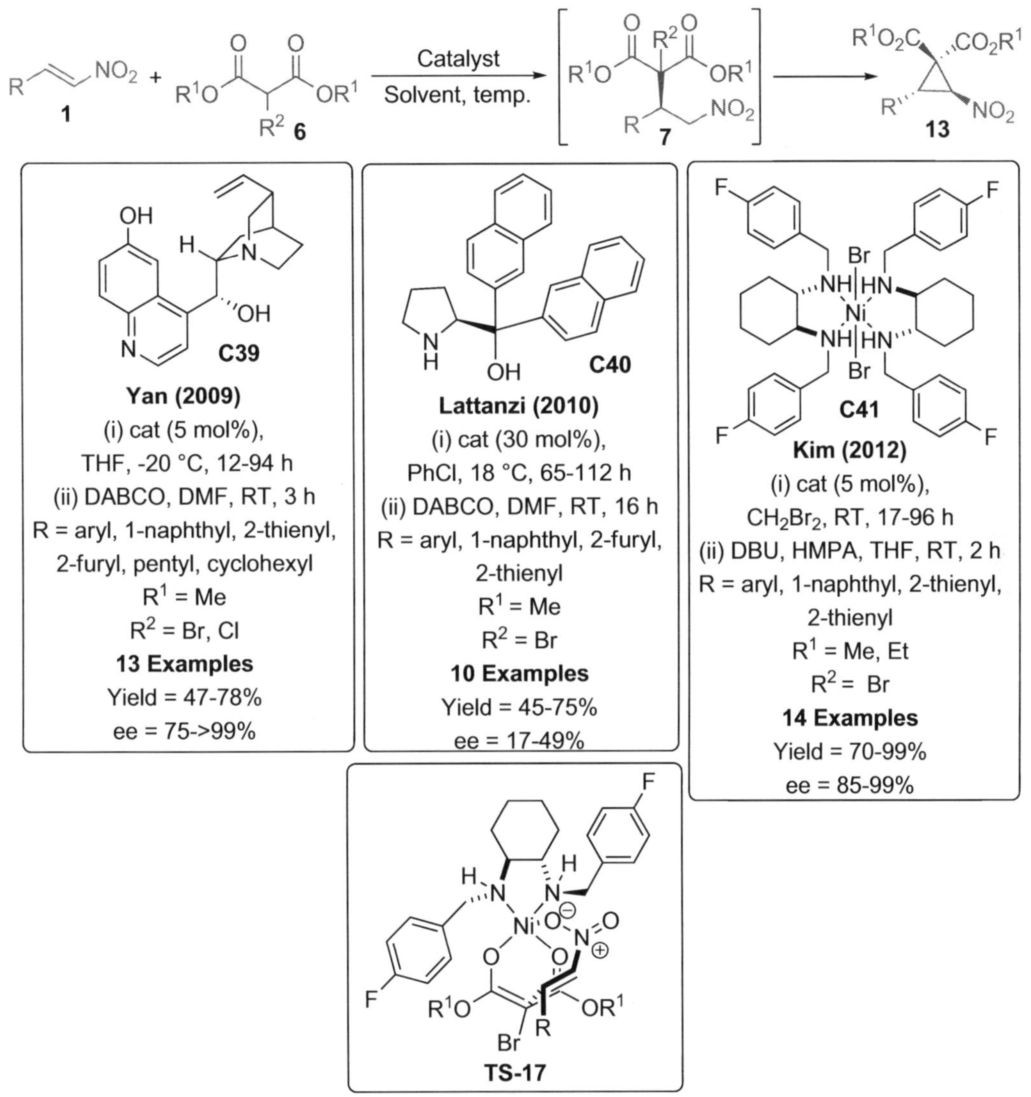

SCHEME 1.23 Asymmetric synthesis of nitrocyclopropanes by organocatalytic conjugate addition of halomalonates to nitroalkenes.

The Bolm group designed and synthesized ephedrine- and pseudoephedrine-derived thiourea catalysts and investigated the asymmetric Michael addition of diketones and β-ketoesters to nitroalkenes **1** (Scheme 1.26).[4] Among these, pseudoephedrine-derived thiourea **C4** with 3,5-trifluoromethylphenyl substituents showed better catalytic efficiency toward the conjugate addition of diketones to nitroalkenes. Using the same catalyst **C4**, the authors studied the asymmetric Michael addition of β-ketoesters **17** to nitrostyrene **1**, which afforded the corresponding adducts **18** in good yields with moderate diastereo and enantioselectivities.

Kim and co-workers reported a catalytic, asymmetric Michael addition of α-fluoro-β-ketoesters **19** with nitroalkenes **1** by employing bifunctional organocatalyst **C1** to afford γ-nitro-α-fluorocarbonyl compounds **20** in high yields and enantioselectivities.[85] A notable feature of this reaction is that a quaternary stereocenter possessing a fluorine atom and a contiguous stereogenic carbon center were generated with a high level of enantioselectivity (Scheme 1.27).

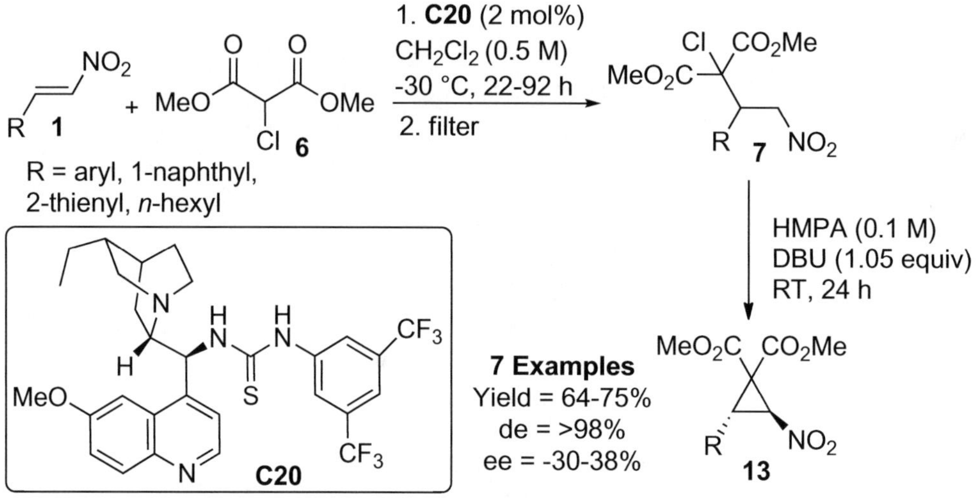

SCHEME 1.24 Asymmetric synthesis of functionalized nitrocyclopropanes by the conjugate addition of dimethyl chloromalonate to nitroalkenes.

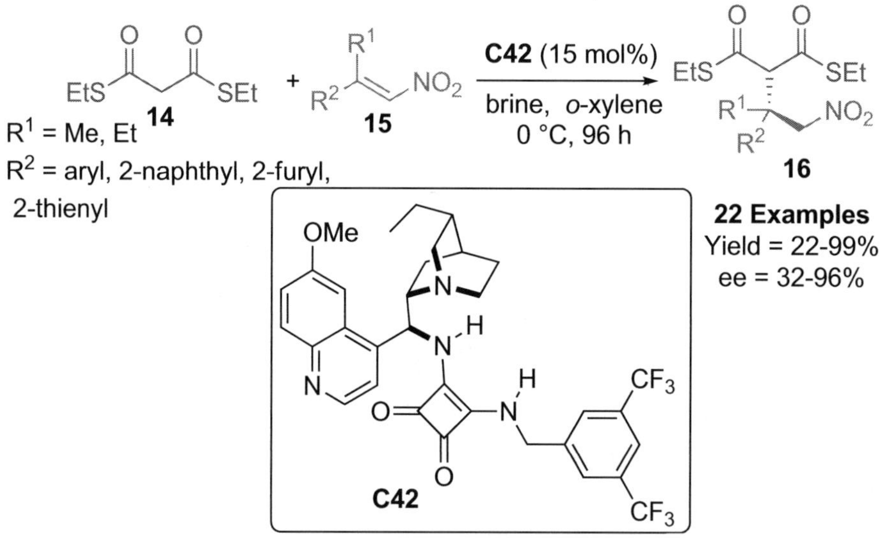

SCHEME 1.25 Michael addition of dithiomalonates to β,β-disubstituted nitroalkenes.

The same authors developed a catalytic, enantioselective conjugate addition of cyclic β-ketoesters **21a** to nitroalkenes **1** for the preparation of γ-nitrocarbonyl compounds **22a** using bifunctional organocatalyst **C1** in good-to-high yields with excellent enantioselectivities (Scheme 1.28).[86]

The Itsuno group introduced a chiral polymer of cinchonidine-based sulfonamide **C44** for the Michael addition of β-ketoesters **21** to nitroalkenes **1** (Scheme 1.29).[87] Even though the polymer catalyst **C44** was insoluble in organic solvents, the reaction provided Michael adducts **22** with high diastereoselectivity (>10:1) and moderate-to-excellent enantioselectivity (>97%). As the catalyst **C44** was insoluble, they recovered the catalyst and reused it several times without losing reactivity.

Peters and co-workers reported the synthesis of a polyfunctional catalyst system **C45** in which an imidazolium-aryloxide betaine moiety cooperates with a Lewis acidic metal center Cu(II) within a chiral catalyst skeleton for the enantioselective Michael addition of β-ketoesters **21** to β-substituted

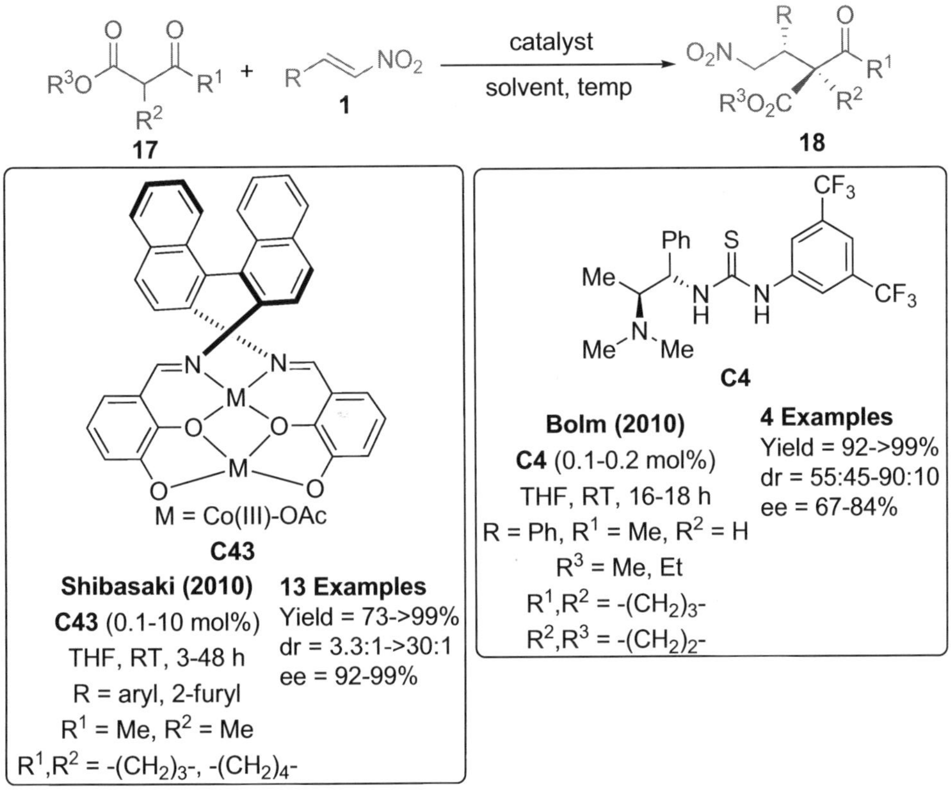

SCHEME 1.26 Asymmetric Michael addition of β-ketoesters to nitroalkenes catalyzed by a Co-Schiff's base complex and pseudoephedrine-derived thiourea.

nitroalkenes **1** (Scheme 1.30).[88] This strategy allowed access to the rare diastereomer arising from the direct 1,4-addition of different 1,3-dicarbonyl substrates to β-substituted nitroalkenes in good yields with high enantioselectivities. The extensive mechanistic investigations revealed that the aryloxide moiety acts as a base to form a Cu(II)-bound enolate, whereas the imidazolium group and the phenolic OH, generated during the proton transfer, together activate the nitroalkene by hydrogen bonding.

Nakano and co-workers designed a chiral azanorbornane-based amino alcohol catalyst **C46**, which was synthesized to mediate the asymmetric Michael addition of β-keto esters **21** to nitroalkenes **1** (Scheme 1.31).[89] It was demonstrated by the authors that the N-phenylethylated azanorbornane bearing a mono-phenylmethanol moiety successfully mediated the asymmetric reaction, and the corresponding Michael adducts **22** possessing a quaternary chiral carbon center were obtained with good yields (up to 99%) and high stereoselectivities (up to 91:9 dr and up to 91% ee).

Nakano and co-workers also reported chiral amino amide organocatalyst **C47** bearing a naphthyl ring at the amido group for the asymmetric version of the Michael reaction among β-ketoesters **21** and nitroalkenes **1** (Scheme 1.32).[90] The corresponding Michael adducts **22** were isolated in good yields and high stereoselectivities (up to 99:1 dr and up to 98% ee). The good catalytic activity was explained by the authors as the consequence of the aid given by the presence of a primary amine for enamine formation, amido group for hydrogen bonding and naphthyl ring as a shielding moiety to enhance the stereochemical output.

Kesavan et al. described an asymmetric synthesis of armeniaspirol analogs. The reaction involves squaramide **C48** catalyzed Michael addition of β-ketoester **23** to nitroalkenes **1** to give corresponding adducts **24** followed by further transformation to natural product analogs **25** and **26** in good yields and excellent diastereoselectivities (Scheme 1.33).[91]

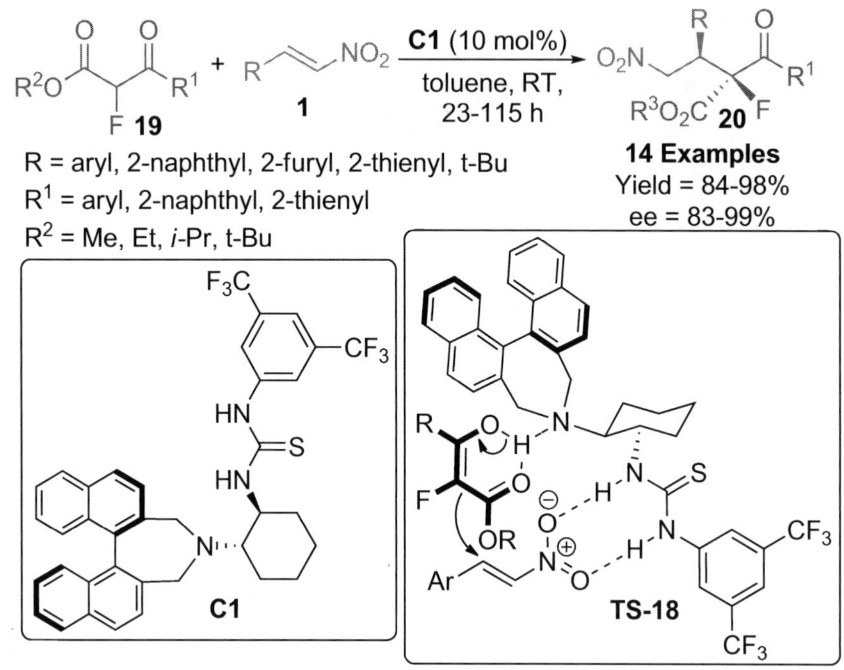

SCHEME 1.27 Asymmetric Michael addition of α-fluoro-β-ketoesters to nitroalkenes catalyzed by bifunctional thiourea.

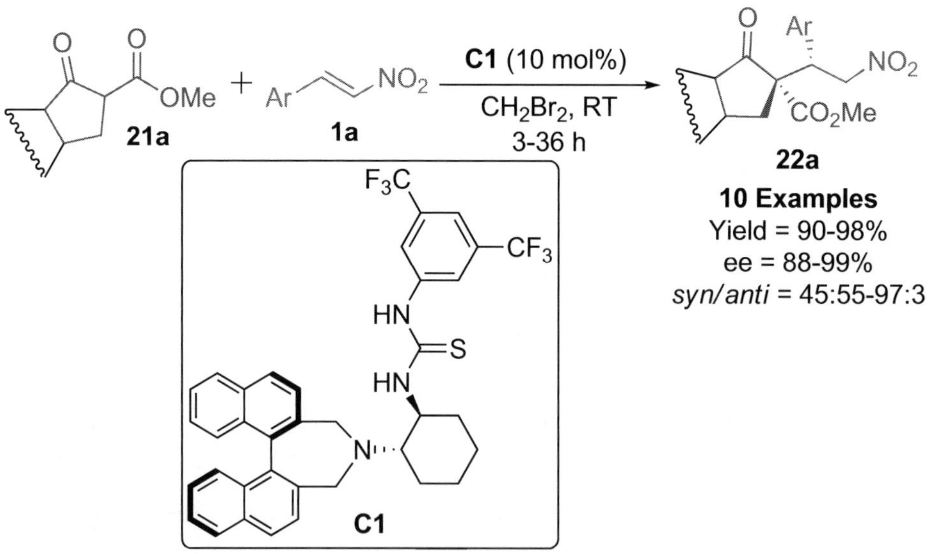

SCHEME 1.28 Bifunctional thiourea catalyzed asymmetric Michael addition of cyclic β-ketoesters to nitroalkenes.

Pan et al. reported an asymmetric Michael addition of dihydroquinolones **27** to nitroalkenes **1** to afford 3,3-disubstituted 3,4-dihydro-2-quinolones **28** in moderate-to-good yields and enantioselectivities (Scheme 1.34).[92] Cinchona alkaloid-derived bifunctional amino-thiourea catalyst **C20** was found to be a suitable catalyst for this reaction.

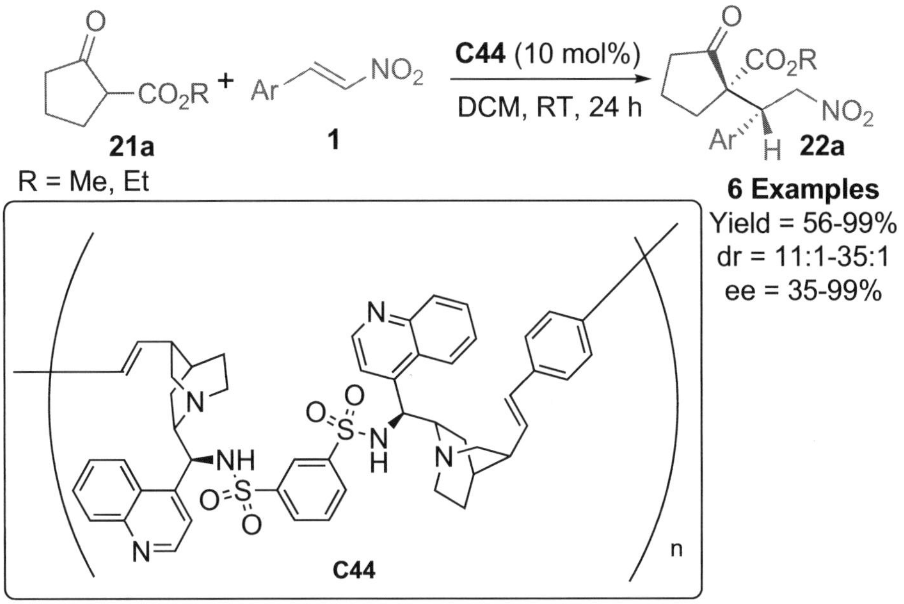

SCHEME 1.29 Michael addition of β-ketoesters to nitroalkenes catalyzed by the chiral polymer of a cinchonidine-based sulfonamide.

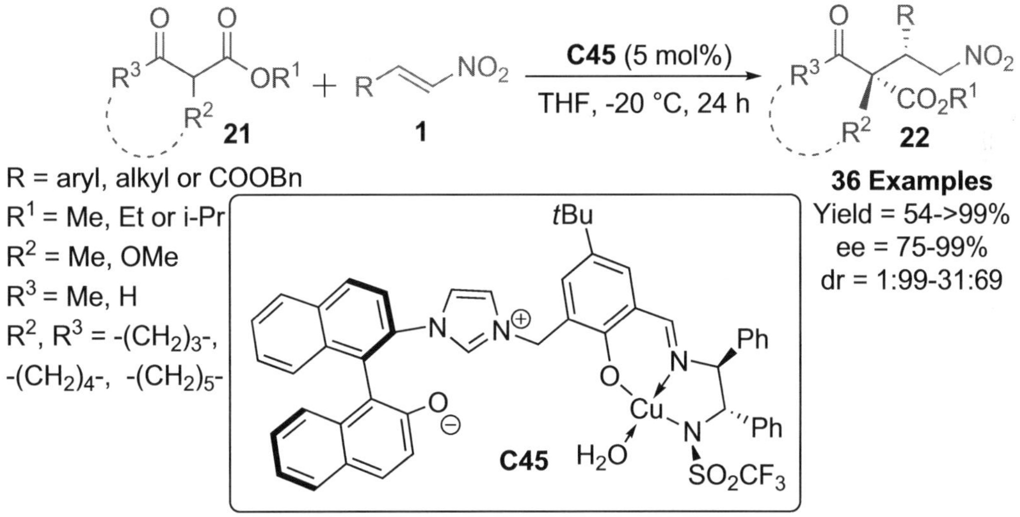

SCHEME 1.30 Asymmetric Michael addition of β-ketoesters to nitroalkenes mediated by polyfunctional imidazolium-aryloxide betaine/Lewis acid catalysts.

The Kim group reported a catalytic asymmetric decarboxylative Michael addition of nitroalkenes **1** and β-ketoacids **29** employing chiral binaphthyl-derived bifunctional organocatalyst **C49** to give corresponding γ-nitroketones **30** in high yields with excellent enantioselectivities (Scheme 1.35).[93] Various aryl, heteroaryl and alkyl-substituted β-ketoacids are well tolerated under these reaction conditions. More importantly, this reaction was performed in gram-scale in the presence of as low as 0.1 mol% catalyst **C49** without affecting the yields and selectivities.

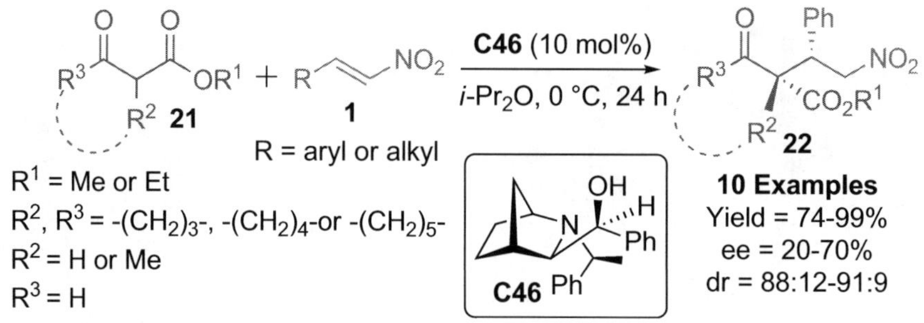

SCHEME 1.31 Asymmetric Michael addition of β-ketoesters to nitroalkenes mediated by chiral azanorbornane-based amino alcohol catalyst.

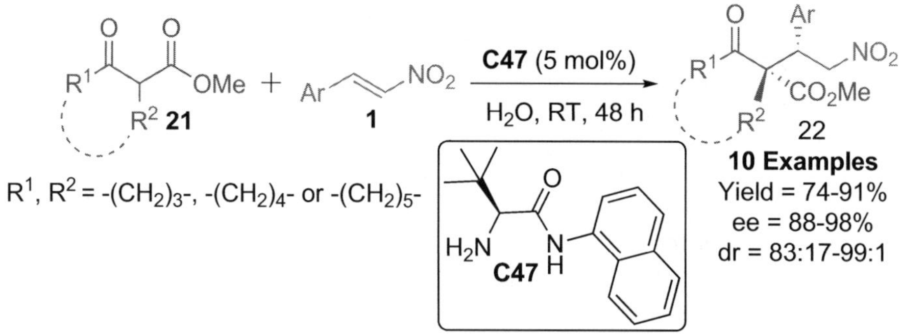

SCHEME 1.32 Asymmetric Michael addition of β-ketoesters to nitroalkenes mediated by amino amide catalyst.

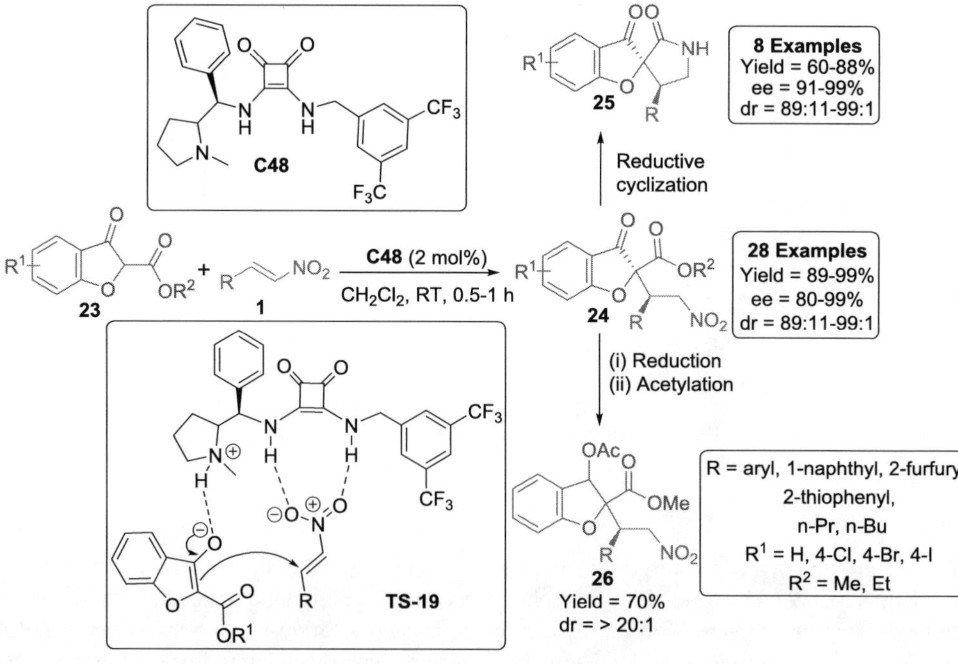

SCHEME 1.33 Enantioselective synthesis of 2,2-disubstituted benzofuranones by a chiral squaramide catalyzed Michael addition of benzofuranone to nitroalkenes.

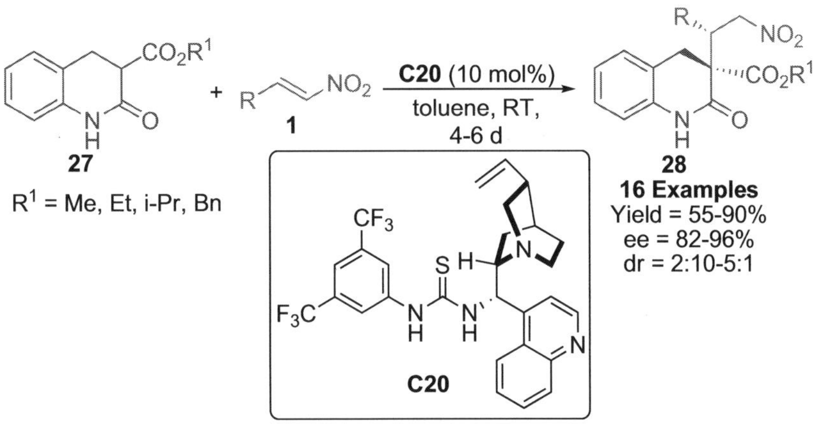

SCHEME 1.34 Asymmetric Michael addition of dihydroquinolones to nitroalkenes in the presence of cinchona alkaloid-derived bifunctional amino-thiourea catalyst.

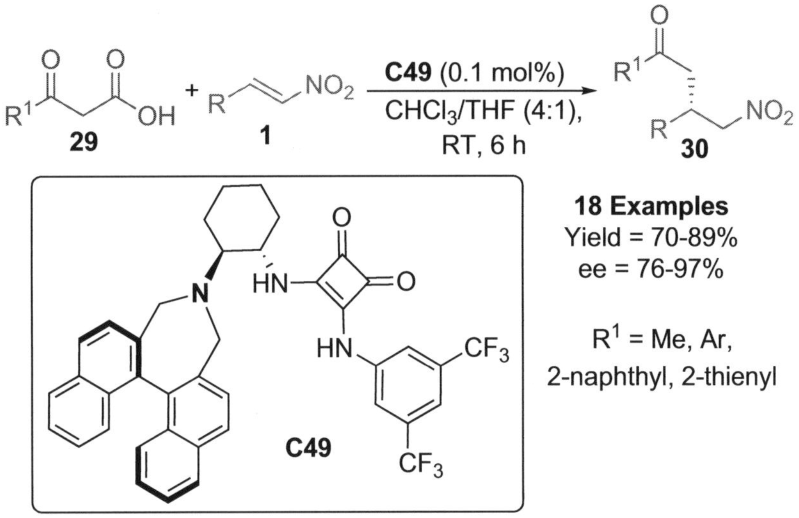

SCHEME 1.35 Asymmetric decarboxylative Michael addition of β-ketoacids to nitroalkenes in the presence of chiral binaphthyl-derived bifunctional organocatalyst.

1.5 ADDITION OF 2-HYDROXYNAPHTHOQUINONES

2-Hydroxy-1,4-naphthoquinone is a stable enol form of a β-diketone, that is, 1,3-dicarbonyl compound. Because of the presence of an enol motif in the structure, it serves as a Michael donor in Michael additions. Over the years, many research groups exploited its Michael donor ability in many Michael reactions to obtain important products and intermediates, which in turn led to the synthesis of biologically important molecules. Of the various Michael acceptors employed with 2-hydroxy-1,4-naphthoquinone **31**, nitroalkenes **1** are of significant interest, and most of the reactions were asymmetric in nature. Our group delineated the asymmetric Michael addition of 2-hydroxynaphthoquinone **31** to nitroalkenes **1** in a recently published review article.[94] Hence, here, we will summarize the entire content as we didn't come across any additional reports other than those included in the review article (Scheme 1.36).[95–102]

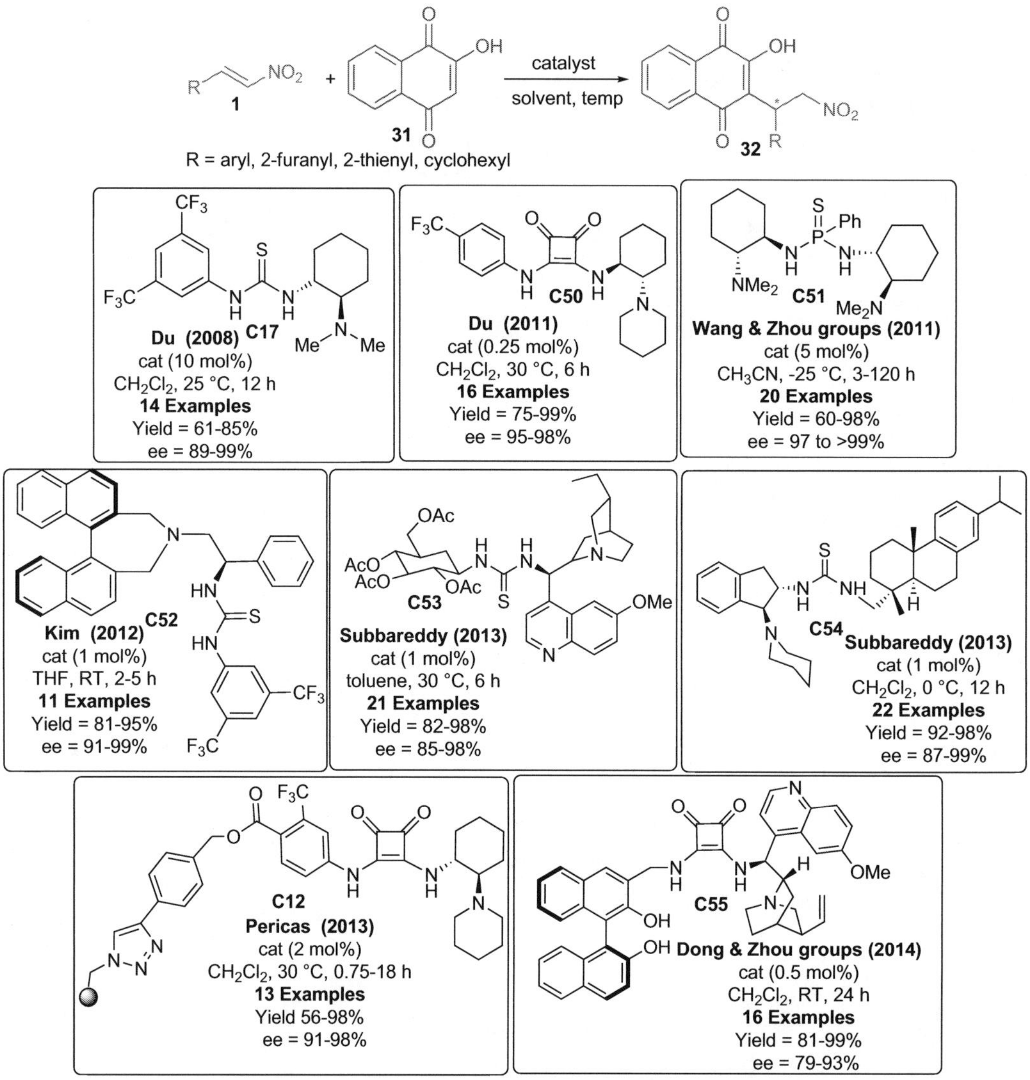

SCHEME 1.36 Asymmetric Michael addition of 2-hydroxynaphthoquinone to nitroalkenes in the presence of various organocatalysts.

Transition states outlined in Figure 1.2 account for the Michael addition of 2-hydroxynaphthoquinone **31** to nitroalkenes **1** in the presence of thiourea-based catalysts **C17**, **C52**, **C53**, **C54**, squaramide-based catalysts **C50**, **C12**, **C55**, and (thio)phosphorodiamide catalyst **C51**. In the case of the thiourea and squaramide catalysts, the nitro group forms hydrogen bond with the NH groups of thiourea/squaramide segment of the catalyst. Deprotonation of the OH group of 2-hydroxy-1,4-naphthoquinone **31** occurs by the tertiary amine of the catalyst, and the enolate is stabilized by H-bonding interactions. As a result of double interaction between the hydrogen and the two vicinal oxygen atoms of hydroxynaphthoquinone **31** in the transition state, the nucleophilic hydroxynaphthoquinone **31** attacks the nitroalkene **1** from *Re*-face to afford the product as (*S*)-isomer **32** (**TS-20** and **21,** Figure 1.2).

When the (thio)phosphorodiamide **C51** was employed as a chiral catalyst, the phosphonothioic diamide group of the catalyst participates in the activation of the nitro group for the nucleophilic

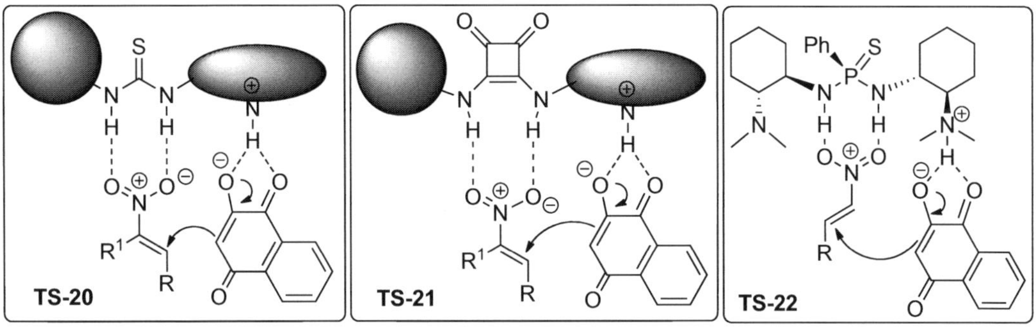

FIGURE 1.2 Transition state models for chiral thiourea, squaramide and phosphorodiamide catalyzed Michael addition of 2-hydroxynaphthoquinone to nitroalkenes.

addition, whereas the tertiary amine group acts as a base and deprotonates 2-hydroxynaphthoquinone **31**. Thus, the deprotonated 2-hydroxynaphthoquinone coordinates to the protonated tertiary amine of the catalyst via H-bonding interactions. Further, nucleophilic addition to nitroalkenes occurs from the *Re*-face to afford the product with (*S*)-configuration as the major product **32** (**TS-22,** Figure 1.2).

The Namboothiri group also employed 2-hydroxynaphthoquinone **31** in the asymmetric Michael addition to nitroallylic acetates **33** using chiral quinine-squaramide catalyst **C56** to afford bioactive pyranonaphthoquinones **34** (α-lapachones) in high yields and excellent enantioselectivities (Scheme 1.37).[103] This reaction sequence proceeds via S_N2' intramolecular oxa-Michael addition. The first step of the reaction involves the formation of enolate from **31,** which then adds to the nitroallylic acetates **33** from the *Si*-face followed by acetate group removal (**TS-23**). Double hydrogen bonding of the squaramide moiety in the catalyst **C56** with a nitro group of **33** facilitates this Michael addition. Further, 6-*endo-trig* intramolecular oxa-Michael addition leads to the formation of respective dihydropyran adduct **34** (Scheme 1.37).

The continuous flow synthesis of the above-described reaction was shown by Pericas and co-workers where different pyranonaphthoquinones were produced in an enantioselective fashion employing polystyrene-supported squaramide **C57** as a catalyst (Scheme 1.38).[104] The reaction was carried out in a sequential two-step process where Michael addition was followed by oxa-Michael cyclization. Here, the authors also emphasized on the significant role played by the linker in the catalytic performance of heterogeneous catalysts.

1.6 ADDITION OF HYDROXYCOUMARIN

Goldfuss and co-workers recently reported an asymmetric Michael addition of 4-hydroxycoumarin **35** to different nitroalkenes **1** to afford 3-substituted 4-hydroxycoumarins **36** in good-to-excellent yields and enantioselectivities (Scheme 1.39).[105] Among the various organocatalysts screened including thiourea and squaramide-based catalysts, high catalytic activity was observed for squaramide catalyst **C56** owing to its higher NH acidities relative to thioureas. A computational study was undertaken to assign the configuration of the major isomer. From the study, it was clear that the favored transition structure shows the coordination of the amino group of squaramide binding to the oxygen atom of the enolate nucleophile through H-bonding interactions. At the same time, nitroalkene **1** gets activated by coordinating through hydrogen bonding with the tertiary amine of the catalyst **C56**. Hence, the orientation of the squaramide favors one-atom binding, and the major enantiomer of the product **36** was assigned as *R* (**TS-25**, Scheme 1.39).

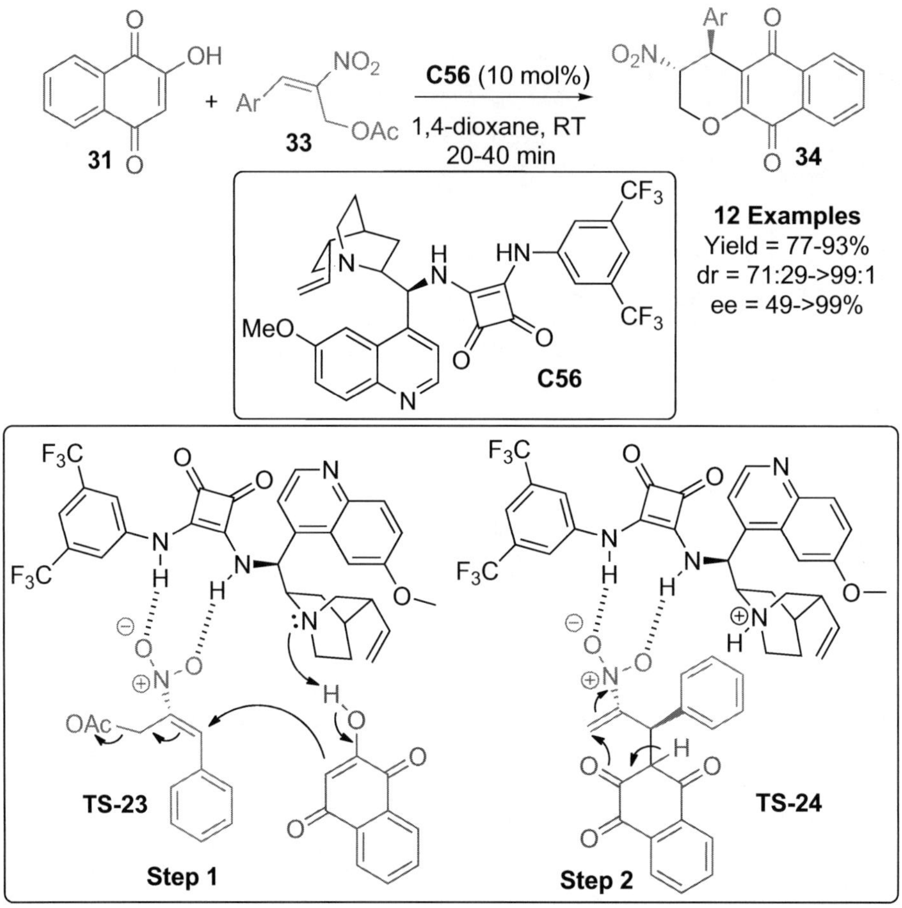

SCHEME 1.37 Synthesis of pyranonaphthoquinones by asymmetric Michael addition of 2-hydroxynaphthoquinone to nitroalkenes followed by intramolecular cyclization.

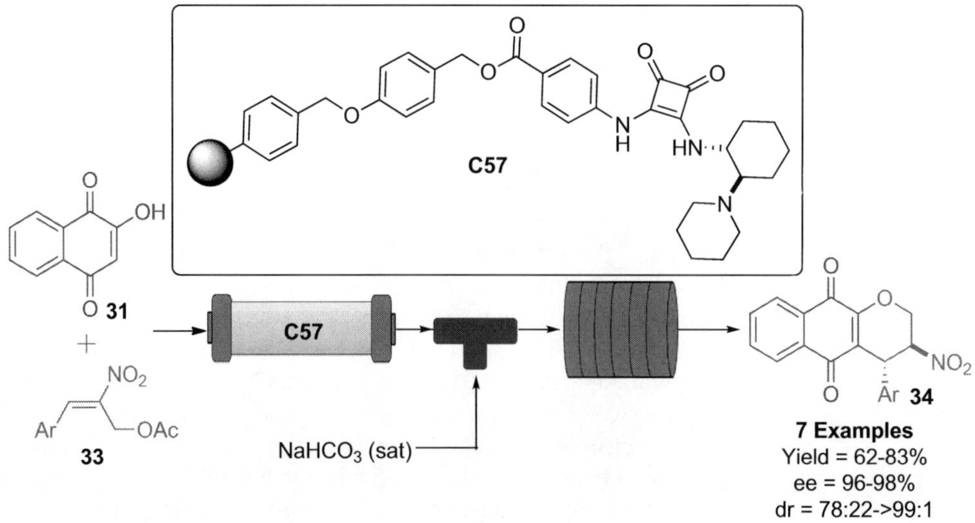

SCHEME 1.38 The continuous flow synthesis of pyranonaphthoquinones.

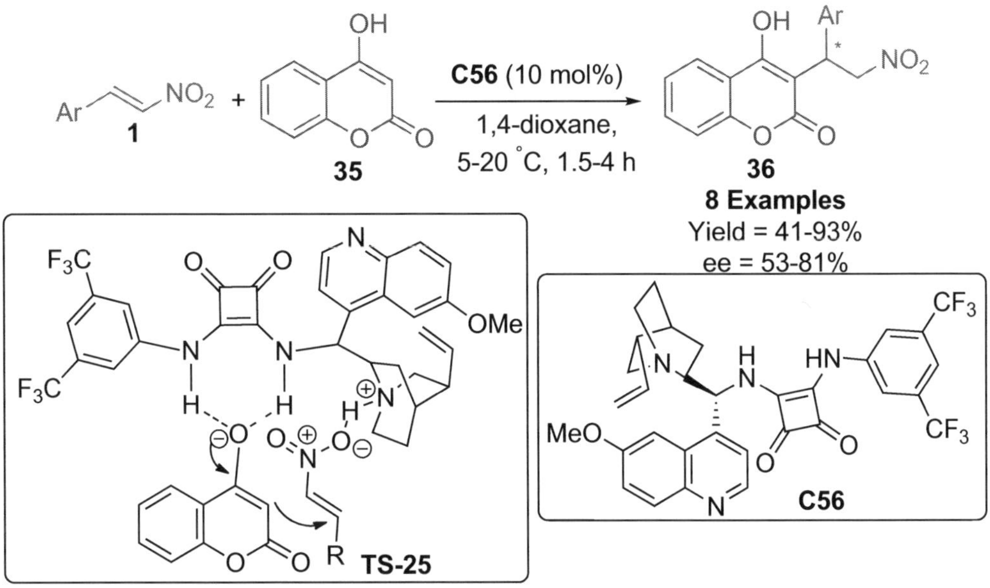

SCHEME 1.39 Chiral squaramide catalyzed Michael addition of hydroxycoumarin to nitroalkenes.

1.7 ADDITION OF MELDRUM'S ACID

Meldrum's acid **37** is a 1,3-dicarbonyl compound possessing an active methylene group which can be deprotonated to yield an anion. This anion can serve as a Michael donor and can participate in Michael additions with various Michael acceptors. The utility of this interesting entity is still at its infancy as far as organocatalysis is concerned. However, the first-ever organocatalytic asymmetric reaction of Meldrum's acid **37** with nitroalkene **15** was reported in 2007 by the Sas group.[106] The group used cinchona-derived catalyst **C22a** to effect the transformation to obtain corresponding adducts **38** in good yields and enantioselectivities of 11%–21% ee (Scheme 1.40).

Followed by the Sas group, many research groups investigated the Michael donor ability of Meldrum's acid **37** with nitroalkenes **15** by using diverse organocatalysts to afford products **38** with good-to-excellent enantioselectivity. Recently, Brière and co-workers published a review article[107] entitled "Meldrum's acid: A useful platform in asymmetric organocatalysis," which dealt with the asymmetric reactions of Meldrum's acid with various Michael acceptors, including nitroalkenes, which is the topic of our present discussion. As the reported review already covered the literature on this compound, we will only summarize it in this book.[108–112]

The asymmetric Michael addition of Meldrum's acid **37** to nitroalkene **15** was assumed to proceed through a mechanism similar to that of Takemoto's catalyst. The transition state model involves the activation of nitroalkene **15** by H-bonding interactions with the squaramide/thiourea NH groups of the catalyst. At the same time, Meldrum's acid **37** gets deprotonated by the tertiary amine of the catalyst, and the resulting enolate gets stabilized by the protonated tertiary amine through H-bonding interaction. This interaction directs the nitroalkene to add from the *Re*-face to Meldrum's acid resulting in (*S*)-configured product **38** (**TS-26/TS-27**, Figure 1.3).

Rawal et al. described an asymmetric Michael addition of barbituric acid **39** to nitroalkenes **1** to afford the chiral barbiturate derivatives **40** in excellent yields and enantioselectivities by employing 0.05 mol% of newly synthesized thiosquaramide catalyst **C63** (Scheme 1.41).[113]

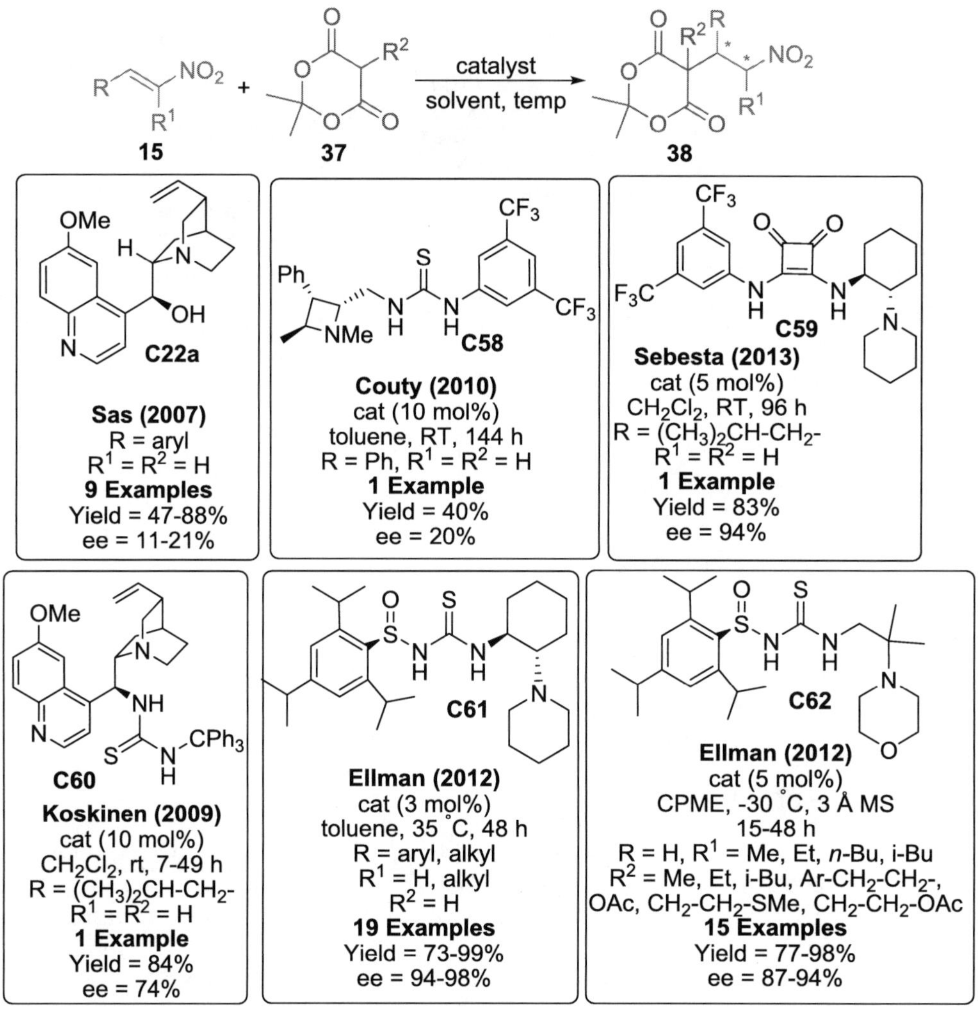

SCHEME 1.40 Asymmetric Michael addition of Meldrum's acid to nitroalkenes.

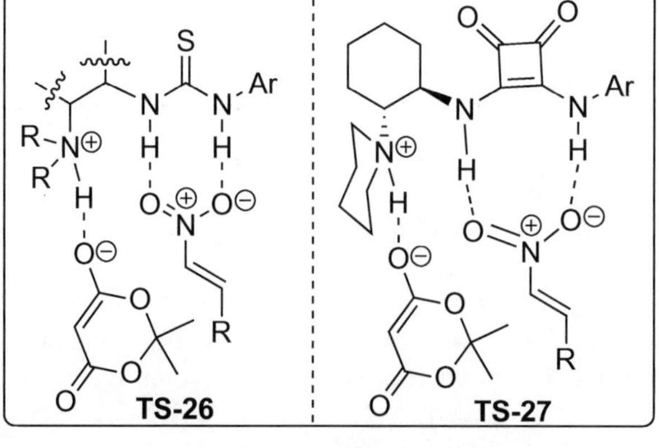

FIGURE 1.3 Transition state models for chiral thiourea and squaramide catalyzed Michael addition of Meldrum's acid to nitroalkenes.

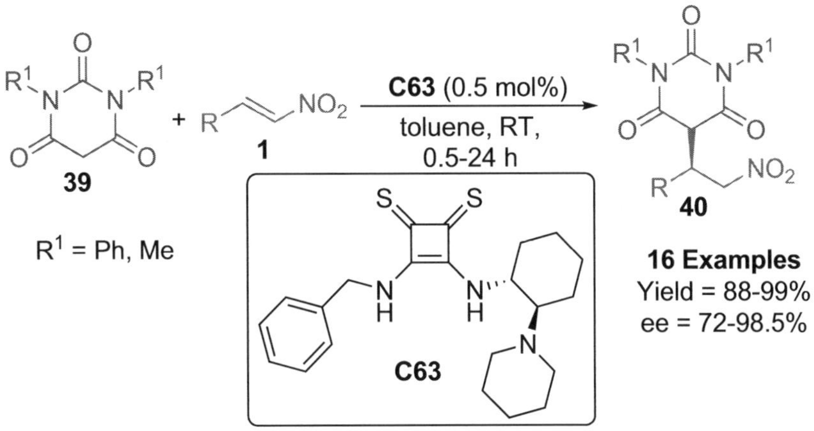

SCHEME 1.41 Asymmetric Michael addition of barbituric acid to nitroalkenes in the presence of a thiosquaramide catalyst.

1.8 CONCLUSIONS

A detailed account of the catalytic stereoselective 1,4-conjugate addition of different 1,3-dicarbonyls as Michael donors to nitroalkenes as Michael acceptors, reported over a period of two decades, is provided in this chapter. Bifunctional organocatalysts bearing a strong H-bonding donor moiety alongside a tertiary amine functionality mounted onto a chiral scaffold have emerged as revolutionary catalysts for such Michael additions to nitroalkenes. Nonetheless, transition metal complexes along with chiral ligands have also been successful in this asymmetric venture. Impressive results have been obtained for the chosen set of reactions in terms of good-to-excellent yields, as well as remarkable enantioselectivities, under minimal catalyst loading and ambient reaction conditions.

REFERENCES

1. Peng, F.-Z.; Shao, Z.-H.; Fan, B.-M.; Song, H.; Li, G.-P.; Zhang, H.-B. *J. Org. Chem.* **2008**, *73*, 5202.
2. Pu, X.; Li, P.; Peng, F.; Li, X.; Zhang, H.; Shao, Z. *Eur. J. Org. Chem.* **2009**, *2009*, 4622.
3. Wang, J.; Li, H.; Duan, W.; Zu, L.; Wang, W. *Org. Lett.* **2005**, *7*, 4713.
4. Flock, A. M.; Krebs, A.; Bolm, C. *Synlett* **2010,** *2010*, 1219.
5. Lai, Q.; Li, Y.; Gong, Z.; Liu, Q.; Wei, C.; Song, Z. *Chirality* **2015**, *27*, 979.
6. Malerich, J. P.; Hagihara, K.; Rawal, V. H. *J. Am. Chem. Soc.* **2008**, *130*, 14416.
7. Bae, H. Y.; Some, S.; Oh, J. S.; Lee, Y. S.; Song, C. E. *Chem. Commun.* **2011**, *47*, 9621.
8. Tukhvatshin, R. S.; Kucherenko, A. S.; Nelyubina, Y. V.; Zlotin, S. G. *ACS Catal.* **2017**, *7*, 2981.
9. Hirashima, S.-i.; Nakashima, K.; Fujino, Y.; Arai, R.; Sakai, T.; Kawada, M.; Koseki, Y.; Murahashi, M.; Tada, N.; Itoh, A.; Miura, T. *Tetrahedron Lett.* **2014**, *55*, 4619.
10. Rao, K. S.; Trivedi, R.; Kantam, M. L. *Synlett* **2015**, *26*, 221.
11. Min, C.; Han, X.; Liao, Z.; Wu, X.; Zhou, H.-B.; Dong, C. *Adv. Synth. Catal.* **2011**, *353*, 2715.
12. Işık, M.; Unver, M. Y.; Tanyeli, C. *J. Org. Chem.* **2015**, *80*, 828.
13. Kasaplar, P.; Riente, P.; Hartmann, C.; Pericàs, M. A. *Adv. Synth. Catal.* **2012**, *354*, 2905.
14. Tan, B.; Zhang, X.; Chua, P. J.; Zhong, G. *Chem. Commun.* **2009**, *2009*, 779.
15. Dong, Z.; Jin, X.; Wang, P.; Min, C.; Zhang, J.; Chen, Z.; Zhou, H.-B.; Dong, C. *ARKIVOC* **2011**, *ix*, 367.
16. Kim, Y.; Kim, Y. J.; Jeong, H. I.; Kim, D. Y. *J. Fluor. Chem.* **2017**, *201*, 43.
17. Kucherenko, A. S.; Kostenko, A. A.; Komogortsev, A. N.; Lichitsky, B. V.; Fedotov, M. Y.; Zlotin, S. G. *J. Org. Chem.* **2019**, *84*, 4304.
18. Bania, N.; Pan, S. C. *Org. Biomol. Chem.* **2019**, *17*, 1718.
19. Okino, T.; Hoashi, Y.; Takemoto, Y. *J. Am. Chem. Soc.* **2003**, *125*, 12672.
20. Liu, J.; Wang, X.; Ge, Z.; Sun, Q.; Cheng, T.; Li, R. *Tetrahedron* **2011**, *67*, 636.

21. Nigmatov, A. G.; Kuchurov, I. V.; Siyutkin, D. E.; Zlotin, S. G. *Tetrahedron Lett.* **2012**, *53*, 3502.
22. Hamza, A.; Schubert, G.; Soos, T.; Papai, I. *J. Am. Chem. Soc.* **2006**, *128*, 13151.
23. Izzo, J. A.; Myshchuk, Y.; Hirschi, J. S.; Vetticatt, M. J. *Org. Biomol. Chem.* **2019**, *17*, 3934.
24. Ye, J.; Dixon, D. J.; Hynes, P. S. *Chem. Commun.* **2005**, *2005*, 4481.
25. Jiang, X.; Zhang, Y.; Liu, X.; Zhang, G.; Lai, L.; Wu, L.; Zhang, J.; Wang, R. *J. Org. Chem.* **2009**, *74*, 5562.
26. McCooey, S. H.; Connon, S. J. *Angew. Chem. Int. Ed.* **2005**, *44*, 6367.
27. Li, X.-J.; Liu, K.; Ma, H.; Nie, J.; Ma, J.-A. *Synlett* **2008**, *2008*, 3242.
28. Yan, L. J.; Liu, Q. Z.; Wang, X. L. *Chin. Chem. Lett.* **2009**, *20*, 310.
29. Kwon, B. K.; Kim, S. M.; Kim, D. Y. *J. Fluor. Chem.* **2009**, *130*, 759.
30. Li, H.; Wang, Y.; Tang, L.; Deng, L. *J. Am. Chem. Soc.* **2004**, *126*, 9906.
31. Chen, F.-X.; Shao, C.; Wang, Q.; Gong, P.; Zhang, D.-Y.; Zhang, B.-Z.; Wang, R. *Tetrahedron Lett.* **2007**, *48*, 8456.
32. Li, H.; Zu, L.; Xie, H.; Wang, W. *Synthesis* **2009**, *9*, 1525.
33. Terada, M.; Ube, H.; Yaguchi, Y. *J. Am. Chem. Soc.* **2006**, *128*, 1454.
34. Suez, G.; Bloch, V.; Nisnevich, G.; Gandelman, M. *Eur. J. Org. Chem.* **2012**, *11*, 2118.
35. Veverková, E.; Bilka, S.; Baran, R.; Šebesta, R. *Synthesis* **2016**, *48*, 1474.
36. Zhang, L.; Lee, M.-M.; Lee, S.-M.; Lee, J.; Cheng, M.; Jeong, B.-S.; Park, H.; Jew, S. *Adv. Synth. Catal.* **2009**, *351*, 3063.
37. Almaşi, D.; Alonso, D. A.; Gómez-Bengoa, E.; Nájera, C. *J. Org. Chem.* **2009**, *74*, 6163.
38. Lee, M.; Zhang, L.; Park, Y.; Park, H. *Tetrahedron* **2012**, *68*, 1452.
39. Li, F.; Li, Y.-Z.; Jia, Z.-S.; Xu, M.-H.; Tian, P.; Lin, G.-Q. *Tetrahedron* **2011**, *67*, 10186.
40. Evans, D. A.; Mito, S.; Seidel, D. J. *Am. Chem. Soc.* **2007**, *129*, 11583.
41. Evans, D. A.; Seidel, D. *J. Am. Chem. Soc.* **2005**, *127*, 9958.
42. Kang, S. H.; Kim, D. Y. *Bull. Korean Chem. Soc.* **2009**, *30*, 1439.
43. Wilckens, K.; Duhs, M.-A.; Lentz, D.; Czekelius, C. *Eur. J. Org. Chem.* **2011**, *2011*, 5441.
44. Tsubogo, T.; Yamashita, Y.; Kobayashi, S. *Angew. Chem. Int. Ed.* **2009**, *48*, 9117.
45. Zhu, Q.; Huang, H.; Shi, D.; Shen, Z.; Xia, C. *Org. Lett.* **2009**, *11*, 4536.
46. Liu, K.; Jin, R.; Cheng, T.; Xu, X.; Gao, F.; Liu, G.; Li, H. *Chem. –Eur. J.* **2012**, *18*, 15546.
47. Jin, R.; Liu, K.; Xia, D.; Qian, Q.; Liu, G.; Li, H. *Adv. Synth. Catal.* **2012**, *354*, 3265.
48. Billault, I.; Launez, R.; Scherrmann, M. C. *RSC Adv.* **2015**, *5*, 29386.
49. Okino, T.; Hoashi, Y.; Furukawa, T.; Xu, X.; Takemoto, Y. *J. Am. Chem. Soc.* **2005**, *127*, 119.
50. Nichols, P. J.; DeMattei, J. A.; Barnett, B. R.; LeFur, N. A.; Chuang, T.-H.; Piscopio, A. D.; Koch, K. *Org. Lett.* **2006**, *8*, 1495.
51. Rapi, Z.; Démuth, B.; Keglevich, G.; Grűn, A.; Drahos, L.; Sóti, P. L.; Bakó, P. *Tetrahedron: Asymmetry* **2014**, *25*, 141.
52. Nie, S.; Hu, Z.; Xuan, Y.; Wang, J.; Li, X.; Yan, M. *Tetrahedron: Asymmetry* **2010**, *21*, 2055.
53. Chen, W.-Y.; Ouyang, L.; Chen, R.-Y.; Li, X.-S. *Tetrahedron Lett.* **2010**, *51*, 3972.
54. Chen, W.-Y.; Li, P.; Xie, J.-w.; Li, X.-S. *Catal. Commun.* **2011**, *12*, 502.
55. Li, X.-J.; Peng, F.-Z.; Li, X.; Wu, W.-T.; Sun, Z.-W.; Li, Y.-M.; Zhang, S.-X.; Shao, Z.-H. *Chem. Asian J.* **2011**, *6*, 220.
56. Li, X.; Li, X.; Peng, F.; Shao, Z. *Adv. Synth. Catal.* **2012**, *354*, 2873.
57. Chowdhury, R.; Vamisetti, G. B.; Ghosh, S. K. *Tetrahedron: Asymmetry* **2014**, *25*, 516.
58. Xuan, Y.; Nie, S.; Dong, L.; Zhang, J.; Yan, M. *Org. Lett.* **2009**, *11*, 1583.
59. Russo, A.; Lattanzi, A. *Tetrahedron: Asymmetry* **2010**, *21*, 1155.
60. Lee, H. J.; Kim, S. M.; Kim, D. Y. *Tetrahedron Lett.* **2012**, *53*, 3437.
61. McCooey, S. H.; McCabe, T.; Connon, S. J. *J. Org. Chem.* **2006**, *71*, 7494.
62. Sim, J. H.; Song, C. E. *Angew. Chem., Int. Ed.* **2017**, *56*, 1835.
63. Govender, T.; Arvidsson, P. I.; Maguire, G. E. M.; Kruger, H. G.; Naicker, T. *Chem. Rev.* **2016**, *116*, 9375.
64. Barnes, D. M.; Ji, J.; Fickes, M. G.; Fitzgerald, M. A.; King, S. A.; Morton, H. E.; Plagge, F. A.; Preskill, M.; Wagaw, S. H.; Wittenberger, S. J.; Zhang, J. *J. Am. Chem. Soc.* **2002**, *124*, 13097.
65. Li, H.; Wang, Y.; Tang, L.; Wu, F.; Liu, X.; Guo, C.; Foxman, B. M.; Deng, L. *Angew. Chem. Int. Ed.* **2005**, *44*, 105.
66. Tan, B.; Chua, P. J.; Li, Y.; Zhong, G. *Org. Lett.* **2008**, *10*, 2437.
67. Tan, B.; Chua, P. J.; Zeng, X.; Lu, M.; Zhong, G. *Org. Lett.* **2008**, *10*, 3489.
68. Zhang, Z.-H.; Dong, X.-Q.; Chen, D.; Wang, C.-J. *Chem. –Eur. J.* **2008**, *14*, 8780.
69. Han, X.; Luo, J.; Liu, C.; Lu, Y. *Chem. Commun.* **2009**, *15*, 2044.

70. Kokotos, C. G.; Kokotos, G. *Adv. Synth. Catal.* **2009**, *351*, 1355.
71. Li, H.; Zhang, S.; Yu, C.; Song, X.; Wang, W. *Chem. Commun.* **2009**, *2009*, 2136.
72. Liu, X.; Lin, L.; Feng, X. *Chem. Commun.* **2009**, *41*, 6145.
73. Luo, J.; Xu, L.-W.; Hay, R. A. S.; Lu, Y. *Org. Lett.* **2009**, *11*, 437.
74. Yu, Z.; Liu, X.; Zhou, L.; Lin, L.; Feng, X. *Angew. Chem. Int. Ed.* **2009**, *48*, 5195.
75. Jiang, H.; Paixão, M. W.; Monge, D.; Jørgensen, K. A. *J. Am. Chem. Soc.* **2010**, *132*, 2775.
76. Jiang, Z.-Y.; Yang, H.-M.; Ju, Y.-D.; Li, L.; Luo, M.-X.; Lai, G.-Q.; Jiang, J.-X.; Xu, L.-W. *Molecules* **2010**, *15*, 2551.
77. Manzano, R.; Andrés, J. M.; Muruzábal, M. D.; Pedrosa, R. *Adv. Synth. Catal.* **2010**, *352*, 3364.
78. Murai, K.; Fukushima, S.; Hayashi, S.; Takahara, Y.; Fujioka, H. *Org. Lett.* **2010**, *12*, 964.
79. Jakubec, P.; Kyle, A. F.; Calleja, J.; Dixon, D. J. *Tetrahedron Lett.* **2011**, *52*, 6094.
80. Dong, Z.; Qiu, G.; Zhou, H.-B.; Dong, C. *Tetrahedron: Asymmetry* **2012**, *23*, 1550.
81. Naicker, T.; Arvidsson, P. I.; Kruger, H. G.; Maguire, G. E. M.; Govender, T. *Eur. J. Org. Chem.* **2012**, *71*, 3331.
82. Liu, B.; Han, X.; Dong, Z.; Lv, H.; Zhou, H.-B.; Dong, C. *Tetrahedron: Asymmetry* **2013**, *24*, 1276.
83. Zhao, Y.; Wang, X.-J.; Lin, Y.; Cai, C.-X.; Liu, J.-T. *Tetrahedron* **2014**, *70*, 2523.
84. Furutachi, M.; Chen, Z.; Matsunaga, S.; Shibasaki, M. *Molecules* **2010**, *15*, 532.
85. Oh, Y.; Kim, S. M.; Kim, D. Y. *Tetrahedron Lett.* **2009**, *50*, 4674.
86. Kwon, B. K.; Kim, D. Y. *Bull. Korean Chem. Soc.* **2009**, *30*, 1441.
87. Endo, Y.; Takata, S.; Kumpuga, B. T.; Itsuno, S. *ChemistrySelect* **2017**, *2*, 10107.
88. Willig, F.; Lang, J.; Hans, A. C.; Ringenberg, M. R.; Pfeffer, D.; Frey, W.; Peters, R. *J. Am. Chem. Soc.* **2019**, *141*, 12029.
89. Togashi, R.; Chennapuram, M.; Seki, C.; Okuyama, Y.; Kwon, E.; Uwai, K.; Tokiwa, M.; Takeshita, M.; Nakano, H. *Eur. J. Org. Chem.* **2019**, *24*, 3882.
90. Owolabi, I. A.; Chennapuram, M.; Seki, C.; Okuyama, Y.; Kwon, E.; Uwai, K.; Tokiwa, M.; Takeshita, M.; Nakano, H. *Bull. Chem. Soc. Jpn.* **2019**, *92*, 696.
91. Sivamuthuraman, K.; Kumarswamyreddy, N.; Kesavan, V. *J. Org. Chem.* **2017**, *82*, 10812.
92. Mukhopadhyay, S.; Nath, U.; Pan, S. C. *Adv. Synth. Catal.* **2017**, *359*, 3911.
93. Moon, H. W.; Kim, D. Y. *Tetrahedron Lett.* **2012**, *53*, 6569.
94. Hosamani, B.; Ribeiro, M. F.; da Silva Junior, E. N.; Namboothiri, I. N. N. *Org. Biomol. Chem.* **2016**, *14*, 6913.
95. Zhou, W.-M.; Liu, H.; Du, D.-M. *Org. Lett.* **2008**, *10*, 2817.
96. Wu, R.; Chang, X.; Lu, A.; Wang, Y.; Wu, G.; Song, H.; Zhou, Z.; Tang, C. *Chem. Commun.* **2011**, *47*, 5034.
97. Yang, W.; Du, D.-M. *Adv. Synth. Catal.* **2011**, *353*, 1241.
98. Kasaplar, P.; Rodríguez-Escrich, C.; Pericàs, M. A. *Org. Lett.* **2013**, *15*, 3498.
99. Reddy, B. V. S.; Reddy, S. M.; Swain, M. *RSC Adv.* **2013**, *3*, 930.
100. Reddy, B. V. S.; Swain, M.; Reddy, S. M.; Yadav, J. S. *RSC Adv.* **2013**, *3*, 8756.
101. Zhou, E.; Liu, B.; Dong, C. *Tetrahedron: Asymmetry* **2014**, *25*, 181.
102. Woo, S. B.; Kim, D. Y. *Beilstein J. Org. Chem.* **2012**, *8*, 699.
103. Nair, D. K.; Menna-Barreto, R. F. S.; da Silva Júnior, E. N.; Mobin, S. M.; Namboothiri, I. N. N. *Chem. Commun.* **2014**, *50*, 6973.
104. Osorio-Planes, L.; Rodríguez-Escrich, C.; Pericàs, M. A. *J. Adv. Catal. Sci. Techno.* **2016**, *6*, 4686.
105. Wolf, F. F.; Klare, H.; Goldfuss, B. *J. Org. Chem.* **2016**, *81*, 1762.
106. Kleczkowska, E.; Sas, W. *Pol. J. Chem.* **2007**, *81*, 1457.
107. Pair, E.; Cadart, T.; Levacher, V.; Brière, J.-F. *ChemCatChem* **2016**, *8*, 1882.
108. Baran, R.; Veverkova, E.; Skvorcova, A.; Sebesta, R. *Org. Biomol. Chem.* **2013**, *11*, 7705.
109. Kimmel, K. L.; Weaver, J. D.; Lee, M.; Ellman, J. A. *J. Am. Chem. Soc.* **2012**, *134*, 9058.
110. Kimmel, K. L.; Weaver, J. D.; Ellman, J. A. *Chem. Sci.* **2012**, *3*, 121.
111. Menguy, L.; Couty, F. *Tetrahedron: Asymmetry* **2010**, *21*, 2385.
112. Bassas, O.; Huuskonen, J.; Rissanen, K.; Koskinen, A. M. P. *Eur. J. Org. Chem.* **2009**, 1340.
113. Rombola, M.; Sumaria, C. S.; Montgomery, T. D.; Rawal, V. H. *J. Am. Chem. Soc.* **2017**, *139*, 9, 5297.

2 Catalytic Asymmetric Michael Addition of Aldehydes to Nitroalkenes

2.1 INTRODUCTION

Having discussed the catalytic asymmetric Michael addition of various 1,3-dicarbonyls to nitroalkenes in the previous chapter, similar addition of enolates and umpolung of aldehydes is covered in this chapter. In 2007, independent reviews by the Najera group[1] and Tsogoeva[2] delineated the asymmetric 1,4-addition of aldehydes and ketones to nitroalkenes reported till then in numerous articles.[3–43] The reader is referred to the above reviews for details, and here we discuss only the subsequent literature on the asymmetric conjugate addition of aldehydes to nitroalkenes. Chiral N-heterocyclic carbene (NHC) catalysts were the exclusive catalysts for the handful of reported Stetter reactions of nitroalkenes with aldehydes, including α,β-unsaturated ones.

2.2 ADDITION OF ALDEHYDE ENOLATES

Asymmetric Michael addition of aldehydes to nitroalkenes is one of the extensively studied C–C-bond forming reactions. In 2001, Barbas and co-workers reported the asymmetric conjugate addition of aliphatic aldehydes **2** to nitroalkenes **1** for the first time using 20 mol% of (*S*)-2-(morpholinomethyl)-pyrrolidine catalyst **C1**, which provided γ-formyl nitro compounds **3** in moderate-to-excellent yields with high enantio and diastereoselectivities (Scheme 2.1).[44] These γ-formyl nitro compounds **3** were successfully converted to trans-3,4-disubstituted pyrrolidines **4** without affecting the enantioselectivity. However, there was no reaction in the case of *tert*-butylacetaldehyde and isopropyl-substituted nitroalkenes.

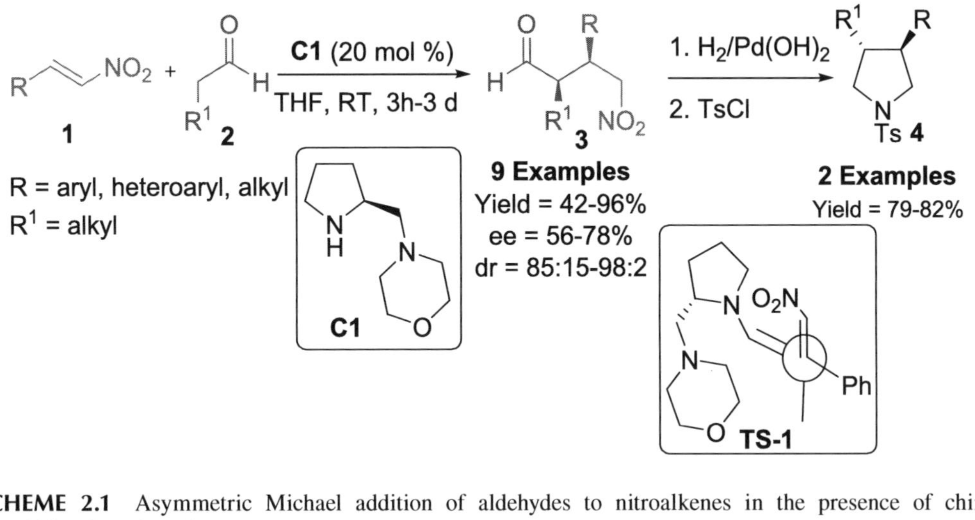

SCHEME 2.1 Asymmetric Michael addition of aldehydes to nitroalkenes in the presence of chiral pyrrolidine-based catalyst.

The Alexakis group also demonstrated (*R,R*)-*N*-^{i}Pr-2,2′-bipyrrolidine **C2** catalyzed asymmetric conjugate addition of aldehydes **2** to nitrostyrene **1** (Scheme 2.2).[4,45] Use of HCl as an additive provided the Michael adducts *ent*-**3** in better enantio and diastereoselectivities in shorter reaction time. The group also investigated the reaction under microwave irradiation, which took 1–2 h to complete the reaction and provided adducts *ent*-**3** in high yields and good-to-excellent enantio and diastereoselectivities (Scheme 2.2).[45]

In 2004, the Barbas group reported the asymmetric Michael addition of α,α-disubstituted aldehydes **5** to nitroalkene **1a** employing L-proline-derived diamine catalyst **C3** and trifluoroacetic acid (TFA) in isopropanol (Scheme 2.3).[46] γ-Nitroaldehydes **6a** containing a quaternary stereocenter were obtained in excellent yields and moderate-to-good enantio and diastereoselectivities.

In 2005, the Wang group reported chiral pyrrolidine sulfonamide **C4** catalyzed enantioselective Michael addition of linear aldehydes and α,α-disubstituted aldehydes **5** with aryl and alkyl-substituted nitroalkenes **1** (Scheme 2.4).[47] Though the reaction with aryl-substituted nitroalkenes

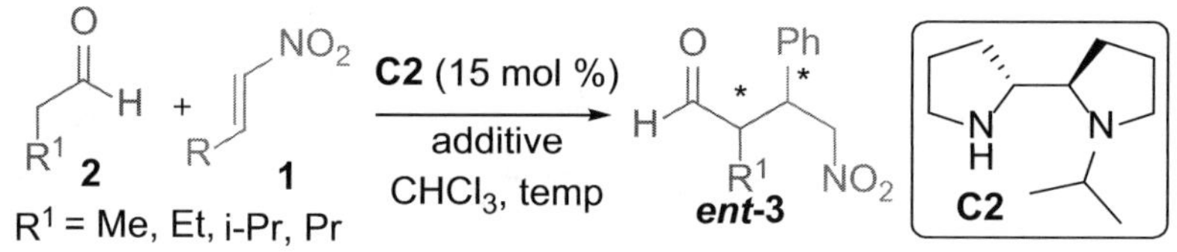

Additive	RT	-25 or 0°C
none	RT, 1 h-2 d, yield = 99% ee = 61-66%, dr = 75:25-87:23%	-25 °C, 2-4 d, Yield = 70-98% ee = 70-83%, dr = 90:10-96:4%
HCl (15 mol%)	RT, 3 h-5 d, Yield = 82-99% ee = 67-79%, dr = 85:15-95:5%	0°C, 2 d, Yield = 82-83% ee = 68-85%, dr = 88:12-96:4%

R^1 = i-Pr, $(Me)_2$

C2 (15 mol %), $CHCl_3$, MW (15 W), 1-2 h

27 °C, Yield = 76-97%, ee = 77-78%, dr = 90:10

SCHEME 2.2 Asymmetric conjugate addition of aldehydes to nitrostyrene catalyzed by chiral bipyrrolidine-based organocatalyst.

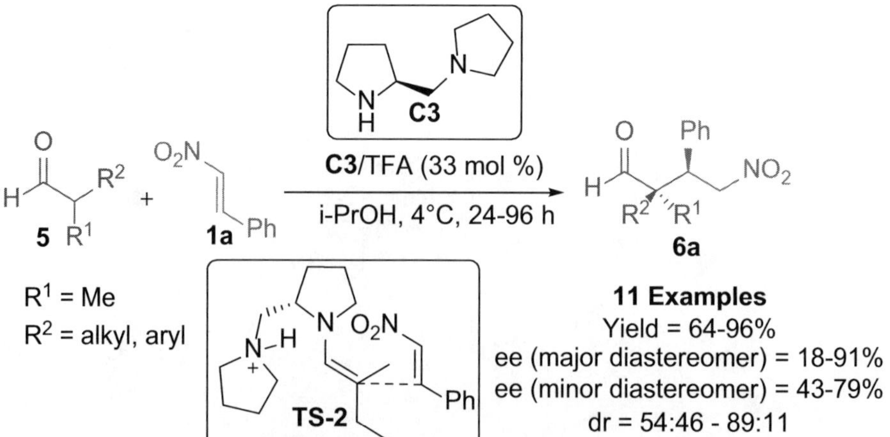

SCHEME 2.3 Asymmetric Michael addition of α,α-disubstituted aldehydes to nitroalkene in the presence of L-proline-derived diamine catalyst.

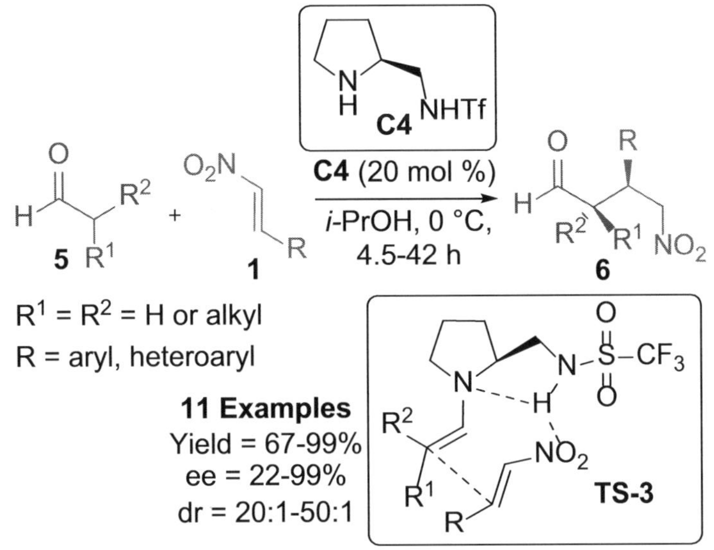

SCHEME 2.4 Asymmetric Michael addition of aldehydes to nitroalkenes catalyzed by chiral pyrrolidine sulfonamide organocatalyst.

afforded corresponding Michael adducts **6** in excellent yields and good diastereo and enantioselectivities, alkyl-substituted nitroalkenes provided the Michael adduct **6** with low enantioselectivity (22%).

In the same year, the Hayashi group developed diphenylprolinol silyl ether **C5a** as a highly efficient catalyst and used it for the asymmetric Michael addition of aldehydes **5** to nitroalkenes **1** (Scheme 2.5).[48] The Michael adducts **6** were obtained in good yields (52%–85%) with good-to-excellent enantioselectivities (68%–99% ee), as well as with good diastereoselectivities (*syn/anti* 94:6). However, the reaction of isobutyraldehyde with nitroalkene under similar reaction conditions gave the Michael adduct with moderate enantioselectivity (68% ee). The diphenylsiloxymethyl group of the pyrrolidine motif accounted for the excellent enantioselectivity because it allowed selective *anti*-enamine formation and shielded the *Re*-face of the enamine, which would react with nitroalkene via an acyclic synclinal transition state **TS-4** (Scheme 2.5).

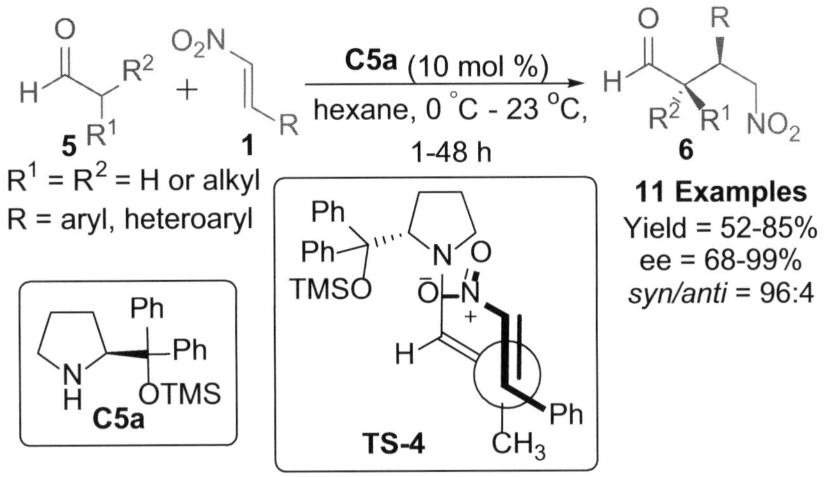

SCHEME 2.5 Asymmetric Michael addition of aldehydes to nitroalkenes catalyzed by diphenylprolinol silyl ether.

However, the Jacobsen group's report on the conjugate addition of α,α-disubstituted aldehydes **5a** to nitroalkenes **1** is superior to the previous reports in terms of the scope of the reaction, yields, as well as diastereo and enantioselectivity of Michael adducts **6b** (Scheme 2.6).[49] Chiral cyclohexane diamine-derived primary amine-thiourea **C6** catalyzed the Michael addition of various α,α-disubstituted aldehydes **5** to various alkyl and aryl-substituted nitroalkenes **1** to provide synthetically versatile nitroaldehyde adducts **6** in excellent yields and enantio and diastereoselectivities.

Following these pioneering works in the asymmetric Michael addition of aldehydes to nitroalkenes and their application in natural product synthesis, several groups were inspired to further investigate this Michael addition to attain excellent stereoselectivities and yields under mild reaction conditions. Thus, various chiral primary and secondary amines, most of them derived from amino acids, were examined in this reaction.

In 2006, Alexakis, Kanger and co-workers reported a new catalyst *N*-isopropyl-3,3′-bimorpholine **C7**, which catalyzed the reaction between various linear alkyl aldehydes **2** and aryl and alkyl-substituted nitroalkenes **1** providing the Michael adducts *ent*-**3** in moderate-to-good yields and enantio and diastereoselectivities (Scheme 2.7).[50] However, isobutyraldehyde and 3,3-dimethylbutyraldehyde failed to undergo Michael addition to nitrostyrene under these reaction conditions. As the synthesis of catalyst involved multiple steps starting from tartaric acid, in search of a short synthetic method for catalyst synthesis, the Kanger group reported the synthesis of (2*R*,2′*R*)-*N*-isopropyl-2,2′-bipiperidine **C8** from 2,2-bipyridine in two steps.[51] This catalyst is also equally efficient to catalyze the Michael addition of aldehydes **2** to nitroalkenes **1** and provided the

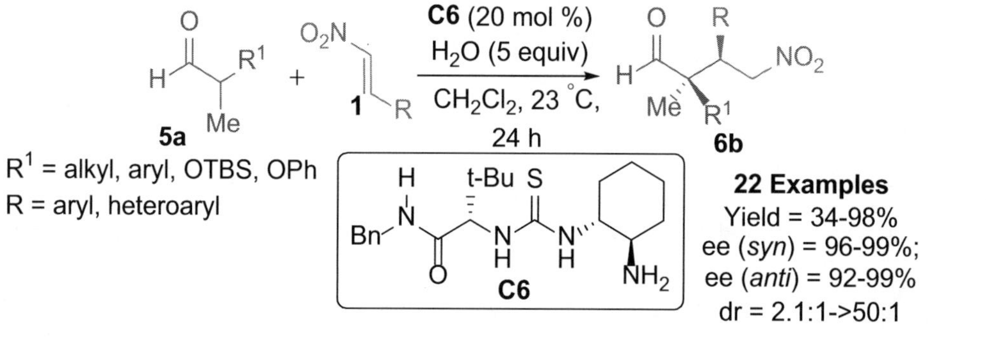

SCHEME 2.6 Asymmetric Michael addition of aldehydes to nitroalkenes catalyzed by chiral cyclohexane diamine-derived primary amine-thiourea.

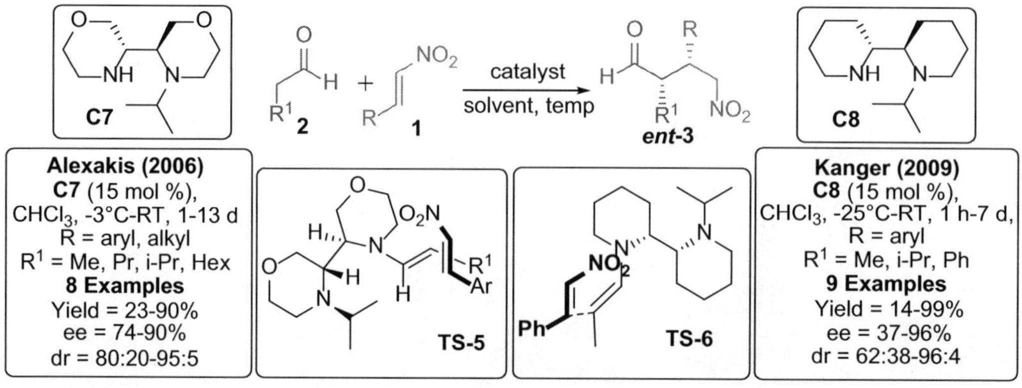

SCHEME 2.7 Asymmetric Michael addition of aldehydes to nitroalkenes catalyzed by chiral bimorpholine and bipiperidine-based organocatalysts.

adducts *ent*-**3** in moderate-to-excellent yields and enantio- and diastereoselectivities, except in the case of isobutyraldehyde which gave the corresponding adduct in poor yield and enantioselectivity.

Palomo et al. observed that the catalyst systems with the hydrogen-bond donor at the α-position of the pyrrolidine nitrogen atom were found to catalyze the self-aldol reaction preferably over Michael addition.[52] They suggested that this may be responsible for the low enantioselectivities. Therefore, it was proposed that if the α-hydrogen-bond donor is removed, the Michael addition should proceed with high enantio and diastereocontrol. Thus, *trans*-4-hydroxyprolylamide **C9** was investigated for the Michael addition of aldehydes **2** to nitroalkenes **1**, and best results were obtained with 5–10 mol% of **C9**, which provided the products **3** in good yields (70%–90%) with high diastereomeric ratios (up to 99:1) and high enantioselectivities (up to 99%, Scheme 2.8).

Ni and co-workers reported an asymmetric Michael addition of α,α-disubstituted aldehydes **5** to nitroalkenes **1** using a diamine-based organocatalyst **C10** in the presence of an acidic co-catalyst benzoic acid to give products **6** with quaternary carbons in high-to-excellent yields and high levels of enantioselectivities under solvent-free reaction conditions (Scheme 2.9).[53] The List group reported an asymmetric Michael addition of acetaldehyde **5** ($R^1 = R^2 = H$) to aromatic as well as aliphatic nitroalkenes **1** using silyl prolinol ether catalyst **C5a** to afford the corresponding Michael products **6** in moderate-to-good yields and high enantioselectivities (Scheme 2.9, **TS-7**).[54] Later, Hayashi et al. carried out Michael reactions of α-unsubstituted aldehyde, that is, acetaldehyde **5** with various nitroalkenes **1** using diphenylprolinol silyl ether catalyst **C5a**. The chiral α-unsubstituted γ-nitroaldehydes **6** were obtained with excellent enantioselectivity (92% to >99% ee) and moderate yields (54%–77%, Scheme 2.9).[55] Alza and Pericas developed a polymer-linked α,α-diarylprolinol silyl ether homogeneous catalyst **C11** for the Michael addition of aldehydes **5** to nitroalkenes **1**. Here, the polymer backbone along with triazole linker and the catalyst provide best results, especially with linear and short chain aldehydes (Scheme 2.9).[56]

Seebach et al. investigated the mechanism of the asymmetric Michael addition of aldehydes **2** to nitroalkenes **1** catalyzed by diphenylprolinol silyl ether **C5a** and identified 4-nitrophenol as a potent acid additive for various aldehydes and nitroalkenes allowing shorter reaction time and lower catalyst loading without affecting the enantio and diastereoselectivities of the process (Scheme 2.10).[57] In-situ nuclear magnetic resonance (NMR) studies led, for the first time, to the identification of cyclobutanes in the amine-catalyzed Michael addition of aldehydes to nitroalkenes. The cyclobutanes may be regarded as "parasitic" components, i.e., an off-cycle resting state of the catalyst, by which the zwitterion intermediate is removed from the catalytic cycle.

The Headley group developed a water-soluble di(methylimidazole)prolinol silyl ether **C5b** as an organocatalyst in combination with sodium bicarbonate as an additive for the asymmetric Michael addition of aldehydes **2** to various nitroalkenes **1** using brine as a reaction medium to obtain the adducts **3** with high stereoselectivities (ee up to 99%; dr (*syn/anti*) up to 98:2) at room temperature

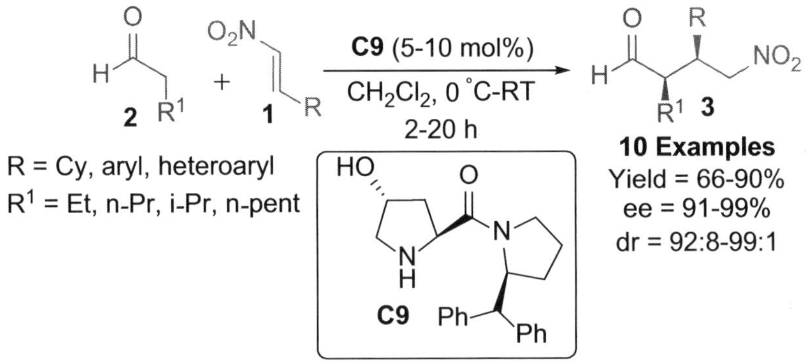

SCHEME 2.8 Asymmetric Michael addition of aldehydes to nitroalkenes catalyzed by *trans*-4-hydroxyprolylamide.

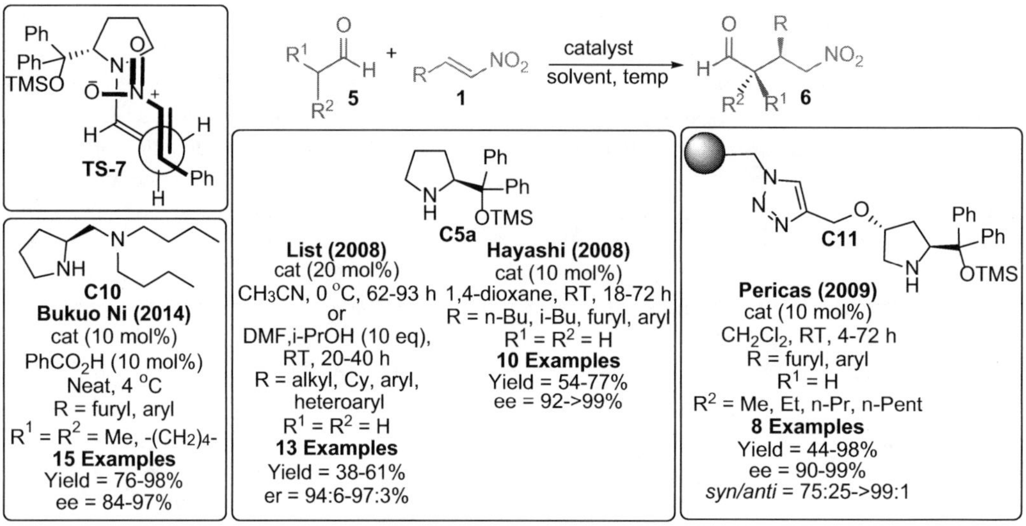

SCHEME 2.9 Asymmetric Michael addition of α,α-disubstituted aldehydes to nitroalkenes catalyzed by chiral pyrrolidine-based diamine and silyl prolinol ether catalysts.

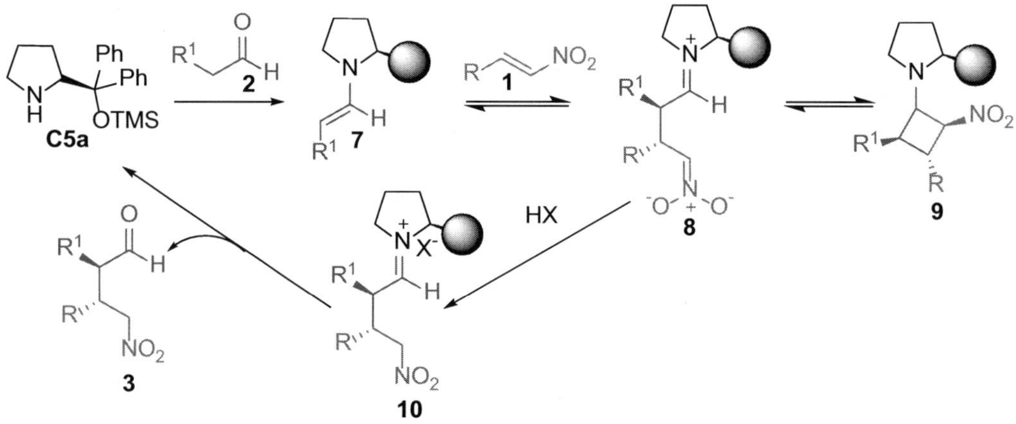

SCHEME 2.10 Generally accepted mechanism for the organocatalyzed Michael addition of aldehydes to nitroalkenes in the presence of a catalytic amount of acid.

(Scheme 2.11).[58] The Huang group conducted an asymmetric Michael addition of aldehydes **2** to various nitroalkenes **1** using only 2 mol% of catalyst **C5c** in the presence of 20 mol% HCO_2H. The reaction was performed at room temperature in an aqueous medium providing the corresponding adducts **3** in good yields with high diastereo and enantioselectivities (up to >99% ee, Scheme 2.11).[59] The Ni group reported an enantioselective Michael addition of aldehydes **2** to nitroalkenes **1** using only 3 mol% of the catalyst **C5d** in water to provide the corresponding Michael adducts **3** in excellent enantio and diastereoselectivities (Scheme 2.11). Moreover, the catalyst **C5d** can be easily recycled and reused without loss of activity.[60]

The Wang group synthesized a novel superparamagnetic nanoparticle-supported (*S*)-diphenylprolinol trimethylsilyl ether catalyst **C12** by immobilizing the Jørgensen–Hayashi catalyst onto the magnetic nanoparticles (MNPs) of $Fe_3O_4@SiO_2$ and employed it for an enantioselective conjugate addition of aldehydes **2** to nitroalkenes **1** in water, yielding adducts **3** in excellent yields and good enantio and diastereoselectivities (Scheme 2.12).[61] Jørgensen–Hayashi catalyst embedded

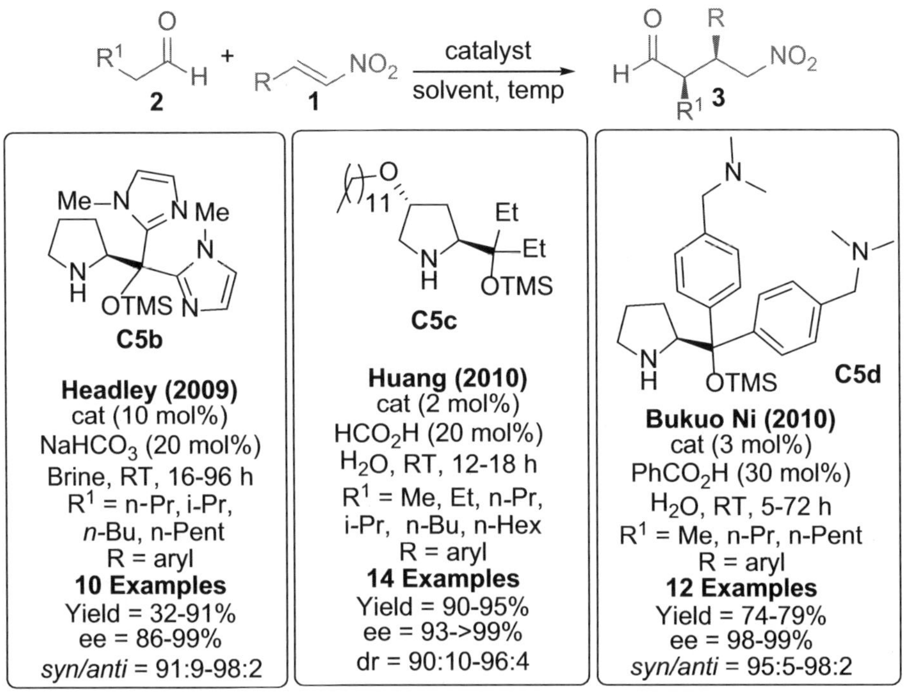

SCHEME 2.11 Asymmetric Michael addition of aldehydes to nitroalkenes catalyzed by various proline-based organocatalysts.

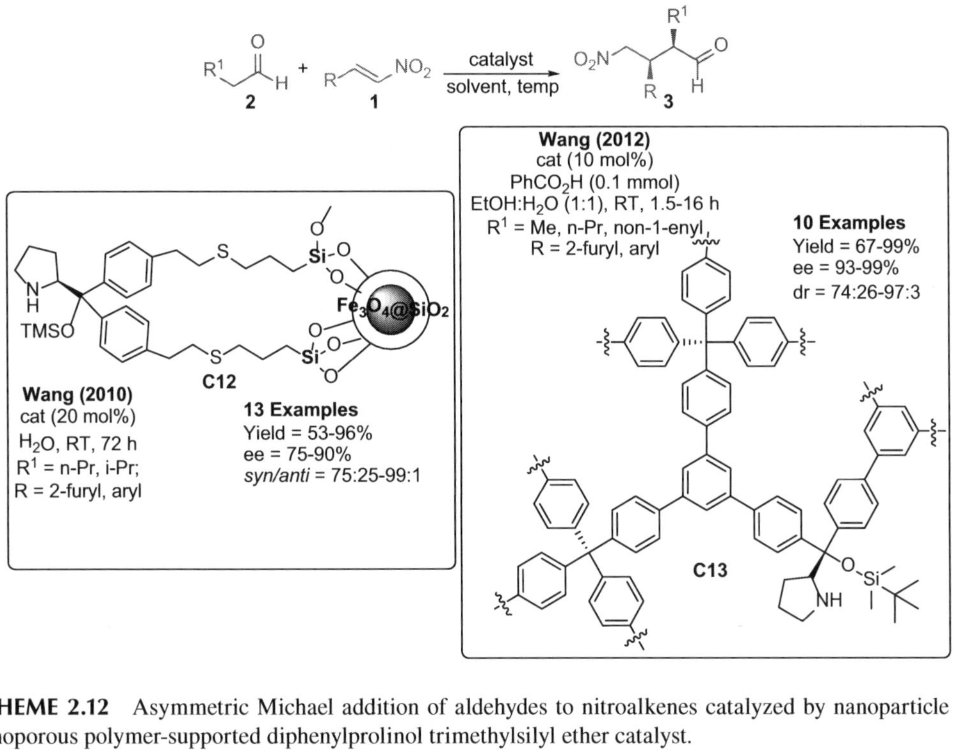

SCHEME 2.12 Asymmetric Michael addition of aldehydes to nitroalkenes catalyzed by nanoparticle and nanoporous polymer-supported diphenylprolinol trimethylsilyl ether catalyst.

into a nanoporous polymer JH-CPP to form catalyst **C13**, which was employed for the asymmetric Michael addition of aldehydes **2** to nitroalkenes **1** to give the corresponding products **3** in good-to-excellent yield (67%–99%), high enantioselectivity (93%–99% ee) and high diastereoselectivity (dr 75:25–99:1).[62] The catalyst can also be reused at least four times without significant loss in enantioselectivity (97%–99% ee) and diastereoselectivity (dr 80:20–86:14, Scheme 2.12).

The Headley group employed 5 mol% of organocatalyst **C5e** in combination with ILS-benzoic acid for the Michael addition of aldehydes **2** to aromatic and aliphatic nitroalkenes **1** in water to give products **3** with excellent stereoselectivity (Scheme 2.13).[63] The Paixao group described an organocatalytic Michael addition of aldehydes **2** to nitroalkenes **1** by employing polyethylene glycol as a recyclable solvent medium wherein it assists in the formation of a host–guest complex. Further, the hydrophobic alkyl side chains of the catalyst **C5f** greatly influences the reaction in yielding the products **3** (Scheme 2.13).[64]

The Ma group employed MNPs (Fe_3O_4)-supported catalyst Fe_3O_4/PVP@SiO_2/ProTMS **C14** in the asymmetric Michael addition of propanal **2a** to various nitroalkenes **1** to obtain the corresponding *syn* adducts **3a** with (2*R*, 3*S*)-configurations (Scheme 2.14).[65] The chemical composition, surface morphology and pore structure, related to the degree of hydrolysis of $Si(OCH_3)_3$, made this catalyst system **C14** to achieve high yields and excellent stereoselectivities.

Lu and co-workers developed novel perhydroindole-derived chiral catalyst **C15** for the asymmetric Michael reaction of aldehydes **5** to nitroalkenes **1**.[66] (2*S*,3′*S*,7′*S*)-Diphenylperhydroindolinol silyl ether **C15** was found to catalyze the reaction of various aldehydes and nitroalkenes, affording corresponding Michael adducts **6** in good-to-excellent yields and enantioselectivities (Scheme 2.15). Bolm et al. designed and synthesized 2-silylated pyrrolidine organocatalyst **C16** and employed it in asymmetric Michael addition of aldehydes **5** to nitroalkenes **1**, which provided the adducts **6** in high yields and stereoselectivities (Scheme 2.15).[67] Later, the Franz group used *tert*-butyldiphenylsilylpyrrolidine catalyst **C17** for the asymmetric Michael reaction of aldehydes **5** to nitroalkenes **1** (Scheme 2.15).[68] In this case, even acetaldehyde donors reacted giving products with yields up to 77% and enantioselectivities up to 96% ee, avoiding common side reactions that were often encountered leading to lower yields. They illustrated the mechanism of the pyrrolidine-based catalyst **C17** by Electrospray ionization (ESI) mass spectrometric analyses of corresponding enamine formation, nitro complexation and Michael adduct formation (**TS-8**, Scheme 2.15).

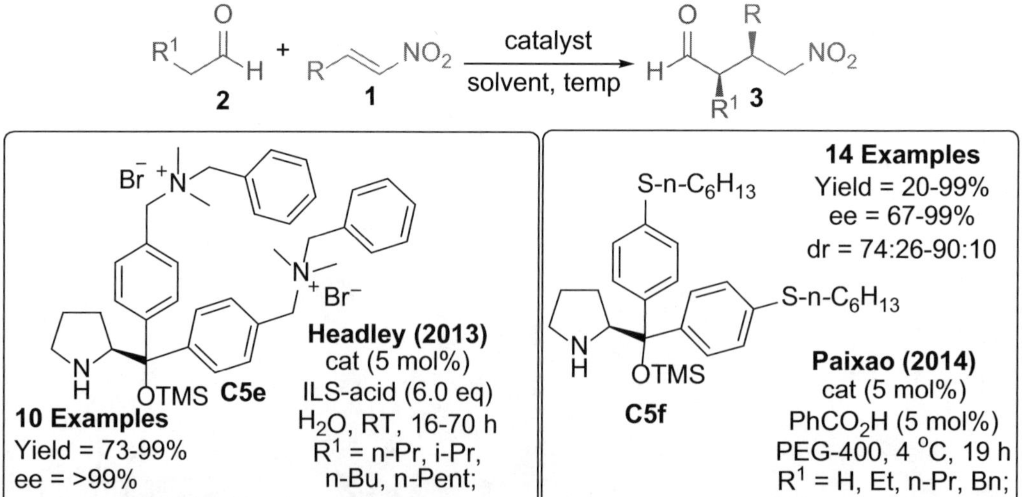

SCHEME 2.13 Asymmetric Michael addition of aldehydes to nitroalkenes catalyzed by diphenylprolinol trimethylsilyl ether-based organocatalysts.

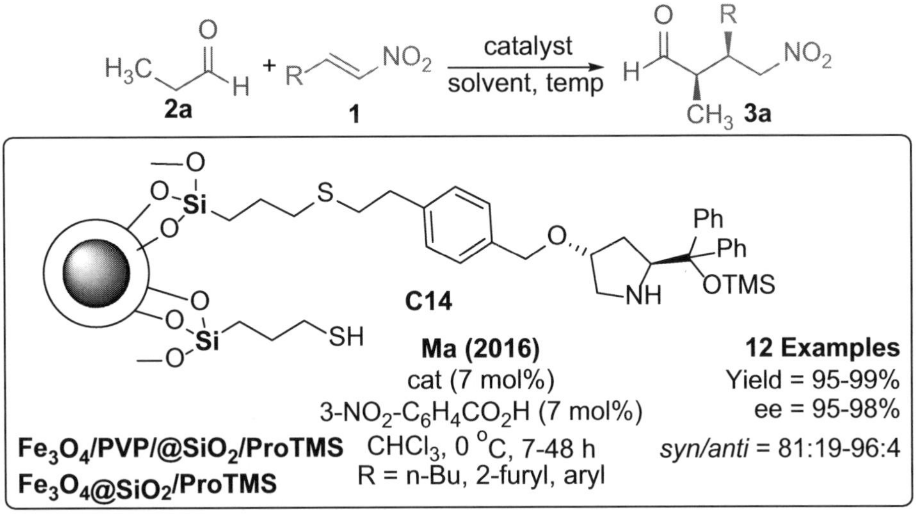

SCHEME 2.14 Asymmetric Michael addition of aldehydes to nitroalkenes catalyzed by supported (*S*)-diphenylprolinol trimethylsilyl ether (Fe_3O_4/PVP@SiO_2/ProTMS).

The Wennemers group introduced highly effective tripeptidic catalysts (H-D-Pro-Pro-Asp-NH_2 **C18** and H-D-Pro-Pro-Glu-NH_2 **C19**) for the conjugate addition of aldehydes **2** to β-substituted nitroalkenes **1**.[69] One mole percent of the peptidic catalyst **C18** effectively catalyzed the conjugate addition to afford γ-nitroaldehydes **3** in excellent yields (65%–98%) and stereoselectivities (*syn/anti* = 4:1 to >99:1, 81%–99% ee, Scheme 2.16). Opposite enantioselectivity of γ-nitroaldehydes **3** was observed when the diastereomer of peptide **C18** was employed and the influence of conformation of peptides was analyzed by molecular modeling study. The proposed transition state structure for the stereoselective formation of γ-nitroaldehyde is shown in Figure 2.1, and the two transition states

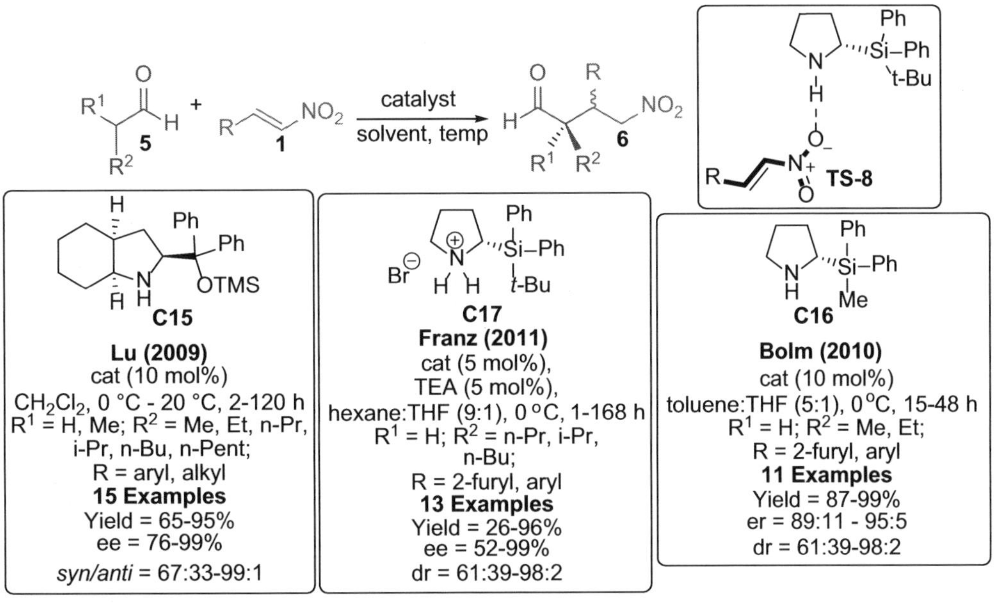

SCHEME 2.15 Asymmetric Michael addition of aldehydes to nitroalkenes catalyzed by perhydroindole and silylated pyrrolidine-derived chiral catalysts.

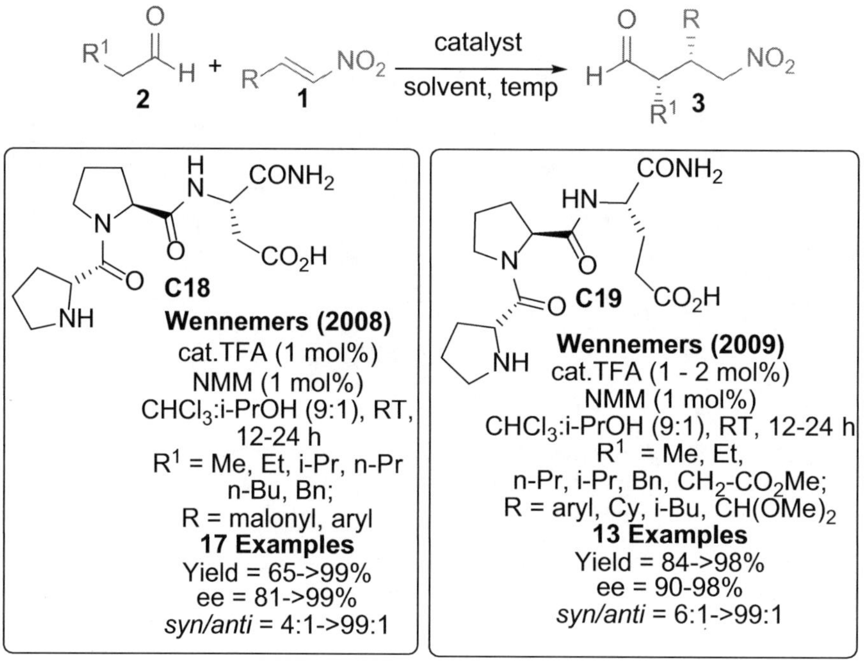

SCHEME 2.16 Tripeptides as asymmetric catalysts for the Michael addition of aldehydes to nitroalkenes.

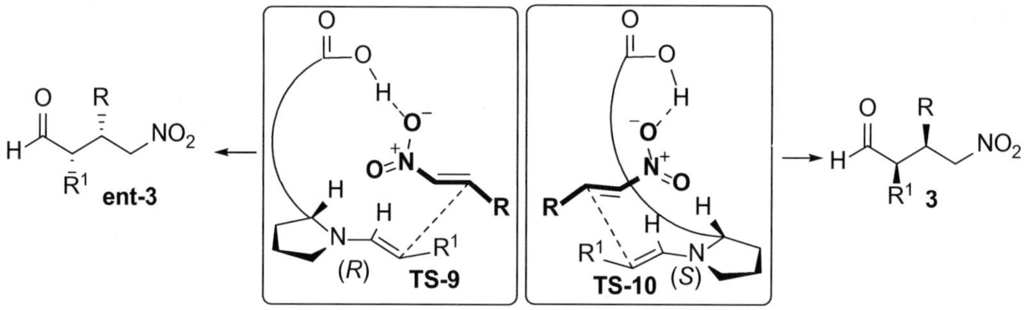

FIGURE 2.1 Proposed transition state structure of tripeptide catalyst with nitroalkene.

TS9 and **TS10** of the diastereomeric peptides behave like pseudoenantiomers. The same group employed 1 mol% of H-D-Pro-Pro-Glu-NH_2 **C19** for the asymmetric Michael addition of aldehydes **2** to nitroalkenes **1** to afford γ-nitroaldehydes **3** with high stereoselectivities (Scheme 2.16). D-Pro-Pro motif in the peptidic catalyst skeleton was responsible for the high asymmetric induction. Whereas the C-terminal amide bond and the spacer length between the terminal carboxylic acid and peptide bond play an important role in fine-tuning the stereoselectivity.[70] Kinetic studies were performed by the same group revealing that both the imine formation and hydrolysis of the corresponding imine are rate-determining steps.

The Lecouvey group developed a combination of both an aminocatalyst and a phosphonic acid **C20** to catalyze the asymmetric conjugate addition of aldehydes **2** to nitroalkenes **1**.[71] Very low catalyst loading of 1 mol% was sufficient to afford the corresponding γ-nitroaldehydes **3** in excellent yields and good stereoselectivities without the use of any external base (Scheme 2.17). Later, a detailed study was conducted to unveil the mechanism of tripeptide catalyzed Michael addition *via* kinetic studies, NMR analyses and structural modifications.[72] Tripeptide catalyst is involved in

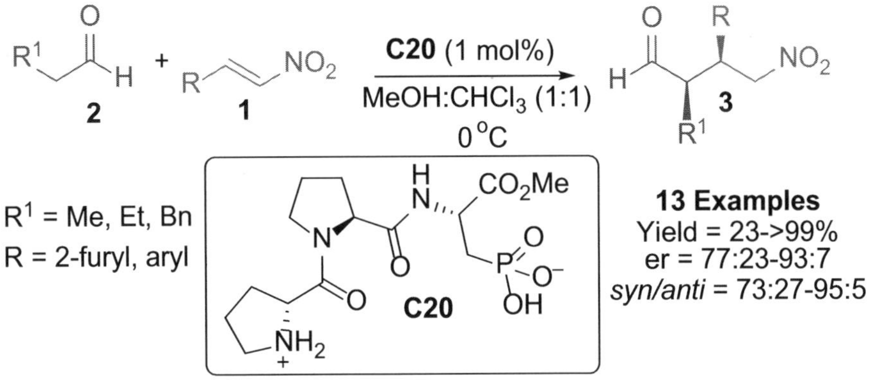

SCHEME 2.17 Phosphonic acid-containing tripeptide as a bifunctional organocatalyst for the stereoselective Michael addition of aldehydes to nitroalkenes.

activating the aldehyde through (*E*)-enamine. It activates nitroalkene through noncovalent acidic interactions. Additionally, the tripeptidic structure plays a significant role in terms of distance and rigidity between both activating sites and influences the reaction selectivities and reaction rates.

The Paixao group designed, synthesized and employed new prolyl peptide-peptoid hybrid catalysts **C21** for the asymmetric conjugate addition of aldehydes **5** to nitroalkenes **1** to give Michael adducts **6** with good-to-excellent yields and diastereo and enantioselectivities (Scheme 2.18).[73] An *anti*-enamine intermediate **I-1** (Scheme 2.18) was proposed for this reaction.

Chen and co-workers designed, synthesized and evaluated pyrrolidine-camphor organocatalysts **C22** and **C23** for the asymmetric Michael reaction of aldehydes **5** to nitroalkenes **1** (Scheme 2.18). The catalyst **C22** was found to be the most effective for α,α-disubstituted aldehydes **5** in the presence of 20 mol% benzoic acid under solvent-free conditions,[74] whereas the catalyst **C23** gave best results with unsubstituted aldehydes providing the desired Michael products **6** in high yields (up to 94%) and stereoselectivities (up to 99% ee).[75] Here, the pyrrolidine unit of the bifunctional organocatalyst

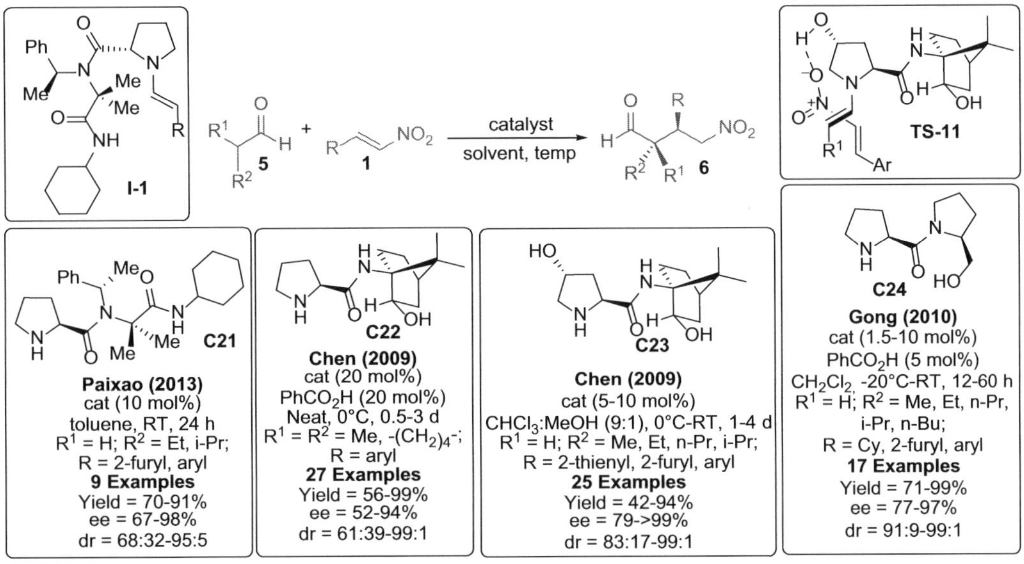

SCHEME 2.18 Asymmetric Michael addition of aldehydes to nitroalkenes catalyzed by various proline-based organocatalysts.

C23 reacts with the aldehyde **5** to form nucleophilic enamine, and the 4-hydroxy group of the catalyst **C23** activates the nitro group through hydrogen bonding to organize a favorable transition state model. The neighboring rigid bicyclic camphor scaffold serves as an efficient stereocontrolling element. The approach of the nitroalkene from the less-hindered *Si*-face of the enamine would produce the observed stereochemistry (**TS-11**, Scheme 2.18).

The Gong group designed and synthesized the novel prolylprolinol catalysts **C24** for promoting the direct addition of unmodified aldehydes **5** to nitroalkenes **1** (Scheme 2.18).[76] The less sterically hindered one was the most effective, affording the products in high enantioselectivities by employing 1.5–5 mol% of the catalyst. A model reaction between propanal and nitrostyrene catalyzed by organocatalyst **24** was studied by density functional theory (DFT) calculations. The involvement of the C–H group to fix the location of nitrostyrene in the DFT studies accounted for the highest diastereo and enantioselectivity exhibited by the catalyst without any bulky group on the prolinol moiety.

The Gao group employed 1 mol% of (*S*)-2-(2′-piperidinyl)pyridine-derived *trans*-4-hydroxy-L-prolinamide **C25** for the asymmetric conjugate addition of aldehyde **5** to nitroalkene **1** in combination with a co-catalyst bearing –OH group, that is, (*S*)-1,1-bi-2-naphthol (Scheme 2.19).[77] During the reaction, pyrrolidine activates the aldehyde **5** by forming an enamine intermediate. Meanwhile,

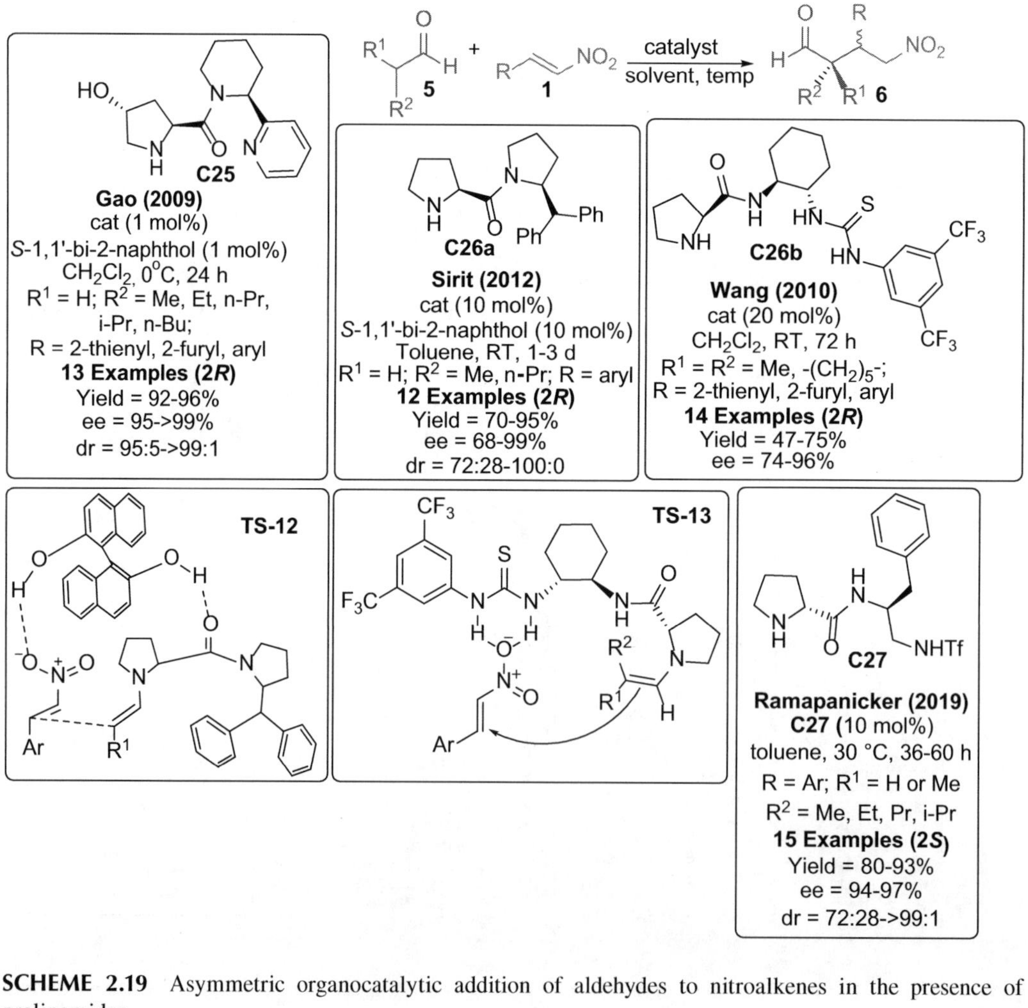

SCHEME 2.19 Asymmetric organocatalytic addition of aldehydes to nitroalkenes in the presence of prolinamides.

hydrogen-bonding interactions between the two OH groups of the (*S*)-1,1-bi-2-naphthol and nitro group of the nitroalkene **1** and the hydroxyl group of the catalyst **C25** not only activate the nitro group of the nitroalkene **1** but also direct the nitroalkene **1** to approach the less-hindered enamine face, affording the adduct **6**.

The Sirit group reported an asymmetric Michael addition of aldehydes **5** to nitroalkenes **1** employing a combination of L-prolinamide derivative **C26a** and an acidic additive such as (*S*)-1,1-bi-2-naphthol to give nitroaldehyde products **6** in excellent yields with excellent enantio and diastereoselectivities (Scheme 2.19).[78] Initially, the pyrrolidine unit of the catalyst **C26a** reacts with aldehyde **5** to form a nucleophilic enamine, and the co-catalyst (*S*)-1,1-bi-2-naphthol activates the nitro group through hydrogen bonding to organize a favorable transition state model **TS-12** (Scheme 2.19). The *anti*-enamine double bond is far from the bulky substituent of the pyrrolidine motif, and the attack of this enamine onto the less-hindered *Si*-face of the nitroalkene **1** results in the formation of the *syn*-diastereomer.

Wang and co-workers synthesized and employed secondary amine-thiourea catalyst **C26b** derived from L-proline and chiral diamine for the asymmetric Michael addition of α,α-disubstituted aldehydes **5** to nitroalkenes **1**.[79] Moderate yields with excellent enantioselectivities (up to 96% ee) were obtained with a wide range of heteroaryl and aryl nitroalkenes (Scheme 2.19). The reaction proceeds through enamine, and a theoretical study revealed that only one oxygen atom of the nitro group was bonded to the thiourea group, which could enhance the electrophilic property of nitroalkene **1** (**TS-13**, Scheme 2.19).

The Ramapanicker group reported the asymmetric Michael addition of aldehyde **5** to nitroalkene **1** mediated by a D-prolinamide-based catalyst **C27** bearing a triflicamide group as a hydrogen bond donor (Scheme 2.19).[80] The corresponding Michael products **6** were formed in good yields (up to 93%) and excellent stereoselectivities (up to 97% ee and up to >99:1 dr). The catalyst was able to perform better than proline itself and was also successful in reactions with hindered α,α-branched aldehydes.

The Yoshida group reported the use of L-phenylalanine lithium salt **C28** as a catalyst system for the Michael addition of isobutyraldehyde **5** to nitroalkenes **1** to form quaternary carbon-containing nitroalkanes **6** in high yields with high enantioselectivity (Scheme 2.20).[81] Michael addition

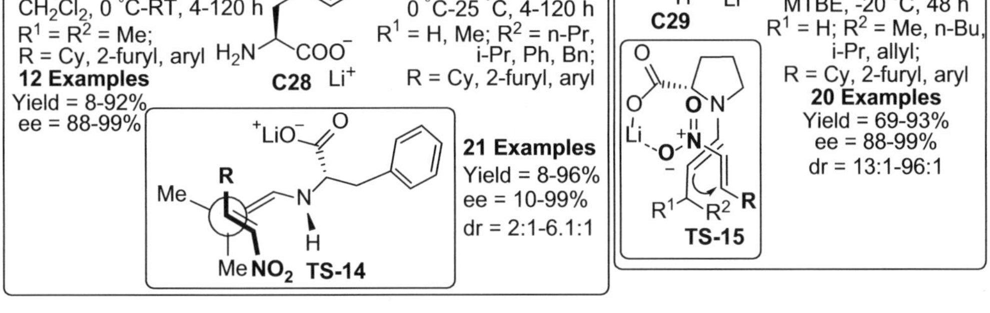

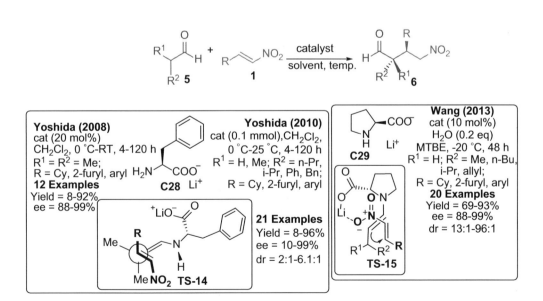

SCHEME 2.20 L-Phenylalanine lithium salt and lithium salt of L-proline as catalysts for the Michael addition of aldehydes to nitroalkenes.

proceeds *via* an enamine mechanism (**TS-14**, Scheme 2.20). In 2010, the same group generalized the method by reacting several aldehydes **5** with different nitroalkenes **1** to produce various substituted γ-nitroaldehydes **6** under the same catalytic conditions (Scheme 2.20).[82] From asymmetric α-branched and unbranched aldehydes **5**, the corresponding Michael adducts **6** were obtained with *syn*-selectivity. Various functionalized aromatic and heteroaromatic nitroalkenes **1** were found to be good Michael acceptors for this reaction. Conjugated nitroalkadienes also gave the corresponding Michael adducts **6** in good yields with high enantioselectivity without generating the 1,6-adducts. According to the possible mechanism, the reaction pathway involves the formation of an enamine of aldehyde and the lithium salt of phenylalanine **C28**. Although two transition states, **TS-16** and **TS-17**, have been proposed, **TS-16**, in which there is no electrostatic repulsion between the carboxylate and the nitro group as well as overall less steric congestion, is favored over **TS-17** despite the fact that there is chelation in the latter. The carboxylate moiety of catalyst **C28** orients the enamine such that it attacks the *Re*-face of nitroalkene to afford the *S*-configured product **6** (Scheme 2.21). The Wang group reported an asymmetric Michael addition of aldehydes **5** to nitroalkenes **1** in the presence of the lithium salt of L-proline **C29**, which proceeded exclusively in *syn*-manner (Scheme 2.21).[83] The reaction follows an enamine pathway, wherein nitroalkene **1** approaches the enamine from the *Si*-face resulting in (2*R*-3*S*) the configured product **6**. Lithium ion could serve as a Lewis acid to ensure the activation of the nitroalkene **1** for the nucleophilic attack in this reaction (**TS-15**, Scheme 2.21).

In 2010, Demir and Eymur showed that a proline-thiourea self-assembled organocatalyst mediates the enantioselective Michael addition of aldehydes **5** to nitroalkenes **1** (Scheme 2.22).[84] It appears that an achiral additive, such as 1,3-bis[3,5-bis(trifluoromethyl)phenyl] thiourea **C30**, which assists in providing effective H-bonding interactions, has an impact on reactivity, selectivity and, presumably, on solubility. The reaction was efficient with only 5 mol% thiourea **C30** in conjunction with L-proline (20 mol%), which afforded corresponding products in good enantioselectivity and high *syn*-selectivity. This reaction represented the first example of its kind to employ the self-assembly of organocatalyst **C30** with L-proline as an additive in an enantioselective Michael addition of branched and unbranched aliphatic aldehydes **5** with nitroalkenes **1**. To explain the higher *syn*-diastereoselectivities and the enantioselectivities with respect to proline, Demir and Eymur proposed a transition state based on Seebach's model. According to this model, the *anti*-enamine formed will orient away from the bulkier group of the catalyst **C30**, allowing the enamine to attack nitroalkene *via* an acyclic synclinal transition state **TS-18** (Scheme 2.22).

Hirose and co-workers described an asymmetric Michael addition of aldehydes **5** to nitroalkenes **1** in the presence of 10 mol% of self-assembled L-proline-amino thiourea **C31** catalyst system to afford the products **6** in excellent yields and stereoselectivities (Scheme 2.22).[85] Bifunctional thiourea increases the solubility of L-proline in less polar solvents and improves the reaction rate and selectivity. In the possible mechanism, the thiourea group of the catalyst **C31** through its H-bonding interactions with proline forms a complex. This on dehydration reaction with the aldehyde **5** forms an iminium intermediate, which subsequently undergoes deprotonation to give the enamine intermediate. Because of its higher highest occupied molecular orbital (HOMO) energy level, this enamine becomes strongly nucleophilic and adds to nitroalkene **1** to give the product.

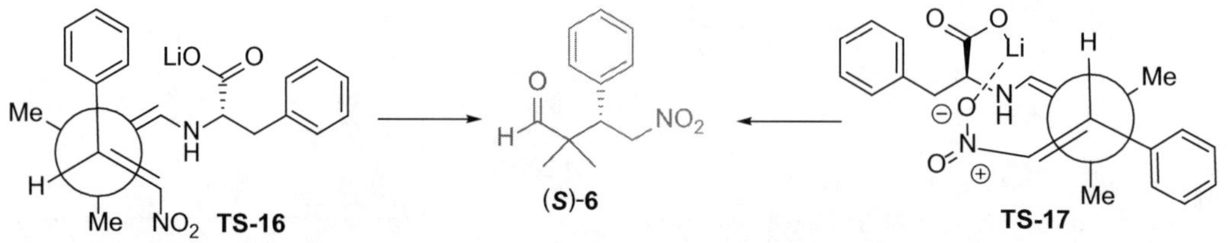

SCHEME 2.21 Plausible transition state for the Michael addition of aldehydes to nitroalkenes in the presence of L-phenylalanine lithium salt.

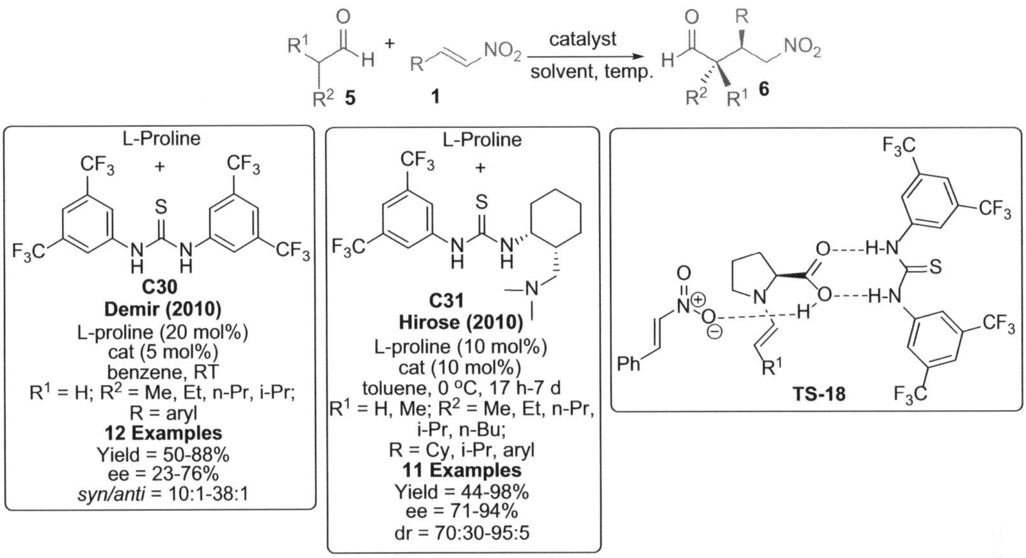

SCHEME 2.22 Self-assembled L-proline-thiourea and L-proline-amino-thiourea-mediated Michael addition of aldehydes to nitroalkenes.

Wu and co-workers reported a catalytic asymmetric Michael reaction of **5** with nitroalkenes **1** using 10 mol% chiral primary amine-thiourea **C32**, and the corresponding products **6** were obtained in excellent enantioselectivity with up to 98% yield (Scheme 2.23).[86] The Yan group used chiral bifunctional sulfamide **C33** as a double hydrogen donor for the asymmetric Michael addition of aldehydes **5** to nitroalkenes **1**. The base additive accelerates the reaction rate to give the products **6** with high stereoselectivities (Scheme 2.23).[87] At first, the imine intermediate is formed from the catalyst **C33** and aldehyde **5**. Subsequently, the base promotes the tautomerization of the imine intermediate to enamine. Hydrogen-bonding interactions between a nitro group of nitroalkene **1** and the sulfamide moiety of the catalyst **C33** activate the nitroalkene **1** to make it more electrophilic and bring it sufficiently close to the reactive enamine. The resulting nucleophilic addition of enamine occurs via the *Si*-face of the double bond of the enamine and forms the intermediate. The subsequent proton transfer followed by hydrolysis yields the product **6** with the regeneration of the catalyst **C33** (**TS-19**, Scheme 2.23).

The Xiao group demonstrated an asymmetric Michael addition of aldehydes **5** to nitroalkenes **1** using a primary amine-thiourea bifunctional catalyst **C34** in the presence of 1,4-diazabicyclo[2.2.2] octane (DABCO) to give products **6** with high enantioselectivities of up to 90%–98% ee (Scheme 2.24).[88] In this reaction, the in-situ generated enamine intermediate from aldehyde **5** and the catalyst attack the *Si*-face of the nitroalkene to give the corresponding products **6** with *R*-configuration. The role of DABCO is to accelerate the tautomerization of imine to the enamine by deprotonation (**TS-21**, Scheme 2.24).

Proline-thiourea self-assembled organocatalysts formed from (*S*)-proline and bipyridine-derived thioureas **C35** were employed for the asymmetric Michael addition of aldehydes **5** to nitroalkenes **1** to furnish the corresponding Michael adducts **6** in excellent yields with high enantio and diastereoselectivities (up to 5:95 dr and 98% ee, Scheme 2.24).[89] The presence of secondary amine functionalities showed high catalytic activity and selectivity (**TS-21**, Scheme 2.24). Later, Lu and co-workers designed, synthesized and investigated several pyrrolidine-camphor-derived organocatalysts for the enantioselective Michael addition of aldehydes **5** to nitroalkenes **1**.[90] Among them, catalyst **C36** (Figure 2.2) was found to be the best which gave adducts in excellent yields and good-to-excellent diastereo and enantioselectivities. The transition state involves hydrogen-bonding interactions between nitroalkenes **1** and in-situ generated nucleophilic enamine (**TS-22**, Figure 2.2).

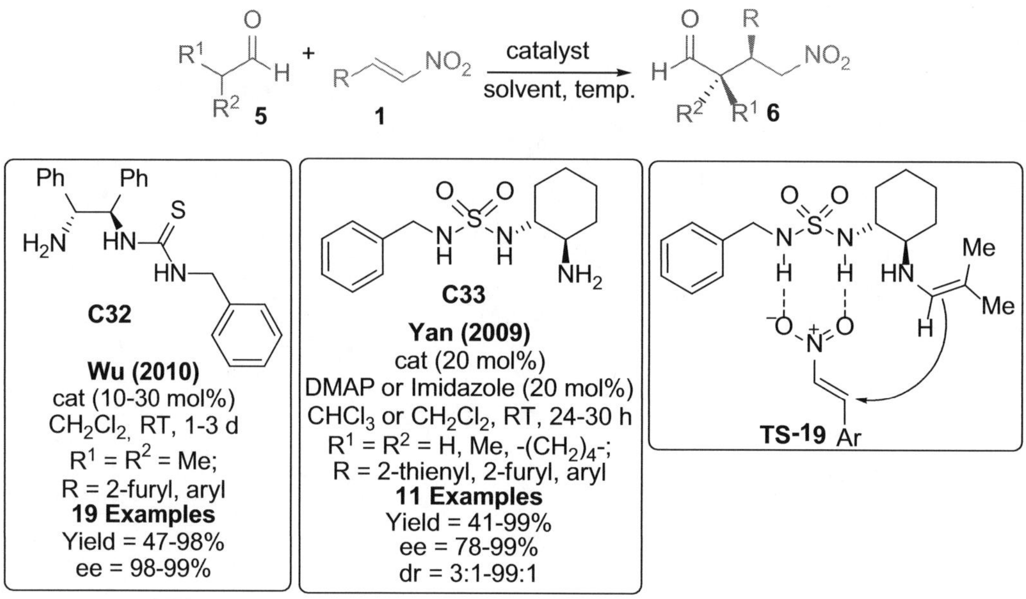

SCHEME 2.23 Primary amine-thiourea and bifunctional sulfamides as organocatalysts for the Michael addition of aldehydes to nitroalkenes.

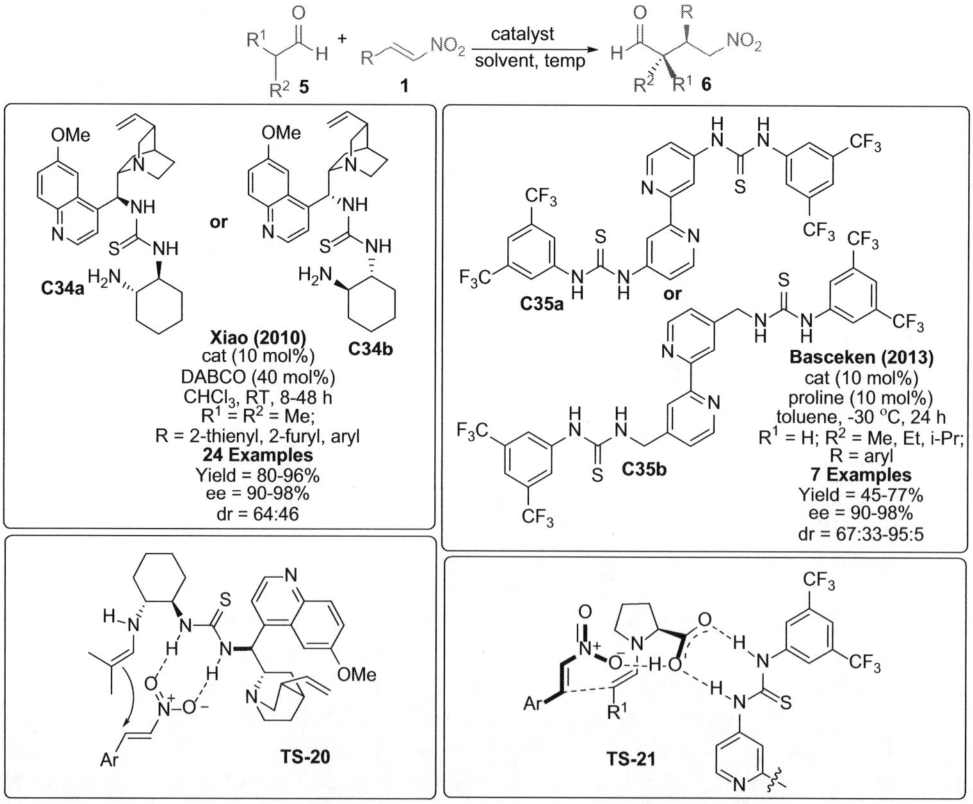

SCHEME 2.24 Primary amine-thiourea and (*S*)-proline and bipyridine-derived thiourea as organocatalysts for the Michael addition of aldehydes to nitroalkenes.

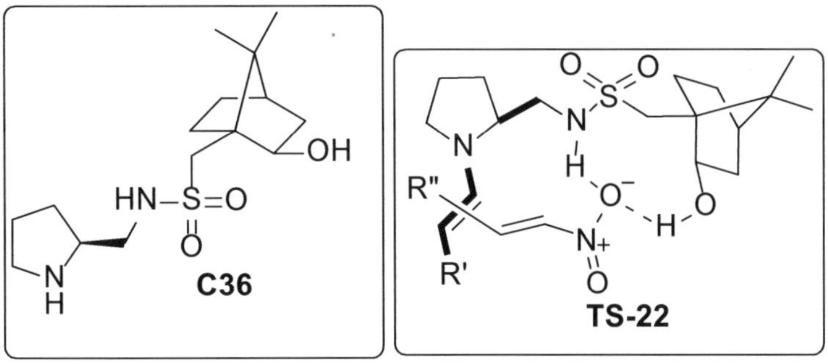

FIGURE 2.2 Pyrrolidine-camphor-derived organocatalyst and its transition state model for the asymmetric Michael addition of aldehydes to nitroalkenes.

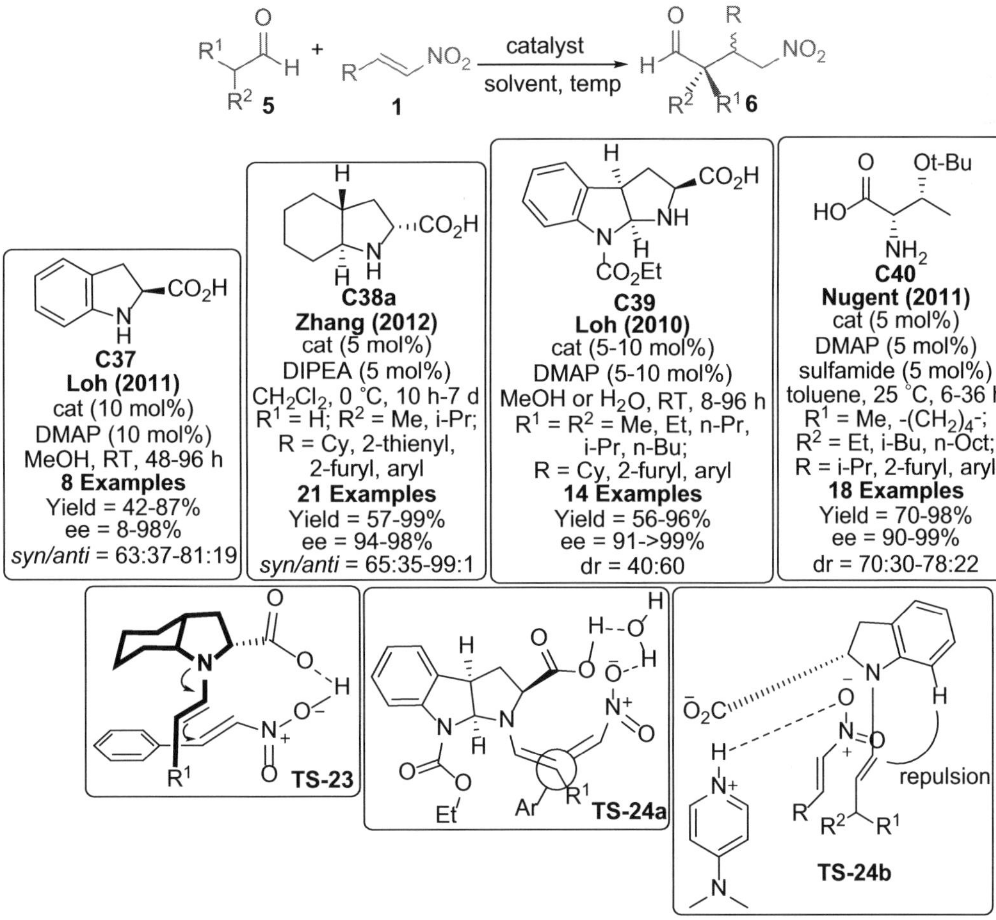

SCHEME 2.25 Asymmetric Michael addition of aldehydes to nitroalkenes catalyzed by L-threonine and indoline-based organocatalysts.

The pyrrolidine structural unit, camphor scaffold, appropriate linker functionality and the free hydroxyl group contributed significantly toward the diastereoselectivity and enantioselectivity of the reaction.

Zhang and co-workers employed perhydroindolic acid **C38a** for the asymmetric Michael addition of aldehydes **5** to nitroalkenes **1** (Scheme 2.25).[91] The proline-like catalyst **C38** possesses a rigid bicyclic structure with two H atoms attached at the bridgehead C-atoms residing on the opposite side of the ring. With 5 mol% of the catalyst **C38**, the reaction afforded the products with enantioselectivities up to 98% ee. According to the stereochemical pathway, the proposed enamine was formed from **C38a** and aldehyde while nitroalkene was activated by hydrogen bonding through carboxylic acid of **C38a** (**TS-23**, Scheme 2.25). The Loh group introduced self-assembly-based organocatalysts for the asymmetric Michael addition of aldehydes **5** to nitroalkenes **1** (Scheme 2.25).[92] A combination of DMAP and pyrrolidine-based organocatalyst **C37** catalyzes the transformation in high yields with excellent stereoselectivities. Hexahydropyrrolo[2,3-*b*]indole-based rigid tricyclic chiral catalyst **C39**-DMAP works well both in organic as well as aqueous media to give the products **6**. DMAP plays a dual role to improve the solubility of the catalyst and can serve as a phase transfer catalyst when the reaction is carried out in water. The rigid tricyclic skeleton provides a site for asymmetric induction, and the hydrophobic pocket enables the reaction to proceed smoothly in aqueous media (**TS-24a**, Scheme 2.25). The organocatalytic system DMAP-**C37** effectively mediated the transformation through the *syn*-enamine intermediate, and DMAP plays a crucial role in the entire synthetic strategy to achieve high yield and enantioselectivity (**TS-24b**).

Nugent et al. developed a three-component catalyst system containing an amino acid (O^{t}Bu-L-threonine) **C40**, a hydrogen bond donor (sulfamide), and an amine base (DMAP) for the asymmetric Michael addition of α-branched aldehydes **5** to nitroalkenes **1** to give Michael adducts **6** bearing a quaternary carbon in good-to-high yield and excellent enantioselectivities (Scheme 2.25).[93] Here, the primary amine of the catalyst **40** forms an enamine with aldehyde **5**, while the carboxylate-sulfamide assembly activates the nitroalkene **1** and directs it into bonding distance with the β-carbon of the enamine via hydrogen bonding.

The Portnoy group developed a polymer-supported bifunctional catalyst **C41** with a primary amine for the asymmetric Michael addition of aldehydes **5** to nitroalkenes **1** (Scheme 2.26). The reaction as usual proceeds through the enamine mechanism (**TS-25**, Scheme 2.26).[94] Luo and co-workers designed and synthesized new perhydroindole derivative **C42** and employed them in asymmetric Michael reaction of aldehydes **5** to nitroalkenes **1** in water providing Michael adducts **6** with high enantioselectivities of up to 98% ee (Scheme 2.26).[95]

Recently, the Lu group described the design and synthesis of 3-monosubstituted and 3,3′-disubstituted chiral amine catalysts **C43** having a binaphthyl skeleton and subsequently utilized them for the asymmetric Michael addition of aldehydes **5** to nitroalkenes **1** (Scheme 2.26).[96] However, 3-monosubstituted amine catalyst **C43** afforded corresponding Michael adducts **6** in excellent yields with high enantio and diastereoselectivities. In this reaction, the azepine group of the catalyst **C43** reacts with the aldehyde **5** to form a nucleophilic *anti*-enamine. The bulky substituent at the 3-position of binaphthyl served as an efficient stereocontrolling element. Nitroalkene **1** might approach from the lower side, that is, *Re*-face of *anti*-enamine to *Re*-face of nitroalkene **1**, to avoid the steric repulsion from the 3-diphenylmethanol silyl ether functionality, resulting in *syn*-Michael adducts **6** with (2*S*,3*R*) absolute configuration (**TS-26**, Scheme 2.26). Lombardo et al. developed an ion-tagged diphenylprolinol silyl ether **C44** as an efficient catalyst for an enantioselective Michael addition of aliphatic aldehydes **5** to nitroalkenes **1** employing as low as 0.25–5 mol% of catalyst to afford the products **6** with >99.5% ee (Scheme 2.26).[97]

Tanchoux and co-workers reported the asymmetric Michael addition of aldehydes **5** to nitroolefins **1** using a heterogeneous catalyst **C45** possessing 9-amino-9-deoxy epi-quinine adsorbed onto alginate gels (Scheme 2.27).[98] The corresponding Michael adducts **6** were obtained in good yields (up to 93%) as well as enantioselectivities (up to 98%). The authors revealed the dual role of the carboxylic moiety of the biopolymer as a co-catalyst as well as a noncovalent binding site by infrared spectroscopic analysis. Moreover, the biopolymer imparted better mechanical support, thus, improving the overall efficiency of the catalyst.

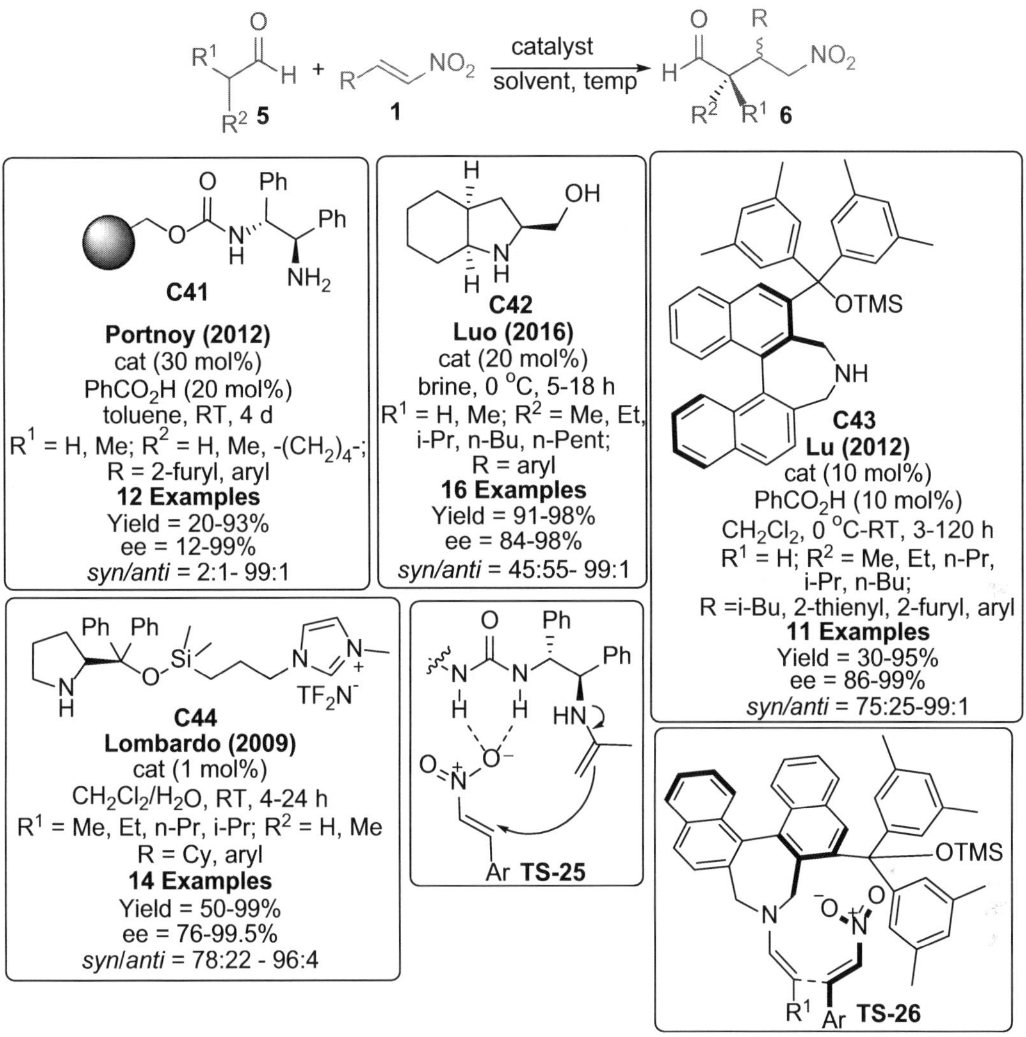

SCHEME 2.26 Polymer-supported amine, perhydroindole, ion-tagged diphenylprolinol silyl ether and binaphthyl-based chiral amine as catalysts for the Michael addition of aldehydes to nitroalkenes.

Of late, Rachwalski and co-workers reported phosphine oxides bearing chiral aziridines subunit **C46** as a catalyst for mediating the asymmetric Michael addition of β-nitrostyrenes **1** to different aldehydes **2** (Scheme 2.28).[99] It was observed that the reactions afforded enantioenriched Michael products **3** with remarkable yields (up to 97%) and excellent stereoselectivities (up to 95:5 dr and up to 98% ee). Here, the authors explained that the dipole interactions of the nitrostyrenes with the phosphine oxide group, as well as the hydrogen-bonding interactions, were responsible for obtaining the stereoselectivity.

Šebesta and co-workers performed diphenylprolinol silyl ether **C5a** catalyzed asymmetric Michael addition of alkyl and aryloxyacetaldehydes **2b** to various nitroalkenes **1** (Scheme 2.29).[100] The addition products 2-(alkoxy)-3-aryl-4-nitrobutanal **3** were obtained in moderate-to-good yield (55%–86%) as diastereomeric mixtures but with high enantiomeric purity (up to 96% ee) (Scheme 2.29). The *syn* and *anti* diastereomers of **3** were formed through **TS-27** and **TS-28**, respectively.

The Kelleher group synthesized a series of α-methyl prolinamide organocatalysts **C47**, **C48** and **C49** as epimers at the α-center from a common starting material, L-proline (Scheme 2.30).[101]

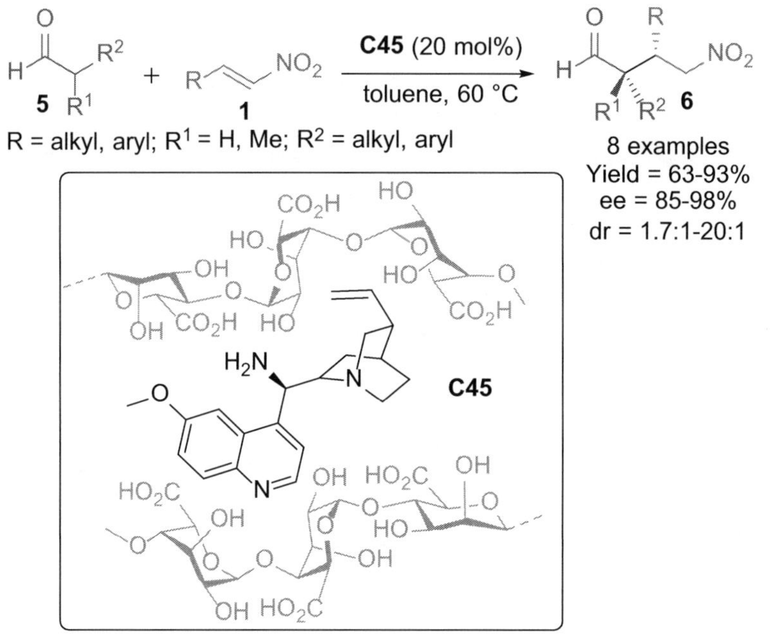

SCHEME 2.27 Asymmetric Michael addition of aldehydes to nitroolefins mediated by heterogeneous alginate-bound epi-quinine as a catalyst.

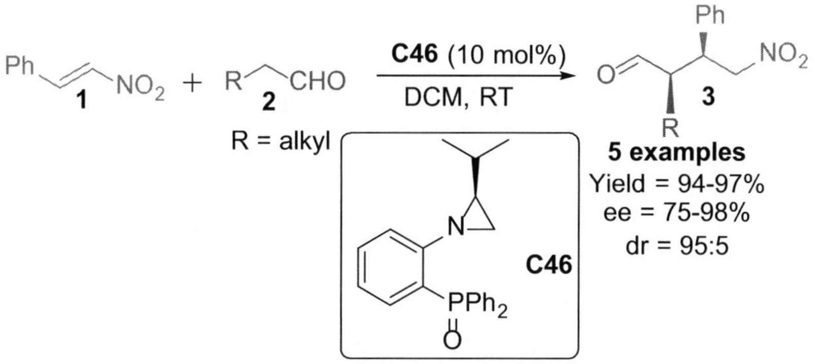

SCHEME 2.28 Enantioselective Michael addition of aliphatic aldehydes to β-nitrostyrene mediated by phosphinoyl-aziridine catalyst.

One can selectively form either an enantiomer of the *syn*-Michael addition product **3b** in excellent yield with good stereocontrol (Scheme 2.30). For other proline-derived catalysts, this would only be possible by separately preparing the catalysts from D-proline. Furthermore, only 5 mol% of catalyst and 1.5 equivalent of aldehyde are required for this reaction.

Primary and secondary amines catalyze the Michael addition of aldehydes **2** to nitroalkenes **1** but α-amination of hindered aldehydes is difficult with a pyrrolidine core. Greck et al. developed double enamine activation of aldehydes **2a** for the synthesis of α,α-disubstituted aldehydes **8** containing a quaternary stereogenic center (Scheme 2.31).[102] A one-pot reaction sequence consisting of a Michael addition of aldehyde **2a** to nitroalkenes **1** catalyzed by diphenylprolinol silyl ether **C5a** and followed by electrophilic α-amination with DBAD **7** catalyzed by a primary amine 9-amino(9-deoxy)*epi*-cinchonine **C50** provided α-hydrazino aldehydes **8** as single diastereoisomers.

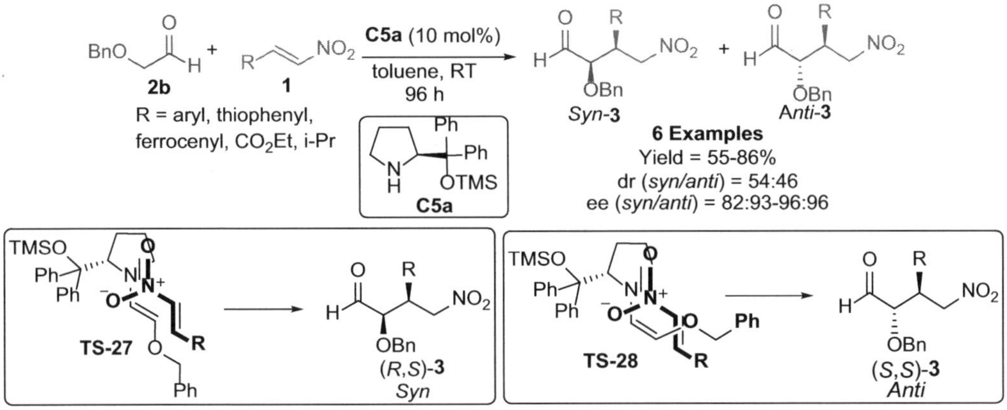

SCHEME 2.29 Diphenylprolinol silyl ether as a catalyst for the Michael addition of aldehydes to nitroalkenes.

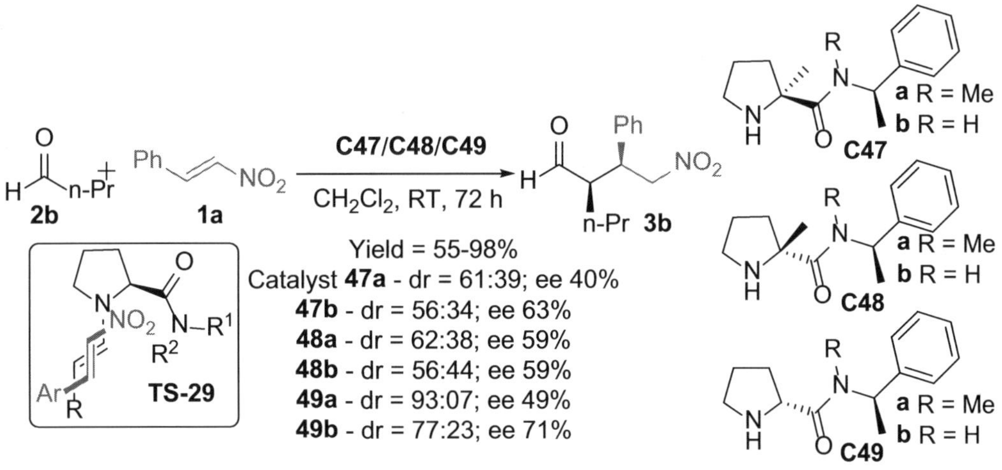

SCHEME 2.30 α-Methyl prolinamide organocatalysts for the Michael addition of aldehydes to nitroalkenes.

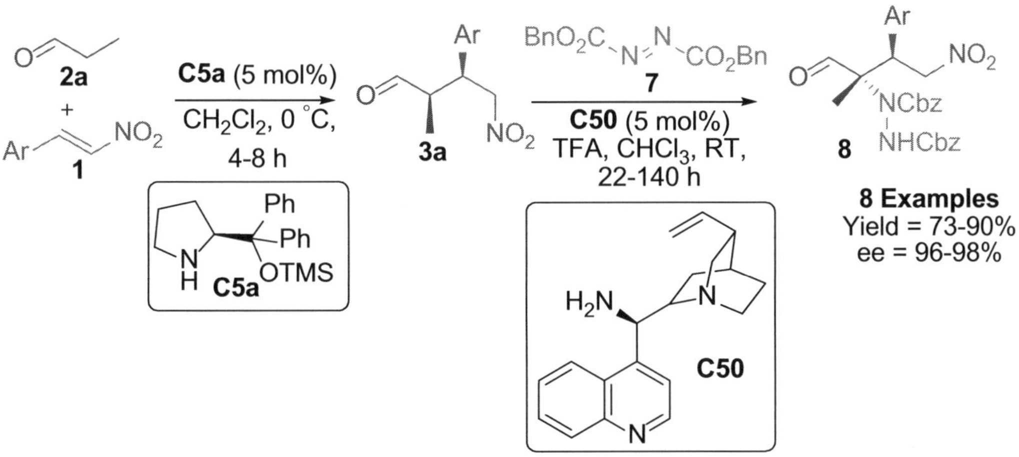

SCHEME 2.31 Synthesis of α,α-disubstituted aldehydes through double enamine activation of aldehydes in the presence of primary amine 9-amino(9-deoxy)*epi*-cinchonine catalyst.

Later, Ma and co-workers investigated the Michael addition of protected 2-nitro-ethenamines **11** and aldehydes **2**.[103] Both diphenylprolinol silyl ether **C5a** and naphthylprolinol silyl ether **C51** were equally efficient to catalyze the Michael addition and provided 1,2-diamine precursors **12** and **14** in excellent yields and selectivities (Scheme 2.32). The phthaloyl-protected 2-nitroethenamine **11** existed in the *E*-form and afforded the Michael adducts with the usual relative and absolute stereochemistry. Acyl-protected 2-nitroethenamine **13** existed in the *Z*-form due to a strong intramolecular hydrogen bond. Thus, with less bulky aldehydes, the reaction proceeded through *E*-enamine intermediate and provided the Michael adducts **12/14** as major *anti*-diastereomer with opposite absolute stereochemistry. In the case of bulky aldehydes, to avoid steric interactions between aldehyde substituent and acetamide group, the reaction proceeded *via* *Z*-enamine intermediate and favored the formation of *syn*-diastereomer (**TS-30**, Scheme 2.32).

Bonne et al. developed a new enantio and diastereoselective synthesis of diversely substituted fused isoxazolines **18**, involving an organocatalytic Michael addition catalyzed by diphenylprolinol silyl ether catalyst **C5a** (resulting into γ-nitroaldehydes **16**), followed by [3+2]-heterocyclization between an in-situ generated silylnitronate and the unactivated double bond (Scheme 2.33).[104]

The synthetic importance of γ-formyl nitro compounds **20** was demonstrated by the enantioselective synthesis of (-)-botryodiplodin **21** (Scheme 2.34).[105] Chiral diamine (*S,S*)-*N*-^{i}Pr-2,2′-bipyrrolidine **C52** catalyzed the Michael addition of propionaldehyde **2** to nitroalkene **19**, affording Michael adduct **20** with 92% yield as a mixture of four diastereomers. Further functional group transformations of adduct **20** provided (-)-botryodiplodin **21** in few steps with good diastereo and enantioselectivities.

The Merino group synthesized optically active five-membered cyclic nitrones **23** in a one-pot procedure via diphenylprolinol silyl ether **C5a** catalyzed Michael addition of aldehydes **22** to nitroalkenes **1** followed by in-situ Zn-mediated reductive cyclization in an aqueous medium

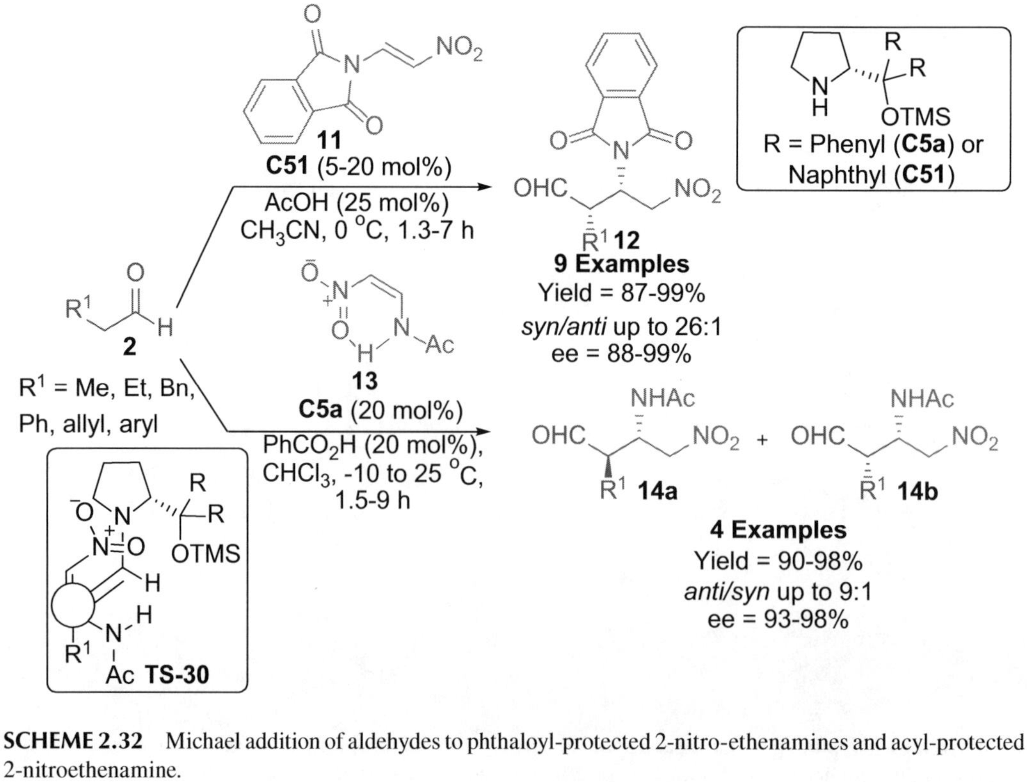

SCHEME 2.32 Michael addition of aldehydes to phthaloyl-protected 2-nitro-ethenamines and acyl-protected 2-nitroethenamine.

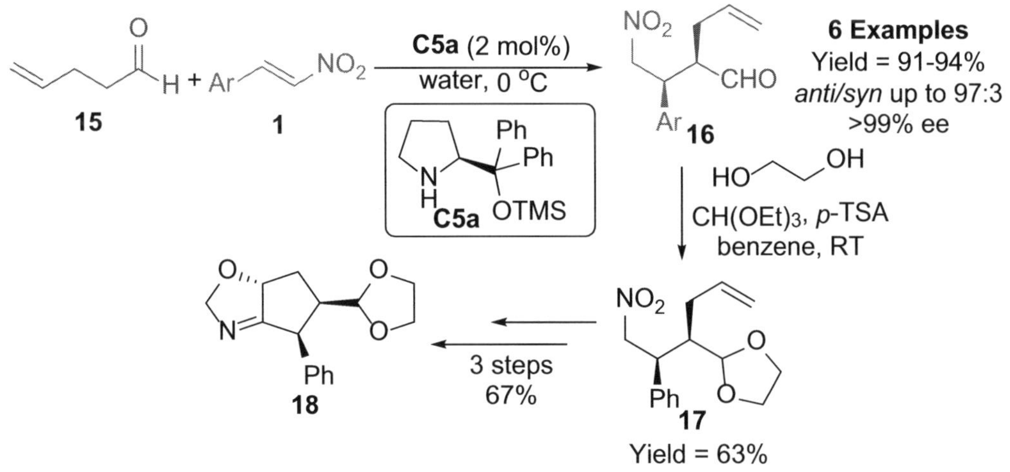

SCHEME 2.33 Organocatalytic asymmetric Michael addition of 4-pentenal to nitroalkenes and further conversion to substituted fused isoxazolines.

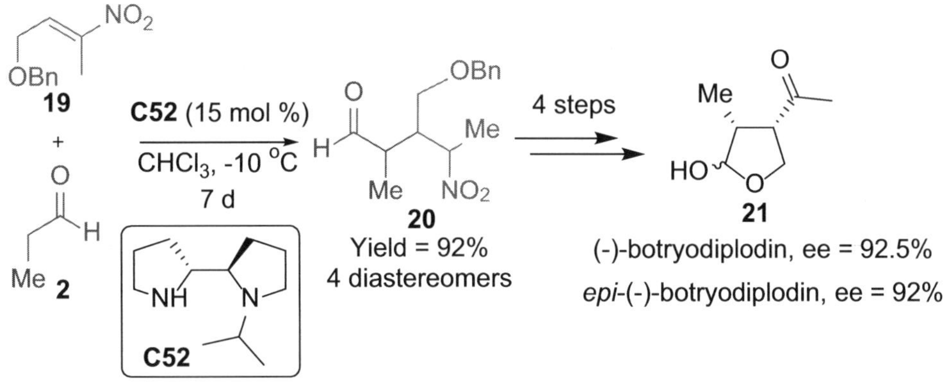

SCHEME 2.34 Michael addition of propionaldehyde to nitroalkene in the presence of bipyrrolidine-based organocatalyst and further transformation to (-)-botryodiplodin.

(Scheme 2.35).[106] This approach was extended to the preparation of alkenyl cyclic nitrones **23** that undergo spontaneous intramolecular 1,3-dipolar cycloaddition to provide tricyclic derivatives **24**.

Kanger et al. developed an asymmetric aminocatalytic approach for the asymmetric synthesis of alkylidenecyclopropane derivatives **26** by an enantioselective Michael addition of aldehydes **25** to nitroalkenes **1** using a proline-based catalyst **C5a** in the presence of an acidic co-catalyst, 4-nitrophenol (Scheme 2.36).[107]

Ma et al. described an asymmetric Michael addition of aldehydes **2** to nitroalkene **27** by using diarylprolinol ether catalyst **C5a** in the presence of benzoic acid, which assisted in the formation of an enamine intermediate (Scheme 2.37).[108]

The catalysts with an appropriately positioned proton donor do not require any additives for high reactivity and stereoselectivity. Therefore, Wennemers et al. developed a peptidic catalyst H-D-Pro-Pro-NHCH(Ph)CH_2_4-Me-C_6H_4 **C53** having an additional acid group for the asymmetric Michael addition of aldehydes **2** to β,β-disubstituted nitroalkenes **29** to afford γ-nitroaldehydes **30** with an all-carbon quaternary stereogenic center adjacent to a tertiary stereocenter in high yields (Scheme 2.38).[109] As the TFA salt of the peptide was used in the reaction, an equivalent amount of *N*-methylmorpholine was used. In this reaction, steric shielding interaction between the aromatic

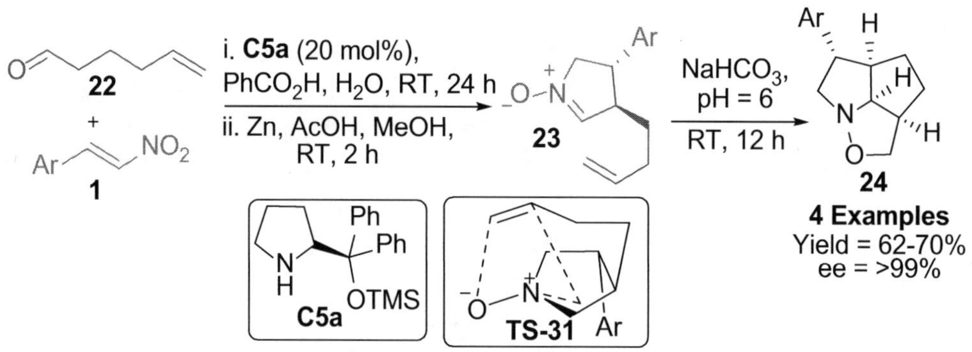

SCHEME 2.35 Synthesis of cyclic nitrones by diphenylprolinol silyl ether catalyzed Michael addition of aldehydes to nitroalkenes and further transformation to tricyclic compounds.

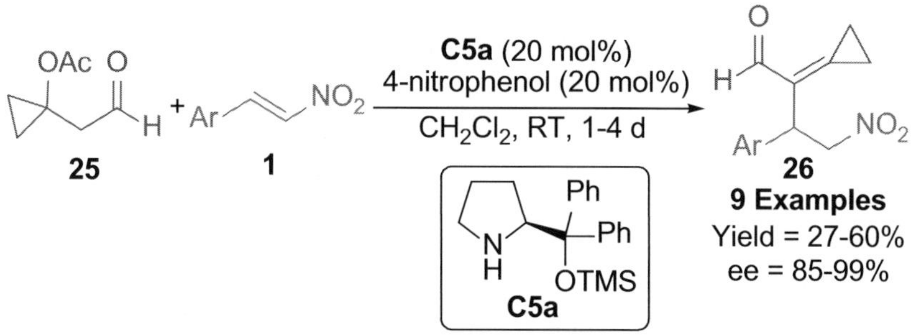

SCHEME 2.36 Synthesis of alkylidenecyclopropanes by the Michael addition of aldehydes to nitroalkenes.

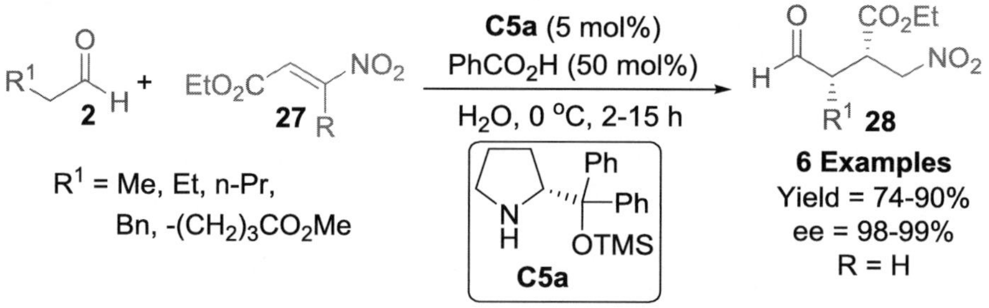

SCHEME 2.37 Asymmetric Michael addition of aldehydes to β-nitroacrylate using a diarylprolinol ether catalyst.

portion of catalyst **C53** and the β,β-disubstituted nitroalkene **29** are crucial for favoring the conjugate addition over the competing aldol reaction.

Wang et al. described an asymmetric Michael addition of aliphatic aldehydes **2** to indole-3-carboxaldehyde-derived nitroalkenes **31** employing (*S*)-diphenylprolinol trimethylsilyl ether catalyst **C5a** to afford optically active *syn*-2-alkyl-3-(1*H*-indol-3-yl)-4-nitrobutanal derivatives **32** with high yields and excellent enantioselectivities (>99% ee, Scheme 2.39).[110]

The Wennemers group employed only 5 mol% of tripeptides H-Pro-Pro-D-Gln-OH **C54** and H-Pro-Pro-Asn-OH **C55** for the asymmetric Michael addition of aldehydes **2** to α,β-disubstituted nitroalkenes **33** to give γ-nitroaldehydes **34** with three contiguous stereocenters in excellent yields and enantiomeric excesses (Scheme 2.40).[111]

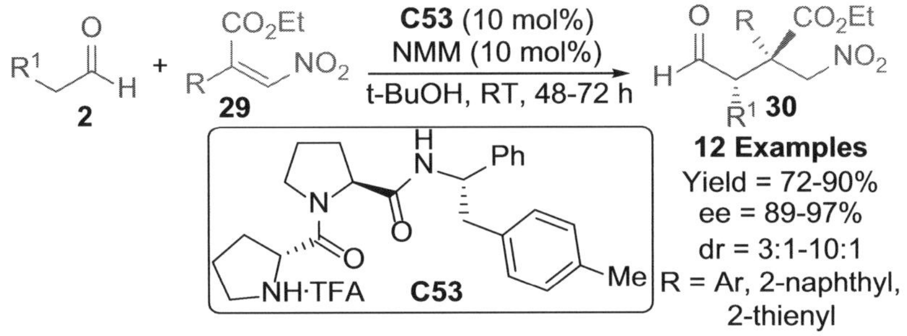

SCHEME 2.38 Peptidic catalyst having an additional acid group for asymmetric Michael addition of aldehydes to β,β-disubstituted nitroalkenes.

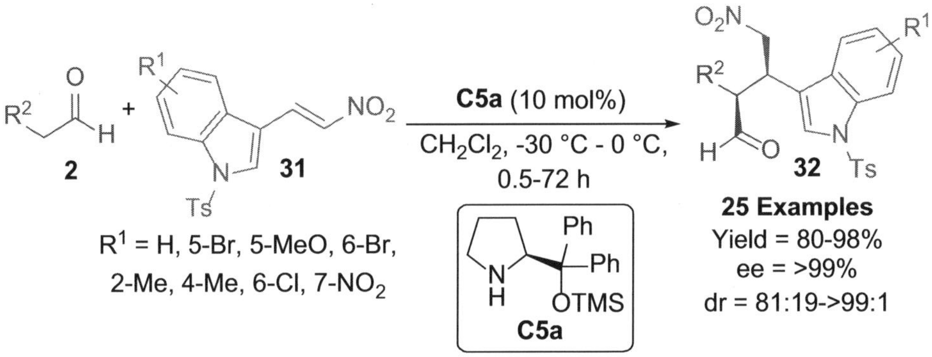

SCHEME 2.39 Asymmetric Michael addition of aliphatic aldehydes to nitroalkenes derived from indole-3-carboxaldehyde.

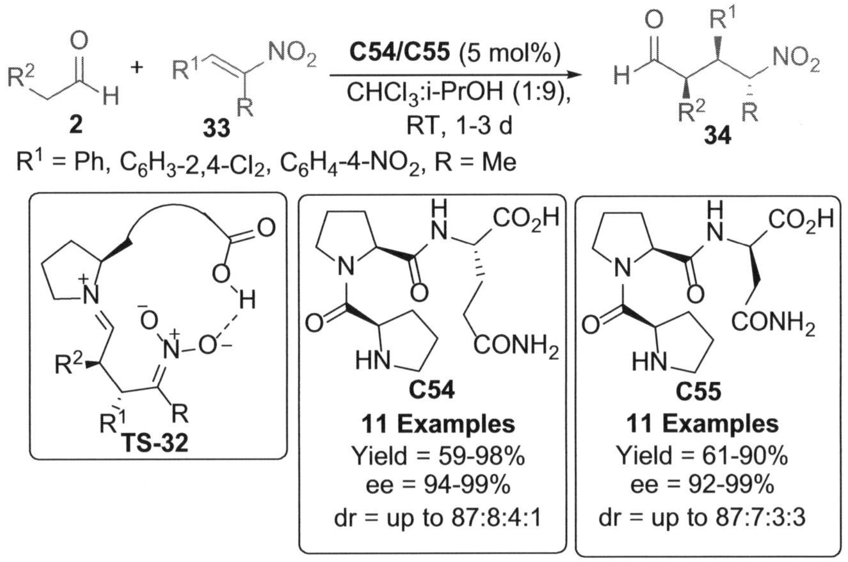

SCHEME 2.40 Tripeptides as catalysts for asymmetric Michael addition of aldehydes to α,β-disubstituted nitroalkenes.

The Peng group developed a catalytic asymmetric Michael addition of aldehydes **2** to α-substituted nitroalkenes **35** using proline-based catalyst **C56** in the presence of an acidic additive, 4-nitrobenzoic acid, to generate γ-nitro carbonyl compounds **36** with high yields and excellent stereoselectivities (Scheme 2.41).[112]

Ma et al. employed a combination of trimethylsilyl(TMS)-protected diphenyl-prolinol **C5a** and benzoic acid to catalyze the asymmetric Michael addition of aldehydes **2** to 3-substituted 3-nitroacrylates **27** to give the products **29** with excellent enantioselectivities (Scheme 2.42).[113] The catalyst reactivity was found to be highly dependent on the electronic nature of the nitroalkenes.

Ramapanicker and Gorde synthesized a small array of four D-prolyl-2(trifluoromethylsulfonamidoalkyl)pyrrolidine organocatalysts and compared their performance in the asymmetric Michael addition of aldehydes **2** to nitroalkenes **1** at ambient conditions (Scheme 2.43).[114] The authors envisaged that the secondary amine would activate the aldehyde by enamine formation while the NHTf group by hydrogen-bonding interactions would direct the approach of the Michael acceptor. It was observed that the D-prolyl-2-(trifluoromethylsulfonamidopropyl)pyrrolidine **C57** performed superior compared to other synthesized organocatalysts in the asymmetric Michael reaction providing the Michael adducts **3** with high yields (up to 95%) and remarkable enantioselectivities (up to 97%) using just 10 mol% of the catalyst, avoiding the use of an additive (Scheme 2.43).[114] Here, the authors suggested the requirement of an optimal distance between the secondary amine functionality and the hydrogen-bond donor unit in the catalyst design.

The Miura group reported an asymmetric conjugate addition of furfurals **29** with nitroalkenes **1** using a pyrrolidine-diaminomethylenemalononitrile organocatalyst **C58** (Scheme 2.44).[115] The pyrrolidine unit present in the catalyst **C58** and the aldehyde moiety of **29** react to form an enamine intermediate, which adds to the nitroalkene substrate **1** activated via hydrogen bonding with amine

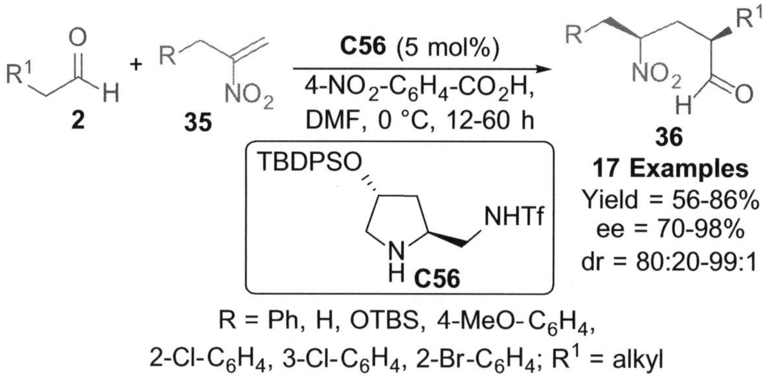

SCHEME 2.41 Michael addition of aldehydes to α-substituted nitroalkenes in the presence of a proline-based catalyst.

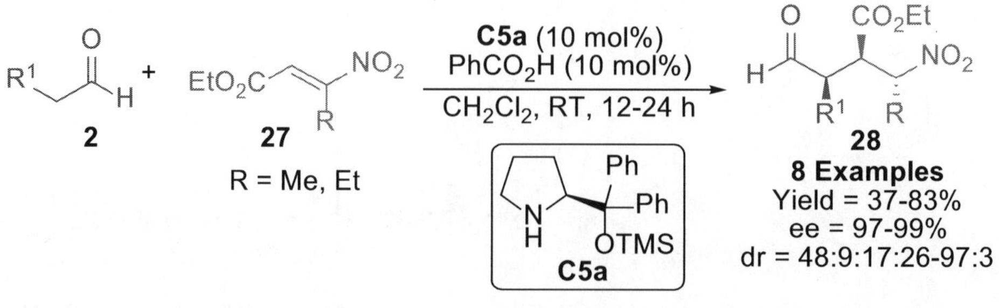

SCHEME 2.42 Asymmetric Michael addition of aldehydes to 3-substituted 3-nitroacrylates.

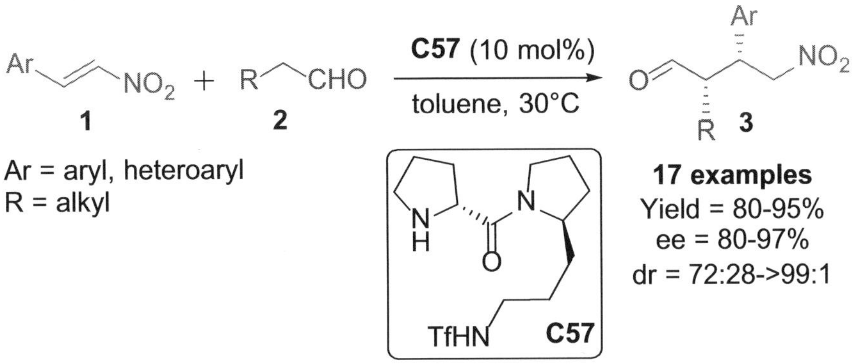

SCHEME 2.43 Asymmetric Michael addition of aldehydes to nitroalkenes mediated by D-prolyl-2-(trifluoromethylsulfonamidopropyl)pyrrolidine organocatalyst.

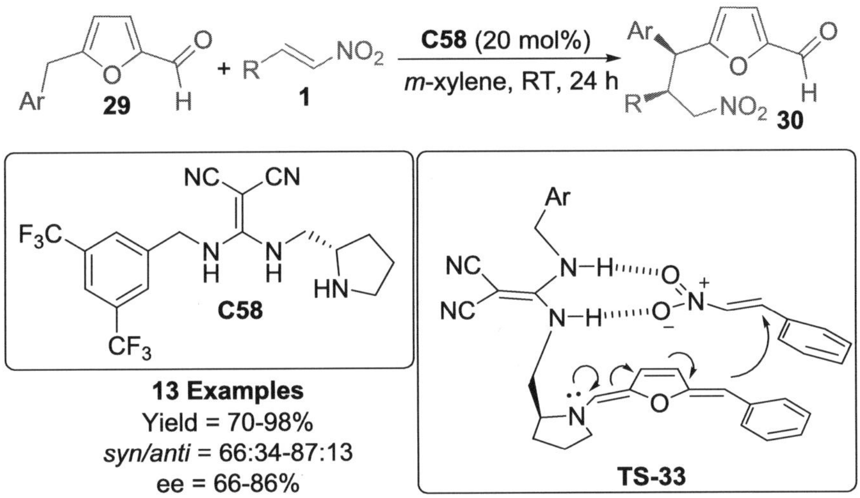

SCHEME 2.44 Asymmetric conjugate addition of furfurals to nitroalkenes.

protons of the catalyst **C58** (**TS-33**, Scheme 2.44) to afford adducts **30** in high yields and enantioselectivities (up to 86%).

Recently, the same group reported the success of a pyrrolidine-thiourea organocatalyst **C59** in efficiently mediating the asymmetric conjugate addition of 5-benzylfurfurals **29** derivatives to nitroalkenes **1**, where the thiourea moiety activated the nitroalkene through H-bonding interactions and the pyrrolidine unit activated the aldehyde through covalent interactions. It was worth observing that the catalyst **C59** could catalyze the transformations with high yields (up to 99%) and enantioselectivities (up to 95%) using just 5 mol% catalyst loading (Scheme 2.45).[116]

Recently, Sirit and co-workers designed a tetraoxacalix[2]arene[2]triazine-based organocatalyst **C60a** for the asymmetric Michael addition of isobutyraldehyde **5a** to various nitrostyrenes **1** using THF as the solvent (Scheme 2.46).[117] It was observed that the Michael adducts were formed smoothly providing remarkable yields (up to 97%) and excellent enantioselectivities (up to 99%). Here the authors emphasized that the positive stereochemical outcome obtained in the reactions was because of the effect of bulky heteroatom-bridged calixarene skeleton linked to the chiral amine group in the catalyst.

Chinchilla and co-workers reported the enantioselective conjugate addition of aldehyde **5a** to nitroalkenes **1** mediated by a chiral primary amine-salicylamide catalyst **C60b** (Scheme 2.46).[118]

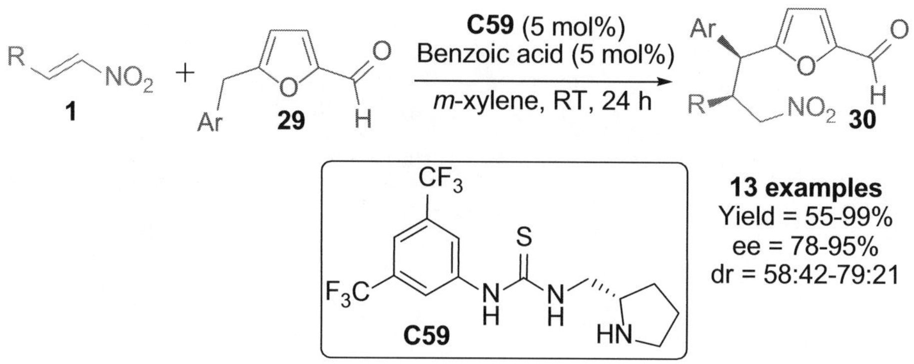

SCHEME 2.45 Asymmetric conjugate addition of 5-benzylfurfurals to nitroalkenes using pyrrolidine-thiourea organocatalyst.

These reactions were performed in deep eutectic mixtures (choline chloride/water) as green solvents and resulted in high yields and good enantioselectivities. The recyclability of the eutectic mixture along with the organocatalyst was also observed with almost similar results.

Kostenko and co-workers reported the Michael addition of aliphatic aldehydes **5a** to nitroalkenes **1** mediated by a C_2-symmetric squaramide **C12** as catalyst (Scheme 2.46).[119] The corresponding Michael adducts **6** were formed with high enantioenrichment. They also possessed the potential to be transformed into diverse biologically active γ-aminobutyric acid derivatives.

Alemán et al. described the asymmetric 1,3-addition of silyl dienol ether **31**, a masked aldehyde functionality, to nitroalkenes **1** in the presence of bifunctional organocatalyst **C61**, followed by double-bond isomerization to afford Rauhut–Currier type products **32** with high yields and enantioselectivities (Scheme 2.47).[120]

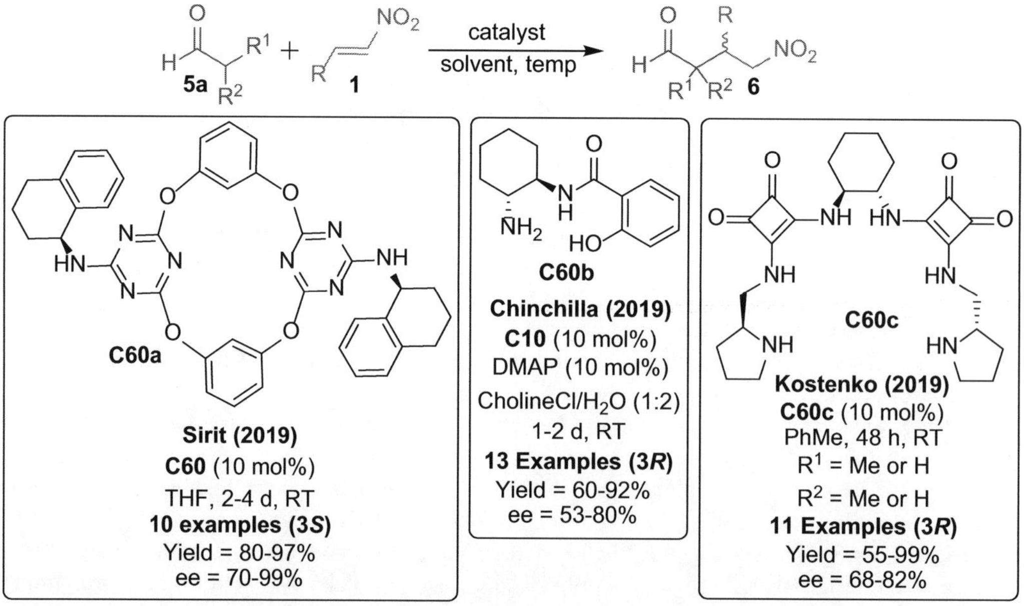

SCHEME 2.46 The Michael addition of aldehyde to nitrostyrenes mediated by chiral calix[2]arene[2]triazine, cyclohexanediamine and squaramide-based catalysts.

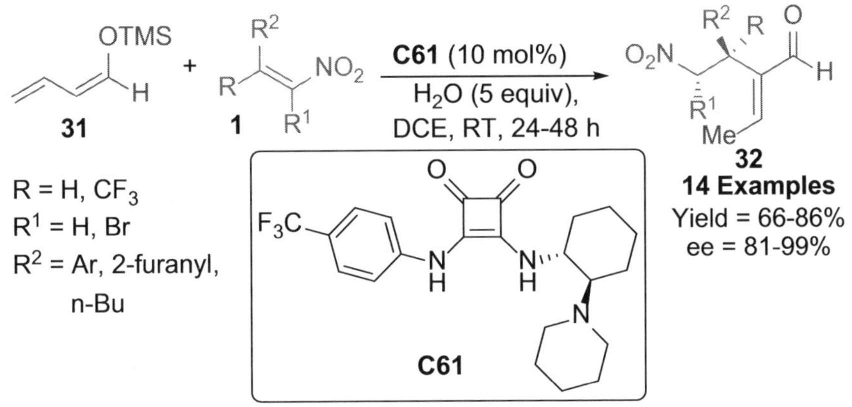

SCHEME 2.47 Asymmetric synthesis of Rauhut–Currier-type products by 1,3-addition of silyl dienol ether to nitroalkenes.

2.3 ADDITION OF UMPOLUNG OF ALDEHYDES: THE STETTER REACTION

The Stetter reaction utilizes the umpolung reactivity of aldehydes, the inversion of their normal mode of reactivity, generating acyl anion equivalents that can participate in 1,4-conjugate additions with various Michael acceptors.

The Rovis group reported an enantioselective intermolecular Stetter reaction of nitroalkenes **1** with heteroarylaldehydes **33** in the presence of an NHC catalyst **C62** to form corresponding Stetter products **34** with high yields and excellent enantioselectivities (Scheme 2.48).[121] The increased enantioselectivity of the products **34** in this reaction is attributed to the conformational change in the bicyclic ring system of the catalyst **C62**, which was accounted for by the stereoelectronic effects. This orients the incoming nitroalkene **1** to improve enantiofacial discrimination.

The same group also reported a diastereo and enantioselective synthesis of *syn* δ-nitroesters **36** with high yields from nitroalkenes **1** and α,β-unsaturated aldehydes **35** in the presence of an NHC catalyst **C63**, which operates exclusively via the homoenolate pathway rather than the typical acyl anion pathway (Scheme 2.49).[122]

An enantioselective intermolecular Stetter reaction of enals **35** with nitroalkenes **1** employing chiral NHC catalyst **C62** along with catechol provides Stetter products **37** in good yields and enantioselectivities (Scheme 2.50).[123] The use of a bifunctional Brønsted acid, that is, catechol in the reaction facilitates proton transfer and assists in the intermolecular reaction. Further, this reaction

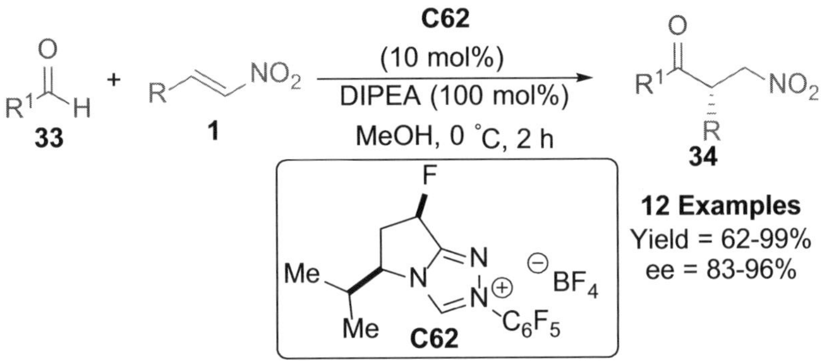

SCHEME 2.48 Enantioselective intermolecular Stetter reaction of nitroalkenes with aldehydes in the presence of an NHC catalyst.

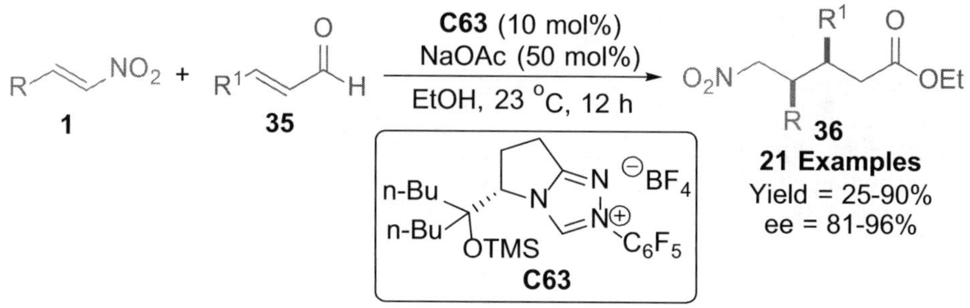

SCHEME 2.49 Asymmetric NHC catalyzed addition of enals to nitroalkenes.

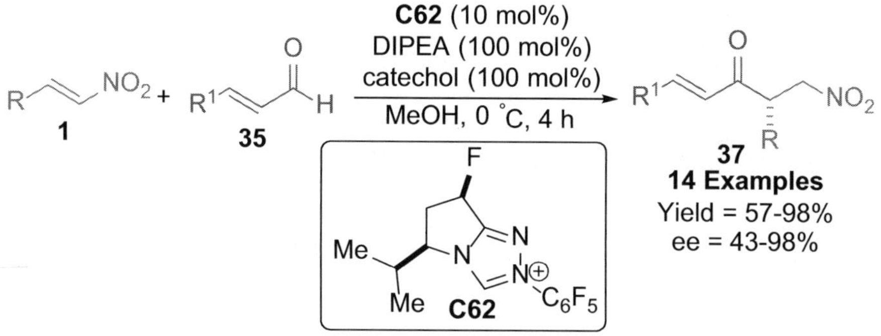

SCHEME 2.50 Enantioselective Stetter reaction of enals with nitroalkenes in the presence of chiral NHC catalyst and catechol.

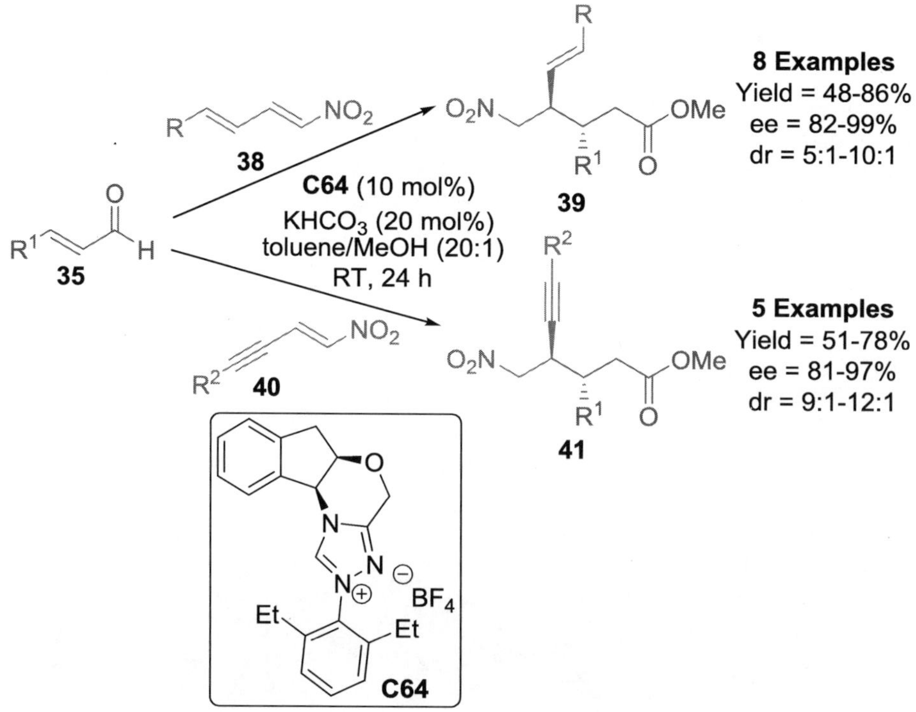

SCHEME 2.51 NHC catalyzed homoenolate addition of enals to conjugated nitroalkenes.

compliments the homoenolate strategy, which was generally observed in the NHC catalyzed reactions of enals.

The Liu group described the chiral NHC **C64** catalyzed asymmetric synthesis of highly enantioenriched and highly functionalized 5-carbon-synthon δ-nitro esters **39/41** by reacting enals **35** with different nitroalkenes **38/40**, including nitrodienes, nitroenynes and nitrostyrenes.[124] The corresponding chiral nitroesters were obtained in high yields with excellent enantioselectivities (Scheme 2.51).

2.4 CONCLUSIONS

Most catalytic asymmetric Michael additions of aldehydes to nitroalkenes reported so far take place under organocatalysis, especially enamine (Lewis base) catalysis. Proline and other amino acid derivatives as well as cinchona-derived catalysts have been extensively employed for this purpose. Several chiral NHCs have been employed for the addition of umpolung of aldehydes in the emerging topic of Stetter reaction, which is a convenient method for the synthesis of β- or δ-ketonitroalkanes. The products formed in high yields and stereoselectivities possess a key carbonyl group and a nitroalkyl moiety, which are amenable for further synthetic elaboration.

REFERENCES

1. Almaşi, D.; Alonso, D. A.; Nájera, C. *Tetrahedron: Asymmetry* **2007**, *18*, 299.
2. Tsogoeva, S. B. *Eur. J. Org. Chem.* **2007**, *2007*, 1701.
3. List, B.; Pojarliev, P.; Martin, H. J. *Org. Lett.* **2001**, *3*, 2423.
4. Alexakis, A.; Andrey, O. *Org. Lett.* **2002**, *4*, 3611.
5. Enders, D.; Seki, A. *Synlett* **2002**, *2002*, 26.
6. Andrey, O.; Alexakis, A.; Bernardinelli, G. *Org. Lett.* **2003**, *5*, 2559.
7. Benaglia, M.; Puglisi, A.; Cozzi, F. *Chem. Rev.* **2003**, *103*, 3401.
8. Andrey, O.; Alexakis, A.; Tomassini, A.; Bernardinelli, G. *Adv. Synth. Catal.* **2004**, *346*, 1147.
9. Ishii, T.; Fujioka, S.; Sekiguchi, Y.; Kotsuki, H. *J. Am. Chem. Soc.* **2004**, *126*, 9558.
10. Kotrusz, P.; Toma, S.; Schmalz, H.-G.; Adler, A. *Eur. J. Org. Chem.* **2004**, *2004*, 1577.
11. Dai, T.; Masako, T.; Takeshi, O. *Chem. Lett.* **2005**, *34*, 962.
12. Mitchell, C. E. T.; Cobb, A. J. A.; Ley, S. V. *Synlett* **2005**, *2005*, 611.
13. Rasalkar, M. S.; Potdar, M. K.; Mohile, S. S.; Salunkhe, M. M. *J. Mol. Catal. A: Chemical* **2005**, *235*, 267.
14. Tsogoeva, S. B.; Yalalov, D. A.; Hateley, M. J.; Weckbecker, C.; Huthmacher, K. *Eur. J. Org. Chem.* **2005**, *2005*, 4995.
15. Almaşi, D.; Alonso, D. A.; Nájera, C. *Tetrahedron: Asymmetry* **2006**, *17*, 2064.
16. Cao, C.-L.; Ye, M.-C.; Sun, X.-L.; Tang, Y. *Org. Lett.* **2006**, *8*, 2901.
17. Cao, Y.-J.; Lu, H.-H.; Lai, Y.-Y.; Lu, L.-Q.; Xiao, W.-J. *Synthesis* **2006**, *2006*, 3795.
18. Corma, A.; Garcia, H. *Adv. Synth. Catal.* **2006**, *348*, 1391.
19. Cozzi, F. *Adv. Synth. Catal.* **2006**, *348*, 1367.
20. Enders, D.; Chow, S. *Eur. J. Org. Chem.* **2006**, *2006*, 4578.
21. Guo, H.-C.; Ma, J.-A. *Angew. Chem. Int. Ed.* **2006**, *45*, 354.
22. Huang, H.; Jacobsen, E. N. *J. Am. Chem. Soc.* **2006**, *128*, 7170.
23. Li, Y.; Liu, X.-Y.; Zhao, G. *Tetrahedron: Asymmetry* **2006**, *17*, 2034.
24. Luo, S.; Mi, X.; Zhang, L.; Liu, S.; Xu, H.; Cheng, J.-P. *Angew. Chem. Int. Ed.* **2006**, *45*, 3093.
25. Luo, S.; Mi, X.; Liu, S.; Xu, H.; Cheng, J.-P. *Chem. Commun.* **2006**, *2006*, 3687.
26. Luo, S.; Xu, H.; Mi, X.; Li, J.; Zheng, X.; Cheng, J.-P. *J. Org. Chem.* **2006**, *71*, 9244.
27. Mase, N.; Watanabe, K.; Yoda, H.; Takabe, K.; Tanaka, F.; Barbas, C. F. *J. Am. Chem. Soc.* **2006**, *128*, 4966.
28. Pansare, S. V.; Pandya, K. *J. Am. Chem. Soc.* **2006**, *128*, 9624.
29. Pellissier, H. *Tetrahedron* **2006**, *62*, 2143.
30. Reyes, E.; Vicario, J. L.; Badía, D.; Carrillo, L. *Org. Lett.* **2006**, *8*, 6135.
31. Trost, B. M.; Jiang, C. *Synthesis* **2006**, *2006*, 369.
32. Tsogoeva, S. B.; Wei, S. *Chem. Commun.* **2006**, *2006*, 1451.

33. Wang, J.; Li, H.; Lou, B.; Zu, L.; Guo, H.; Wang, W. *Chem. - Eur. J.* **2006**, *12*, 4321.
34. Xu, Y.; Cordova, A. *Chem. Commun.* **2006**, *2006*, 460.
35. Xu, Y.; Zou, W.; Sundén, H.; Ibrahem, I.; Córdova, A. *Adv. Synth. Catal.* **2006**, *348*, 418.
36. Yalalov, D. A.; Tsogoeva, S. B.; Schmatz, S. *Adv. Synth. Catal.* **2006**, *348*, 826.
37. Zhu, M.-K.; Cun, L.-F.; Mi, A.-Q.; Jiang, Y.-Z.; Gong, L.-Z. *Tetrahedron: Asymmetry* **2006**, *17*, 491.
38. Zu, L.; Li, H.; Wang, J.; Yu, X.; Wang, W. *Tetrahedron Lett.* **2006**, *47*, 5131.
39. Zu, L.; Wang, J.; Li, H.; Wang, W. *Org. Lett.* **2006**, *8*, 3077.
40. Barros, M. T.; Faísca Phillips, A. M. *Eur. J. Org. Chem.* **2007**, *2007*, 178.
41. Chapman, C. J.; Frost, C. G. *Synthesis* **2007**, *2007*, 1.
42. Clarke, M. L.; Fuentes, J. A. *Angew. Chem. Int. Ed.* **2007**, *46*, 930.
43. Diez, D.; Gil, J. M.; Moro, R. F.; Marcos, I. S.; García, P.; Basabe, P.; Garrido, N. M.; Broughton, H. B.; Urones, J. G. *Tetrahedron* **2007**, *63*, 740.
44. Betancort, J. M.; Barbas, C. F. *Org. Lett.* **2001**, *3*, 3737.
45. Mossé, S.; Alexakis, A. *Org. Lett.* **2006**, *8*, 3577.
46. Mase, N.; Thayumanavan, R.; Tanaka, F.; Barbas, C. F. *Org. Lett.* **2004**, *6*, 2527.
47. Wang, W.; Wang, J.; Li, H. *Angew. Chem. Int. Ed.* **2005**, *44*, 1369.
48. Hayashi, Y.; Gotoh, H.; Hayashi, T.; Shoji, M. *Angew. Chem. Int. Ed.* **2005**, *44*, 4212.
49. Lalonde, M. P.; Chen, Y.; Jacobsen, E. N. *Angew. Chem. Int. Ed.* **2006**, *45*, 6366.
50. Mossé, S.; Laars, M.; Kriis, K.; Kanger, T.; Alexakis, A. *Org. Lett.* **2006**, *8*, 2559.
51. Laars, M.; Ausmees, K.; Uudsemaa, M.; Tamm, T.; Kanger, T.; Lopp, M. *J. Org. Chem.* **2009**, *74*, 3772.
52. Palomo, C.; Vera, S.; Mielgo, A.; Gómez-Bengoa, E. *Angew. Chem. Int. Ed.* **2006**, *45*, 5984.
53. He, J.; Chen, Q.; Ni, B. *Tetrahedron Lett.* **2014**, *55*, 3030.
54. García-García, P.; Ladépêche, A.; Halder, R.; List, B. *Angew. Chem. Int. Ed.* **2008**, *47*, 4719.
55. Hayashi, Y.; Itoh, T.; Ohkubo, M.; Ishikawa, H. *Angew. Chem. Int. Ed.* **2008**, *47*, 4722.
56. Alza, E.; Pericàs, M. A. *Adv. Synth. Catal.* **2009**, *351*, 3051.
57. Patora-Komisarska, K.; Benohoud, M.; Ishikawa, H.; Seebach, D.; Hayashi, Y. *Helv. Chimica Acta* **2011**, *94*, 719.
58. Wu, J.; Ni, B.; Headley, A. D. *Org. Lett.* **2009**, *11*, 3354.
59. Hu, F.; Guo, C.-S.; Xie, J.; Zhu, H.-L.; Huang, Z.-Z. *Chem. Lett.* **2010**, *39*, 412.
60. Zheng, Z.; Perkins, B. L.; Ni, B. *J. Am. Chem. Soc.* **2010**, *132*, 50.
61. Wang, B. G.; Ma, B. C.; Wang, Q.; Wang, W. *Adv. Synth. Catal.* **2010**, *352*, 2923.
62. Wang, C. A.; Zhang, Z. K.; Yue, T.; Sun, Y. L.; Wang, L.; Wang, W. D.; Zhang, Y.; Liu, C.; Wang, W. *Chem. - Eur. J.* **2012**, *18*, 6718.
63. Ghosh, S. K.; Qiao, Y.; Ni, B.; Headley, A. D. *Org. Biomol. Chem.* **2013**, *11*, 1801.
64. Feu, K. S.; de la Torre, A. F.; Silva, S.; de Moraes Junior, M. A. F.; Correa, A. G.; Paixao, M. W. *Green Chem.* **2014**, *16*, 3169.
65. Wu, T.; Feng, D.; Xie, B.; Ma, X. *RSC Adv.* **2016**, *6*, 25246.
66. Luo, R.-S.; Weng, J.; Ai, H.-B.; Lu, G.; Chan, A. S. C. *Adv. Synth. Catal.* **2009**, *351*, 2449.
67. Husmann, R.; Joerres, M.; Raabe, G.; Bolm, C. *Chem. - Eur. J.* **2010**, *16*, 12549.
68. Jentzsch, K. I.; Min, T.; Etcheson, J. I.; Fettinger, J. C.; Franz, A. K. *J. Org. Chem.* **2011**, *76*, 7065.
69. Wiesner, M.; Revell, J. D.; Wennemers, H. *Angew. Chem. Int. Ed.* **2008**, *47*, 1871.
70. Wiesner, M.; Neuburger, M.; Wennemers, H. *Chem. - Eur. J.* **2009**, *15*, 10103.
71. Cortes-Clerget, M.; Gager, O.; Monteil, M.; Pirat, J.-L.; Migianu-Griffoni, E.; Deschamp, J.; Lecouvey, M. *Adv. Synth. Catal.* **2016**, *358*, 34.
72. Cortes-Clerget, M.; Jover, J.; Dussart, J.; Kolodziej, E.; Monteil, M.; Migianu-Griffoni, E.; Gager, O.; Deschamp, J.; Lecouvey, M. *Chem. - Eur. J.* **2017**, *23*, 6654.
73. de la Torre, A. F.; Rivera, D. G.; Ferreira, M. A. B.; Corrêa, A. G.; Paixão, M. W. *J. Org. Chem.* **2013**, *78*, 10221.
74. Chang, C.; Li, S.-H.; Reddy, R. J.; Chen, K. *Adv. Synth. Catal.* **2009**, *351*, 1273.
75. Reddy, R. J.; Kuan, H.-H.; Chou, T.-Y.; Chen, K. *Chem. - Eur. J.* **2009**, *15*, 9294.
76. Lu, D.; Gong, Y.; Wang, W. *Adv. Synth. Catal.* **2010**, *352*, 644.
77. Cheng, Y.-Q.; Bian, Z.; He, Y.-B.; Han, F.-S.; Kang, C.-Q.; Ning, Z.-L.; Gao, L.-X. *Tetrahedron: Asymmetry* **2009**, *20*, 1753.
78. Naziroglu, H. N.; Durmaz, M.; Bozkurt, S.; Demir, A. S.; Sirit, A. *Tetrahedron: Asymmetry* **2012**, *23*, 164.
79. Bai, J.-F.; Xu, X.-Y.; Huang, Q.-C.; Peng, L.; Wang, L.-X. *Tetrahedron Lett.* **2010**, *51*, 2803.
80. Gorde, A. B.; Ramapanicker, R. *Eur. J. Org. Chem.* **2019**, *2019*, 4745.
81. Sato, A.; Yoshida, M.; Hara, S. *Chem. Commun.* **2008**, *2008*, 6242.

82. Yoshida, M.; Sato, A.; Hara, S. *Org. Biomol. Chem.* **2010**, *8*, 3031.
83. Xu, K.; Zhang, S.; Hu, Y.; Zha, Z.; Wang, Z. *Chem. - Eur. J.* **2013**, *19*, 3573.
84. Demir, A. S.; Eymur, S. *Tetrahedron: Asymmetry* **2010**, *21*, 112.
85. Wang, W.-H.; Abe, T.; Wang, X.-B.; Kodama, K.; Hirose, T.; Zhang, G.-Y. *Tetrahedron: Asymmetry* **2010**, *21*, 2925.
86. He, T.; Gu, Q.; Wu, X.-Y. *Tetrahedron* **2010**, *66*, 3195.
87. Zhang, X.; Liu, S.; Li, X.; Yan, M.; Chan, A. S. C. *Chem. Commun.* **2009**, *7*, 833.
88. Chen, J.-R.; Zou, Y.-Q.; Fu, L.; Ren, F.; Tan, F.; Xiao, W.-J. *Tetrahedron* **2010**, *66*, 5367.
89. Demir, A. S.; Basceken, S. *Tetrahedron: Asymmetry* **2013**, *24*, 1218.
90. Weng, J.; Ai, H.-B.; Luo, R.-S.; Lu, G. *Chirality* **2012**, *24*, 271.
91. Zhao, L.; Shen, J.; Liu, D.; Liu, Y.; Zhang, W. *Org. Biomol. Chem.* **2012**, *10*, 2840.
92. (a) Xiao, J.; Xu, F.-X.; Lu, Y.-P.; Loh, T.-P. *Org. Lett.* **2010**, *12*, 1220. (b) Xiao, J.; Xu, F.-X.; Lu, Y.-P.; Liu, Y.-L.; Loh, T.-P. *Synthesis* **2011**, *2011*, 1912.
93. Nugent, T. C.; Shoaib, M.; Shoaib, A. *Org. Biomol. Chem.* **2011**, *9*, 52.
94. Tuchman-Shukron, L.; Miller, S. J.; Portnoy, M. *Chem. - Eur. J.* **2012**, *18*, 2290.
95. Hu, X.; Wei, Y.-F.; Wu, N.; Jiang, Z.; Liu, C.; Luo, R.-S. *Tetrahedron: Asymmetry* **2016**, *27*, 420.
96. Liang, D.-C.; Luo, R.-S.; Yin, L.-H.; Chan, A. S. C.; Lu, G. *Org. Biomol. Chem.* **2012**, *10*, 3071.
97. Lombardo, M.; Chiarucci, M.; Quintavalla, A.; Trombini, C. *Adv. Synth. Catal.* **2009**, *351*, 2801.
98. Aguilera, D. A.; Spinozzi Di Sante, L.; Pettignano, A.; Riccioli, R.; Roeske, J.; Albergati, L.; Corti, V.; Fochi, M.; Bernardi, L.; Quignard, F.; Tanchoux, N. *Eur. J. Org. Chem.* **2019**, 24, *3842*.
99. Wujkowska, Z.; Zawisza, A.; Leśniak, S.; Rachwalski, M. *Tetrahedron* **2019**, *75*, 230.
100. Huťka, M.; Poláčková, V.; Marák, J.; Kaniansky, D.; Šebesta, R.; Toma, Š. *Eur. J. Org. Chem.* **2010**, *2010*, 6430.
101. Kelleher, F.; Kelly, S.; Watts, J.; McKee, V. *Tetrahedron* **2010**, *66*, 3525.
102. Desmarchelier, A.; Marrot, J.; Moreau, X.; Greck, C. *Org. Biomol. Chem.* **2011**, *9*, 994.
103. Zhu, S.; Yu, S.; Wang, Y.; Ma, D. *Angew. Chem. Int. Ed.* **2010**, *49*, 4656.
104. Bonne, D.; Salat, L.; Dulcère, J.-P.; Rodriguez, J. *Org. Lett.* **2008**, *10*, 5409.
105. Andrey, O.; Vidonne, A.; Alexakis, A. *Tetrahedron Lett.* **2003**, *44*, 7901.
106. Sádaba, D.; Delso, I.; Tejero, T.; Merino, P. *Tetrahedron Lett.* **2011**, *52*, 5976.
107. Reitel, K.; Lippur, K.; Järving, I.; Kudrjašova, M.; Lopp, M.; Kanger, T. *Synthesis* **2013**, *45*, 2679.
108. Zhu, S.; Yu, S.; Ma, D. *Angew. Chem. Int. Ed.* **2008**, *47*, 545.
109. Kastl, R.; Wennemers, H. *Angew. Chem. Int. Ed.* **2013**, *52*, 7228.
110. Chen, J.; Geng, Z.-C.; Li, N.; Huang, X.-F.; Pan, F.-F.; Wang, X.-W. *J. Org. Chem.* **2013**, *78*, 2362.
111. Duschmalé, J.; Wennemers, H. *Chem. - Eur. J.* **2012**, *18*, 1111.
112. Zheng, B.; Wang, H.; Han, Y.; Liu, C.; Peng, Y. *Chem. Commun.* **2013**, *49*, 4561.
113. Wang, L.; Zhang, X.; Ma, D. *Tetrahedron* **2012**, *68*, 7675.
114. Gorde, A. B.; Ramapanicker, R. *J. Org. Chem.* **2019**, *84*, 1523.
115. Akutsu, H.; Nakashima, K.; Hirashima, S.-i.; Kitahara, M.; Koseki, Y.; Miura, T. *Tetrahedron Lett.* **2017**, *58*, 4759.
116. Akutsu, H.; Nakashima, K.; Hirashima, S.-i.; Matsumoto, H.; Koseki, Y.; Miura, T. *Tetrahedron* **2019**, *75*, 2431.
117. Genc, H. N.; Ozgun, U.; Sirit, A. *Chirality* **2019**, *31*, 293.
118. Torregrosa-Chinillach, A.; Sánchez-Laó, A.; Santagostino, E.; Chinchilla, R. *Molecules* **2019**, *24*, 4058.
119. Bykova, K. A.; Kostenko, A. A.; Kucherenko, A. S.; Zlotin, S. G. *Russ. Chem. Bull.* **2019**, *68*, 1402.
120. Frias, M.; Mas-Ballesté, R.; Arias, S.; Alvarado, C.; Alemán, J. *J. Am. Chem. Soc.* **2017**, *139*, 672.
121. DiRocco, D. A.; Oberg, K. M.; Dalton, D. M.; Rovis, T. *J. Am. Chem. Soc.* **2009**, *131*, 10872.
122. White, N. A.; DiRocco, D. A.; Rovis, T. *J. Am. Chem. Soc.* **2013**, *135*, 8504.
123. DiRocco, D. A.; Rovis, T. *J. Am. Chem. Soc.* **2011**, *133*, 10402.
124. Maji, B.; Ji, L.; Wang, S.; Vedachalam, S.; Ganguly, R.; Liu, X.-W. *Angew. Chem. Int. Ed.* **2012**, *51*, 8276.

3 Catalytic Asymmetric Michael Addition of Ketones to Nitroalkenes

3.1 INTRODUCTION

The reactivity of 1,3-dicarbonyl compounds and aldehydes with nitroalkenes in the presence of various chiral catalysts has been described in the last two chapters. This chapter deals with the similar reaction of ketones as carbon-centered nucleophiles under the influence of a wide variety of chiral organo- and metal-ligand complex as catalysts. Thus, enolates derived from various alkyl and aryl ketones as well as activated ketones, such as α-ketoesters/amides/sulfones/phosphonates and β-ketosulfones, participate in the Michael addition to nitroalkenes to afford synthetically useful δ-nitroketones.

Although ketones are less reactive than aldehydes, maintenance of configurational stability of the products, the δ-nitroketones, at the chiral center α- to the carbonyl group is challenging and the appropriate choice of the catalyst is crucial to the success of the outcome. Due to greater structural diversity compared to aldehyde derivatives, a broad spectrum of reactivity is observed concerning cyclic and acyclic ketones, alkyl and aryl ketones and various other activated ketones, which requires considerable insights and experimentation. The ketones that have been employed as Michael donors with nitroalkenes in the presence of chiral catalysts have been broadly classified in this chapter as (i) aliphatic, including cyclic ketones and aryl ketones and (ii) ketones activated by an ester, amide, sulfone or phosphonate moiety.

3.2 ADDITION OF ALIPHATIC AND AROMATIC KETONES

Zhao and co-workers demonstrated that organocatalytic self-assemblies **C1b** and phenylglycine (**C1a**) could be established via ionic interactions between ammonium and carboxylate ions from natural α-amino acids and alkaloids. These self-assembled organocatalysts **C1b** and phenylglycine are excellent catalysts for the asymmetric nitro-Michael addition of ketones **2** to nitroalkenes **1** (Scheme 3.1).[1] With L-phenylglycine, *Z*-enamine was formed exclusively, and the activated nitroalkene adds to the enamine double bond by *Re*, *Si*-attack resulting in (3*R*, 4*R*)-configured *anti*-product **3** (**TS-1**, Scheme 3.1).

The Xu group used the chiral proline amide-thiourea bifunctional catalyst **C2** to catalyze the Michael addition of ketone **2** to nitroalkenes **1** to provide products **3** in excellent yields and enantioselectivities up to 91% ee (Scheme 3.1).[2] The Portnoy group in 2012 developed a polymer-supported bifunctional urea–primary amine catalyst **C3** using L-amino acid as a spacer. They employed the same catalyst **C3** for the asymmetric Michael addition of ketones **2** to nitroalkenes **1** to give the products **3** with good enantioselectivities (Scheme 3.1).[3]

Yan and co-workers employed chiral bifunctional sulfamide **C4** for the asymmetric Michael addition of ketones **2** to nitroalkenes **1**. In this reaction, along with the catalyst, additive boosts the yield and enantioselectivity of the product **3a** (Scheme 3.2).[4] Interestingly, base additives enhanced the reaction yields and were involved in proton transfer by abstracting the proton from imine cation. In the plausible mechanism, ketone **2** and catalyst **C4** form imine which in the presence of base tautomerizes to enamine. This enamine attacks from the *Si*-face of nitroalkene **1** which is activated by the catalyst **C4** through H-bonding interaction and on subsequent proton transfer and hydrolysis affords product **3a** with the regeneration of the catalyst **C4** (**TS-2**, Scheme 3.2).

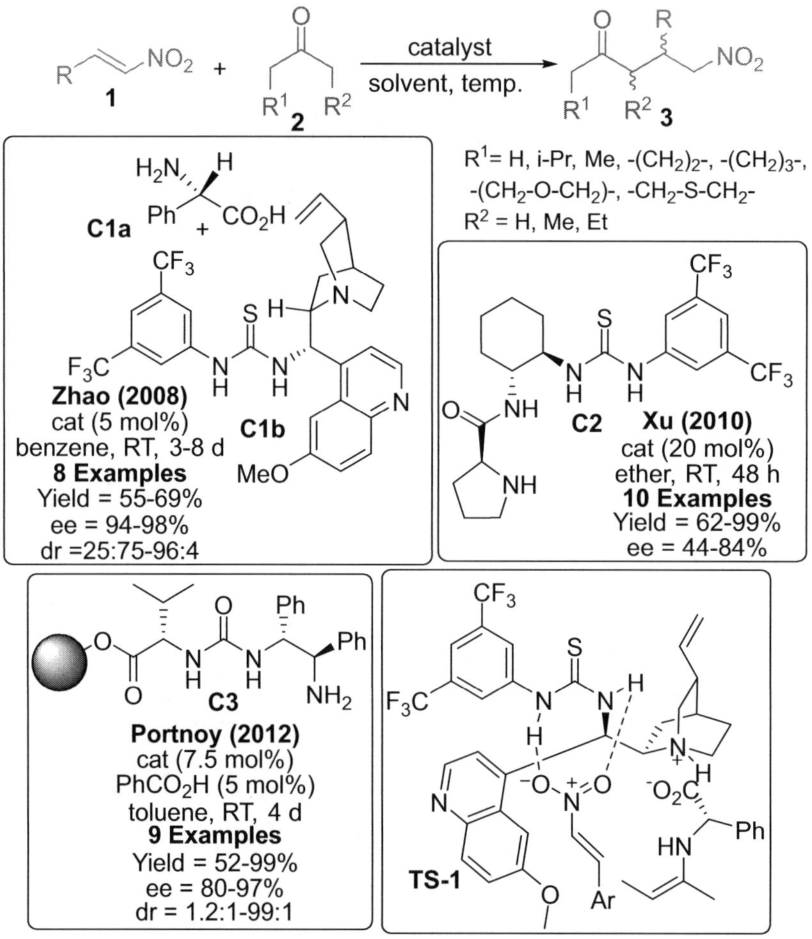

SCHEME 3.1 Asymmetric Michael addition of ketones to nitroalkenes catalyzed by self-assembled thiourea, proline amide-thiourea and polymer-supported bifunctional urea–primary amine organocatalysts.

Wang and co-workers reported an asymmetric Michael addition of acetone **2** ($R^1 = R^2 = H$) to various nitroalkenes **1** by employing simple chiral primary amine bifunctional catalyst **C5** in the presence of terephthalic acid as an additive to give products **3a** with high yields and enantioselectivities (**TS-3**, Scheme 3.2).[5] The Wang group developed a chiral *trans*-cyclohexane diamine-based thiophosphoramide catalyst **C6** for the asymmetric Michael reaction of acetone **2** to nitroalkenes **1** to afford corresponding adducts **3a** in excellent yields with excellent enantioselectivities (Scheme 3.2).[6] Nuclear magnetic resonance (NMR) studies of this reaction revealed the weak H-bonding interaction between the nitro group of nitroalkene and the thiophosphoramide moiety of the catalyst **C6**. In this reaction, the primary amine of the catalyst **C6** reacts with a carbonyl compound to form an enamine intermediate in the presence of an acidic co-catalyst. Meanwhile, nitroalkene **1** gets activated by the catalyst **C6** and enamine attacks this nitroalkene **1** from the *Si*-face to give the adducts **3a** (**TS-4**, Scheme 3.2).

The Kokotos group in their report described the asymmetric Michael reaction between aryl methyl ketones **2a** and nitroalkenes **1** for the synthesis of corresponding adducts **3b**, which were further converted to (*S*)-baclofen, (*R*)-baclofen and (*S*)-phenibut in good yields and enantioselectivity.[7] The catalyst **C7** derived from (*S*)-di-*tert*-butyl aspartate and (1*R*, 2*R*)-diphenyl-ethylenediamine was used as a chiral catalyst for the above transformation in yielding the adducts **3b** with good

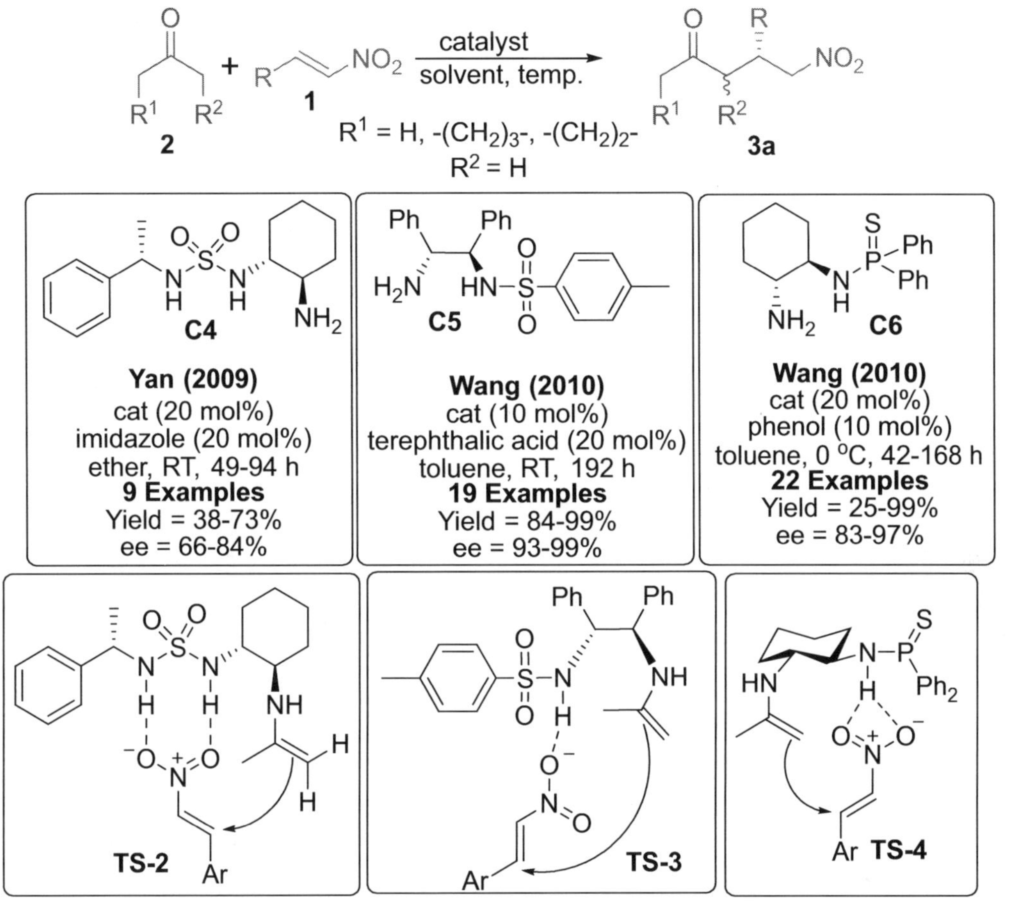

SCHEME 3.2 Asymmetric Michael addition of ketones to nitroalkenes catalyzed by sulfamides and *trans*-cyclohexane diamine-based thiophosphoramide.

yields and enantioselectivity (Scheme 3.3). The primary amine moiety of catalyst **C7** initially forms enamine by reacting with a ketone. Simultaneously, the nitroalkene gets activated by hydrogen bonding between the thiourea moiety of the catalyst and the nitro group and guides the enantioselective Michael addition from one face (**TS-5**, Scheme 3.3). Wang and co-workers developed a new class of dehydroabietic amine-substituted primary amine-thiourea bifunctional catalyst **C8** for an organocatalytic conjugate addition of various heterocycle-bearing ketones **2a** to nitroalkenes **1**, affording (*S*)- or (*R*)-γ-nitro heteroaromatic ketones **3b** with excellent enantioselectivities (up to ee >99%, Scheme 3.3).[8]

The Ley group performed Michael reaction of a range of ketones **2b** with aromatic nitroalkenes **1** catalyzed by L-proline-derived 5-pyrrolidin-2-yl tetrazoles **C9** (Scheme 3.4).[9] It was found that the L-proline-derived tetrazole **C9** exhibited better results compared to L-proline both in terms of yield and enantioselectivity (96% and 78% yield, 62% and 47% ee, respectively). This could be ascribed either to the difference in H-bonding strengths between the tetrazole and the carboxylic acid group or to the increased size of the tetrazole moiety.

Headley and co-workers designed and prepared chiral pyrrolidine-based pyridinium ionic liquids **C10a**, **C10b** and **C10c** with varying anions such as Cl^-, BF_4^- and NTf_2^-, respectively.[10] They also utilized them for the enantioselective Michael addition of ketones **2b,** especially cyclic ketones, to nitroalkenes **1** and afforded the corresponding adducts **3c** in good-to-excellent yields (up to 99%)

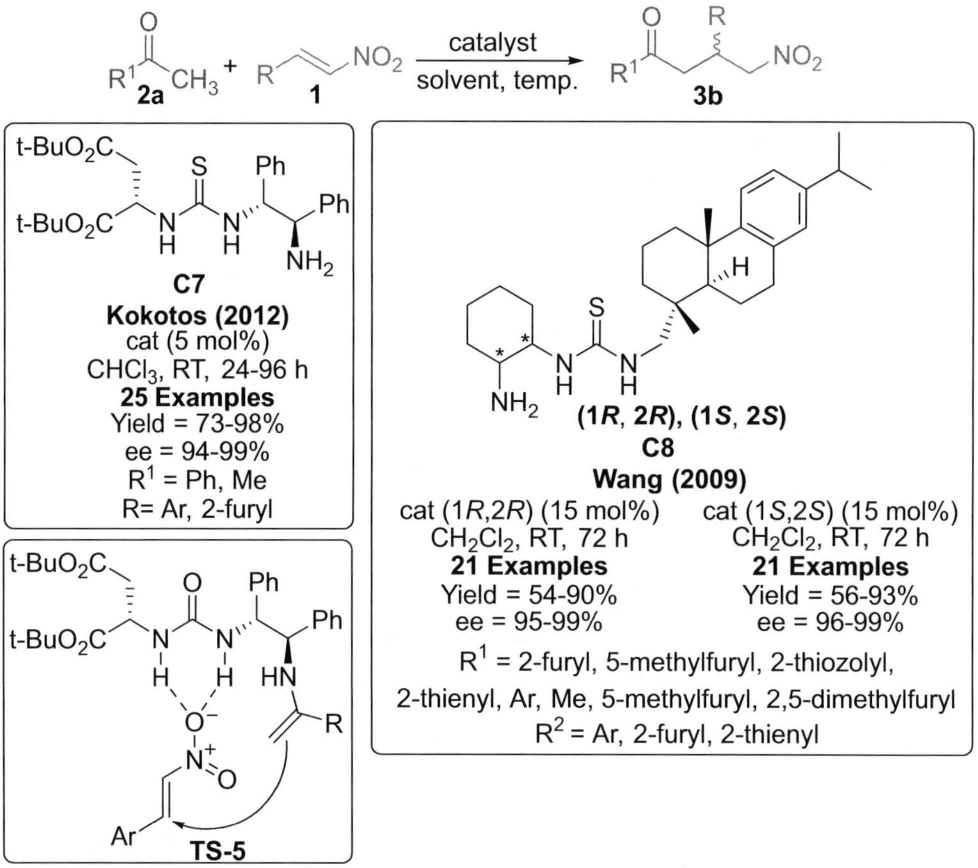

SCHEME 3.3 Asymmetric Michael addition of ketones to nitroalkenes in the presence of thiourea organocatalysts.

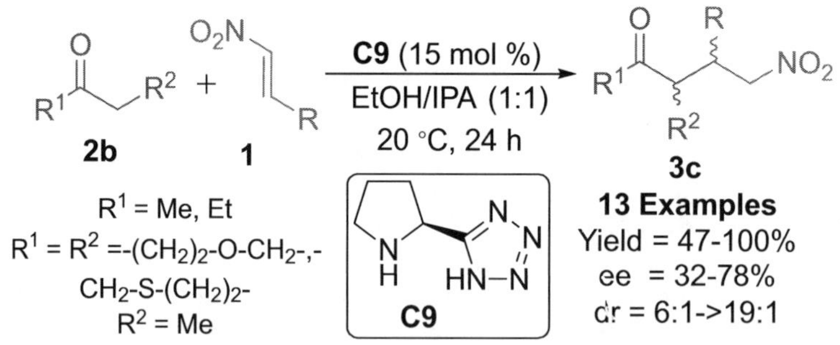

SCHEME 3.4 Asymmetric Michael addition of ketones to nitroalkenes catalyzed by L-proline-derived 5-pyrrolidin-2-yl tetrazoles.

and enantioselectivities (up to 99% ee, Scheme 3.5). The catalysts can be successfully recycled and reused without significant deterioration in the activity.

The Wang group synthesized a pyrrolidine-1,4-diazabicyclo[2.2.2]octane (DABCO)-type chiral quaternary ammonium ionic liquid **C11** from Boc-L-prolinol and DABCO and employed it for the asymmetric Michael addition of ketones **2b** to nitroalkenes **1** to afford the products **3c** with excellent yields with excellent enantioselectivities (Scheme 3.5).[11] The catalyst **C11** can be recycled and

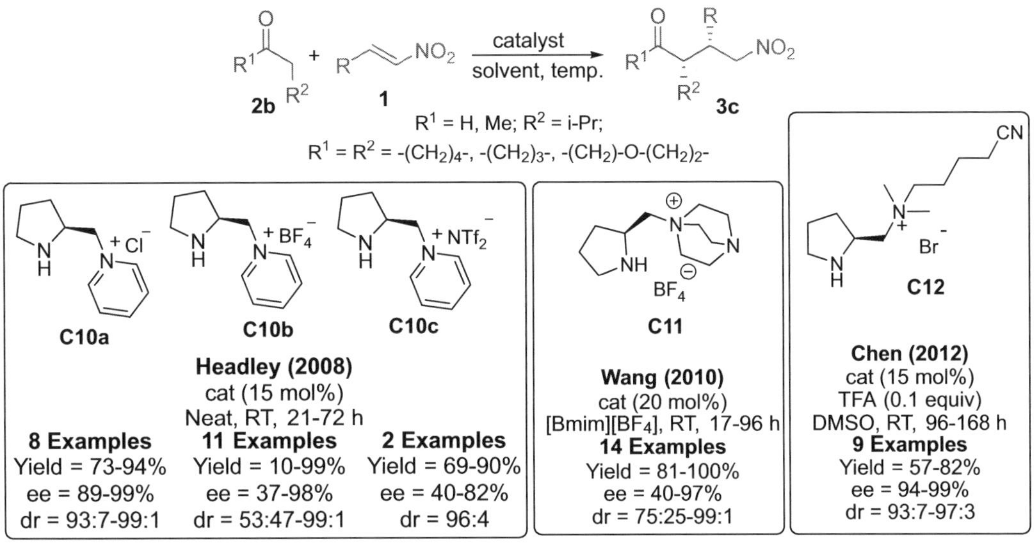

SCHEME 3.5 Asymmetric Michael addition of ketones to nitroalkenes catalyzed by pyrrolidine-pyridinium and pyrrolidine-DABCO-type chiral quaternary ammonium ionic liquids.

reused five times without loss of catalytic activity. Chen and co-workers used chiral pyrrolidine-based quaternary ammonium bromide **C12** as a chiral catalyst for the asymmetric Michael addition of ketones **2b** to nitroalkenes **1** to give products **3c** in high yields (up to 96%) with high enantio and diastereoselectivities (up to 97:3 dr, Scheme 3.5).[12]

Zhong and co-workers demonstrated the L-prolinol **C13** catalyzed enantioselective Michael addition of cyclohexanones **2b** to nitroalkenes **1** in the presence of an acid additive, such as benzoic acid, to obtain Michael adducts **3c** in excellent yields with excellent enantio and diastereoselectivities (Scheme 3.6).[13] This Michael addition proceeds through transition state **TS-6** (Scheme 3.6). the Peng group employed 5 mol% of 4-trifluoromethanesulfonamidyl prolinol *tert*-butyldiphenylsilyl ether bifunctional organocatalyst **C14** for the asymmetric Michael addition of ketones **2b** to nitroalkenes **1**, resulting in *syn*-selective adducts **3c** in excellent yields (up to 99%) with excellent enantio and diastereoselectivities (up to 99% ee and up to 98:2 dr, respectively, Scheme 3.6).[14] In this reaction, nitroalkene **1** gets activated by the H-bonding interaction between the sulfonamide proton and the nitro group. Meanwhile, the bulky group ($CH_2OTBDPS$) shields the *Si*-face of enamine and makes the nitroalkene to attack from the nonshielded side to afford the products **3c** (**TS-7**, Scheme 3.6).

Peng and co-workers developed homodiphenylprolinol methyl ether **C15** which catalyzed asymmetric Michael reaction of ketones **2b** to nitroalkenes **1** in the presence of an acid additive such as *o*-methylbenzoic acid (Scheme 3.6).[15] Five mole percent of catalyst was used for the reaction to give the products **3c** at high yields with excellent enantioselectivities at room temperature. In this reaction, the bulky group $CH_2C(OMe)(Ph)_2$ shields the *Si*-face of enamine and makes nitroalkene **1** to attack from the nonshielded side of the enamine to afford the product (**TS-8,** Scheme 3.6).

Liang et al. developed tunable pyrrolidine-based triazole organocatalysts **C16** *via* Cu(I)-catalyzed 1,3-dipolar "click" azide-alkyne cycloaddition. The triazole **C16** catalyzed the reactions of cyclohexanones **2b** with substituted nitroalkenes **1** giving very high diastereoselectivity (up to 99:1 dr) and excellent ee values (up to 99%, Scheme 3.7).[16] The reactions were more efficient in water (89%–99%) than in $CHCl_3$ (74%–91%). In general, the Michael additions of aldehydes proceed with poor enantioselectivities (22%–37% ee) and those of aliphatic nitroalkenes with moderate yields, that is, 50% with lower enantioselectivity (22% ee).

New chiral pyrrolidine-pyridine-based organocatalyst **C17** was synthesized and used for the asymmetric Michael addition of cyclic/acyclic/aromatic ketones **2b** with nitroalkenes **1** to give

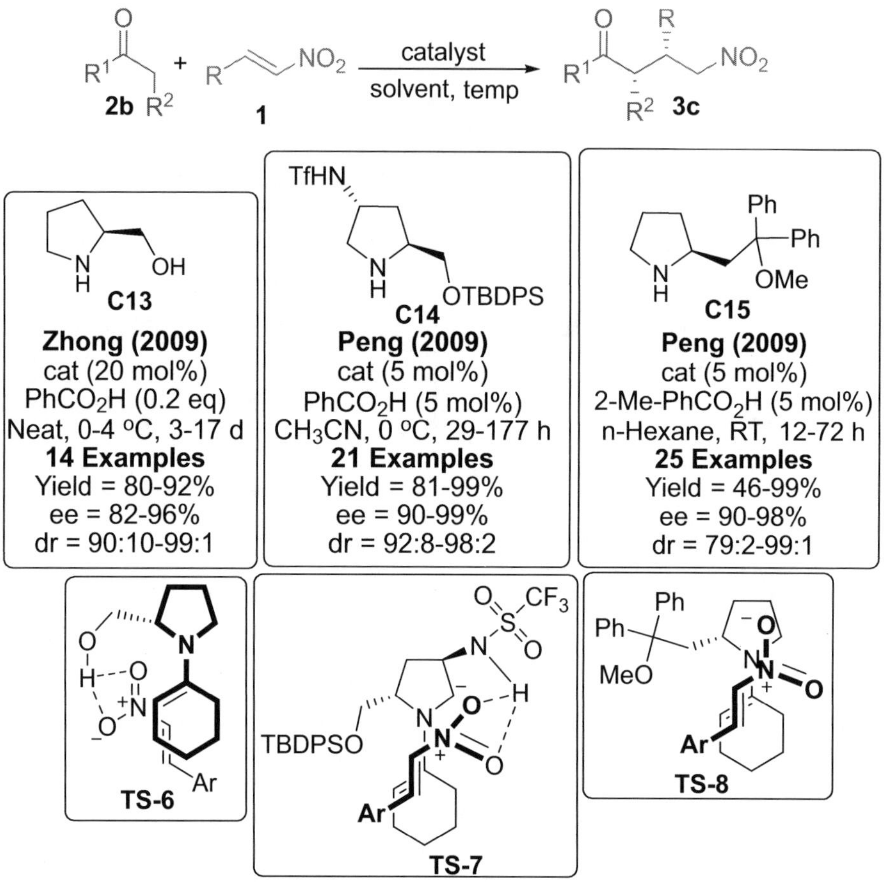

SCHEME 3.6 Asymmetric Michael addition of ketones to nitroalkenes catalyzed by L-prolinol and protected prolinol/homoprolinol-derived organocatalysts.

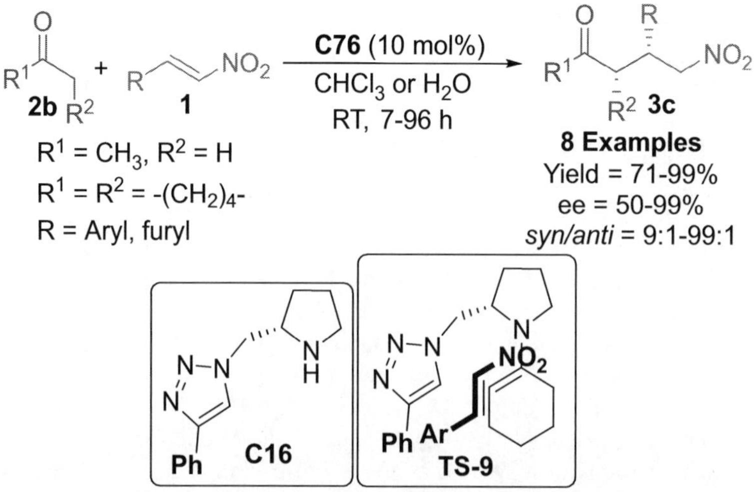

SCHEME 3.7 Asymmetric Michael addition of ketones to nitroalkenes catalyzed by pyrrolidine-based triazole organocatalyst.

corresponding adducts **3c** in high yields and enantioselectivities (up to 99% ee, Scheme 3.8).[17] The reaction proceeds via an enamine pathway. First, the catalyst **C17** forms a chiral enamine with ketone **2b**, and then a Michael reaction between the activated enamine and the nitroalkene **1** results in the formation of the corresponding product **3c** (**TS-10**, Scheme 3.8). The catalyst **C17** is regenerated for use in the subsequent catalytic cycle. The presence of the intermediates in this plausible reaction pathway was confirmed by electrospray ionization mass spectroscopy (ESI-MS).

Diez et al. also synthesized new chiral pyrrolidine-based organocatalyst **C18** and successfully applied for the Michael addition of cyclohexanones **2b** [R^1-R^2 = $(CH_2)_4$] to nitroalkenes **1**.[18] The Peng group designed and synthesized MP-sulfonyl chloride resin-supported sulfonamidyl prolinol *tert*-butyldiphenylsilyl ether catalyst **C19**. Ten mole percent of the catalyst was used for the asymmetric Michael addition of ketones **2b** to nitroalkenes **1** in water to afford the corresponding products in high yields and excellent enantioselectivities (Scheme 3.8).[19] The catalyst can be recycled and reused in subsequent reactions. They proposed an acyclic synclinal transition state model **TS-11** (Scheme 3.8) to explain the high stereoselectivity.

Recently, Pramanik and co-workers synthesized a pyrrolidine-oxadiazolone-based catalyst **C20** to mediate asymmetric Michael addition of differently substituted ketones **2c** to nitroalkenes **1** (Scheme 3.9).[20] The method was able to furnish the Michael adducts **3d** in good yields (up to 97%) and remarkable stereoselectivity (up to 99% ee and >97:3 dr). Here, the authors demonstrated

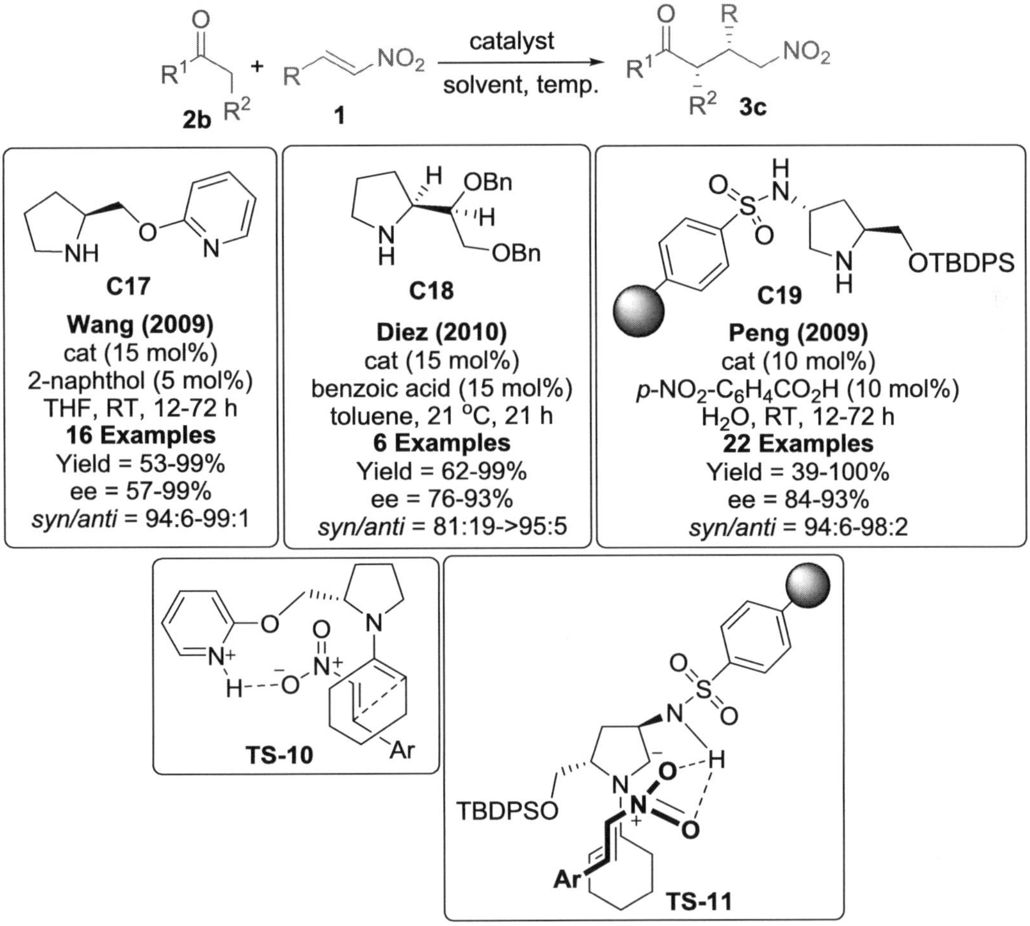

SCHEME 3.8 Asymmetric Michael addition of ketones to nitroalkenes catalyzed by protected prolinol-based organocatalysts.

covalent interactions between ketone and N–H of pyrrolidine, whereas noncovalent interactions were demonstrated between the nitroalkene and oxadiazolone moiety in the transition state (**TS-12**, Scheme 3.9).

Wang and co-workers reported a pyrrolidine-based chiral porous polymer **C21**, which was further used as an organocatalyst in the asymmetric Michael addition of cyclohexanones **2d** to nitroolefins **1a** in water under heterogeneous conditions (Scheme 3.10).[21] Owing to the highly stable, porous and tunable nature of the employed heterogeneous organocatalysts, the adducts **3e** were obtained with high yields (up to 98%) and excellent enantioselectivities (up to 99%).

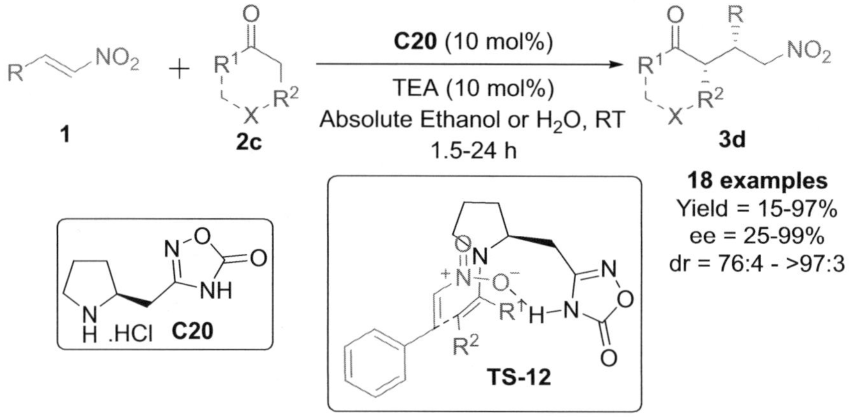

SCHEME 3.9 Asymmetric Michael addition of various ketones to nitrostyrenes using a pyrrolidine-oxadiazolone catalyst.

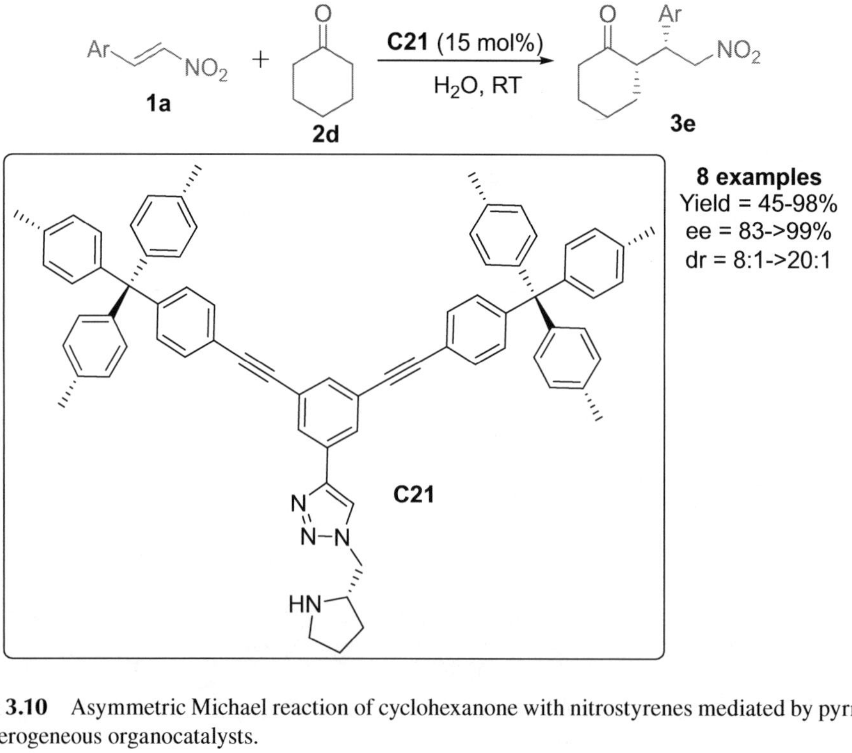

SCHEME 3.10 Asymmetric Michael reaction of cyclohexanone with nitrostyrenes mediated by pyrrolidine-based heterogeneous organocatalysts.

Juaristi and co-workers reported the asymmetric Michael addition of cyclohexanone **2d** to nitroalkenes **1a** mediated by a novel prolinamide-based organocatalyst **C22** bearing a bis-amidophosphoryl fragment attached to a (2-naphthyl) ethyl group (Scheme 3.11).[22] The designed catalyst **C22** could successfully mediate the asymmetric addition of the cyclohexanone **2d** to the nitroalkene **1a** using benzoic acid as an additive in the presence of water, providing the products **3f** with remarkable yields (up to 99%) and a good stereoselectivity (up to 98:2 er and up to 96:4 dr). This superior catalytic efficiency of the designed catalyst over the parent amino acid was credited to the enhanced steric influences created by the presence of naphthyl moieties in the catalytic framework.

Palomo and co-workers reported the use of a chiral tertiary amine catalyst **C23** for activating the α-branched allylic ketones **2e** to generate dienolates, which reacted specifically at the Cα site with nitroalkenes **1a** to give the corresponding products **3g** with high stereoselectivities (99% ee and >98:2 dr) (Scheme 3.12).[23] Here the authors also mentioned that under the employed catalytic conditions, the aliphatic dienolates had very low reactivity for the same reaction.

In 2011, Yu and co-workers demonstrated the use of newly synthesized secondary–secondary–tertiary triamine bifunctional organocatalyst (*S*)-*N*-(pyrrolidin-2-ylmethyl)pyridin-2-amine **C24** for the asymmetric Michael addition of ketones **2b** to nitroalkenes **1** at room temperature (Scheme 3.13).[24] The corresponding adducts **3c** were obtained at high yields and good diastereoselectivities with excellent enantioselectivities in the presence of a catalyst **C24** (10 mol%). In the transition state, the bulky group (–NH–Py) plays a crucial role in shielding the *Si*-face of enamine double bond, which would make nitroalkene **1** to approach from the *Re*-face to afford the product **3c**. The secondary amine (R_2N–H) and the adjacent protonated pyridine (R_3N+…..H) offer double H-bonding interactions to activate the nitroalkene **1** (**TS-13**, Scheme 3.13).

Chandrasekhar et al. developed a new chiral pyrrolidine-pyrazole catalyst **C25** having a pyrazole unit attached to chiral pyrrolidine.[25] It gives high enantio and diastereoselectivities in the asymmetric Michael reaction under solvent-free conditions (Scheme 3.13). Here, the catalyst **C25** forms an *anti*-enamine with cyclohexanone **2b**, and then a Michael reaction between the activated enamine and nitroalkene **1** in anti-*Re* addition fashion results in the formation of corresponding Michael product **3c** via an acyclic synclinal transition state (**TS-14**, Scheme 3.13). Li and co-workers developed pyrrolidine-based phthalimide catalyst **C26** in the presence of an acidic co-catalyst, 2,4-dichlorobenzoic acid (Scheme 3.13).[26]

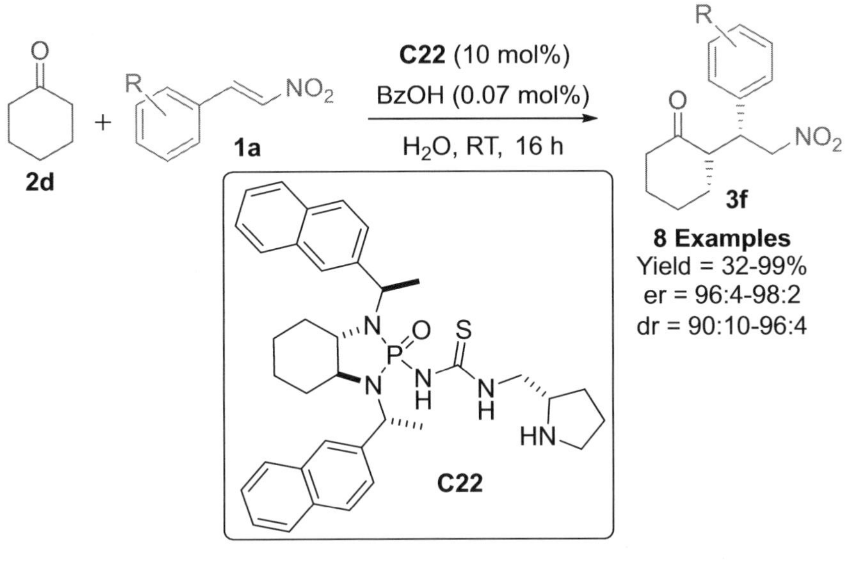

SCHEME 3.11 Asymmetric Michael addition of cyclohexanone to nitroalkenes mediated by phosphoramide-prolinamide organocatalyst.

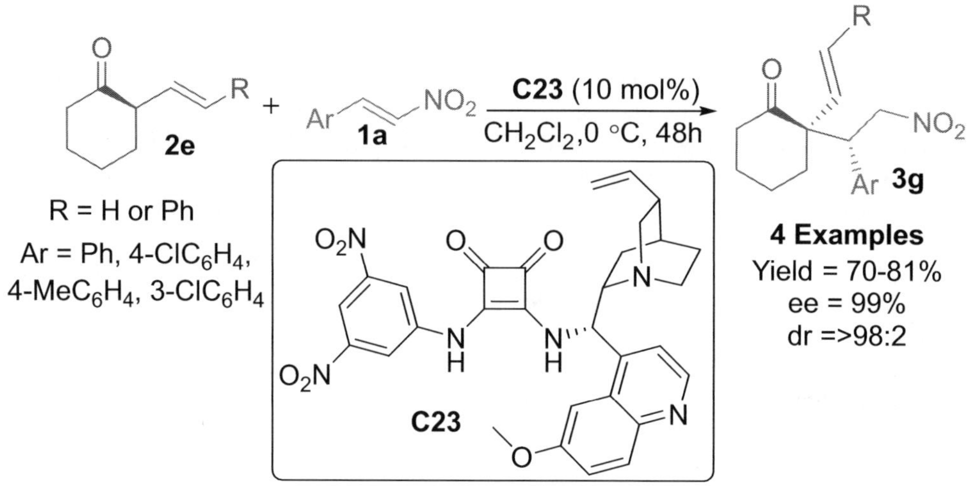

SCHEME 3.12 Asymmetric addition of unsaturated cycloalkanones to nitroalkenes mediated by chiral tertiary amine-squaramide catalyst.

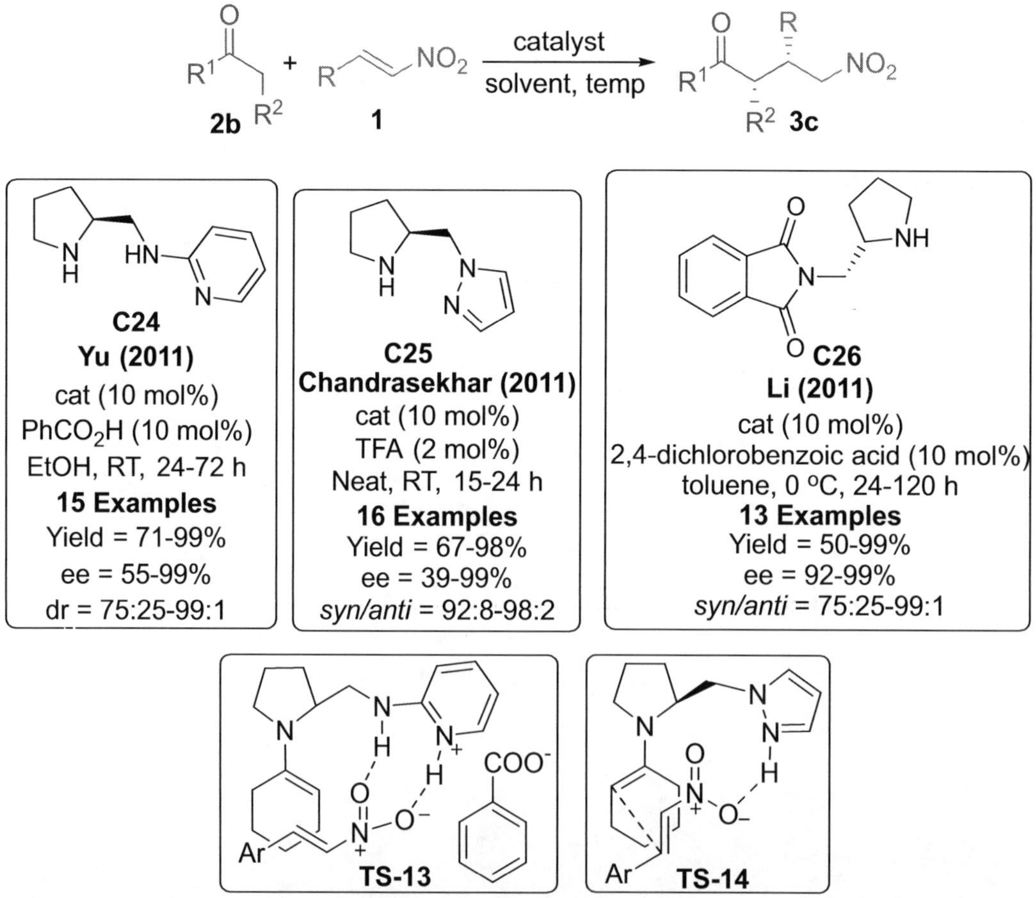

SCHEME 3.13 Asymmetric Michael addition of ketones to nitroalkenes catalyzed by pyrrolidine organocatalysts tethered with pyridine, pyrazole and phthalimide.

The Chen group synthesized and employed a new type of pyrrolidinyl-camphor-based bifunctional organocatalyst **C27** for the diastereo and enantioselective Michael addition of ketones **2b** to nitroalkenes **1** (Scheme 3.14).[27] Here, the C2 hydroxy group of camphor provides additional H-bonding interaction to stabilize the transition state (**TS-15**, Scheme 3.14). The rigid and bulky bicyclic camphor selectively shields the approach of the nitroalkene **1** from the enamine *Si*-face. At the same time, the secondary amine linker plays a key role in the stereochemical outcome via H-bonding with the assistance of the *exo*-hydroxyl group of camphor. The resulting enamine, therefore, attacks the nitroalkene **1** from the *Re*-face to afford the products **3c**. The Zlotin group synthesized natural pinane-derived bifunctional catalyst (1*R*, 2*R*, 3*R*, 5*R*)-2-hydroxy-3-((*S*)-pyrrolidin-2-ylmethylamino) pinane **C28** containing a pyrrolidine unit (Scheme 3.14).[28] Only 10 mol% of the catalyst **C28** was sufficient to catalyze the reaction to afford the products **3c** with stereoselectivities of up to 88% ee.

The Ni group used a combination of water-compatible chiral pyrrolidine-based catalyst **C29** in combination with ionic-liquid-supported Brønsted acid **L1** for the asymmetric Michael addition of ketones **2b** to nitroalkenes **1** to afford products **3c** in excellent yields with excellent enantio and diastereoselectivities (Scheme 3.15).[29] Employing newly synthesized polymer immobilized pyrrolidine-based chiral ionic liquid **C30**, Wang and co-workers performed the Michael addition of ketones **2b** to nitroalkenes **1** to give corresponding adducts **3c** in high yields (up to 97%) and excellent enantio and diastereoselectivities under solvent-free reaction conditions (Scheme 3.15).[30] Ma and co-workers developed a new silica-supported organocatalyst **C31** for the asymmetric Michael addition of ketones **2b** to nitroalkenes **1** (Scheme 3.15).[31] The reactions proceeded well at room temperature to give high yields (up to 98%) and excellent diastereo and enantioselectivities (ee up to 93%). Moreover, the catalyst **C31** can be recycled without substantial loss of activity. 1,2,3-Triazole ring of **C31** plays an important role in grafting the pyrrolidine onto the silica surface.

The Xiao group designed and synthesized chiral pyrrolidinyl-sulfamide derivative **C32** and used them as bifunctional organocatalysts for an asymmetric Michael addition of cyclohexanones **2b** to various nitroalkenes **1**. The corresponding Michael adducts **3c** were obtained in excellent yields and stereoselectivities (Scheme 3.16).[32] Initially, cyclohexanones **2b** forms enamine with catalyst **C32**. The sulfamide group offers H-bonding interactions with the nitro group, thus directing the *Re*-face

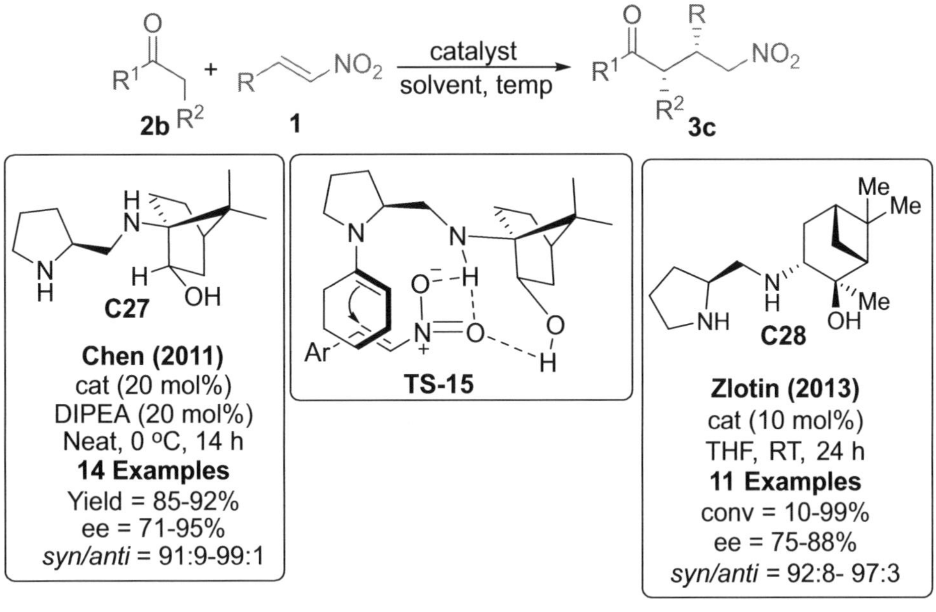

SCHEME 3.14 Pyrrolidinyl-camphor and pyrrolidinylmethylamino pinane as organocatalysts in Michael addition of ketones to nitroalkenes.

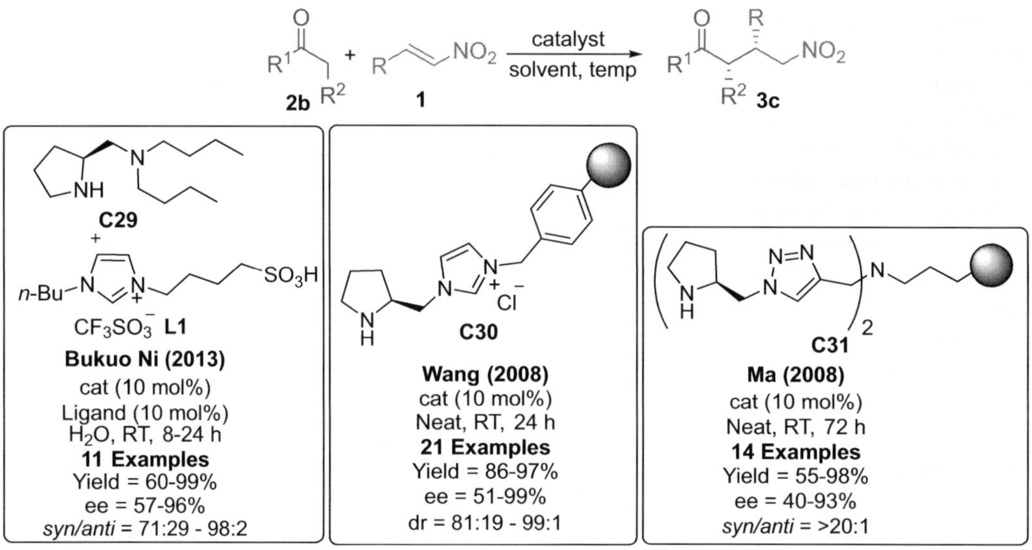

SCHEME 3.15 Asymmetric Michael addition of ketones to nitroalkenes catalyzed by a pyrrolidine-based catalyst, polymer immobilized pyrrolidine-based chiral ionic liquid and silica-supported pyrrolidine catalyst.

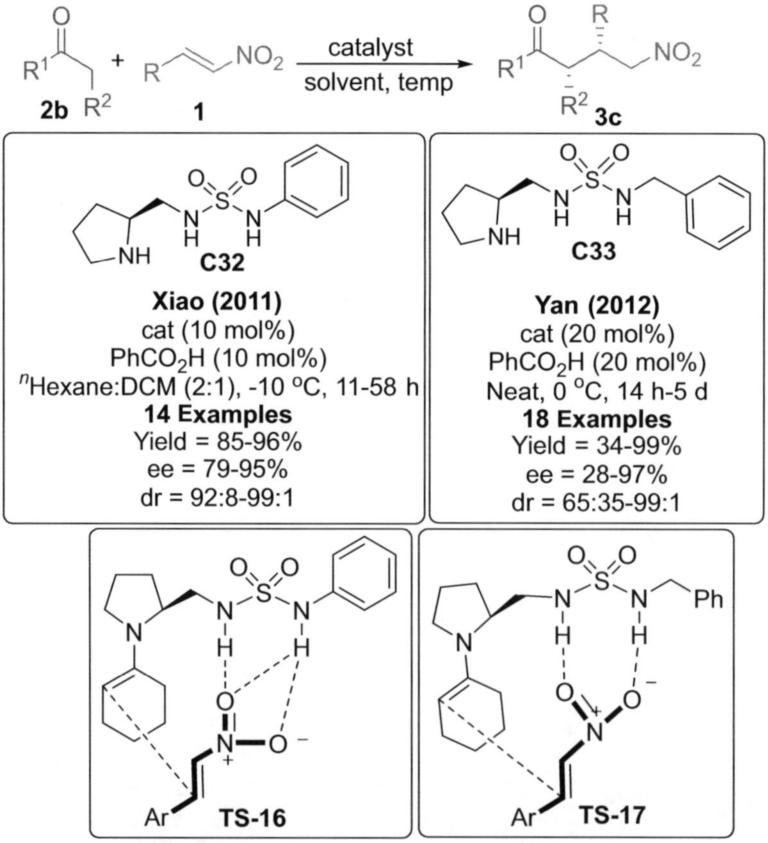

SCHEME 3.16 Pyrrolidinyl-sulfamide derivatives as organocatalysts in the Michael addition of ketones to nitroalkenes.

of nitroalkene **1** to be attacked by the *Re*-face of the enamine to give the corresponding *syn*-adduct **3c** (**TS-16**, Scheme 3.16). Chiral pyrrolidinesulfamide **C33** as an organocatalyst reacted under neat conditions (Scheme 3.16).[33] The reaction proceeds through the formation of a well-organized transition state **TS-17** (Scheme 3.16) formed by H-bonding interaction between the nitro group and sulfamide functionality.

Tang and co-workers prepared pyrrolidine-urea bifunctional organocatalyst **C34** for the asymmetric Michael addition of ketone **2b** to nitroalkene **1** to carry out the theoretical study on the origin of their different activities (Scheme 3.17).[34] The study revealed that the reaction with reduced enantioselectivity was ascribed to the rigid structure formed between catalyst **C34** and nitroalkene **1** through double H-bonding interactions, which retard the approach of nucleophilic enamine intermediate (**TS-18**, Scheme 3.17). The Li group demonstrated the application of new benzoylthiourea–pyrrolidine organocatalyst **C35** to afford Michael product **3c** with moderate-to-excellent diastereo and enantioselectivities under mild conditions (Scheme 3.17).[35] During the reaction, pyrrolidine reacts with the ketone **2b** to form an enamine and, at the same time, benzoylthiourea activates the nitroalkene **1** through H-bonding. Then, the enamine attacks the nitroalkene **1** from the *Re*-face to afford the product **3c** (**TS-19**, Scheme 3.17). Miura and co-workers developed a new pyrrolidine-diaminomethylenemalononitrile (DMM) organocatalyst **C36** that gave the corresponding adducts **3c** in excellent yields with up to 99% ee under solvent-free conditions (Scheme 3.17).[36] Here, the DMM motif serves as a double hydrogen-bond donor site for Michael additions to nitroalkenes **1** to give the product **3c** (**TS-20**, Scheme 3.17).

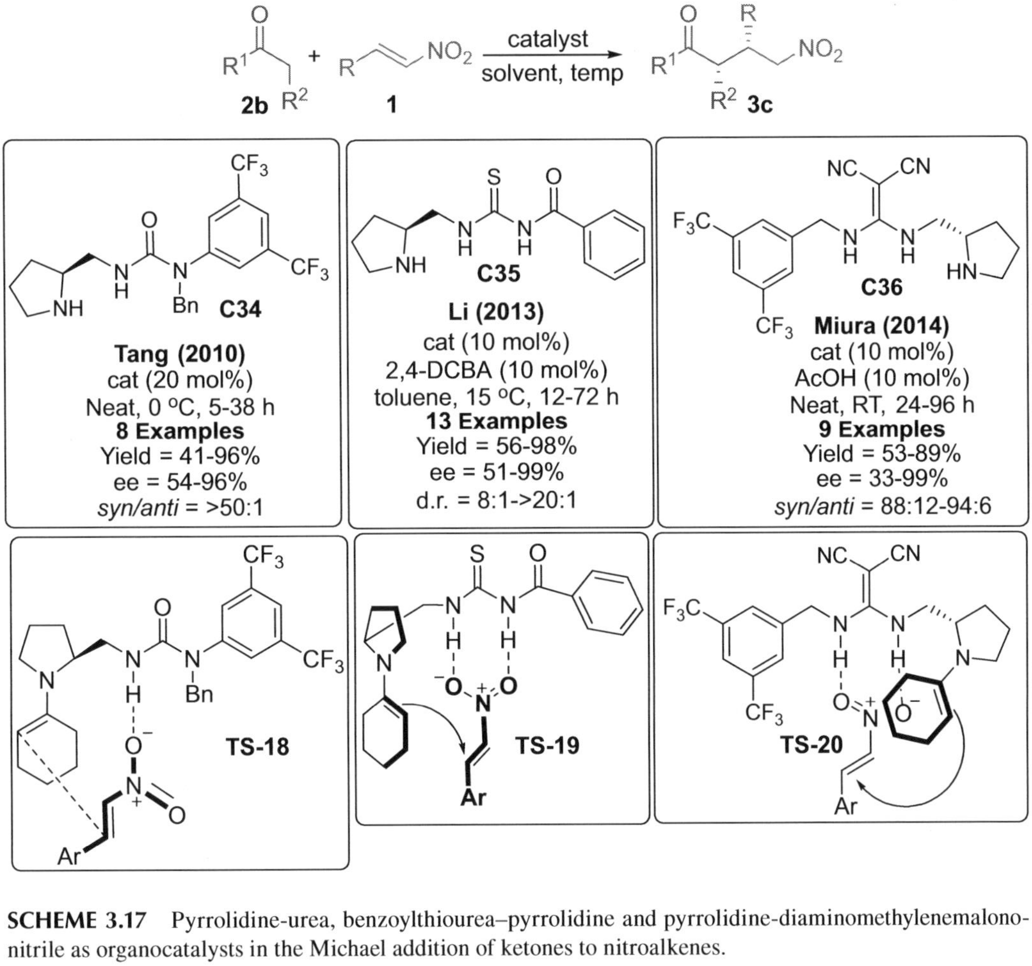

SCHEME 3.17 Pyrrolidine-urea, benzoylthiourea–pyrrolidine and pyrrolidine-diaminomethylenemalononitrile as organocatalysts in the Michael addition of ketones to nitroalkenes.

Proline-based reduced dipeptide **C37** was synthesized and employed for a direct Michael addition of ketones **2b** to nitroalkenes **1** with low catalyst loading (5 mol%) to give products **3c** with high stereoselectivities (ee up to 98%, *syn/anti* up to 99/1) at room temperature (Scheme 3.18).[37] Here, the hydrogen bond donor, amine and carboxyl group direct the nitroalkene **1** to attack the *Re*-face of the enamine to form the adduct **3c** (**TS-21**, Scheme 3.18). The Herrera group reported a new catalyst (2*R*, 3a*S*, 7a*S*)-octahydroindole-2-carboxylic acid **C38** to obtain modest selectivity of up to 58% ee (Scheme 3.18).[38] Such variation on the pyrrolidine ring structure at positions four and five remained unexplored. The substituents on the proline ring appear to influence both the yield and enantioselectivity.

The Moorthy group employed proline-based organocatalyst **C39** for the asymmetric Michael addition of ketones **2b** to nitroalkenes **1** in a brine solution to give the corresponding adducts **3c** with high diastereo and enantioselectivities (Scheme 3.19).[39] The high performance of the catalyst **C39** in this reaction is attributed to the π,π-stacking interactions between the tosyl group of the catalyst **C39** and the aromatic ring of the nitroalkene **1** (**TS-22**, Scheme 3.19). Headley and co-workers reported a new type of ionic-liquid-supported (*S*)-pyrrolidine sulfonamide organocatalyst **C40** that can be recycled and reused at least five times (Scheme 3.19).[40] Further, L-proline-based binaphthyl sulfonimide **C41** was employed in the enantioselective Michael addition of ketones **2b** to nitroalkenes **1** to afford enantiopure γ-nitroketones **3c** in the presence of acidic additive such as benzoic acid (Scheme 3.19).[41] In the proposed mechanism, the pyrrolidine ring first reacts with the ketone **2b** to form an enamine with the aid of the acidic co-catalyst. Subsequently, the oxygen atom of the sulfonimide and Brønsted acid additive direct the nitro group of nitroalkene via H-bonding interaction. Then the enamine attacks the nitroalkene **1** from the *Re*-face to afford the final product **3c** (**TS-23**, Scheme 3.19).

The Peng group designed and synthesized a new type of chiral pyrrolidine-based bifunctional organocatalyst 4-aminothiourea-prolinal dithioacetal **C42** and employed it in an asymmetric

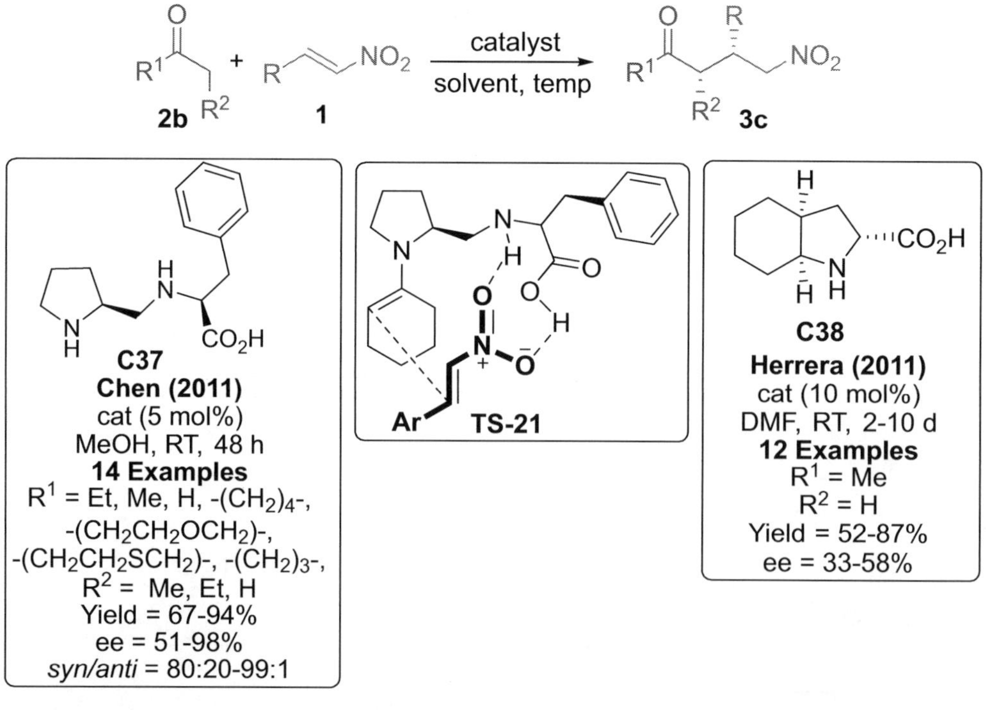

SCHEME 3.18 Proline-based reduced dipeptide and octahydroindole-2-carboxylic acid as organocatalysts in the Michael addition of ketones to nitroalkenes.

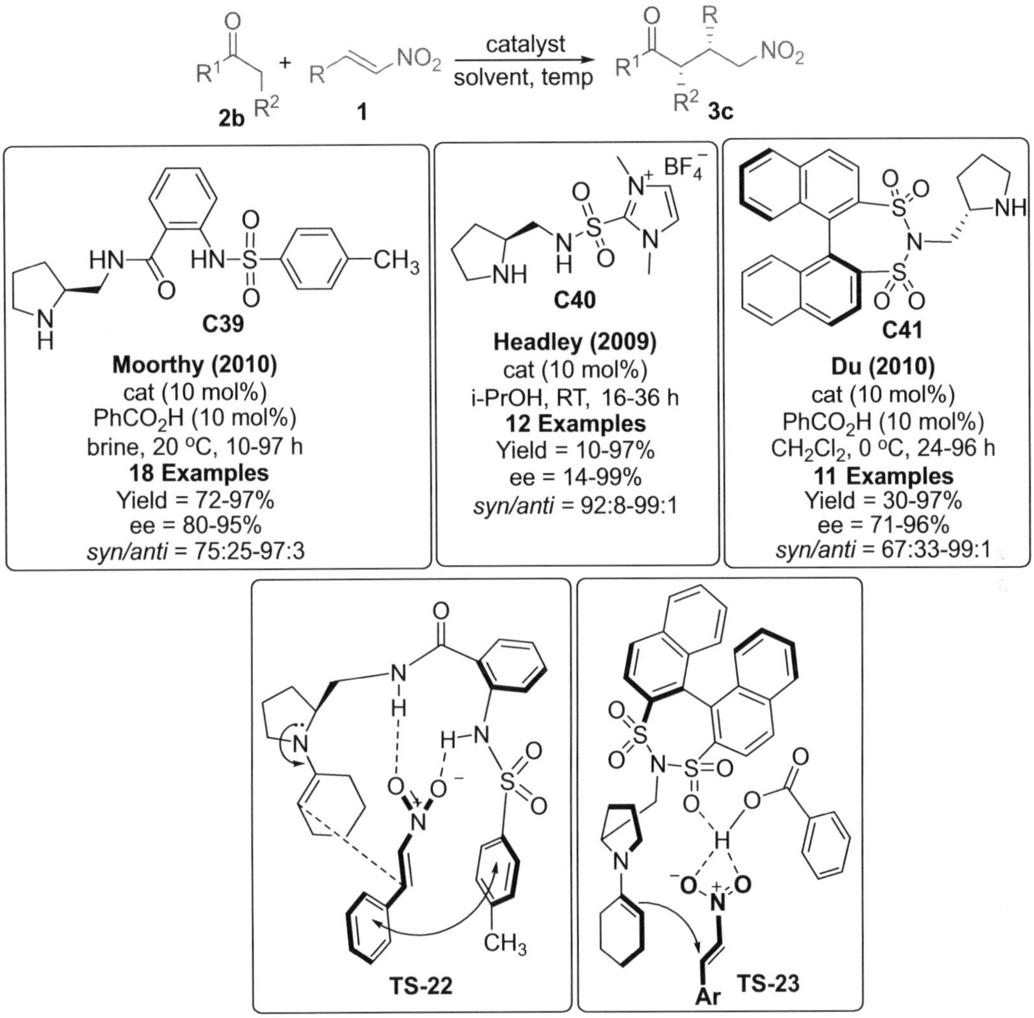

SCHEME 3.19 Proline-based organocatalyst, ionic-liquid-supported (*S*)-pyrrolidine sulfonamide and L-proline-based binaphthyl sulfonimide as catalysts in the asymmetric Michael addition of ketones to nitroalkenes.

Michael addition of ketones **2b** to nitroalkenes **1** (Scheme 3.20).[42] Only 3 mol% of the catalyst **C42** was used to catalyze this reaction to give *syn*-selective adducts under solvent-free conditions at room temperature. In the case of unsymmetrical ketone **2b** such as 2-butanone, the reaction occurred at a more substituted site because the enamine intermediate was formed under thermodynamic control. In this reaction, pyrrolidines react with a ketone to form an enamine. At the same time, thiourea activates the nitroalkene **1** by H-bonding interaction and facilitates the enamine to attack the nitroalkene **1** from the *Re*-face to afford the *syn*-product **3c** (**TS-24**, Scheme 3.20).

The Ma group described an asymmetric Michael addition of aryl ketones **2b** by employing newly synthesized thiourea catalyst **C43** bearing saccharide and primary amine moieties (Scheme 3.20).[43] In the proposed mechanism, the nitroalkene was activated by the thiourea group of the catalyst through H-bonding interactions. At the same time, the primary amine forms an enamine intermediate with ketone **2b.** The thus-formed enamine intermediate approaches the nitroalkene from the

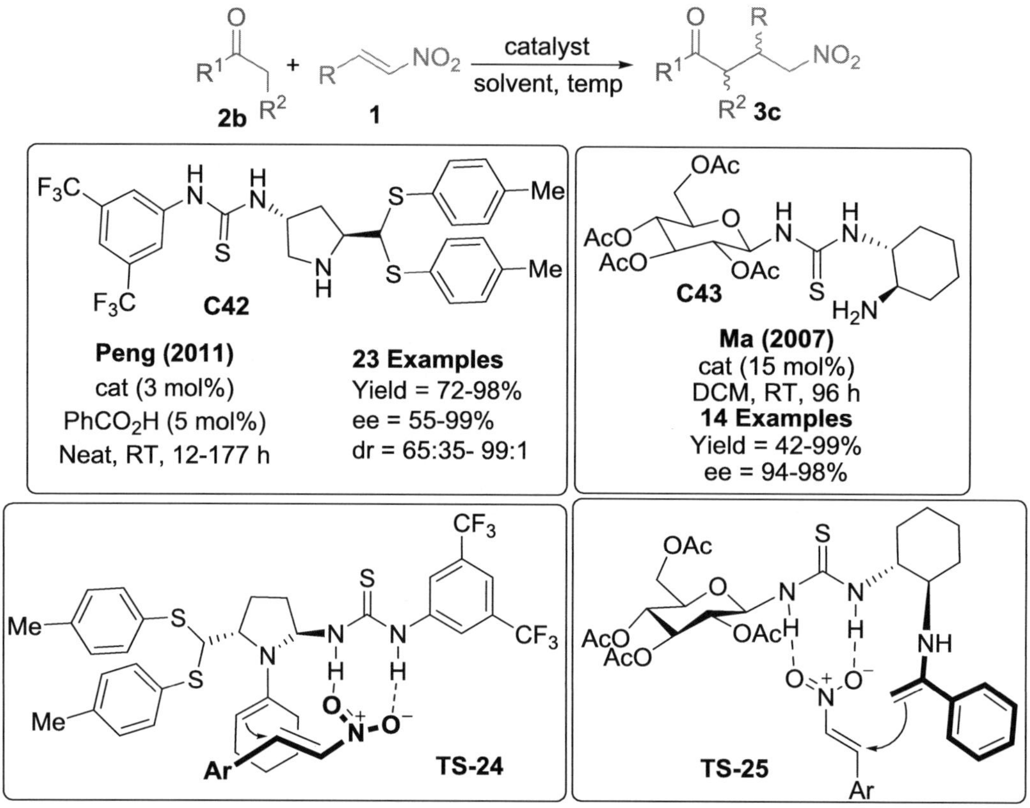

SCHEME 3.20 4-Aminothiourea-prolinal dithioacetal and thiourea-bearing saccharide as organocatalysts in the asymmetric Michael addition of ketones to nitroalkenes.

Si-face as the *Re*-face was shielded by the cyclohexyl group of the catalyst **C43** to afford the product **3c** (**TS-25**, Scheme 3.20).

Xu and co-workers reported 2-[(imidazol-2-ylthio)methyl]pyrrolidine catalyst **C44** which was obtained by L-proline and imidazolylthio compound (Scheme 3.21).[44] The salicylic acid is required as a co-catalyst to deliver the Michael adducts **3c** in good yields and enantioselectivities. The catalytic activity of these reactions can be explained by the dual activation of the substrate by the catalyst **C44**, as well as a co-catalyst salicylic acid through the formation of a stable transition state **TS-26** (Scheme 3.21) formed by the synergistic effect of H-bonding and electrostatic interactions. In another report, Xu and co-workers replaced the co-catalyst salicylic acid by a thioureido acid **C45** in combination with a pyrrolidine-based catalyst **C44**.[45] Here, the plausible reaction pathway was assumed to be through the dual activation of the substrate by H-bonding interactions of the –NH groups of thioureido acid **C45**. The Zhang group reported a new type of pyrrolidine-aminobenzimidazole bifunctional organocatalyst **C46** for the asymmetric Michael addition of cyclohexanones **2b** to various nitroalkenes **1** in brine as the reaction medium.[46] Michael addition products **3c** were obtained with excellent yields and high stereoselectivities (up to 99/1 dr and 99% ee) at room temperature (Scheme 3.21).

The Zhong group developed a proline-based catalyst **C47** with a hypothesis that pyrrolidines would form enamine and phosphine oxide would activate nitroalkene by dipole interaction to achieve the highest stereoselectivity.[47] Thus, the Michael addition of six-membered cyclic ketones **2b** to nitroalkenes **1** in the presence of 15 mol% catalyst **C47** provided Michael adducts **3c** in high yields and stereoselectivities (Scheme 3.22). Singh et al. described the application of newly designed sulfoxide-bearing pyrrolidine organocatalyst **C48** in the asymmetric Michael addition of ketones **2b** to nitroalkenes **1** to obtain the corresponding Michael adducts **3c** with good enantio

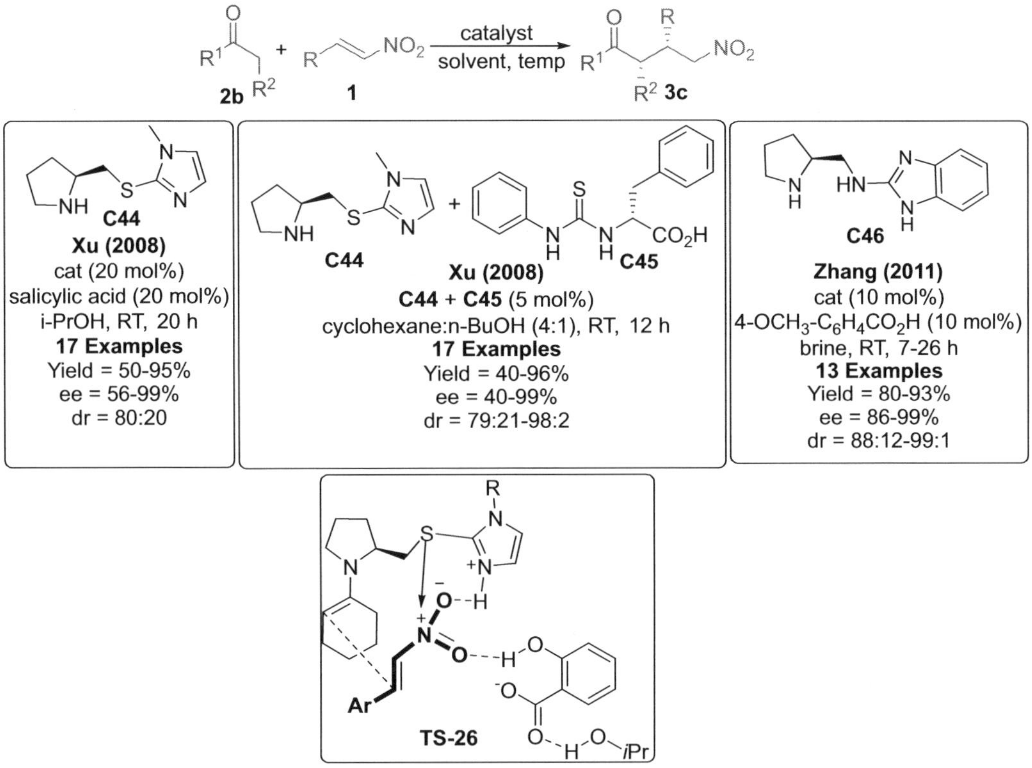

SCHEME 3.21 Imidazolylthiomethylpyrrolidine and pyrrolidine-aminobenzimidazole as organocatalysts in the asymmetric Michael addition of ketones to nitroalkenes.

and diastereoselectivity as well as high yields (Scheme 3.22).[48] The reaction was conducted in wet toluene with a catalyst loading of 20 mol%. To account for the stereochemical results, the authors proposed a transition state (**TS-27**, Scheme 3.22) based on Seebach's Model. (*S*)-2-((Naphthalen-2-ylsulfonyl)methyl)pyrrolidines **C49** bearing a pyrrolidine and a sulfone moiety was employed by Lin and co-workers for the asymmetric Michael addition of cyclohexanone **2b** to nitroalkenes **1** in water without any acid additive. The 15 mol% catalyst loading led to the formation of Michael adducts **3c** in excellent yields with high stereoselectivities (Scheme 3.22). In this reaction, the pyrrolidine ring of the catalyst **C49** reacts with cyclohexanone **3c** and forms enamine. Meanwhile, the oxygen atom of the sulfone group through hydrogen bond with a water molecule directs the nitro group and allows the resulting enamine to attack the nitroalkene **1** from the *Re*-face yielding the adduct **3c** (**TS-28**, Scheme 3.22).[49]

Gao and co-workers reported the use of pyrrolidine-triazole-based dendritic catalyst **C50** in the enantioselective Michael addition of ketones **2b** to nitroalkenes **1** under solvent-free conditions.[50] Michael adducts **3c** were obtained in high yields (up to 99%) and excellent diastereo (up to 45:1) and enantioselectivities (up to 95% ee, Scheme 3.23). The Schmitzer group reported catalytic asymmetric Michael addition of ketones **2b** to nitroalkenes **1** employing 30 mol% of anion-ionic liquid catalyst **C51**.[51] Various Michael adducts **3c** were obtained with excellent yields and modest-to-very good stereoselectivities (Scheme 3.23). In this reaction, the bulkiness of the imidazolium cation located near the negative charge of the proline derivative prevents the ketone **2b** to approach from the sulfonate side and allows the nitroalkene **1** to attack from the other side resulting in *syn*-enantiomer as the major product **3c** (**TS-29**, Scheme 3.23).

Connon and co-workers employed cinchona alkaloid-derived primary amine **C52** for asymmetric organocatalytic Michael reaction of ketones **2f** to nitroalkenes **1** (Scheme 3.24).[52] The use of a

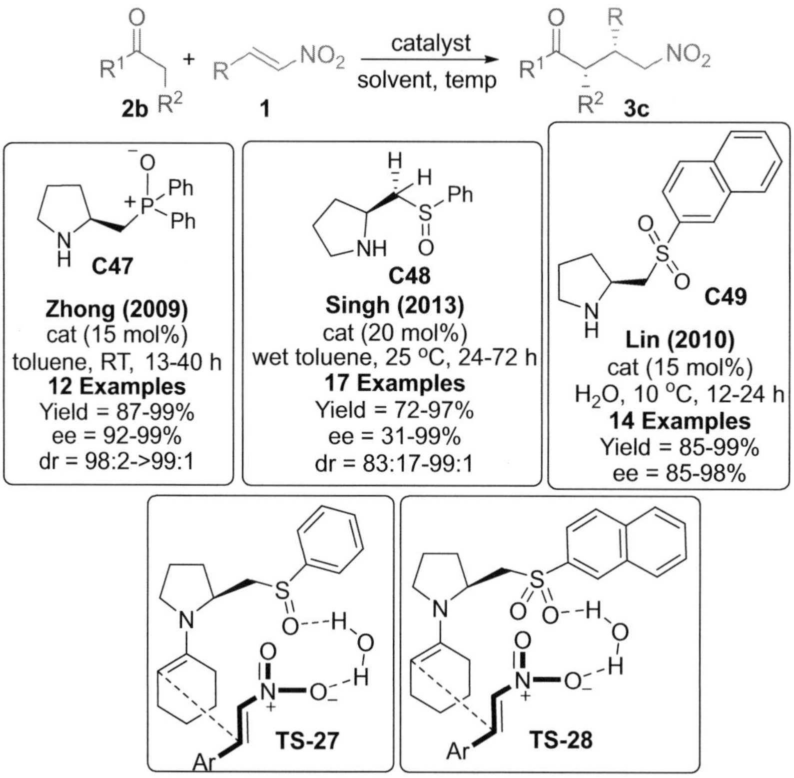

SCHEME 3.22 Phosphine oxide, sulfoxide and sulfone-bearing pyrrolidines as catalysts in the asymmetric Michael addition of ketones to nitroalkenes.

Brønsted acid, such as benzoic acid, as an additive resulted in clean reactions. The 9-amino derivative of dihydroquinine (DHQA) and C-9 inverted analog 9-epi-DHQA promoted the formation of (*R*)-product **3h**. 9-Epi-DHQA **C52** was found to be more active than DHQA. Under solvent-free conditions, the quinidine-derived 9-epi-DHQA **C52** promoted the efficient formation of (*S*)-product **3h** with high enantioselectivity (>95%).

Johnson et al. described the enantioselective synthesis of trisubstituted 2-trifluoromethyl pyrrolidines **4** with three contiguous stereocenters by the Michael addition/hydrogenative cyclization of 1,1,1-trifluoromethylketones **2g** and nitroalkenes **1** using thiourea catalyst **C1** at low catalyst loading (Scheme 3.25).[53]

Simpson and Lam reported the asymmetric Michael addition of 2-acetylazarenes **2h** to nitroalkenes **1** using a chiral Ni(II)-bisoxazoline complex [**L2** and $Ni(OAc)_2$] to afford the corresponding products **3j** in good yields with excellent enantioselectivities (Scheme 3.26).[54] A variety of azines/azoles **2h** including pyridine, pyrimidine, thiazole and quinoline have also been employed in this reaction to afford the corresponding adducts with high enantioselectivities.

The Song group reported the enantioselective Michael addition of 2-acetyl azaarenes **2i** to β-CF_3-β-disubstituted nitroalkenes **1b** by employing Co(II) in combination with imidazoline-oxazolone ligand **L3** to afford the corresponding adducts **3k** with all-carbon quaternary stereogenic centers in excellent yields and enantioselectivities (Scheme 3.27).[55] Recently, the same group reported a novel enantioselective Michael addition of 2-acetyl azaarenes **2i** to β-CF_3-β-(3-indolyl)nitroalkenes **1c** in the presence of a Co complex and (imidazoline-oxazoline) ligand **L3** as the catalyst (Scheme 3.28).[56] Here, a series of 3-substituted indole derivatives **3l** bearing a trifluoromethylated all-carbon quaternary stereocenter was synthesized with remarkable yields (up to 99%) and excellent enantioselectivities (up to 96%).

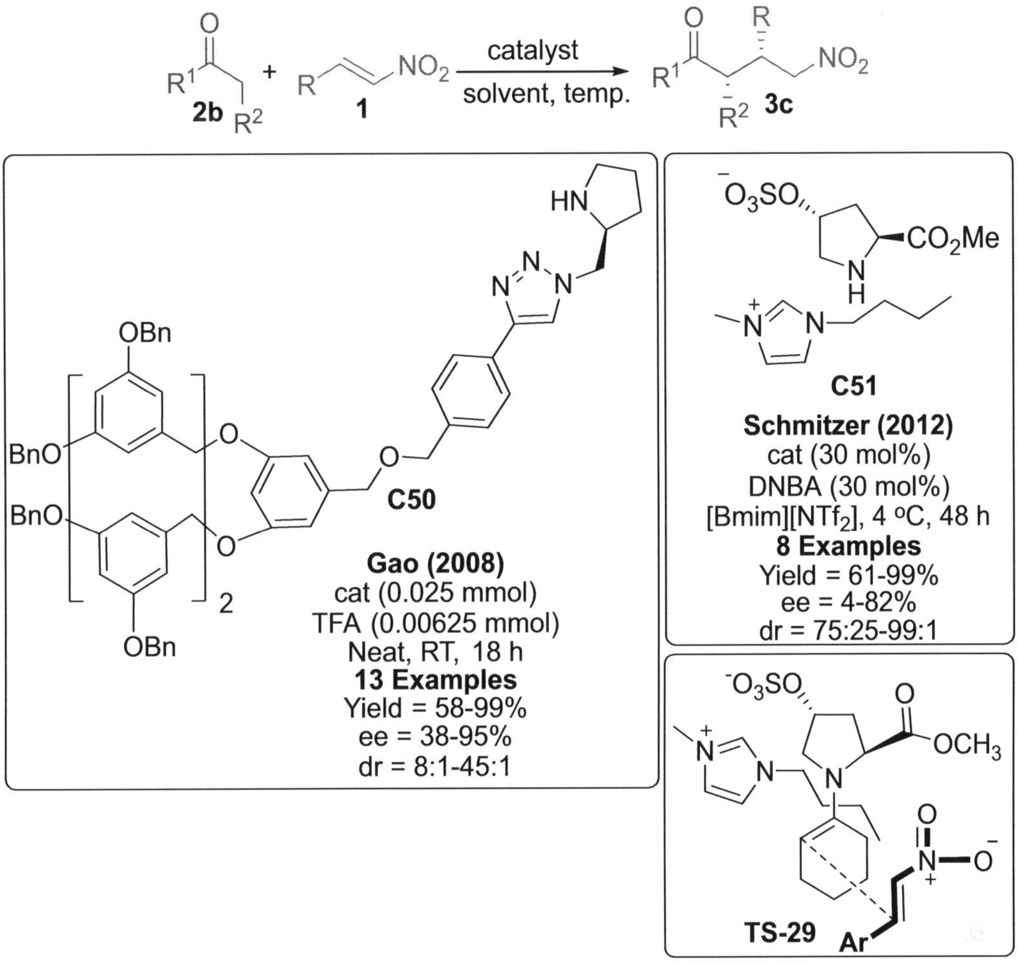

SCHEME 3.23 Pyrrolidine-triazole-based dendritic and an imidazolium salt as catalysts in the asymmetric Michael addition of ketones to nitroalkenes.

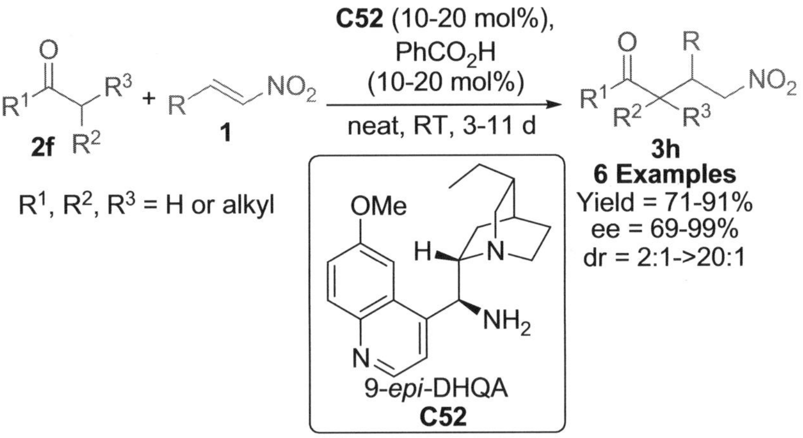

SCHEME 3.24 Cinchona alkaloid-derived primary amine as organocatalyst in the asymmetric Michael addition of ketones to nitroalkenes.

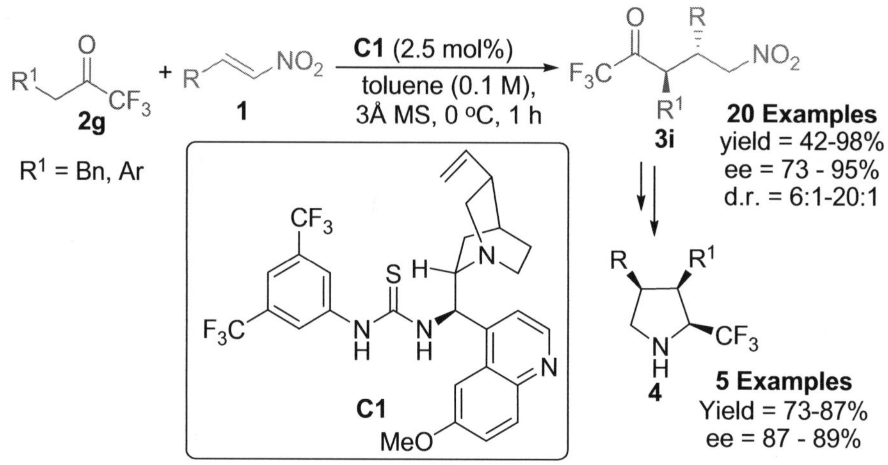

SCHEME 3.25 Asymmetric Michael addition of trifluoromethylketones to nitroalkenes and further transformation to trisubstituted trifluoromethyl pyrrolidines.

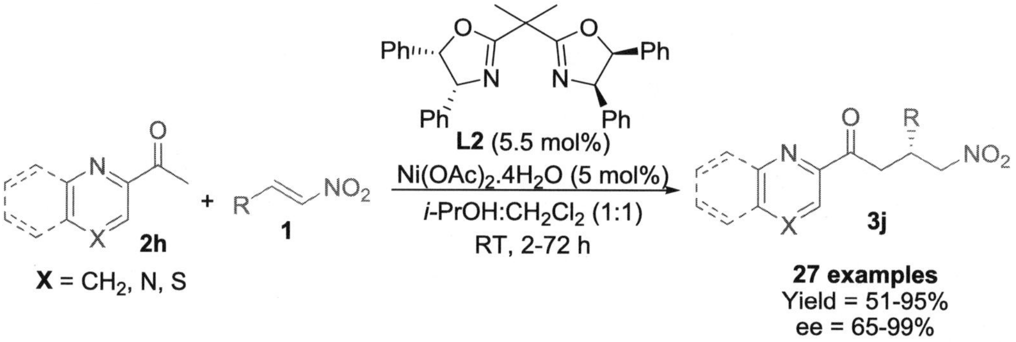

SCHEME 3.26 Michael addition of 2-acetylazarenes to nitroalkenes catalyzed by chiral Ni(II)-bisoxazoline complex.

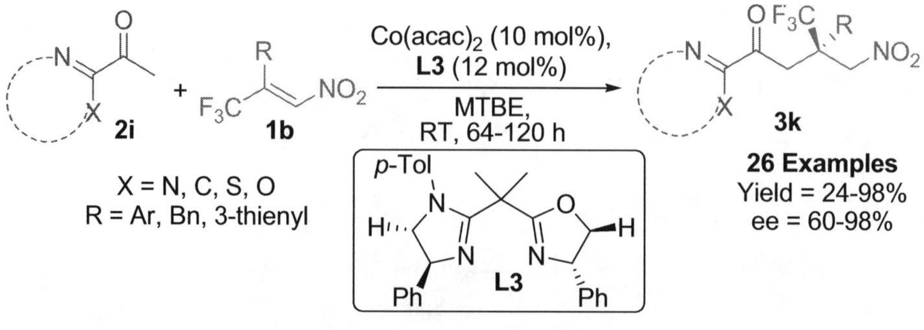

SCHEME 3.27 Michael addition of 2-acetyl azaarenes to β-CF_3-β-disubstituted nitroalkenes in the presence of Co(II)/ imidazoline-oxazolone catalyst system.

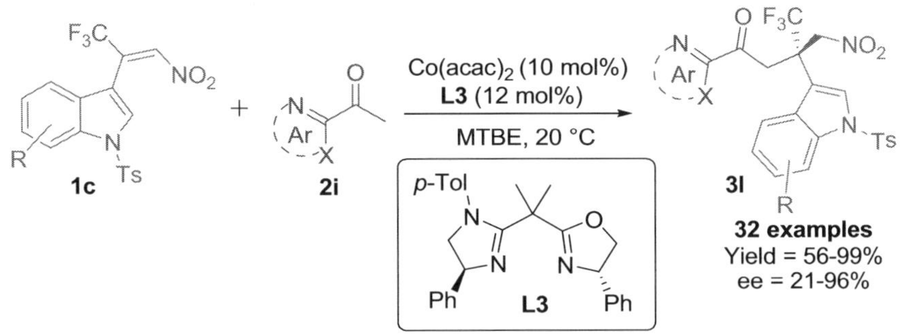

SCHEME 3.28 Asymmetric Michael Addition of 2-acetyl azaarenes to substituted (indolyl)nitroalkenes mediated by Cobalt(II)/(imidazolineoxazoline) as the catalyst.

The Fu group reported the asymmetric Michael addition of 2-acetyl azaarenes **2i** to α-substituted β-nitroacrylates **1d** to give corresponding adducts **3m**.[57] The reaction was performed at room temperature in the presence of Ni(II) catalyst and BOX-ligand **L4**. A variety of azaarenes were employed to give the corresponding products **3m** at high yields with moderate-to-excellent enantioselectivities (Scheme 3.29).

Later, the same group extended the asymmetric Michael addition of 2-acetylazaarenes **2i** with β-difluoromethyl-substituted nitroalkenes **1e** to produce CF_2H containing compounds **3n** with all quaternary stereogenic centers with moderate-to-good enantioselectivities (Scheme 3.30).[58] The in-situ generated Ni-BOX complex activates nitroalkenes **1e** as well as 2-acetylazaarenes **2h** for the Michael addition. Adducts are formed through the predominant *Re*-face attack of **2h** to **1e**, and the configuration of the adduct **3n** was assigned as *S* (Scheme 3.30).

Enders and co-workers reported an organocatalytic asymmetric Michael addition of bis(phenylthio)propan-2-one **2j** to various aromatic nitroalkenes **1** using Takemoto's thiourea catalyst **C53** (Scheme 3.31).[59] The reaction provided the corresponding addition products **3o** at reasonably high yields and enantioselectivities. Moreover, the simple recrystallization of the products further improved the enantioselectivity to 98% ee.

In 2010, Wang and co-workers reported the enantioselective Michael addition of ketones **2k** to nitroalkenes **1**. Chiral amine-thiourea **C54** was used as a bifunctional organocatalyst for this

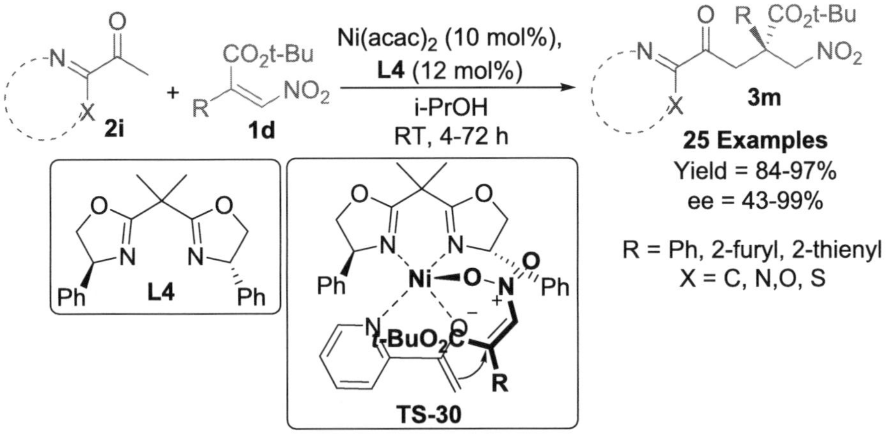

SCHEME 3.29 Michael addition of 2-acetyl azaarenes to α-substituted β-nitroacrylates in the presence of Ni(II)-BOX catalyst system.

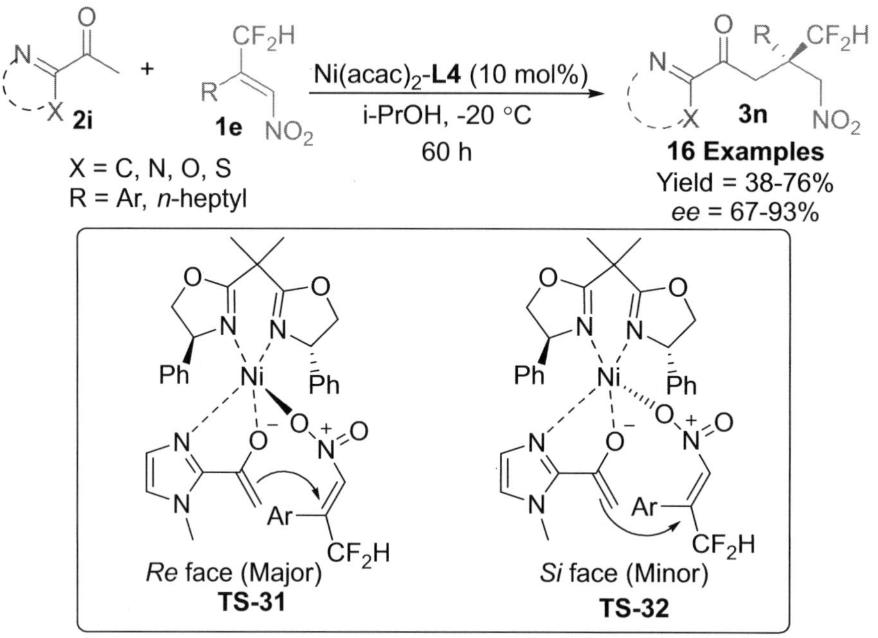

SCHEME 3.30 Michael addition of 2-acetylazaarenes to β-difluoromethyl-substituted nitroalkenes in the presence of Ni(II)-BOX catalyst system.

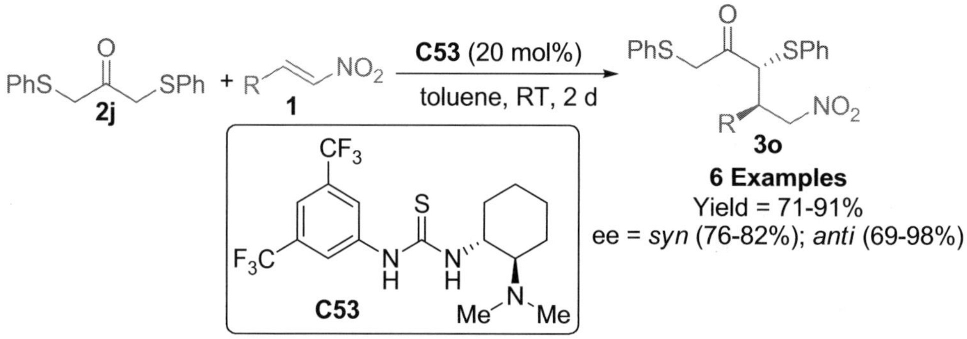

SCHEME 3.31 Michael addition of bis(phenylthio)propan-2-one to nitroalkenes using Takemoto's thiourea catalyst.

reaction along with benzoic acid as an additive to give functionalized Michael adducts **3p** with excellent yields and enantioselectivities (Scheme 3.32).[60]

The Fang group reported an organocatalytic asymmetric Michael addition of 2,4,4,4-tetraflu oro-3,3-dihydroxy-1-phenylbutan-1-ones **2l** to nitroalkenes **1** in the presence of catalyst **C53** to form corresponding decarboxylated γ-nitro-α-fluorocarbonyl compounds **3q** with good yields and enantioselectivities (Scheme 3.33).[61] Interestingly, trifluoroacetate is released with the cleavage of the C–C bond to give the chiral product **3q**.

In the possible reaction mechanism, at first, the primary Michael adduct with *R*-configuration is formed and then the tertiary amine of the catalyst **C53** abstracts the proton from the adduct with the release of trifluoroacetate. In the subsequent step, the thus-formed enol anion abstracts the proton from the protonated amino group of the catalyst to furnish the final product **3q** (**TS-34**, Scheme 3.34).

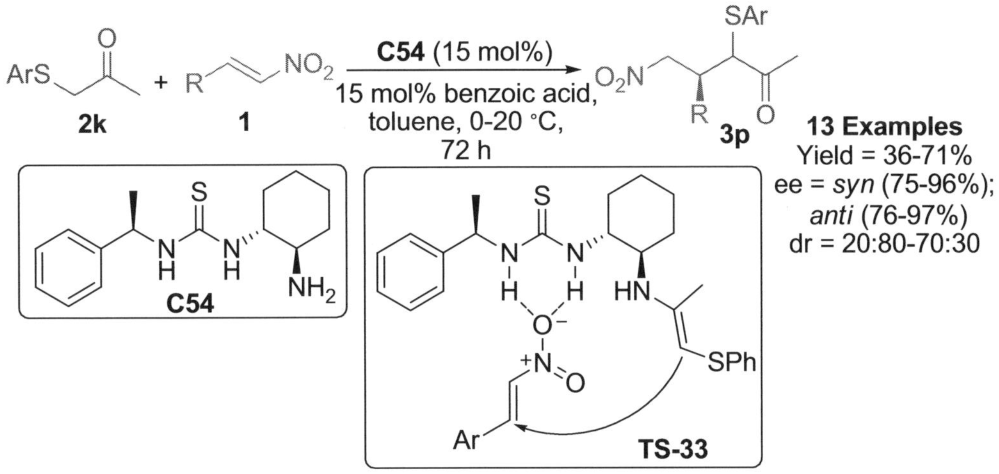

SCHEME 3.32 Michael addition of ketones to nitroalkenes using chiral amine-thiourea bifunctional organocatalyst.

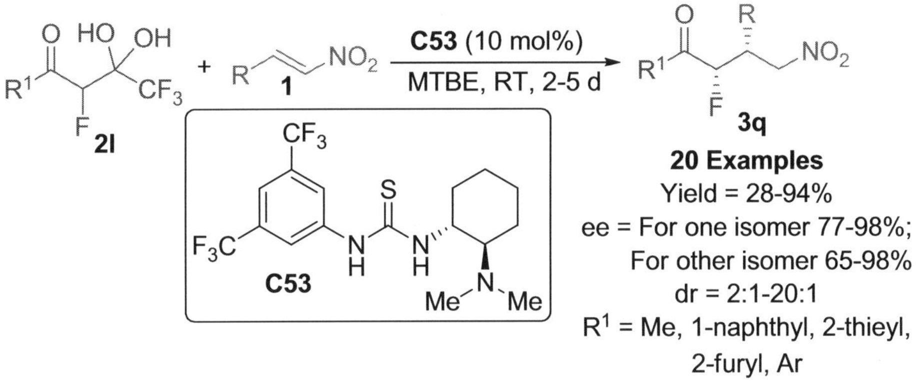

SCHEME 3.33 Michael addition of 2,4,4,4-tetrafluoro-3,3-dihydroxy-1-phenylbutan-1-ones to nitroalkenes using chiral thiourea catalyst.

The Wang group reported an asymmetric Michael addition of imidazole-modified ketones **2m** to nitroalkenes **1** using chiral diamine ligand **L5** and $Ni(OAc)_2$ to form *anti*-selective products **3r** (Scheme 3.35).[62] These *anti*-selective adducts can be transformed into *syn*-selective products through enolization followed by protonation by treating with an appropriate base.

The Wang group also employed imidazole-modified α-heteroatom ketones **2n** and nitroalkenes **1** to demonstrate the *anti*-selective Michael reaction in the presence of chiral diamine ligand **L5** and $Ni(OAc)_2$ to give the corresponding adducts **3s** at high yields and enantioselectivities (Scheme 3.36).[63] In the proposed transition state, the Ni-enolate formed from the imidazolylketone **2n** adds to nitroalkene **1** in a face-selective fashion (**TS-35**, Scheme 3.36).

The same group then disclosed a nickel-catalyzed asymmetric allylic alkylation reaction between nitroallylic acetates **1f** and imidazole-modified ketones **2o** in the presence of chiral diamine ligand **L6** to form enantioenriched α-allylic adducts **3t** in moderate-to-good yields (Scheme 3.37).[64] γ-Nitro-imidazolyl ketones 3t with three contiguous stereocenters can be easily obtained from α-allylic adducts.

Recently, Kang and co-workers presented rhodium-catalyzed asymmetric Michael addition of imidazole-modified ketones **2p** to nitroalkenes **1** to provide γ-nitroketones **3u** (Scheme 3.38).[65]

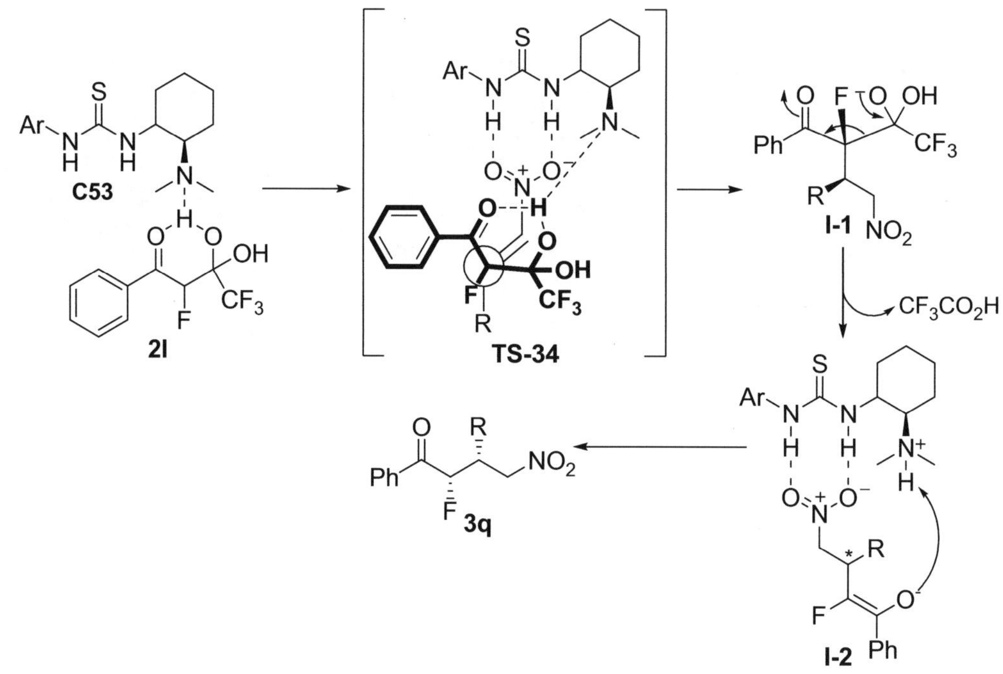

SCHEME 3.34 Proposed reaction mechanism for the Michael addition of 2,4,4,4-tetrafluoro-3,3-dihydroxy-1-phenylbutan-1-ones to nitroalkenes.

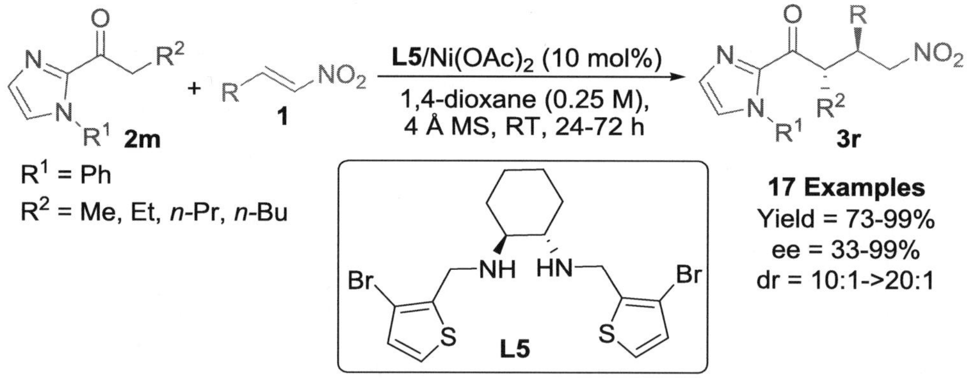

SCHEME 3.35 Michael addition of imidazole-modified ketones to nitroalkenes using chiral diamine ligand and $Ni(OAc)_2$.

This protocol was efficient even with 0.2 mol% of the chiral Rh(III) complex **C55** for the synthesis of adducts **3u** at high yield and excellent ee. According to the proposed mechanism, the chiral Rh(III) complex activates 2-acylimidazoles **2p** through ligand exchange and enolate is generated. Considering the steric factors, the less hindered *Re*-face attack of the enolate is preferred to form the product **3u** (**TS-36**, Scheme 3.38).

The Pan group reported an asymmetric Michael addition of α-branched enones **2q** to nitroalkenes **1** using chiral amine catalyst **C56**.[66] The reaction holds good only for cyclic α-branched enones **2q** such as 1-acetylcyclohexene, 1-acetylcyclopentene and 1-acetylcyclobutene

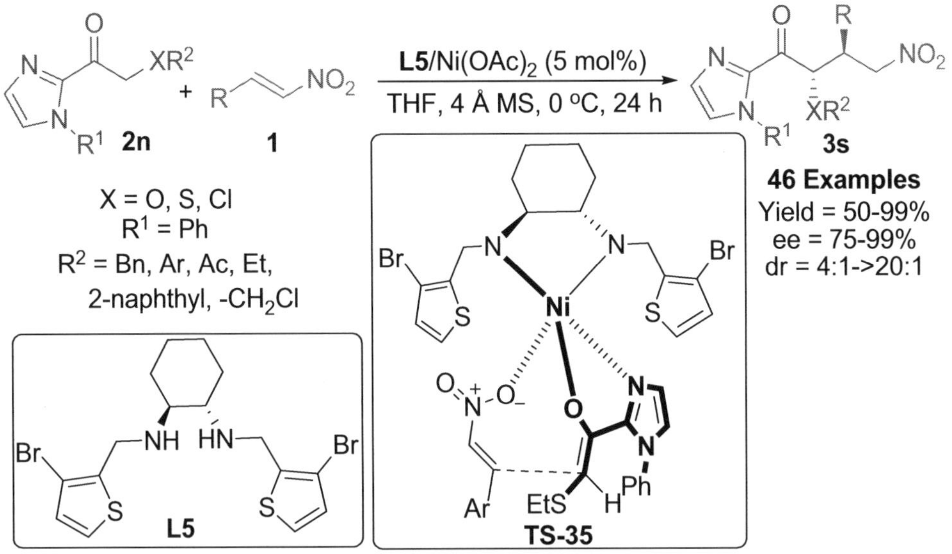

SCHEME 3.36 *Anti*-selective Michael reaction of imidazole-modified α-heteroatom ketones to nitroalkenes using chiral diamine ligand and $Ni(OAc)_2$.

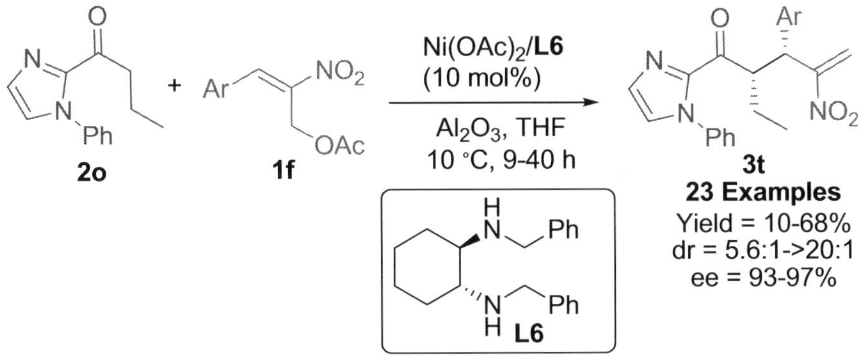

SCHEME 3.37 Asymmetric allylic alkylation of acyl imidazoles using nitroallylic acetates in the presence of diamine-Ni(II) catalyst system.

to give the corresponding adducts **3v** in high yields and excellent stereoselectivities, whereas acyclic α-branched enones remained unreactive (Scheme 3.39). During the reaction, at first, ketone **2q** forms an iminium ion by reacting with the amino group of the catalyst **C56** and then isomerizes to enamine (**TS-37**, Scheme 3.39). In the next step, the tertiary amine gets protonated and interacts with the nitro group to assist in stereocontrol to furnish the product **3v**.

Palomo and co-workers explored α-hydroxyketones **2r** as ester equivalents in Michael addition with nitroalkenes **1** to afford the corresponding Michael adducts **3w** with high diastereo (dr 95:5) and enantioselectivity (up to 99% ee) using **C1** as a catalyst (Scheme 3.40).[67] The stereo and enantioselectivity of the conjugate addition were controlled by the size of the *gem*-R substituents, and bulky groups like benzyl lead to the product with high enantioselectivity. The authors also carried out a synthetic transformation of the formed Michael adducts **3w** to enantioenriched α-substituted carboxylic acids, aldehydes and aryl acetic acids, as well as densely functionalized carbocycles. The stereochemical model **TS-38** (Scheme 3.40) accounts for the reaction outcome.

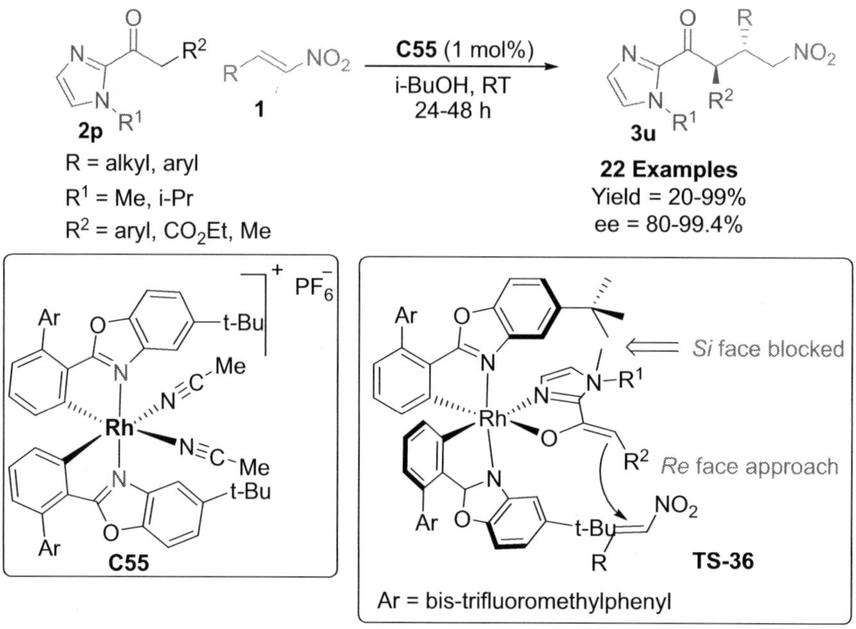

SCHEME 3.38 Rhodium-catalyzed asymmetric Michael addition of imidazole-modified ketones to nitroalkenes.

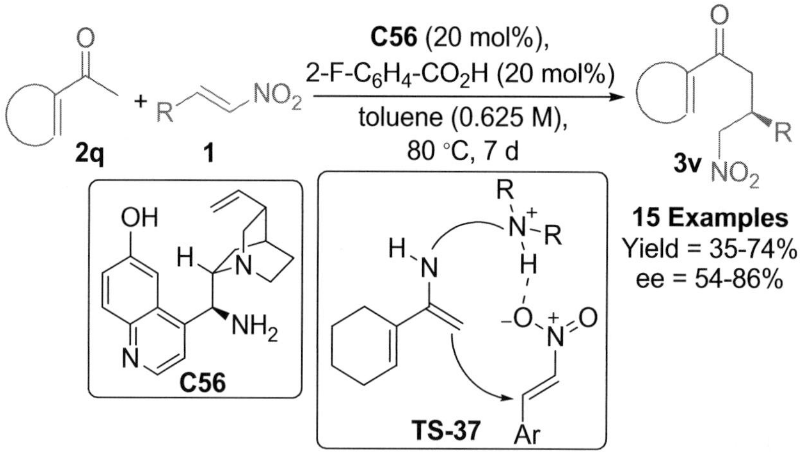

SCHEME 3.39 Michael addition of α-branched enones to nitroalkenes using a chiral amine catalyst.

Recently, the same group reported the asymmetric conjugate addition of enolzable α,β-ynones **2s** to nitroolefins **1** mediated by a bifunctional squaramide-tertiary amine catalyst **C57** (Scheme 3.41).[68] Here, it was observed that the adducts **3x** were formed with excellent yields (up to 99%) and remarkable stereoselectivities (up to 99% ee and up to >20:1 dr). The substrate scope employed was not limited to allyl ynones. Moreover, alkoxymethyl ynones and benzylic ynones were also used as donor ketones.

Zhou and co-workers reported the use of a chiral thiophosphinamide-1,2-diphenylethane-1,2-diamine-based catalyst **C58** for the effective kinetic resolution of axially chiral 2-nitrovinyl biaryls **1g** through asymmetric Michael reaction with acetone **2t** (Scheme 3.42).[69] Using the catalyst, the racemic substrate was separated as highly enantioenriched 2-nitrovinyl biaryls **1g** in considerable

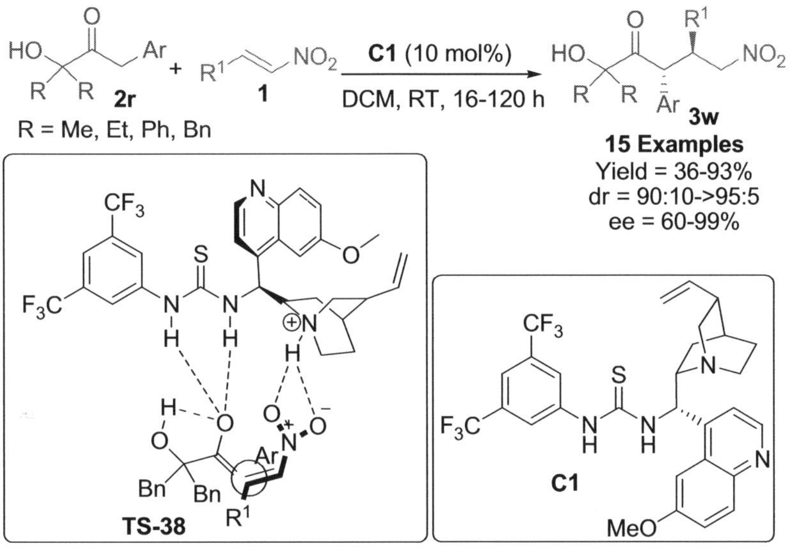

SCHEME 3.40 Michael addition of α-hydroxyketones to nitroalkenes in the presence of bifunctional Brønsted base-Brønsted acid catalyst.

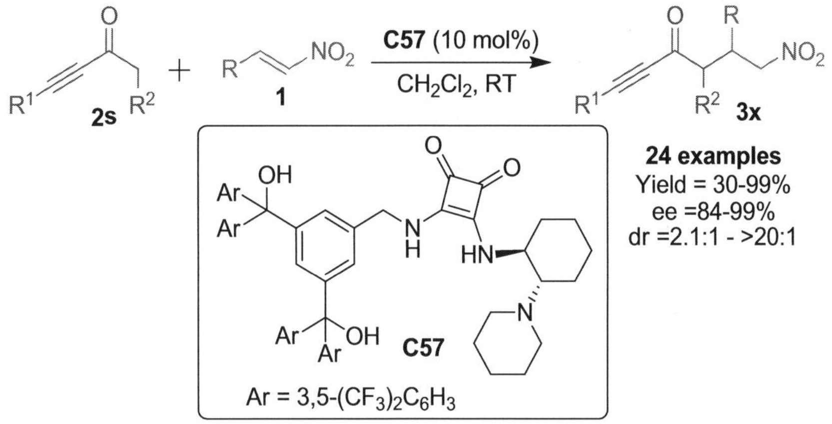

SCHEME 3.41 Brønsted base/H-bonding catalyst mediated by the asymmetric addition of alkynyl ketones to nitroalkenes.

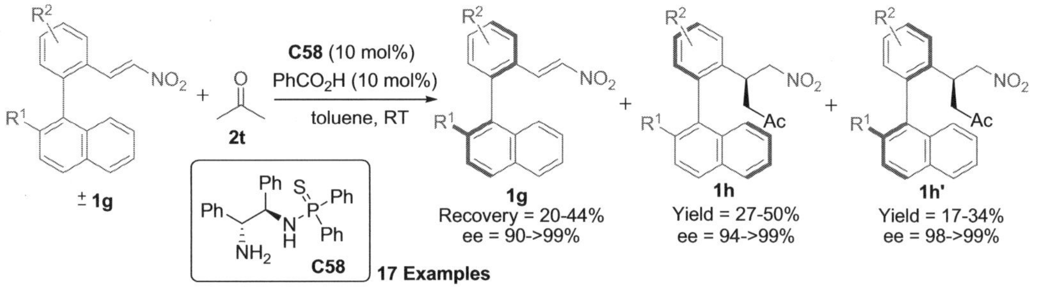

SCHEME 3.42 Kinetic resolution of axially chiral 2 nitrovinyl aryls through asymmetric Michael reaction with acetone mediated by a chiral thiophosphinamide-1,2-diphenylethane-1,2-diamine-based catalyst.

yields (20%–44%) along with two separate Michael adducts **1h** and **1h'** possessing both axial and central chirality with remarkable stereoselectivity (up to >99% ee).

The Palomo group reported an asymmetric Michael addition of β-tetralones **5** with both α-substituted and α-unsubstituted nitroalkenes **1** to afford the corresponding adducts **6**.[70] The reaction uses 10 mol% squaramide-derived catalysts **C59a** or **C59b** to afford the products at high yields and enantioselectivities (Scheme 3.43).

Yuan et al. delineated an asymmetric Michael addition of anthrone **7** to a wide variety of nitroalkenes **1**.[71] The reaction was carried out using bifunctional thiourea-tertiary amine catalyst **C53** to give the corresponding adducts **8** in excellent yields with high enantioselectivities of up to 94% (Scheme 3.44). In the postulated mechanism, the nitroalkene **1** was activated through double H-bonding interactions between thiourea moieties, and the nitro group and anthrone **7** were activated by the tertiary amine moiety of the catalyst **C53** (**TS-39**, Scheme 3.44).

Itsuno and Kumpuga reported the Michael addition of anthrones **7** to nitroalkenes **1** using cinchona-based polyurethanes **C60** as a chiral catalyst (Scheme 3.45).[72] It was observed that the performed reactions provided the Michael adducts **8** in good yields (up to 92%) and enantioselectivities (up to 84%). The catalyst **C60** was recycled for up to five cycles without losing its catalytic efficiency owing to the insolubility of the catalyst in the reaction. Here, the authors emphasized the significant role of the phenolic hydroxyl group of the catalyst in controlling the stereoselectivity of the so-obtained products.

Recently, Genc reported the chiral tetraoxacalix[2]arene[2]triazine **C61** as a catalyst for Michael addition reaction of anthrones **7** to nitroalkenes **1** in an asymmetric fashion (Scheme 3.46).[73] The corresponding Michael adducts **8** were formed in good yields (up to 96%) as well as high enantioselectivities (up to 97%), employing just 10 mol% of the catalyst.

Ma and co-workers reported enantioselective conjugate addition of different ketones **2u-w** to nitrodienes **1i** mediated by bifunctional organocatalysts **C62a-b** derived from saccharides and chiral diamines where the corresponding adducts were formed with excellent enantioselectivities

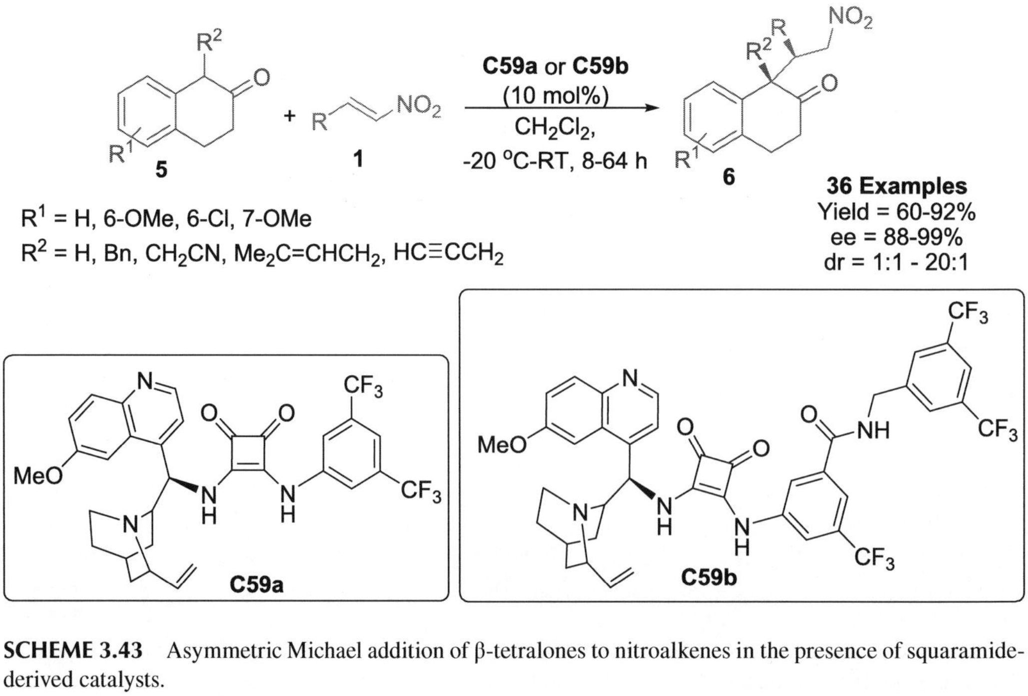

SCHEME 3.43 Asymmetric Michael addition of β-tetralones to nitroalkenes in the presence of squaramide-derived catalysts.

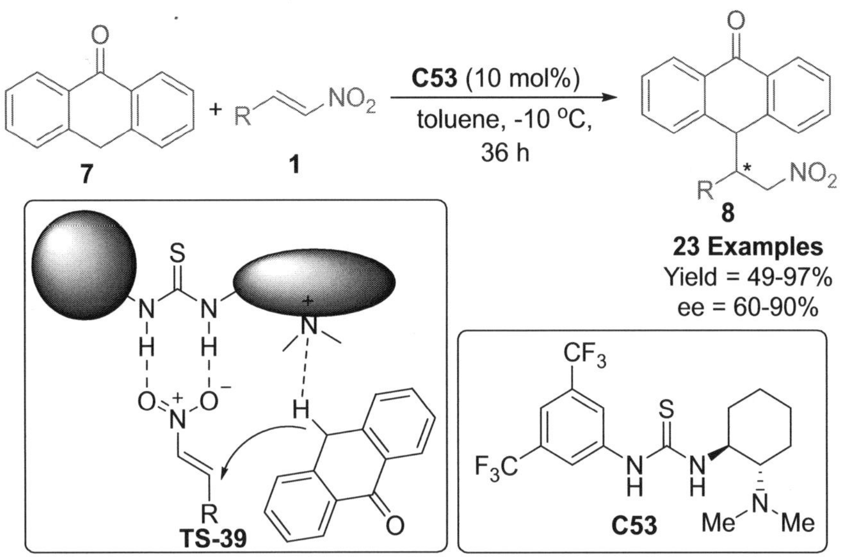

SCHEME 3.44 Asymmetric Michael addition of anthrone to nitroalkenes catalyzed by bifunctional thiourea-tertiary amine catalyst.

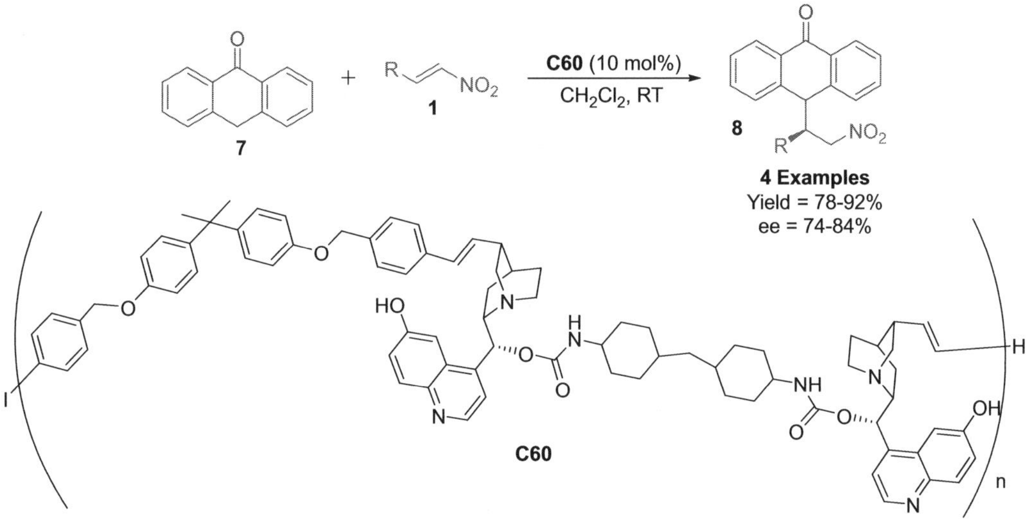

SCHEME 3.45 Asymmetric Michael addition of anthrones to nitroalkenes mediated by cinchona-based polyurethane catalyst.

(Scheme 3.47).[74] 1,4-Addition products **9** and **10** were formed exclusively without any trace of 1,6-addition products.

Wu and co-workers reported highly stereoselective Michael addition of aromatic ketones **2u** to nitrodienes **1i** using a chiral primary amine-thiourea based on dehydroabietic amine as catalyst **C63**, which resulted in the formation of γ-nitroketones **9** in high yields as well as excellent stereoselectivities (Scheme 3.48).[75] Moreover, this protocol is further used for the asymmetric synthesis of chiral 3-(aminomethyl)-5-phenylpentanoic acid. Wu and co-workers reported asymmetric Michael addition

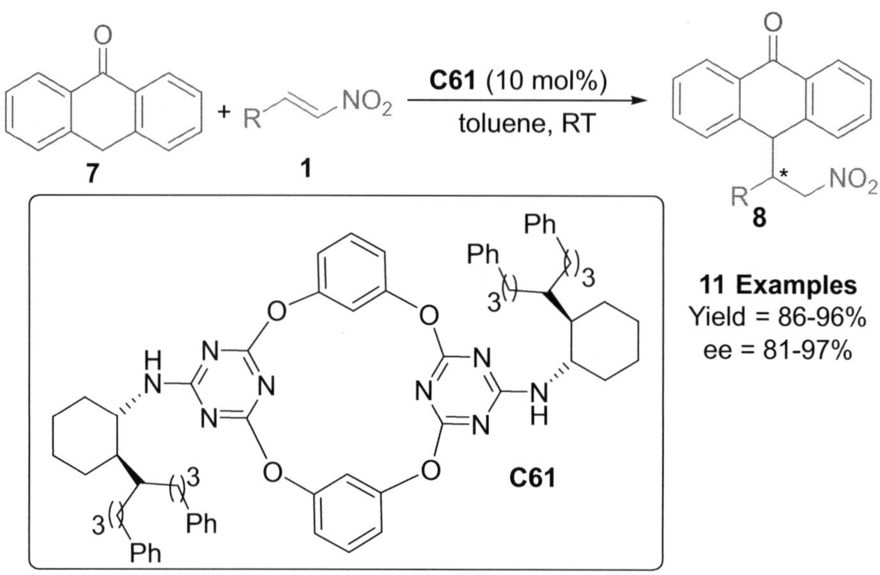

SCHEME 3.46 Michael addition of anthrone to nitroolefins mediated by chiral tetraoxacalix[2]arene[2] triazine.

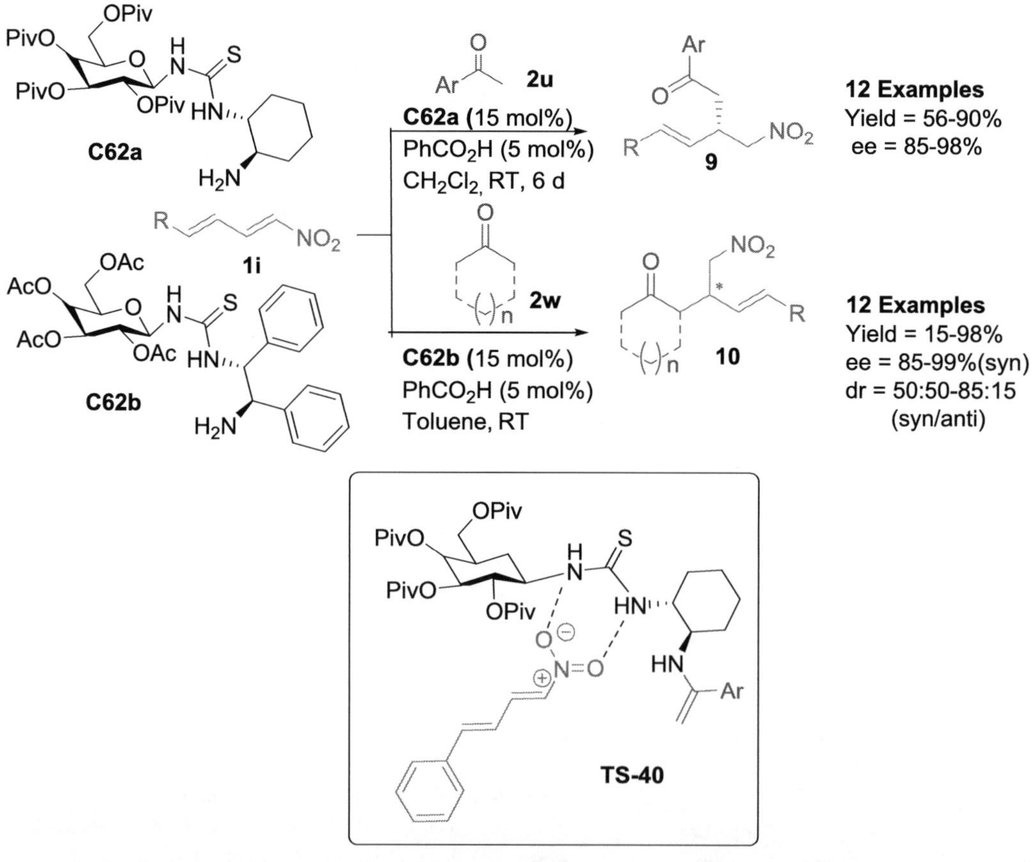

SCHEME 3.47 Enantioselective conjugate addition of different ketones to nitrodienes in the the presence of sugar-derived thiourea catalysts.

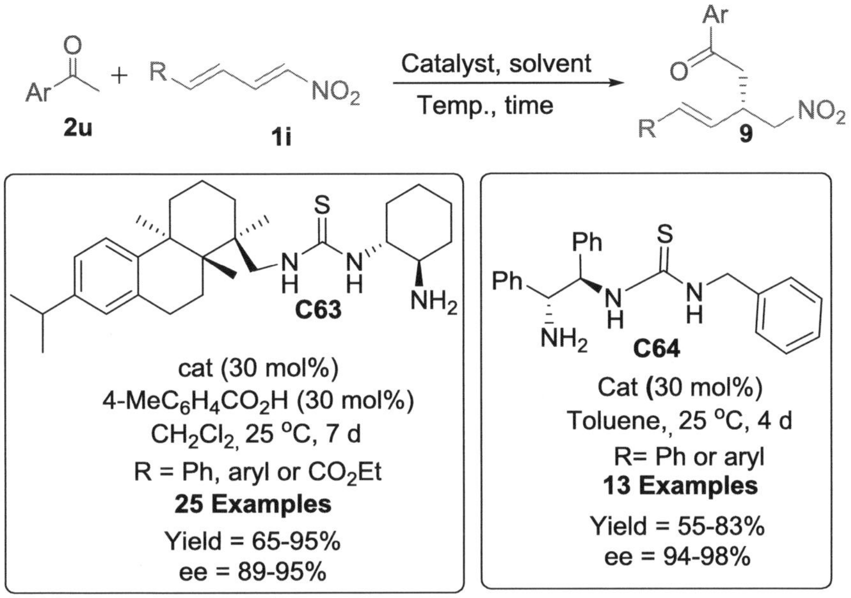

SCHEME 3.48 Stereoselective Michael addition of aryl ketones to nitrodienes mediated by chiral primary amine thioureas.

of aryl ketones **2u** to nitrodienes **1i** mediated by chiral primary amine thioureas as catalysts **C64** (Scheme 3.48).[76] It was observed that good yields and high enantioselectivities in the corresponding Michael adducts **9** were obtained using 30 mol% catalyst loading. The formation of enamine between the catalyst and ketone as well as hydrogen bonding of the nitrodiene and thiourea led to the formation of S-Michael adducts **9** by selective *Si*-facial attack of the enamine to nitrodiene.

3.3 ADDITION OF α-KETOESTERS, α-KETOAMIDES AND α-KETOPHOSPHONATES

Unlike 1,3-dicarbonyl compounds, 1,2-dicarbonyls were only sparingly employed as pronucleophiles. Their use in intermolecular reactions is curtailed by their high electrophilic character, their potential for self-condensation and lower acidity compared to 1,3-dicarbonyl compounds.

Sodeoka et al. described the enantio and diastereoselective conjugate addition of α-ketoesters **11a** to nitroalkenes **1** employing a combination of endogenous and exogenous bases in the presence of a

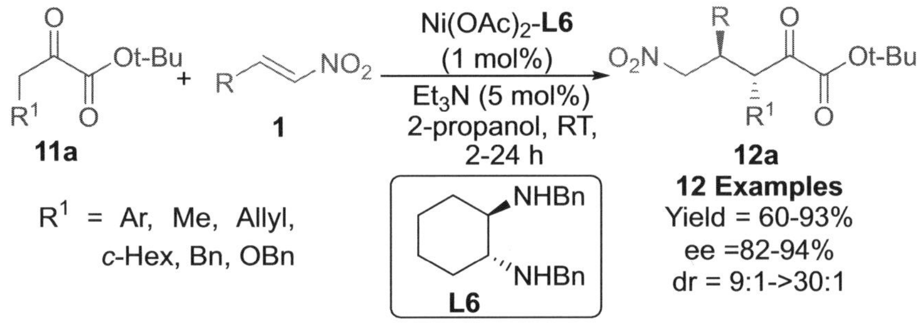

SCHEME 3.49 Conjugate addition of α-ketoesters to nitroalkenes catalyzed by chiral $Ni(OAc)_2$ complex.

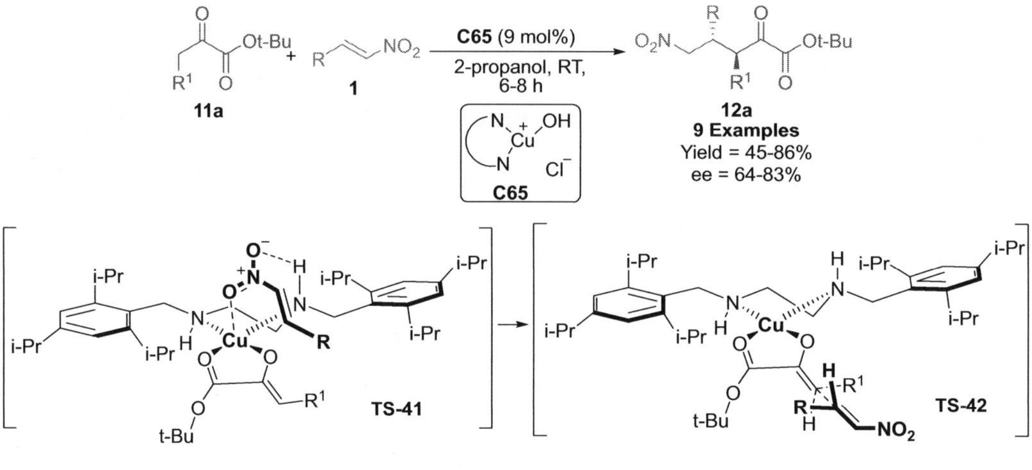

SCHEME 3.50 Chiral Cu(II) hydroxo complex as a catalyst in the asymmetric conjugate addition of α-ketoesters to nitroalkenes.

catalyst-containing ligand **L6** (Scheme 3.49).[77] Because of the usage of a combination of endogenous and exogenous bases, a small amount of the catalyst was sufficient to carry out the reaction efficiently.

The same group in another report described the Cu hydroxo complex containing N-substituted diaminocyclohexane derivative **C65** as a mild acid–base catalyst for the catalytic asymmetric conjugate addition of α-ketoesters **11a** to nitroalkenes **1** (Scheme 3.50).[78] The reactions proceeded smoothly to afford the products **12a** preferentially with *anti*-stereochemistry in good yield with up to 83% ee. At first, Cu hydroxo complex **C65** reacts with α-ketoesters **11a** to form the corresponding bidentate chiral Cu enolate and this Cu ion coordinates to the nitroalkenes **1** (**TS-41**, Scheme 3.50). Moreover, it was speculated that the aryl groups of the diamine ligands are located at the pseudo-equatorial position, which creates a C_2 symmetric chiral environment. This allows the ester group of the substrate to point away to avoid the steric repulsion with the aryl group. Because the *Re*-face of the enolate is blocked by the *tert*-butyl ester and the aryl group of the ligand, the reaction occurs at the *Si*-face of the enolate preferentially (**TS-42**, Scheme 3.50).

Rodriguez and co-workers developed an organocatalytic asymmetric conjugate addition of α-ketoesters **11b** to nitroalkenes **1** employing Takemoto's catalyst **C53** and gave the corresponding

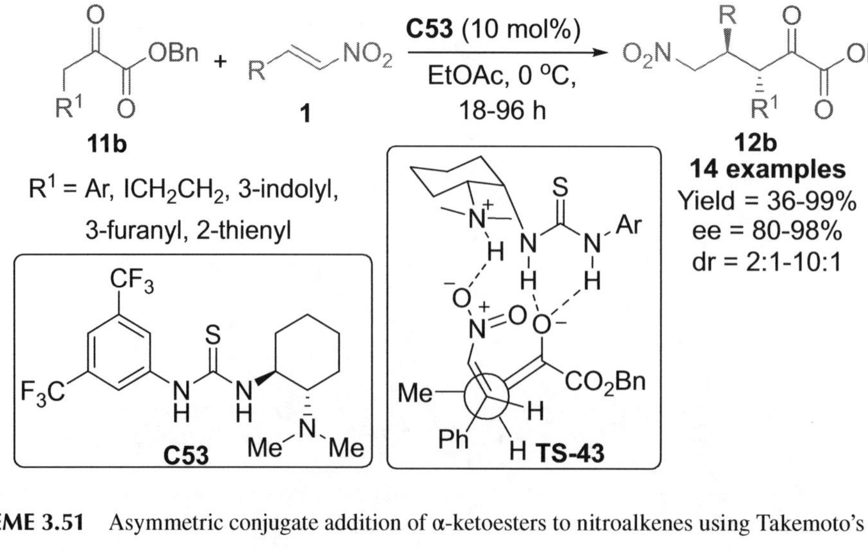

SCHEME 3.51 Asymmetric conjugate addition of α-ketoesters to nitroalkenes using Takemoto's catalyst.

adducts **12b** with excellent yields and enantioselectivities of up to 98% (Scheme 3.51).[79] In the proposed mechanism, the electrophilic character of the nitroalkene **1** is enhanced by the protonated tertiary amine of catalyst **C53**, which binds to the nitro group, whereas the strongly polarized enolate is bound by the thiourea motif. This conformationally restrained environment results in a preferential attack of the *Si*-face of the thermodynamic (*Z*)-enolate on the *Re*-face of the nitroalkene **1** to form the *anti*-(*R*, *R*)-Michael adduct **12b** (**TS-43**, Scheme 3.51). Further transformation of the Michael adducts to five-membered carbo and heterocycles with the creation and control of additional stereocenters was also described.

Organocatalytic asymmetric conjugate addition of α-ketoamides **11c** to nitroalkenes **1** was also demonstrated by the same group employing tertiary amine-thiourea catalyst **C53**. This catalyst with the cooperative effect of the amide proton afforded the corresponding Michael adducts **12c** at high yields and excellent *anti*-selectivity and enantioselectivity (Scheme 3.52).[80] The proposed mechanism involves activation of the nitroalkene **1** *via* a bidentate H-bond interaction with the thiourea moiety and deprotonation of the ketoamide **11c** by the tertiary amine moiety. The resulting ion-pair then adopts a multiple H-bonded transition state, which allows a preferential approach of the *Si*-face of (*Z*)-enolate on the *Re*-face of the nitroalkene **1** (**TS-44**, Scheme 3.52).

Shibasaki and co-workers delineated the catalytic asymmetric conjugate addition of α-ketoanilides **11c** to nitroalkenes **1** using the dinuclear nickel-Schiff base catalyst **C66**, and the complimentary *syn*-selectivity was observed to give the products **12c** in excellent yields and enantioselectivity (Scheme 3.53).[81] In the postulated mechanism, one of the Ni–O bonds in the outer dioxygen cavity is thought to work as a Brønsted base to generate Ni-enolate in situ (**TS-45**, Scheme 3.53). The other Ni in the inner N_2O_2 cavity serves as a Lewis acid to bind with nitroalkene **1** (**TS-46**, Scheme 3.53). Then the C–C bond formation followed by protonation yields the corresponding *syn*-adduct **12c** and regenerates the catalyst **C66**.

Yuan et al. described an asymmetric Michael addition reaction of arylacetyl phosphonates **11d** to nitroalkenes **1** using a chiral bifunctional amine-thiourea catalyst **C67** bearing multiple hydrogen bond donors to afford a range of α-substituted carboxylic esters **12d** with contiguous tertiary stereocenters in good yields with excellent enantio and diastereoselectivities (95% ee and up to >99:1 dr, Scheme 3.54).[82]

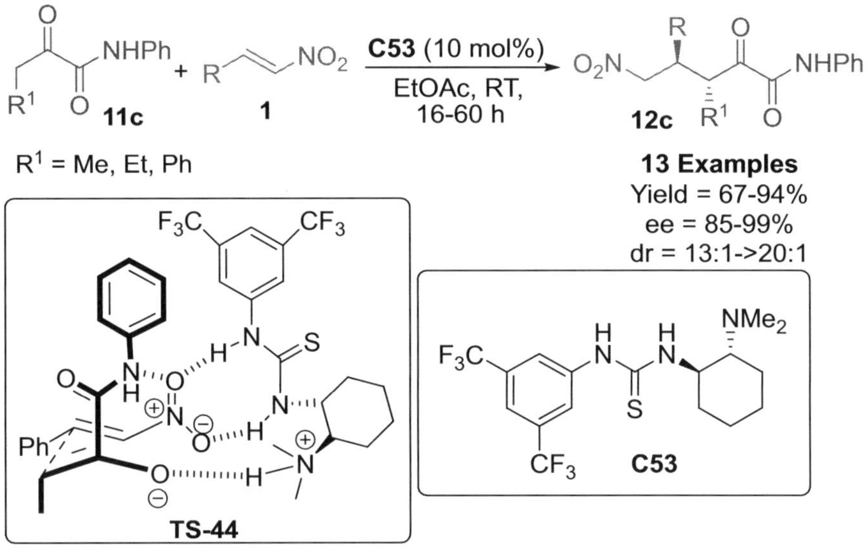

SCHEME 3.52 Asymmetric conjugate addition of α-ketoamides to nitroalkenes in the presence of tertiary amine-thiourea catalyst.

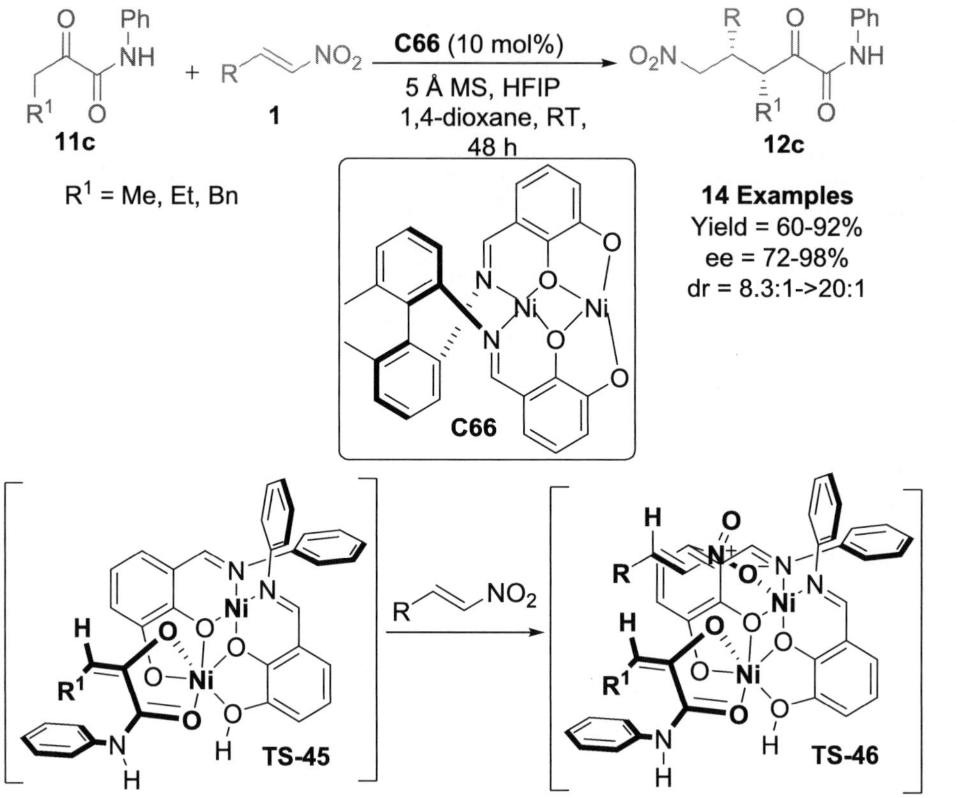

SCHEME 3.53 Asymmetric conjugate addition of α-ketoanilides to nitroalkenes using the dinuclear nickel-Schiff base catalyst.

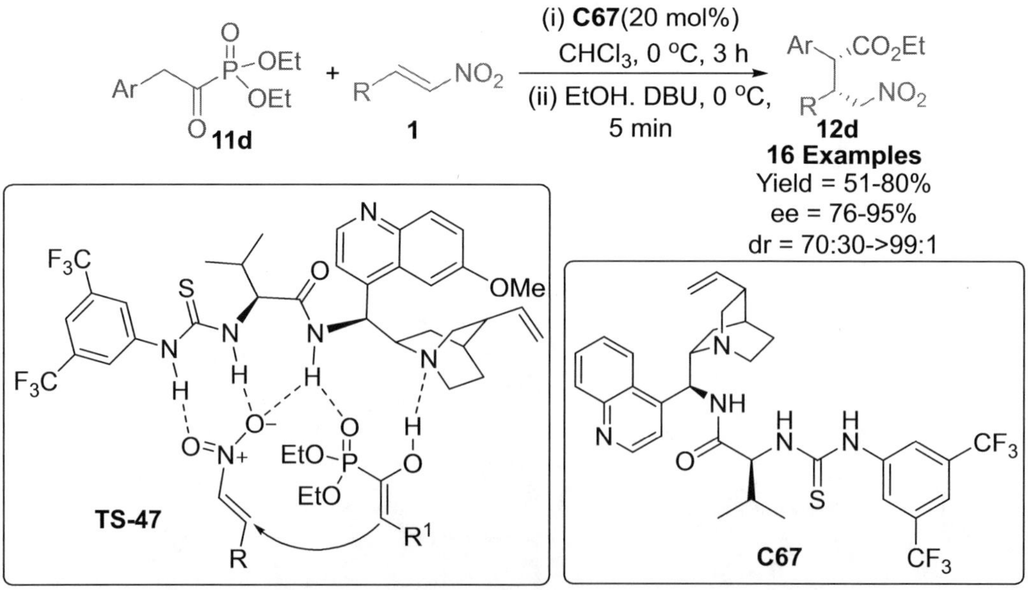

SCHEME 3.54 Asymmetric Michael addition of arylacetyl phosphonates to nitroalkenes in the presence of chiral bifunctional amine-thiourea catalyst.

3.4 ADDITION OF β-KETOSULFONES

Dong and co-workers described C_3-symmetric cinchonine-squaramide **C68** catalyzed asymmetric Michael addition of ketosulfones **11e** to nitroalkenes **1** to afford in-situ generated Michael adduct **12e**. Treatment with zinc in one pot affords chiral cyclic nitrones **13** at high yields and good enantioselectivities (Scheme 3.55).[83] The catalyst employed was recovered and recycled for six runs of Michael addition without any significant loss in its activity and selectivity.

Zhao et al. reacted α-fluoro-α-phenylsulfonyl ketone **11f** with nitroalkene **1** in the presence of a bifunctional chiral thiourea catalyst **C69** to obtain the corresponding Michael adducts **12f** with excellent yields and enantioselectivity (Scheme 3.56).[84] The catalyst **C69** activates the nitroalkene

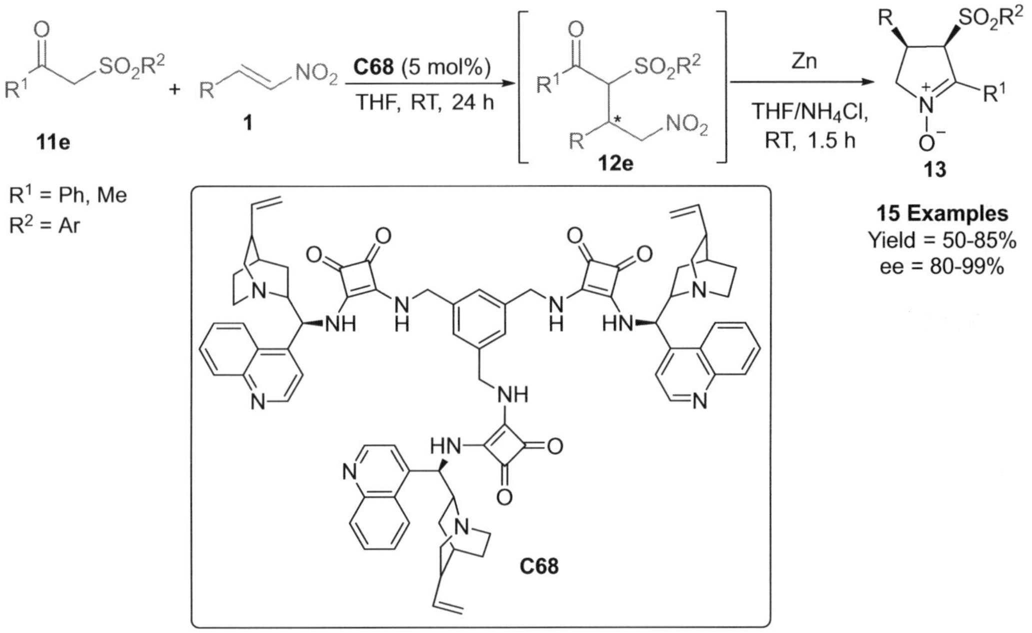

SCHEME 3.55 C_3-symmetric cinchonine-squaramide as catalyst in the asymmetric Michael addition of ketosulfones to nitroalkenes.

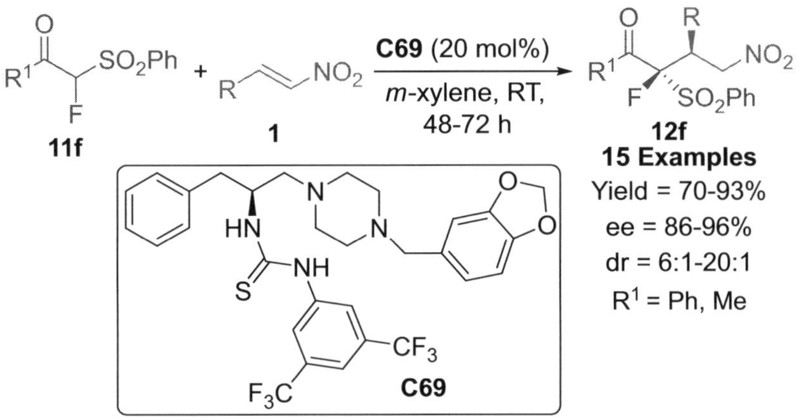

SCHEME 3.56 Asymmetric Michael addition of α-fluoro-α-phenylsulfonyl ketone with nitroalkene **1** in the presence of bifunctional chiral thiourea catalyst.

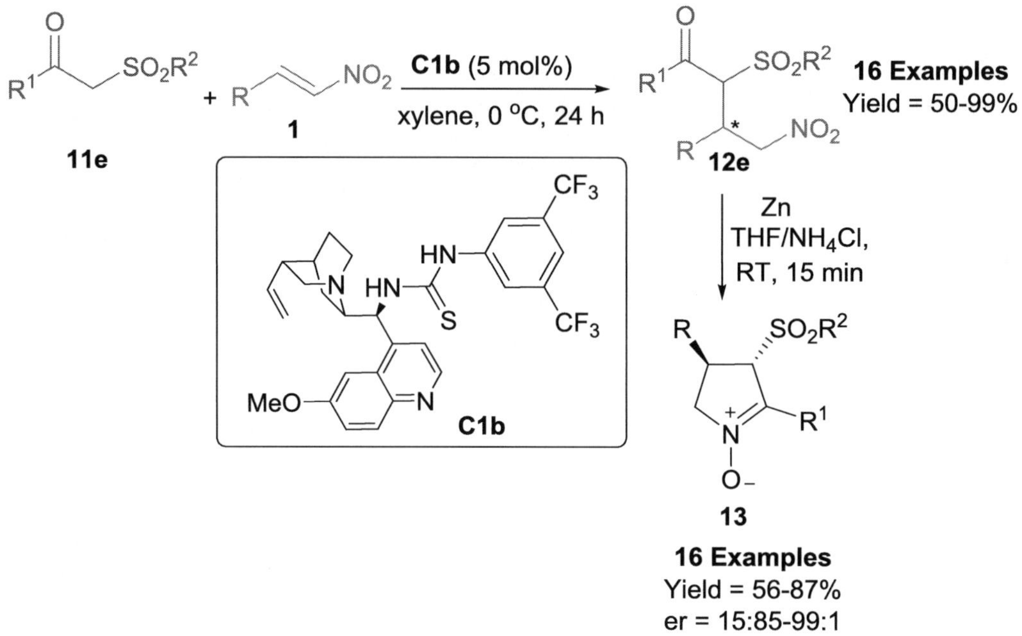

SCHEME 3.57 Michael addition of β-ketosulfones to nitroalkenes catalyzed by thiourea catalyst.

1 through hydrogen bonding, and the α-fluoro-α-phenylsulfonyl ketone **11f** attacks the nitroalkene **1** from the *Si*-face to form the Michael product **12f** with *2S,3S* configuration.

The Alemán group reported the Michael addition of β-ketosulfones **11e** to nitroalkenes **1** catalyzed by thiourea catalyst **C1b**.[85] This catalytic system allows the synthesis of optically enriched ketosulfone derivatives **12e** possessing two chiral centers in excellent yields (Scheme 3.57). The er values were determined by high-performance liquid chromatography analysis of the corresponding nitrone derivatives **13**.

3.5 CONCLUSIONS

Asymmetric Michael addition of enolizable ketones, including alkyl and aryl ketones, α-ketoesters, amides and phosphonates, as well as β-ketosulfones occurs in the presence of various chiral organo- and metal-ligand catalysts. Diverse chiral Lewis/Brønsted bases, Lewis/Brønsted acids and different combinations of bifunctional catalysts efficiently catalyze these reactions to afford the products with high stereoselectivity.

REFERENCES

1. Mandal, T.; Zhao, C.-G. *Angew. Chem. Int. Ed.* **2008**, *47*, 7714.
2. Wang, Q.-W.; Peng, L.; Fu, J.-Y.; Huang, Q.-C.; Wang, L.-X.; Xu, X.-Y. *ARKIVOC* **2010**, *ii*, 340.
3. Tuchman-Shukron, L.; Miller, S. J.; Portnoy, M. *Chem. - Eur. J.* **2012**, *18*, 2290.
4. Liu, S.-P.; Zhang, X.-j.; Lao, J.-h.; Yan, M. *ARKIVOC* **2009**, *vii*, 268.
5. Peng, L.; Xu, X.-Y.; Wang, L.-L.; Huang, J.; Bai, J.-F.; Huang, Q.-C.; Wang, L.-X. *Eur. J. Org. Chem.* **2010**, 2010, 1849.
6. Lu, A.; Liu, T.; Wu, R.; Wang, Y.; Zhou, Z.; Wu, G.; Fang, J.; Tang, C. *Eur. J. Org. Chem.* **2010**, *2010*, 5777.
7. Tsakos, M.; Kokotos, C. G.; Kokotos, G. *Adv. Synth. Catal.* **2012**, *354*, 740.
8. Jiang, X.; Zhang, Y.; Chan, A. S. C.; Wang, R. *Org. Lett.* **2009**, *11*, 153.
9. Cobb, A. J. A.; Longbottom, D. A.; Shaw, D. M.; Ley, S. V. *Chem. Commun.* **2004**, 2004, 1808.

10. Ni, B.; Zhang, Q.; Headley, A. D. *Tetrahedron Lett.* **2008**, *49*, 1249.
11. Xu, D.-Z.; Liu, Y.; Shi, S.; Wang, Y. *Tetrahedron: Asymmetry* **2010**, *21*, 2530.
12. Sun, H.; Wang, G.; Yan, X.; Chen, L. *Res. Chem. Intermed.* **2012**, *38*, 1501.
13. Chua, P. J.; Tan, B.; Zeng, X.; Zhong, G. *Bioorg. Med. Chem. Lett.* **2009**, *19*, 3915.
14. Wang, C.; Yu, C.; Liu, C.; Peng, Y. *Tetrahedron Lett.* **2009**, *50*, 2363.
15. Wang, S.-W.; Chen, J.; Chen, G.-H.; Peng, Y.-G. *Synlett* ***2009***, 1457.
16. Yan, Z.-Y.; Niu, Y.-N.; Wei, H.-L.; Wu, L.-Y.; Zhao, Y.-B.; Liang, Y.-M. *Tetrahedron: Asymmetry* **2006**, *17*, 3288.
17. Xu, D.-Z.; Shi, S.; Wang, Y. *Eur. J. Org. Chem.* **2009**, 2009, 4848.
18. Díez, D.; Antón, A. B.; Peña, J.; García, P.; Garrido, N. M.; Marcos, I. S.; Sanz, F.; Basabe, P.; Urones, J. G. *Tetrahedron: Asymmetry* **2010**, *21*, 786.
19. Chuan, Y.; Chen, G.; Peng, Y. *Tetrahedron Lett.* **2009**, *50*, 3054.
20. Mahato, C. K.; Mukherjee, S.; Kundu, M.; Pramanik, A. *J. Org. Chem.* **2019**, *84*, 1053.
21. Lan, Y.; Yang, C.; Zhang, Y.; An, W.; Xue, H.; Ding, S.; Zhou, P.; Wang, W. *Polym. Chem.* **2019**, *10*, 3298.
22. Cruz-Hernández, C.; Landeros, J. M.; Juaristi, E. *New J. Chem.* **2019**, *43*, 5455.
23. Urruzuno, I.; Mugica, O.; Zanella, G.; Vera, S.; Gómez-Bengoa, E.; Oiarbide, M.; Palomo, C. *Chem. - Eur. J.* **2019**, *25*, 1.
24. Yu, C.; Qiu, J.; Zheng, F.; Zhong, W. *Tetrahedron Lett.* **2011**, *52*, 3298.
25. Chandrasekhar, S.; Kumar, T. P.; Haribabu, K.; Reddy, C. R.; Kumar, C. R. *Tetrahedron: Asymmetry* **2011**, *22*, 697.
26. Ban, S.; Xie, H.; Zhu, X.; Li, Q. *Eur. J. Org. Chem.* **2011**, 2011, 6413.
27. Anwar, S.; Lee, P.-H.; Chou, T.-Y.; Chang, C.; Chen, K. *Tetrahedron* **2011**, *67*, 1171.
28. Siyutkin, D. E.; Kucherenko, A. S.; Frolova, L. L.; Kuchin, A. V.; Zlotin, S. G. *Tetrahedron: Asymmetry* **2013**, *24*, 776.
29. Chen, Q.; Qiao, Y.; Ni, B. *Synlett* **2013**, *24*, 839.
30. Li, P.; Wang, L.; Wang, M.; Zhang, Y. *Eur. J. Org. Chem.* **2008**, 2008, 1157.
31. Zhao, Y.-B.; Zhang, L.-W.; Wu, L.-Y.; Zhong, X.; Li, R.; Ma, J.-T. *Tetrahedron: Asymmetry* **2008**, *19*, 1352.
32. Chen, J.-R.; Fu, L.; Zou, Y.-Q.; Chang, N.-J.; Rong, J.; Xiao, W.-J. *Org. Biomol. Chem.* **2011**, *9*, 5280.
33. Wang, J.; Lao, J.; Du, Q.; Nie, S.; Hu, Z.; Yan, M. *Chirality* **2012**, *24*, 232.
34. Cao, X.-Y.; Zheng, J.-C.; Li, Y.-X.; Shu, Z.-C.; Sun, X.-L.; Wang, B.-Q.; Tang, Y. *Tetrahedron* **2010**, *66*, 9703.
35. Ban, S.; Zhu, X.; Zhang, Z.; Xie, H.; Li, Q. *Eur. J. Org. Chem.* **2013**, *2013*, 2977.
36. Nakashima, K.; Hirashima, S.; Kawada, M.; Koseki, Y.; Tada, N.; Itoh, A.; Miura, T. *Tetrahedron Lett.* **2014**, *55*, 2703.
37. Cao, X.; Wang, G.; Zhang, R.; Wei, Y.; Wang, W.; Sun, H.; Chen, L. *Org. Biomol. Chem.* **2011**, *9*, 6487.
38. Roca-López, D.; Merino, P.; Sayago, F. J.; Cativiela, C.; Herrera, R. P. *Synlett* **2011**, *2011*, 249.
39. Saha, S.; Seth, S.; Moorthy, J. N. *Tetrahedron Lett.* **2010**, *51*, 5281.
40. Ni, B.; Zhang, Q.; Dhungana, K.; Headley, A. D. *Org. Lett.* **2009**, *11*, 1037.
41. Ban, S.; Du, D.-M.; Liu, H.; Yang, W. *Eur. J. Org. Chem.* **2010**, *2010*, 5160.
42. Chuan, Y.-M.; Yin, L.-Y.; Zhang, Y.-M.; Peng, Y.-G. *Eur. J. Org. Chem.* **2011**, *2011*, 578.
43. Liu, K.; Cui, H.-F.; Nie, J.; Dong, K.-Y.; Li, X.-J.; Ma, J.-A. *Org. Lett.* **2007**, *9*, 923.
44. Xu, D.-Q.; Wang, L.-P.; Luo, S.-P.; Wang, Y.-F.; Zhang, S.; Xu, Z.-Y. *Eur. J. Org. Chem.* **2008**, *2008*, 1049.
45. Xu, D.-Q.; Yue, H.-D.; Luo, S.-P.; Xia, A.-B.; Zhang, S.; Xu, Z.-Y. *Org. Biomol. Chem.* **2008**, *6*, 2054.
46. Lin, J.; Tian, H.; Jiang, Y.-J.; Huang, W.-B.; Zheng, L.-Y.; Zhang, S.-Q. *Tetrahedron: Asymmetry* **2011**, *22*, 1434.
47. Tan, B.; Zeng, X.; Lu, Y.; Chua, P. J.; Zhong, G. *Org. Lett.* **2009**, *11*, 1927.
48. Singh, K. N.; Singh, P.; Kaur, A.; Singh, P.; Sharma, S. K.; Khullar, S.; Mandal, S. K. *Synthesis* **2013**, *45*, 1406.
49. Syu, S.-e.; Kao, T.-T.; Lin, W. *Tetrahedron* **2010**, *66*, 891.
50. Lv, G.; Jin, R.; Mai, W.; Gao, L. *Tetrahedron: Asymmetry* **2008**, *19*, 2568.
51. Gauchot, V.; Gravel, J.; Schmitzer, A. R. *Eur. J. Org. Chem.* **2012**, *2012*, 6280.
52. McCooey, S. H.; Connon, S. J. *Org. Lett.* **2007**, *9*, 599.
53. Corbett, M. T.; Xu, Q.; Johnson, J. S. *Org. Lett.* **2014**, *16*, 2362.
54. Simpson, A. J.; Lam, H. W. *Org. Lett.* **2013**, *15*, 2586.

55. Hao, X.-Q.; Wang, C.; Liu, S.-L.; Wang, X.; Wang, L.; Gong, J.-F.; Song, M.-P. *Org. Chem. Front.* **2017**, *4*, 308.
56. Wang, C.; Li, N.; Zhu, W.-J.; Gong, J.-F.; Song, M.-P. *J. Org. Chem.* **2019**, *84*, 191.
57. Ma, H.; Xie, L.; Zhang, Z.; Wu, L.-g.; Fu, B.; Qin, Z. *J. Org. Chem.* **2017**, *82*, 7353.
58. Yu, X.; Bai, H.; Wang, D.; Qin, Z.; Li, J.-Q.; Fu, B. *RSC Adv.* **2018**, *8*, 19402.
59. Ansari, S.; Raabe, G.; Enders, D. *Monatsh. Chem.* **2013**, *144*, 641.
60. Jiang, X.; Zhang, B.; Zhang, Y.; Lin, L.; Yan, W.; Wang, R. *Chirality* **2010**, *22*, 625.
61. Saidalimu, I.; Fang, X.; Lv, W.; Yang, X.; He, X.; Zhang, J.; Wu, F. *Adv. Synth. Catal.* **2013**, *355*, 857.
62. Yang, D.; Wang, L.; Li, D.; Han, F.; Zhao, D.; Wang, R. *Chem. - Eur. J.* **2015**, *21*, 1458.
63. Yang, D.; Li, D.; Wang, L.; Zhao, D.; Wang, R. *J. Org. Chem.* **2015**, *80*, 4336.
64. Wang, J.; Wang, P.; Wang, L.; Li, D.; Wang, K.; Wang, Y.; Zhu, H.; Yang, D.; Wang, R. *Org. Lett.* **2017**, *19*, 4826.
65. Thota, G. K.; Sun, G.-J.; Deng, T.; Li, Y.; Kang, Q. *Adv. Synth. Catal.* **2018**, *360*, 1094.
66. Nath, U.; Banerjee, A.; Ghosh, B.; Pan, S. C. *Org. Biomol. Chem.* **2015**, *13*, 7076.
67. Olaizola, I.; Campano, T. E.; Iriarte, I.; Vera, S.; Mielgo, A.; García, J. M.; Odriozola, J. M.; Oiarbide, M.; Palomo, C. *Chem. - Eur. J.* **2018**, *24*, 3893.
68. Campano, T. E.; Iriarte, I.; Olaizola, O.; Etxabe, J.; Mielgo, A.; Ganboa, I.; Odriozola, J. M.; García, J. M.; Oiarbide, M.; Palomo, C. *Chem. - Eur. J.* **2019**, *25*, 4390.
69. Cui, L.; Wang, Y.; Fan, Z.; Li, Z.; Zhou, Z. *Adv. Synth. Catal.* **2019**, *361*, 3575.
70. Urruzuno, I.; Mugica, O.; Oiarbide, M.; Palomo, C. *Angew. Chem. Int. Ed.* **2017**, *56*, 2059.
71. Liao, Y.-H.; Zhang, H.; Wu, Z.-J.; Cun, L.-F.; Zhang, X.-M.; Yuan, W.-C. *Tetrahedron: Asymmetry* **2009**, *20*, 2397.
72. Kumpuga, B. T.; Itsuno, S. *Catal. Commun.* **2019**, *118*, 5.
73. Genc, H. N. *RSC Adv.* **2019**, *9*, 21063.
74. Ma, H.; Liu, K.; Zhang, F.-G.; Zhu, C.-L.; Nie, J.; Ma, J.-A. J. *Org. Chem.* **2010**, *75*, 1402.
75. Guo, X.-T.; Sha, F.; Wu, X.-Y. *Synthesis* 2017, *49*, 647.
76. He, T.; Qian, J.-Y.; Song, H.-L.; Wu, X.-Y. *Synlett* **2009**, *2009*, 3195.
77. Nakamura, A.; Lectard, S.; Hashizume, D.; Hamashima, Y.; Sodeoka, M. *J. Am. Chem. Soc.* **2010**, *132*, 4036.
78. Nakamura, A.; Lectard, S.; Shimizu, R.; Hamashima, Y.; Sodeoka, M. *Tetrahedron: Asymmetry* **2010**, *21*, 1682.
79. Raimondi, W.; Baslé, O.; Constantieux, T.; Bonne, D.; Rodriguez, J. *Adv. Synth. Catal.* **2012**, *354*, 563.
80. Baslé, O.; Raimondi, W.; Duque, M. M. S.; Bonne, D.; Constantieux, T.; Rodriguez, J. *Org. Lett.* **2010**, *12*, 5246.
81. Xu, Y.; Matsunaga, S.; Shibasaki, M. *Org. Lett.* **2010**, *12*, 3246.
82. Zhang, M.-L.; Chen, L.; You, Y.; Wang, Z.-H.; Yue, D.-F.; Zhang, X.-M.; Xu, X.-Y.; Yuan, W.-C. *Tetrahedron* **2016**, *72*, 2677.
83. Han, X.; Wu, X.; Min, C.; Zhou, H.-B.; Dong, C. *RSC Adv.* **2012**, *2*, 7501.
84. Cui, H.-F.; Li, P.; Wang, X.-W.; Zhu, S.-Z.; Zhao, G. *J. Fluor. Chem.* **2012**, *133*, 120.
85. García Mancheño, O.; Tangen, P.; Rohlmann, R.; Fröhlich, R.; Alemán, J. *Chem. - Eur. J.* **2011**, *17*, 984.

4 Catalytic Asymmetric Michael Addition of Miscellaneous Carbonyl Compounds to Nitroalkenes

4.1 INTRODUCTION

Catalytic asymmetric Michael addition of 1,3-dicarbonyls, aldehydes and ketones to nitroalkenes has been discussed in the previous chapters. However, similar addition of other carbon-centered nucleophiles derived from esters and amides which are traditionally weak nucleophiles also received considerable attention. In particular, the addition of enolates derived from iminoesters, cyclic esters (lactones) and cyclic amides (lactams) to nitroalkenes are discussed in detail in this chapter. While lactones include furanones, oxazolones and isoxazolones, lactams include oxindoles, pyrazolones and pyrrolidones. Presumably because of the poor reactivity of these enolates, many ligand-metal complexes based on chiral amino/imino alcohols and diamines have been employed to achieve high yields and stereoselectivities.

4.2 ADDITION OF ESTERS/LACTONES: IMINOESTERS, FURANONES, OXAZOLONES AND ISOXAZOLONES

Hou et al. described a catalytic asymmetric Michael addition of glycine derivatives **2a** to nitroalkenes **1** using Cu/1,2-*P*, *N*-ferrocene ligand **L1** as a catalyst to afford corresponding adducts **3** in high enantio and diastereoselectivities (Scheme 4.1).[1] The electronic properties of the substituents on the nitroalkenes **1** exerted a limited impact on the diastereo and enantioselectivity. Moreover, α,γ-diaminobutanoate **4** was synthesized in good yield from the adduct **3** in three steps without losing optical activity.

Alemán and co-workers demonstrated the Michael addition of monoactivated glycine ketimine ylides **2b** to nitroalkenes using Takemoto's catalyst **C1**. This is a successful route to various α,γ-diamino acid derivatives with excellent stereoselectivity (Scheme 4.2).[2] Intramolecular hydrogen bonding by the *ortho*-hydroxyl group enhances the acidity of methylene hydrogen atoms and activates the ylide substrate for the conjugate addition to nitroalkenes **1**. Reaction proceeds through two steps: ylide formation through proton transfer from ketimine **2b** to catalyst **C1** (**TS-1**, Scheme 4.2), which is the rate-determining step, and C–C bond formation (**TS-2**, Scheme 4.2).

The Fukuzawa group reported AgOAc/ThioClickFerrophos **L2** complex catalyzed conjugate addition of glycine iminoesters **2c** to nitroalkenes **1** in the presence of triethylamine to afford mainly *anti*-α-imino-γ-nitrobutyrates **3c** in high yields and excellent enantioselectivities (Scheme 4.3).[3] The pyrrolidine cycloadduct **5** was also obtained as the major product in good yields with high enantioselectivities when *tert*-butyl imino ester **2c** was employed in the absence of triethylamine. In the presence of trimethylamine, iminoester **2c** yielded the silver-bound enolate **6a**. This can undergo Michael addition to afford **6b**, which could undergo either protonation to produce the conjugate adduct **3c** or cyclization to produce the proline ester **5** (Scheme 4.4).

The Lam group demonstrated that azaarylacetates and acetamides **2d** participate in highly enantioselective Michael addition to nitroalkenes **1** in the presence of a chiral nickel(II)-bis(diamine)

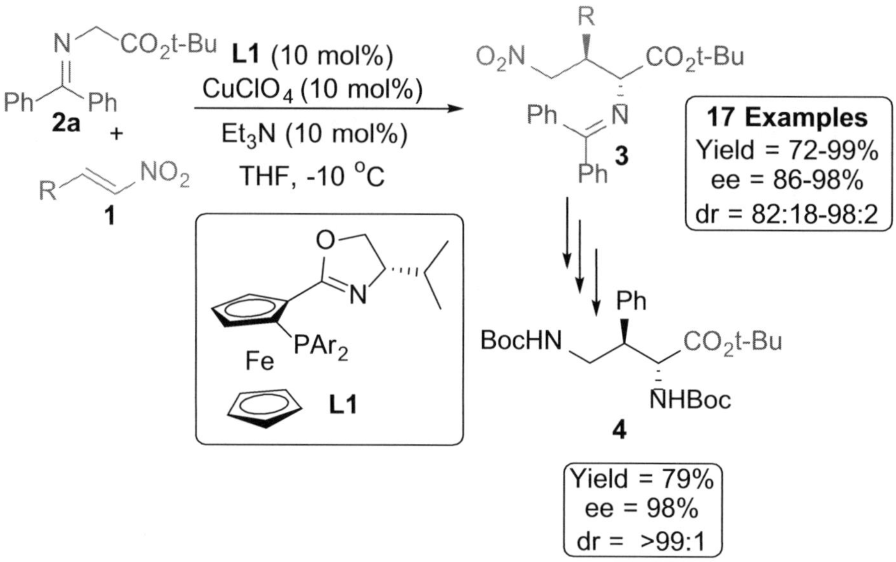

SCHEME 4.1 Asymmetric Michael addition of glycine derivatives to nitroalkenes catalyzed by Cu-ferrocene catalyst system.

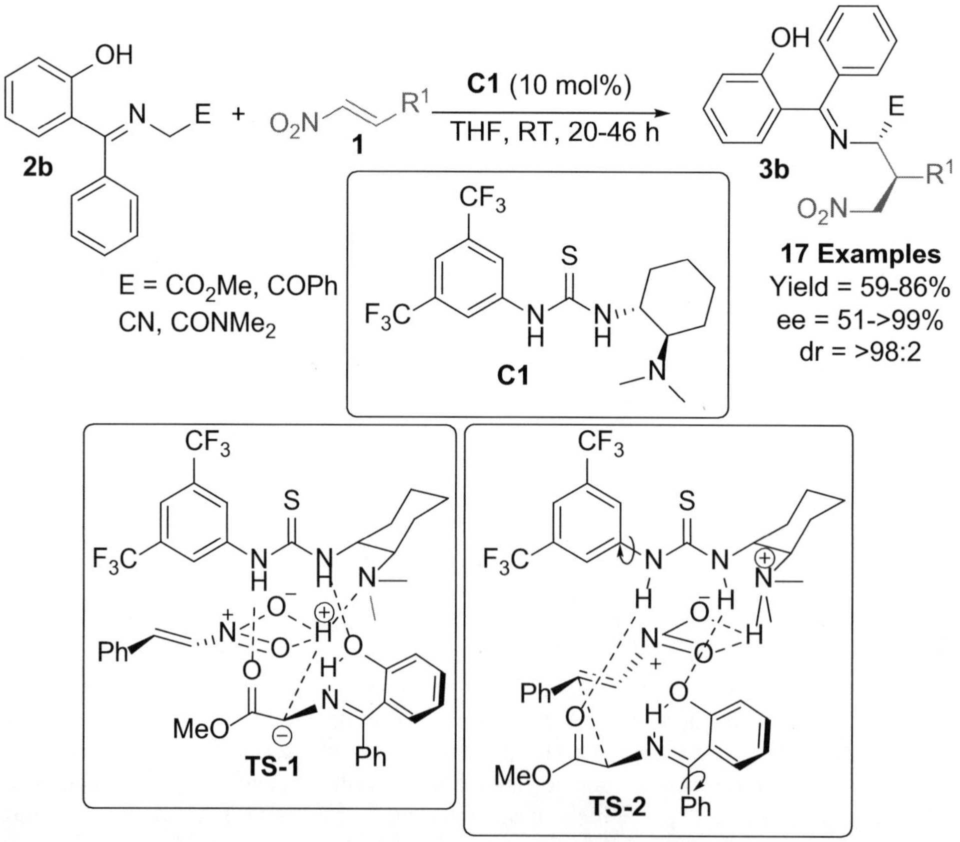

SCHEME 4.2 Michael addition of monoactivated glycine ketimine ylides to nitroalkenes in the presence of Takemoto's catalyst.

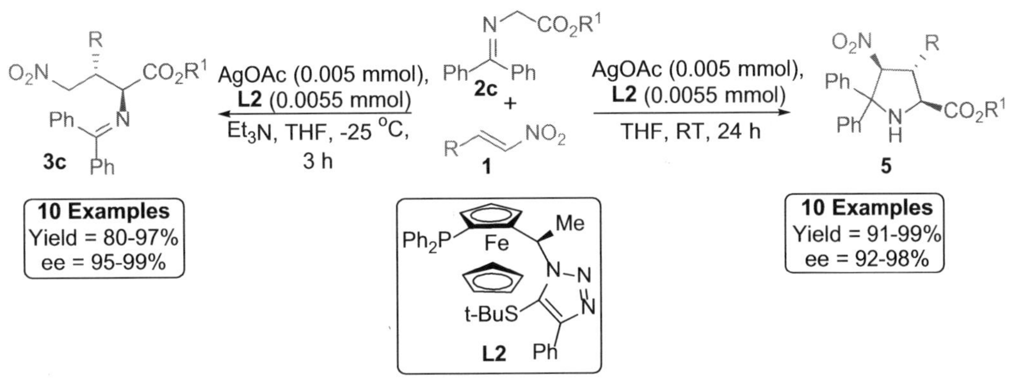

SCHEME 4.3 Conjugate addition of iminomethyl esters to nitroalkenes in the presence of AgOAc/ThioClickFerrophos catalyst system.

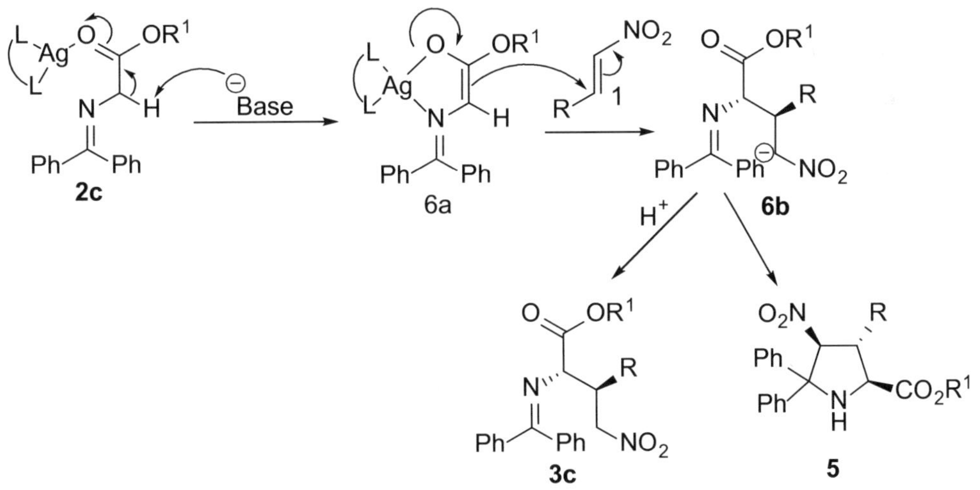

SCHEME 4.4 Proposed mechanism for the formation of α-imino-γ-nitrobutyrates and pyrrolidine.

complex **C2** (Scheme 4.5).[4] A wide array of azaarenes **2d** that includes pyridines, pyrazines, triazines, isoquinolines, quinazolines, benzothiazoles and benzisoxazoles were employed to furnish enantioenriched chiral azaarene containing building blocks **3d** in moderate-to-high yields and enantioselectivities.

The Fukuzawa group also demonstrated a silver-catalyzed asymmetric Michael addition of oxazolines and thioxazolines **2e** with nitroalkenes **1** in the presence of ThioClickFerrophos ligand **L3** (Scheme 4.6).[5] The conjugate addition yields predominantly *anti*-adduct **3e** with high enantiomeric excess. This reaction provides a stereodivergent synthetic route to nonproteinogenic α-quaternary amino acids.

Ma et al. developed a new one-pot sequential conjugate addition/dearomative fluorination of isoxazol-5(4H)-ones **2f** with nitroalkenes **1** and N-fluorobenzenesulfonimide (NFSI) (Scheme 4.7).[6] This reaction was promoted by saccharide-based bifunctional organocatalyst **C3** that contains tertiary amine and thiourea moieties which cooperatively and simultaneously activate the substrates. The aromatic isoxazol-5-ol intermediates **6c** could be fluorinated with concurrent dearomatization to afford the fluorinated chiral isoxazol-5(4H)-ones **3f** in excellent yields and diastereoselectivities with good-to-excellent enantioselectivities.

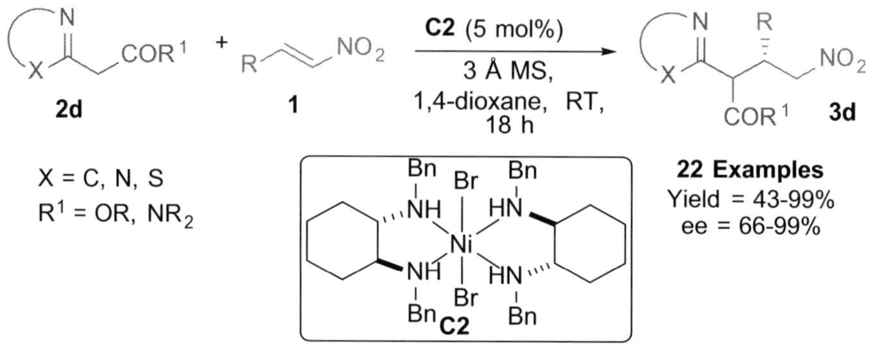

SCHEME 4.5 Michael addition of azaarylacetates and acetamides to nitroalkenes in the presence of chiral nickel(II)-bis(diamine) complex.

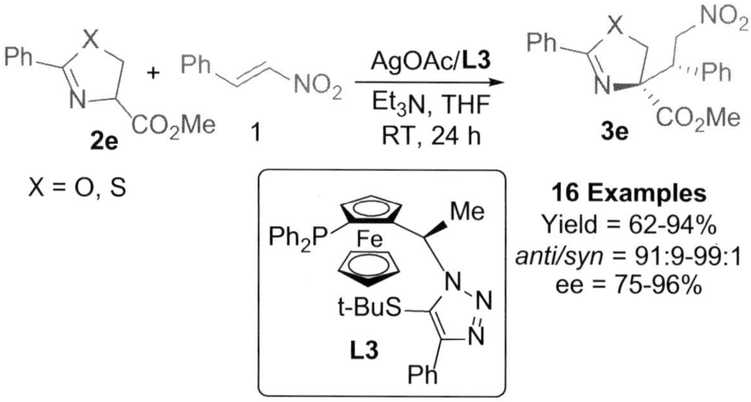

SCHEME 4.6 Asymmetric Michael addition of oxazolines and thioxazolines with nitroalkenes in the presence of silver-ThioClickFerrophos catalyst system.

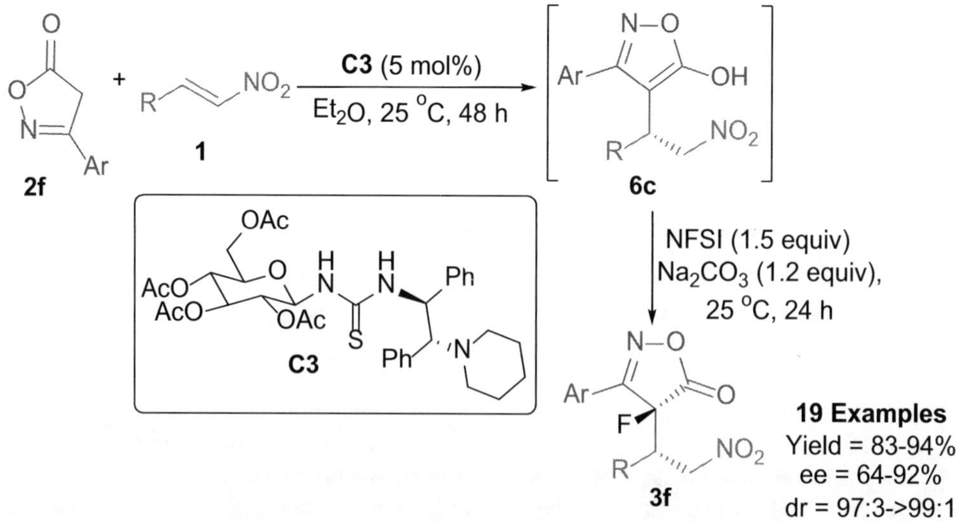

SCHEME 4.7 One-pot sequential conjugate addition/dearomative fluorination of isoxazol-5(4H)-ones on reaction with nitroalkenes and NFSI.

The Terada group employed α-*tert*-butylthio substituted furanones **2g** as vinylogous pronucleophiles in the direct vinylogous Michael addition to nitroalkenes **1** employing an axially chiral guanidine-based catalyst **C4** (Scheme 4.8).[7] The method provides access to functionalized γ-butenolides **3g** as *syn*-diastereomers exclusively with fairly good enantioselectivities.

The addition of oxazolones **2h** to nitroalkenes **1** catalyzed by thiourea cinchona derivatives **C5** involves an asymmetric noncovalent organocatalysis and yielded optically active adducts **3h** in good yields with excellent diastereoselectivities, as well as moderate-to-good enantioselectivities of up to 92% ee (Scheme 4.9).[8]

Wang et al. demonstrated the addition of thiazolones **2i** to nitroalkenes **1** catalyzed by the tertiary amine-thiourea catalyst **C6** with catalyst loadings as low as 1 mol% to obtain optically active 2,4-disubstituted thiazolones **3i** with excellent enantio and diastereoselectivities (up to 96% ee and 10:1 dr, respectively, Scheme 4.10).[9]

Sekikawa et al. reported a highly *anti*-selective nitro-Michael reaction of furanones **2j** by a catalyst-controlled switching of diastereoselectivity from the normal *syn*-selectivity.[10] The reaction proceeded smoothly with 0.1–5 mol% loadings of an *epi*-quinine catalyst **C7** to

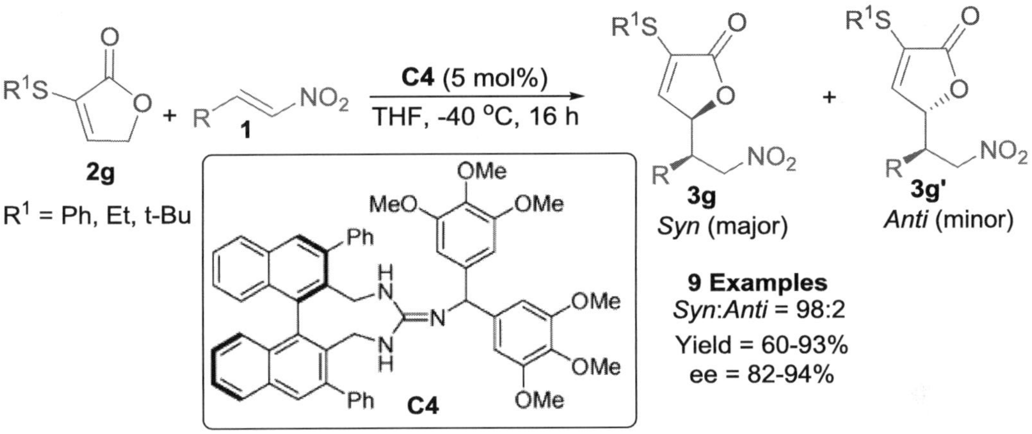

SCHEME 4.8 Vinylogous Michael addition of functionalized furanones to nitroalkenes in the presence of axially chiral guanidine catalyst.

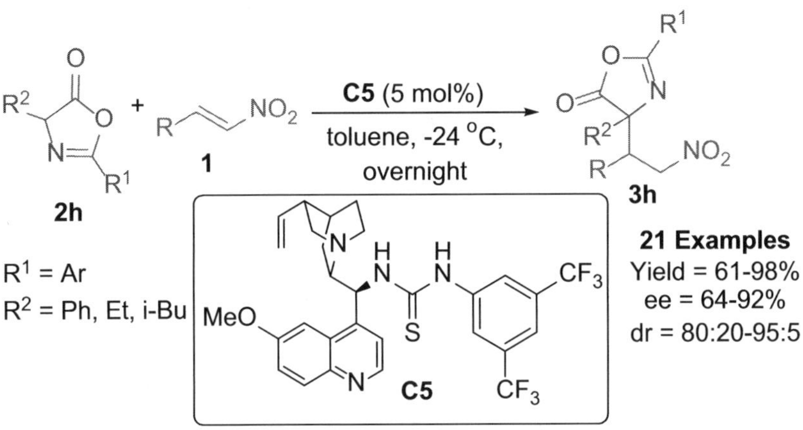

SCHEME 4.9 Asymmetric 1,4-addition of oxazolones to nitroalkenes catalyzed by bifunctional cinchona alkaloid thiourea.

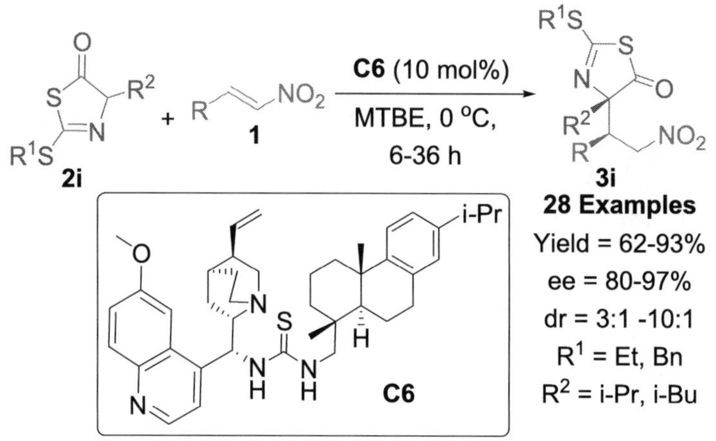

SCHEME 4.10 Michael addition of thiazolones and nitroalkenes in the presence of tertiary amine-thiourea catalyst.

afford the *anti*-Michael adducts **3j** in high yields with excellent enantio and diastereoselectivities (Scheme 4.11). *Anti*-diastereoselectivity of the reaction was further supported by density functional theory (DFT) calculations. DFT calculations suggest the formation of nitroammonium intermediate (**TS-5**, Scheme 4.11) by the conjugate addition of quinuclidine at the *Re*-face of nitroalkene (**TS-3**, Scheme 4.11). The configuration of products **3j** was determined to be (5*R*, 1′*S*), which confirmed that the nucleophilic attack of dienolates derived from **2j** to intermediate **TS-5** occurred from the *Si*-face.

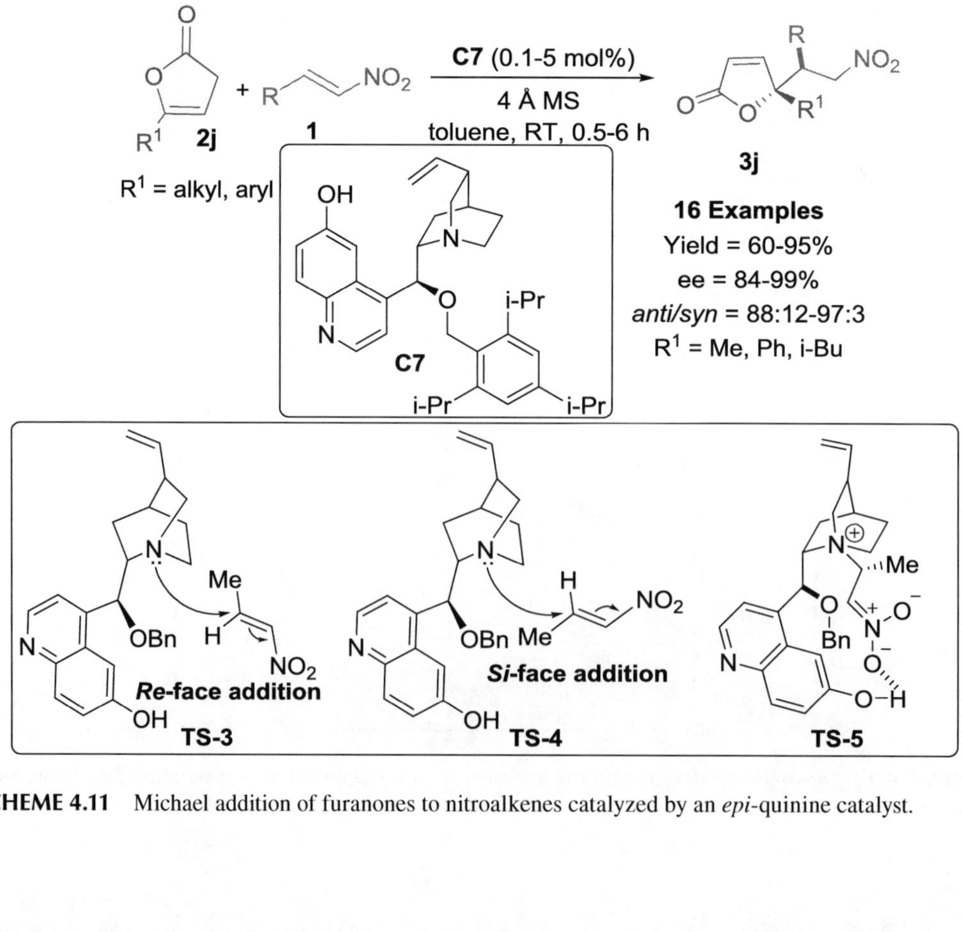

SCHEME 4.11 Michael addition of furanones to nitroalkenes catalyzed by an *epi*-quinine catalyst.

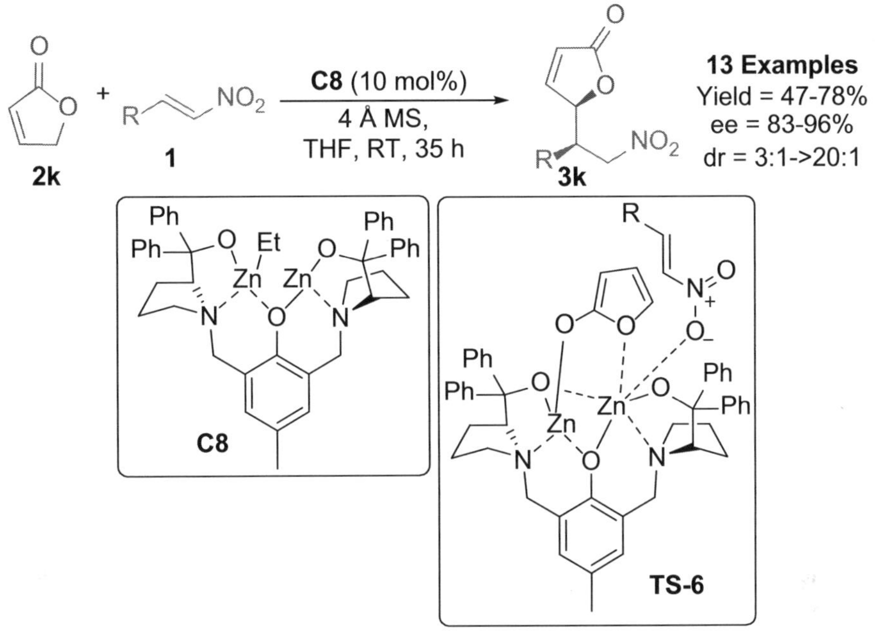

SCHEME 4.12 Asymmetric Michael addition of furanone to nitroalkenes in the presence of a dinuclear zinc complex catalyst.

The Trost group described the synthesis of γ-substituted butenolides **3k** stereoselectively by the direct asymmetric Michael addition of 2(5H)-furanone **2k** to nitroalkenes **1** in the presence of dinuclear zinc catalyst **C8** and molecular sieves, which promote vinylogous nucleophilicity (Scheme 4.12).[11]

Ban and co-workers reported asymmetric double Michael addition of 2-(*3H*)-furanones **2j** to nitroolefins **1** catalyzed by chiral thiosquaramide catalyst **C9**, which easily constructed the enantioenriched 2,4,4-trisubstituted butenolides **3j** with high enantioselectivities (up to 95%) and remarkable diastereoselectivities (up to >99:1) (Scheme 4.13).[12] It was proposed that the

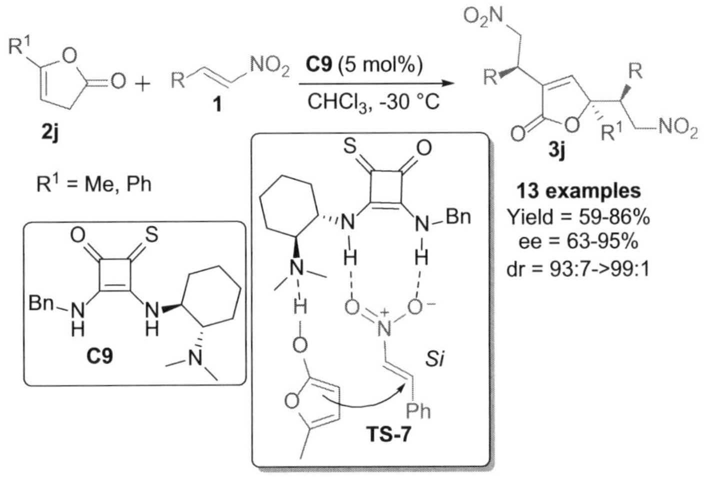

SCHEME 4.13 Asymmetric double Michael addition of 2-(3*H*)-furanones to nitroalkenes mediated by a thiosquaramide catalyst.

chiral bifunctional organocatalyst **C9** catalyzed the two Michael addition steps sequentially, where the Michael acceptors were activated by the NH group of the thiosquaramide by hydrogen bonding, whereas the tertiary amine functionality of the catalyst deprotonated the acidic protons of furanones to generate the dienolates as nucleophiles for the Michael addition (**TS-7**, Scheme 4.13).

4.3 ADDITION OF AMIDES/LACTAMS: OXINDOLES, PYRAZOLONES AND PYRROLIDONES

Zhou et al. employed cinchonidine-derived bifunctional phosphoramide catalyst **C10** for the addition of unprotected 3-prochiral oxindoles **7a** to nitroalkenes **1**.[13] Both unprotected 3-aryl and 3-alkyloxindoles **7a** reacted well with various β-substituted nitroalkenes **1** to deliver the product **8a** bearing C_3 quaternary chiral carbon center with an adjacent tertiary chiral center in high diastereo and enantioselectivities (Scheme 4.14). Unprotected 3-alkyloxindoles were much less reactive, presumably because of their substantially higher pKa value when compared to that of oxindole.

Du et al. described an efficient squaramide **C11a** catalyzed the enantioselective Michael addition of N-Boc-protected 3-substituted oxindoles **7b** to nitroalkenes **1**.[14] The reaction involves low catalyst loading (2 mol%) to afford the adducts **8b** in high yields with good enantio and diastereoselectivities (**TS-8**, Scheme 4.15). Recently, similar addition of 3-oxindoles **7b** to nitrostyrenes **1** has been reported by Pedrosa and co-workers, which was mediated by novel L-*tert*-Leucine-derived thiosquaramide catalyst **C11b** (Scheme 4.15).[15] The catalyst performed better in comparison with its thiourea and squaramide analogs, providing the corresponding products **3** with high yields and excellent stereoselectivities.

Enders et al. developed a secondary amine-catalyzed asymmetric Michael addition of N-Boc-protected oxindoles **7b** to nitroalkenes **1** (Scheme 4.16).[16] The reaction was promoted by didodecyl prolinol TMS-ether **C12** via Brønsted base activation to furnish the corresponding products **8b** in excellent yields (88%–98%), diastereoselectivities (98:2 to 99:1), and enantioselectivities. In the proposed mechanism, the pyrrolidine catalyst **C12** activates both the enol form of the oxindole and the nitro group of the Michael acceptor through hydrogen bonding. The nitroalkene **1** prefers to approach the oxindole **7b** from its *Si*-face due to shielding of the *Re*-face of the prochiral carbon of the oxindole by the bulky side chain of the catalyst (**TS-9**, Scheme 4.16).

Shibasaki et al. developed a homodinuclear $Mn_2(OAc)_2$-Schiff base complex **C13** as a bimetallic catalyst with a loading of 1–5 mol% for the 1,4-addition of 3-substituted oxindoles **7b** to various

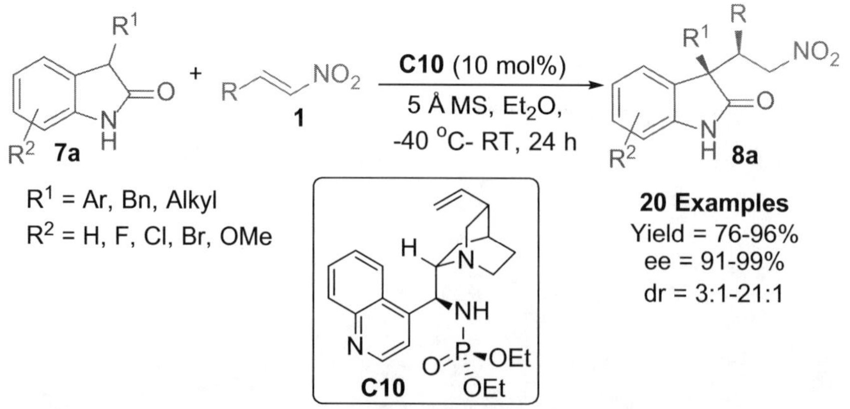

SCHEME 4.14 Conjugate addition of unprotected 3-prochiral oxindoles to nitroalkenes in the presence of cinchonidine-derived bifunctional phosphoramide catalyst.

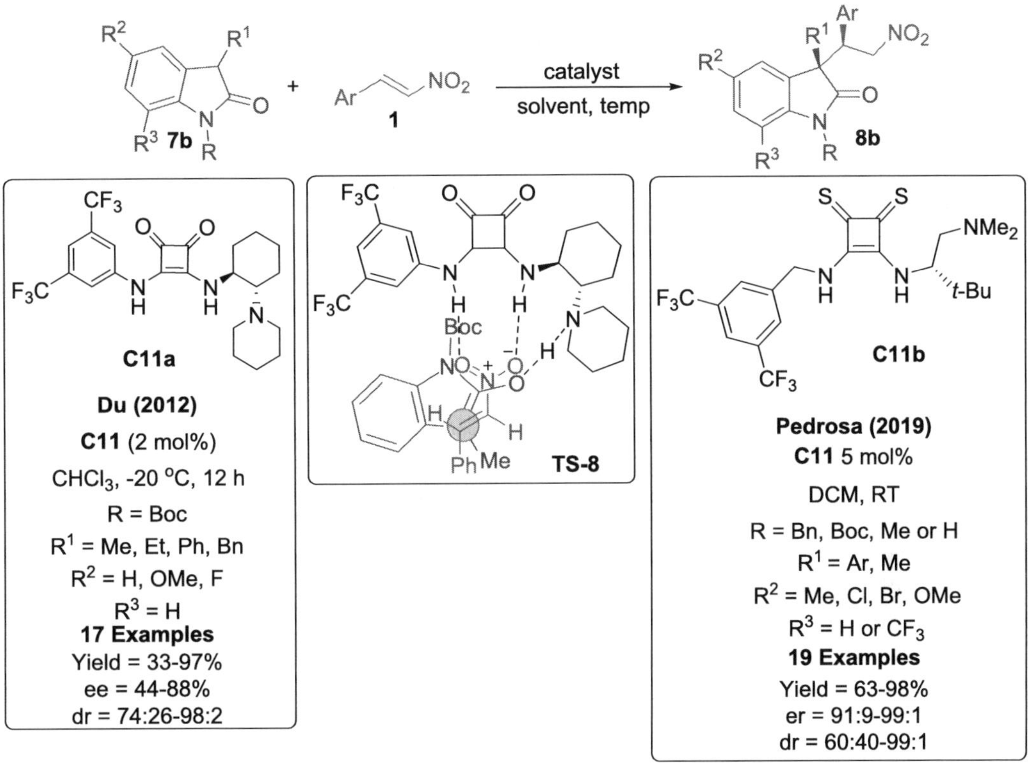

SCHEME 4.15 Squaramide and thiosquaramide-catalyzed enantioselective Michael addition of 3-substituted oxindoles to nitroalkenes.

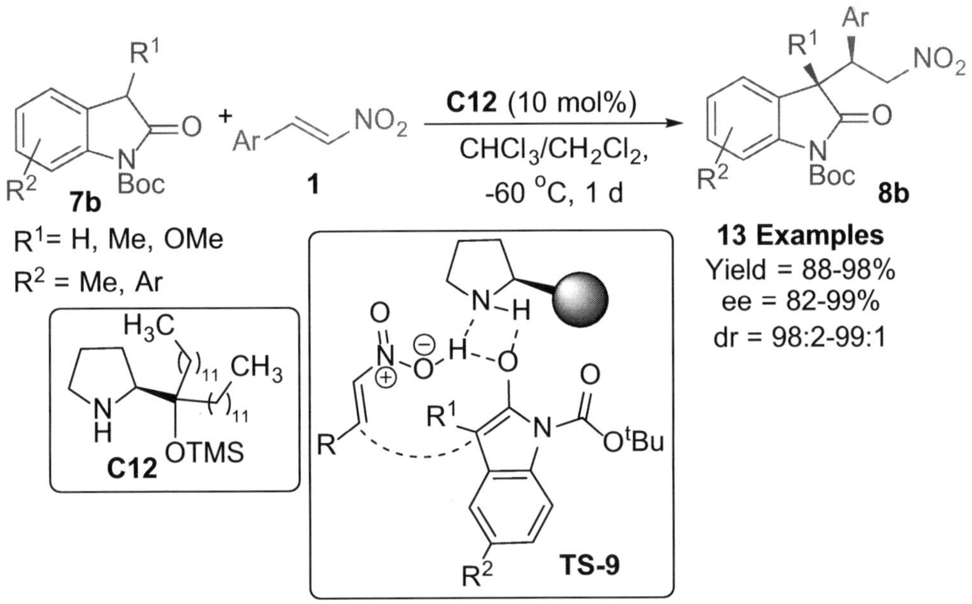

SCHEME 4.16 Secondary amine-catalyzed asymmetric Michael addition of N-Boc-protected oxindoles to nitroalkenes.

nitroalkenes **1.** The reaction afforded the corresponding conjugate addition products **8b** in high yields with excellent enantio and diastereoselectivities (Scheme 4.17).[17] The importance of the outer Mn center for the present reaction was evident from the modest reactivity and/or selectivity observed when monometallic Mn-salen complexes were employed. The heterobimetallic Cu/Mn and Pd/Mn complexes **C13** resulted in poor diastereo and enantioselectivities despite showing good reactivity. This suggested the requirement of two Mn metal centers for the high reactivity and stereoselectivity.

Organocatalytic enantioselective vinylogous Michael-type reaction between 3-alkylidene oxindoles **7c** and trifluoromethylated nitroalkenes **1a** was developed by Wang et al. (Scheme 4.18).[18] The reaction proceeded with very high site selectivity and excellent enantioselectivities in the presence of an organocatalyst **C5** for the formation of trifluoromethylated chiral oxindoles **8c** with all-carbon quaternary stereocenters.

The Kanger group reported the synthesis of 3,3-disubstituted 3-chlorooxindoles **8d** with high diastereoselectivity and enantioselectivity by a conjugate addition of 3-chlorooxindoles **7d** to nitroalkenes **1** under organocatalytic conditions (Scheme 4.19).[19] In the plausible mechanism, nitroalkene **1** was activated by hydrogen bonding to catalyst **C14**, and the tertiary amine of the catalyst **C14** facilitates deprotonation/enolization (**TS-10**, Scheme 4.19). Further, the adduct **8d** was formed by the *Re*-face attack of the enolate at the *Re*-face of the nitroalkene, and the product contains adjacent quaternary and tertiary chiral centers. The presence of chlorine at the quaternary center of oxindole is advantageous in nucleophilic substitution as chlorine can function as a leaving group. Oxindoles also serve as Michael acceptors in the synthesis of spiro-bisoxindoles with excellent diastereo and enantioselectivities.

The Wolf group demonstrated an organocatalytic synthesis of α-oxetanyl and α-azetidinyl alkyl halides **8e** with a tetrasubstituted chiral carbon center from N-Boc-3-fluorooxindole **7e** and strained nitroalkenes **1b** attached to oxetane or azetidine ring (Scheme 4.20).[20] Similar asymmetric induction was achieved by squaramide (**C11** and **C15**) and thiourea (**C16**)-based catalyst system with excellent yield and enantioselectivity. In the proposed mechanism, the reaction proceeds through transition state **TS-11** (Scheme 4.20). The authors have also explored the synthetic applicability of the formed Michael adduct for the preparation of medicinally relevant scaffolds, such as spirocyclopropyl oxindoles, fluorinated pyrrolidines and β-fluoro-β-prolines at high yields and diastereoselectivity.

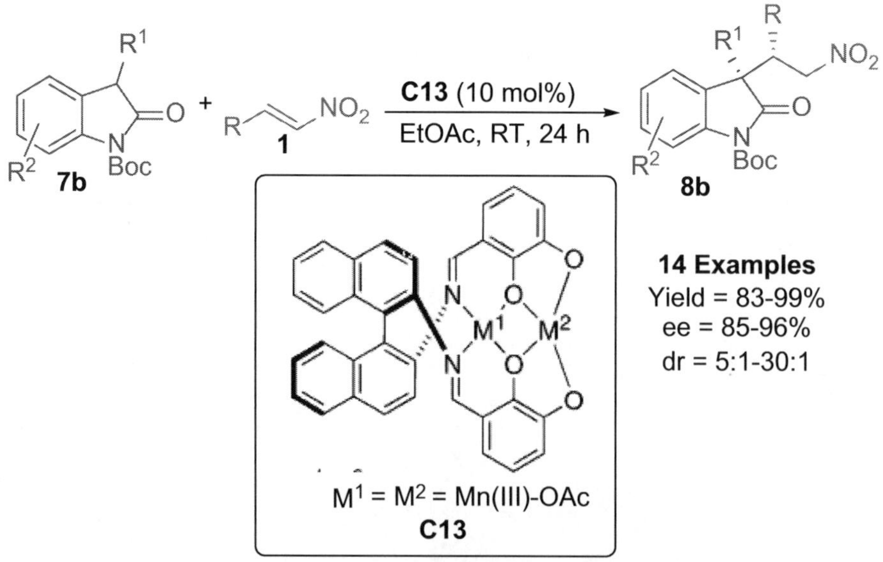

SCHEME 4.17 $Mn_2(OAc)_2$-Schiff base complex as a bimetallic catalyst in the 1,4-addition of 3-substituted oxindoles to nitroalkenes.

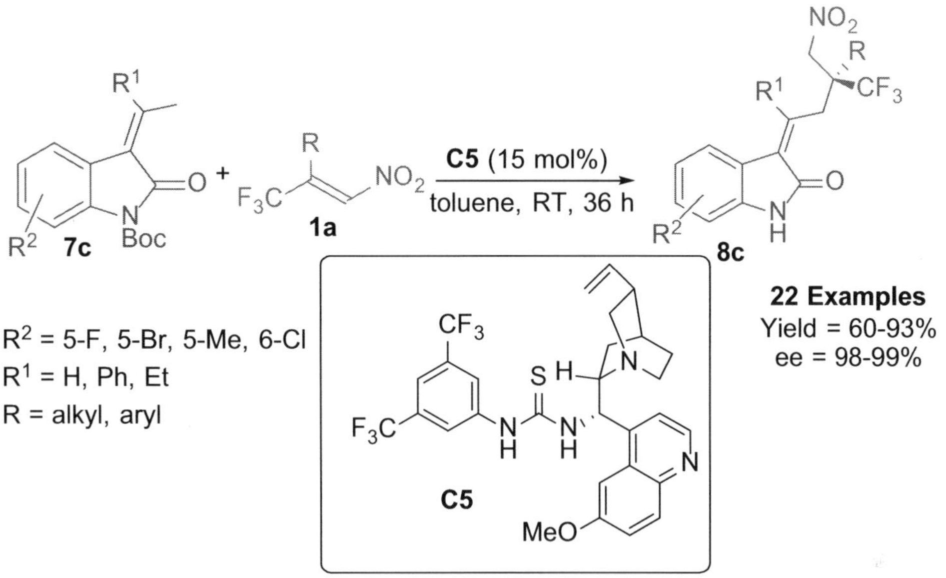

SCHEME 4.18 Enantioselective vinylogous Michael-type reaction between 3-alkylidene oxindoles and trifluoromethylated nitroalkenes in the presence of a quinine-derived thiourea catalyst.

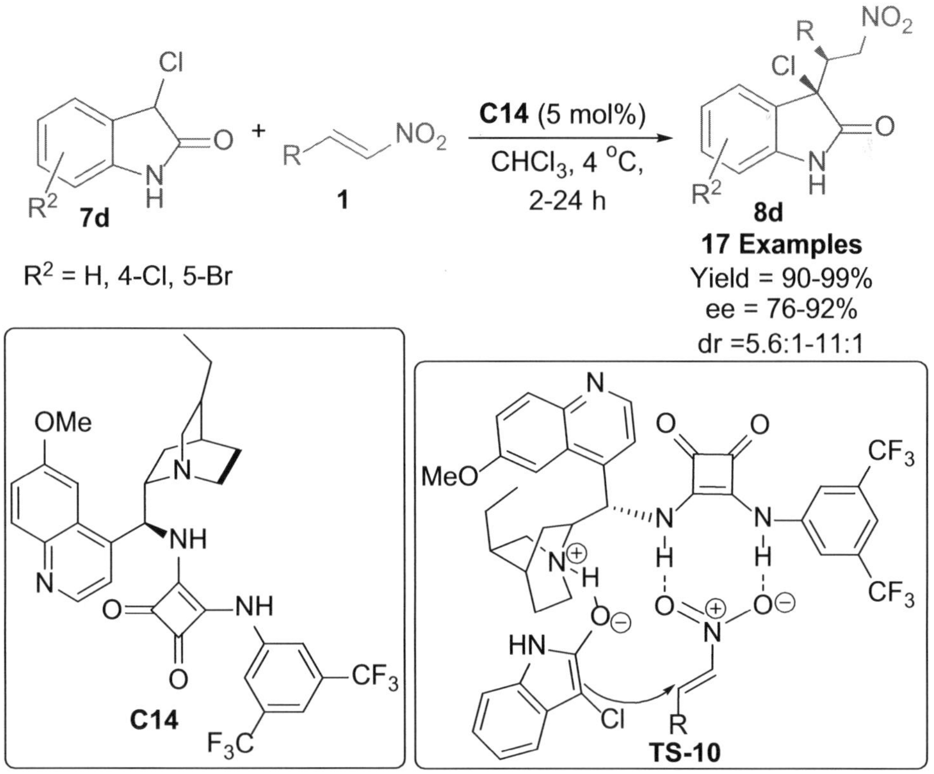

SCHEME 4.19 Synthesis of 3,3-disubstituted 3-chlorooxindoles by the conjugated addition of 3-chlorooxindoles to nitroalkenes.

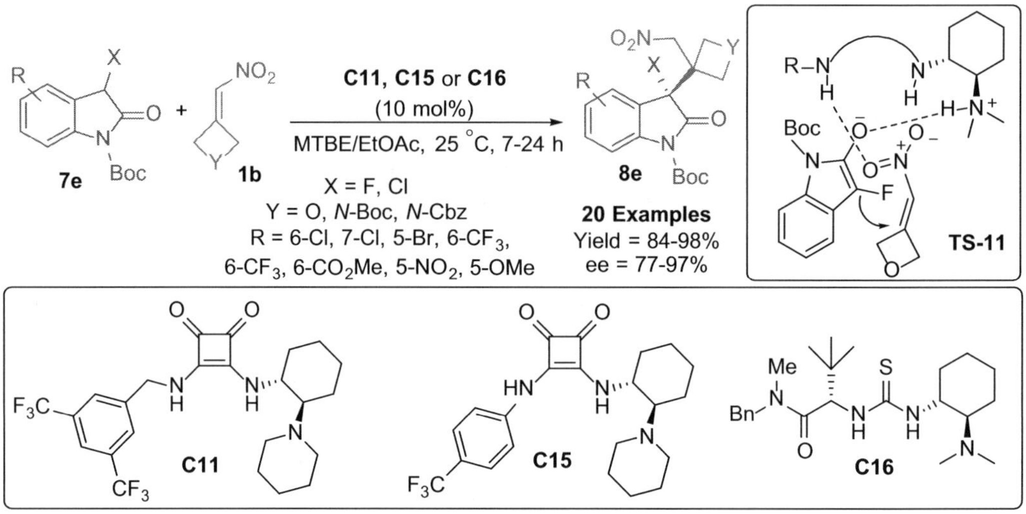

SCHEME 4.20 Synthesis of α-oxetanyl and α-azetidinyl alkyl halides by the conjugate addition of N-Boc-3-halo-oxindole to strained nitroalkenes.

Melchiorre et al. reported an asymmetric Michael addition of dioxindole **7f** to various nitroalkenes **1** employing thiourea-based organocatalyst **C17** to afford corresponding Michael adducts **8f** in relatively good yields with excellent enantioselectivity (Scheme 4.21).[21] This reaction involves an unusual mechanistic scenario wherein the primary amino moiety is not operating as a Brønsted base. In the first step of the speculated mechanism, the primary amine-thiourea catalyst **C17** stabilizes the enol form of the dioxindole **7f** through hydrogen bonding rather than promoting the formation of an enolate intermediate. This is followed by the asymmetric addition of dioxindoles **7f** to nitroalkenes **1**.

Yuan et al. described the synthesis of optically active quaternary 3-aminooxindoles **8g** from 3-monosubstituted 3-aminooxindoles **7g** and nitroalkenes **1** by employing bifunctional thiourea-tertiary amine **C18** catalyzed Michael addition in high yields with excellent enantio and diastereoselectivities (Scheme 4.22).[22] In the proposed mechanism, nitroalkene **1** was activated by hydrogen bonding to the thiourea moiety of the catalyst **C18**. Meanwhile, the tertiary amine moiety of the catalyst **C18** facilitated the deprotonation/enolization of 3-monosubstituted 3-aminooxindole **7g**. The *Si*-face attack of enolate to the *Si*-face of nitroalkene resulted in the Michael adduct **8g** with an (*R*,*R*)-configuration (**TS-12**, Scheme 4.22).

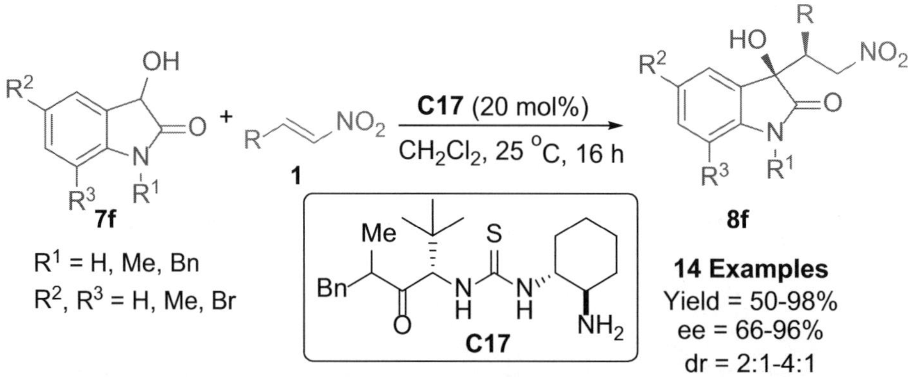

SCHEME 4.21 Asymmetric Michael addition of dioxindole to nitroalkenes in the presence of a thiourea-based organocatalyst.

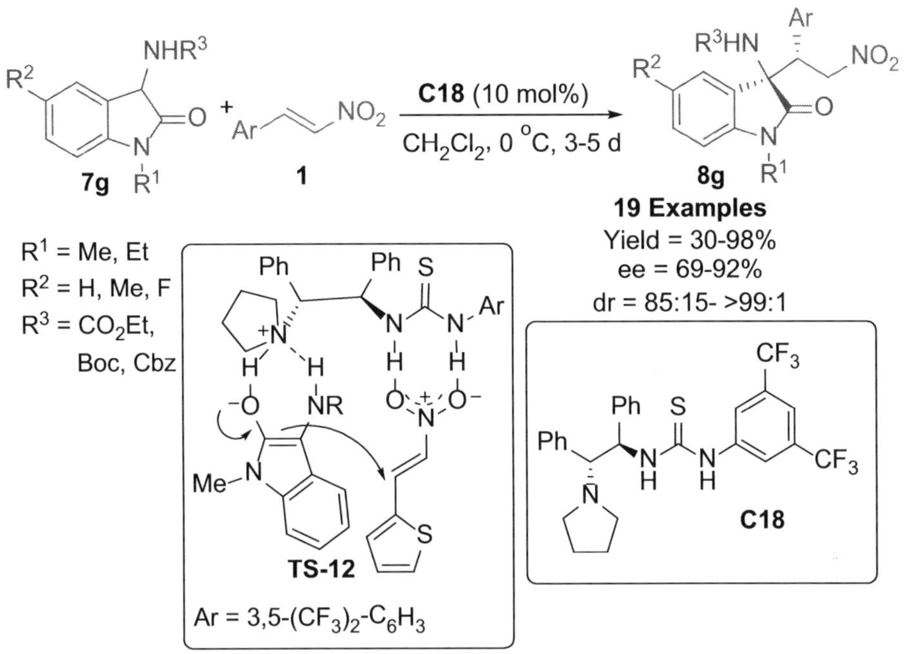

SCHEME 4.22 Asymmetric synthesis of 3-substituted 3-aminooxindoles by the conjugated addition of 3-monosubstituted 3-aminooxindoles to nitroalkenes in the presence of bifunctional thiourea-tertiary amine.

Lu et al. reported the Michael addition of 3-sulfenyloxindoles **7h** to nitroalkenes **1** employing threonine-incorporated multifunctional catalyst **C19**, and the reaction proceeded in a highly stereoselective manner, affording oxindoles **8h** with a 3-sulfenyl-substituted quaternary chiral center in excellent yields and selectivities (Scheme 4.23).[23]

Chen et al. reported an enantioselective *aza*-vinylogous-type reaction of nitrones **7i** derived from isatin to nitroalkenes **1**. The reaction afforded chiral nitrone derivatives **8i** in excellent yields and

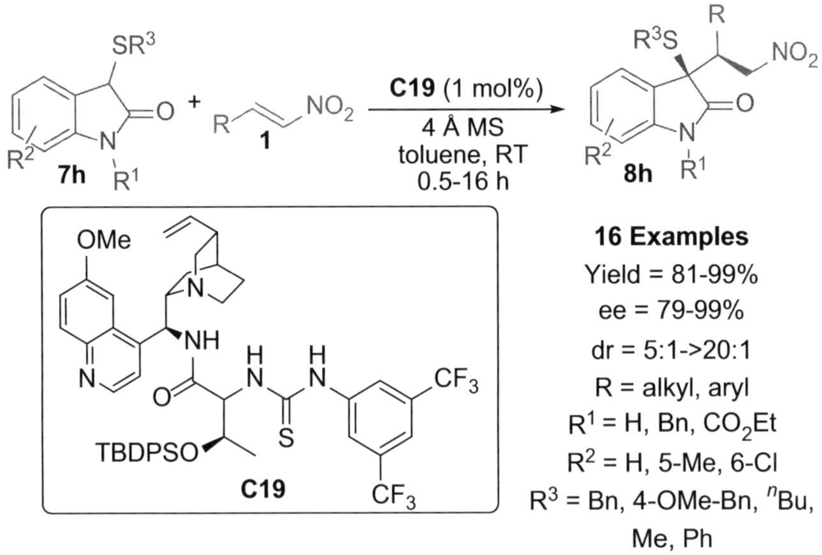

SCHEME 4.23 Michael addition of 3-sulfenyloxindoles to nitroalkenes in the presence of a threonine-incorporated multifunctional catalyst.

stereoselectivities (Scheme 4.24).[24] Various nitroalkenes yielded the products with high efficiency in the presence of thiourea catalyst **C5**.

The reactivity of pyrazol-5-one **9a** has been extensively investigated due to its acidic nature and the formation of an aromatic system upon deprotonation. It appeared suitable for enantioselective nitronate protonation. Phelan and Ellman developed a catalytic enantioselective addition of pyrazolones **9a** to trisubstituted nitroalkenes **1c** in the presence of a bifunctional N-sulfinylurea **C20** as an organocatalyst to afford the corresponding Michael adducts **10a** with high yields and enantioselectivities (Scheme 4.25).[25]

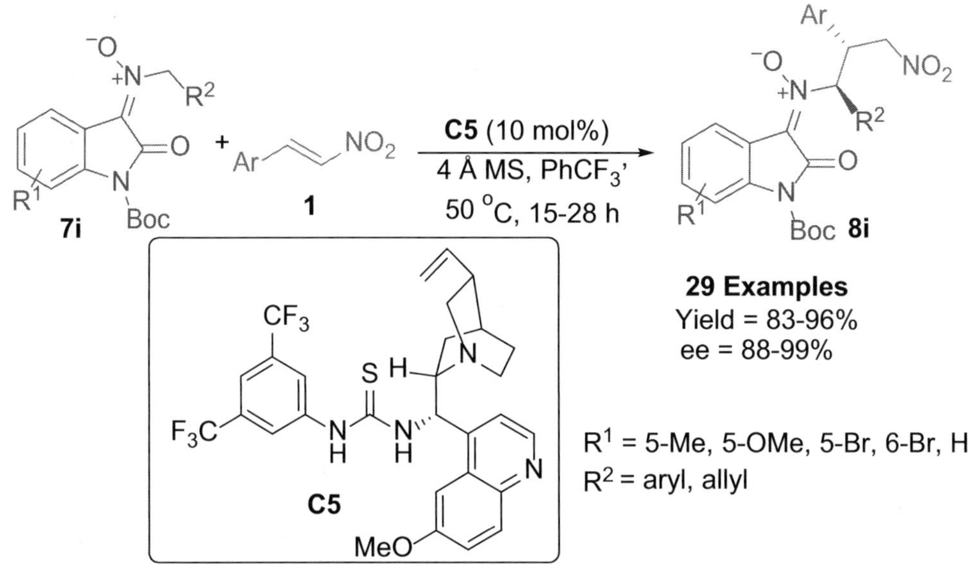

SCHEME 4.24 Enantioselective *aza*-vinylogous-type reaction of nitrones derived from isatin to nitroalkenes in the presence of a thiourea-based catalyst.

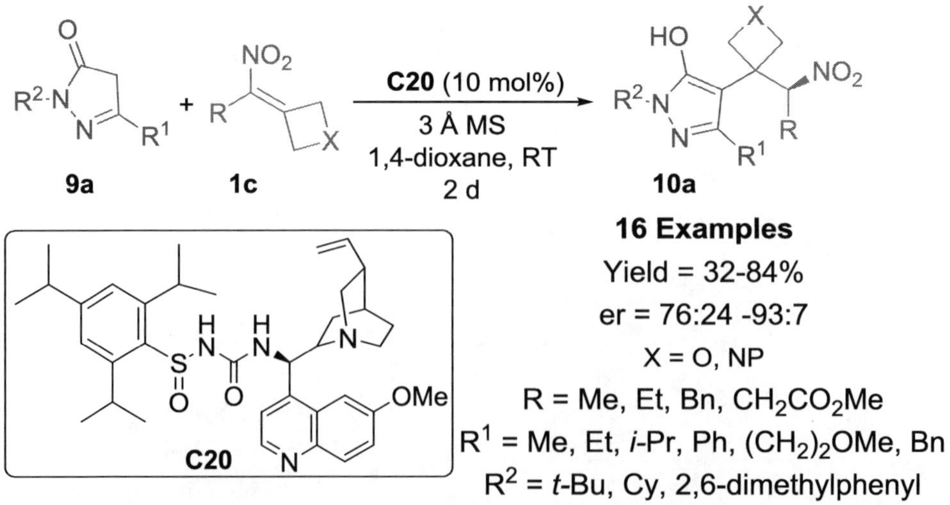

SCHEME 4.25 Enantioselective addition of pyrazolones to highly substituted nitroalkenes in the presence of a bifunctional N-sulfinylurea catalyst.

Du et al. developed an organocatalytic asymmetric Michael addition of pyrazolin-5-ones **9a** to nitroalkenes **1** in the presence of squaramide catalyst **C11** with low catalyst loading (0.25 mol%).[26] The corresponding chiral pyrazol-3-ol derivatives **10b** were obtained with high yields (up to 99%) and enantioselectivities (up to 94% ee) (Scheme 4.26). In general, the 3-methyl-1-phenyl-2-pyrazolin-5-one unit in product **10b** exists in the enol form. In the proposed mechanism, 3-methyl-1-phenyl-2-pyrazolin-5-one **9a** is deprotonated by the tertiary amine of the catalyst via tautomerization. At the same time, the squaramide moiety of catalyst **C11** activates nitroalkene **1** through double hydrogen bonding with the nitro group. The product **10b** with *R*- configuration was formed by attack of the deprotonated 3-methyl-1-phenyl-2-pyrazolin-5-one on the activated nitroalkene **1** from the *Si*-face (**TS-13**, Scheme 4.26).

Ma et al. developed an organocatalytic sequential 1,4-addition/dearomative fluorination of pyrazolones **9a** and nitroalkenes **1** in the presence of NFSI.[27] The transformation is catalyzed by a chiral tertiary amine-thiourea catalyst **C21** in the presence of benzoic acid as an additive to afford the products **10c**, with adjacent tertiary and fluorine-bearing quaternary stereocenters (Scheme 4.27). The postulated mechanism involves two catalytic cycles. The first cycle involves the formation of a complex between nitroalkene **1** and the thiourea group of the chiral catalyst **C21** via hydrogen bonding. The enol form of the pyrazolone **9a** coordinates to the ammonium center of the chiral catalyst **C21** through hydrogen bonding. A highly organized open transition state **TS-14a** has been proposed for the general base-assisted addition of the enol form of the pyrazolone to the nitroalkene (Scheme 4.27). In the subsequent electrophilic-fluorination step, the pyrazol-5-ol intermediate, which is formed in the first step, re-associates with the catalyst **C21** via hydrogen bonding and then undergoes general base-assisted reaction with the fluorinating reagent (**TS-14b**, Scheme 4.27).

Shen et al. described the asymmetric Michael addition of pyrazolin-5-ones **9a** to (*E*)-β-CF_3-β-disubstituted nitroalkenes **1d** by employing a chiral tertiary amine squaramide **C11** as an organocatalyst to afford trifluoromethylated products **10d** with all-carbon quaternary stereocenter (Scheme 4.28).[28] The mechanism involves the deprotonation of 3-methyl-1-phenyl-2-pyrazolin-5-one **9a** by the tertiary amine via tautomerization. Simultaneously, nitroalkene was activated by the H-bonding interaction between the NO_2 group of nitroalkene and squaramide group of the

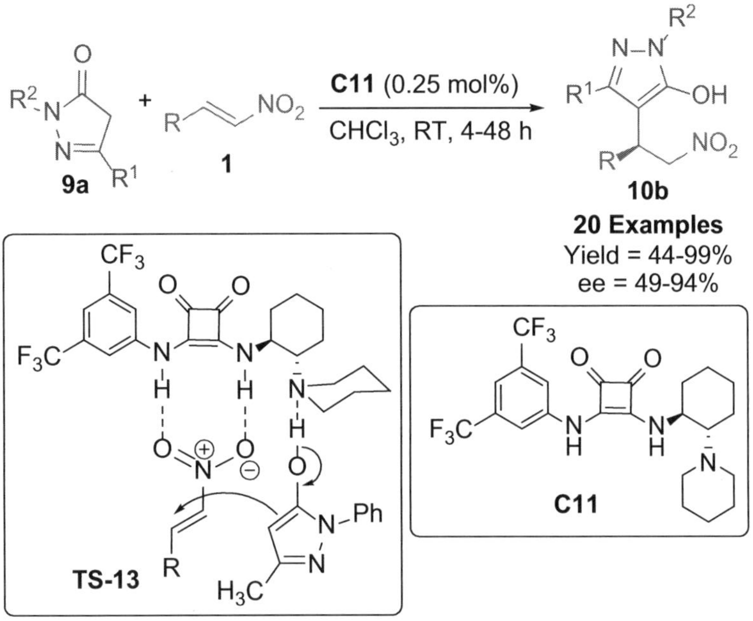

SCHEME 4.26 Asymmetric Michael addition of pyrazolin-5-ones to nitroalkenes in the presence of a cyclohexanediamine-derived squaramide catalyst.

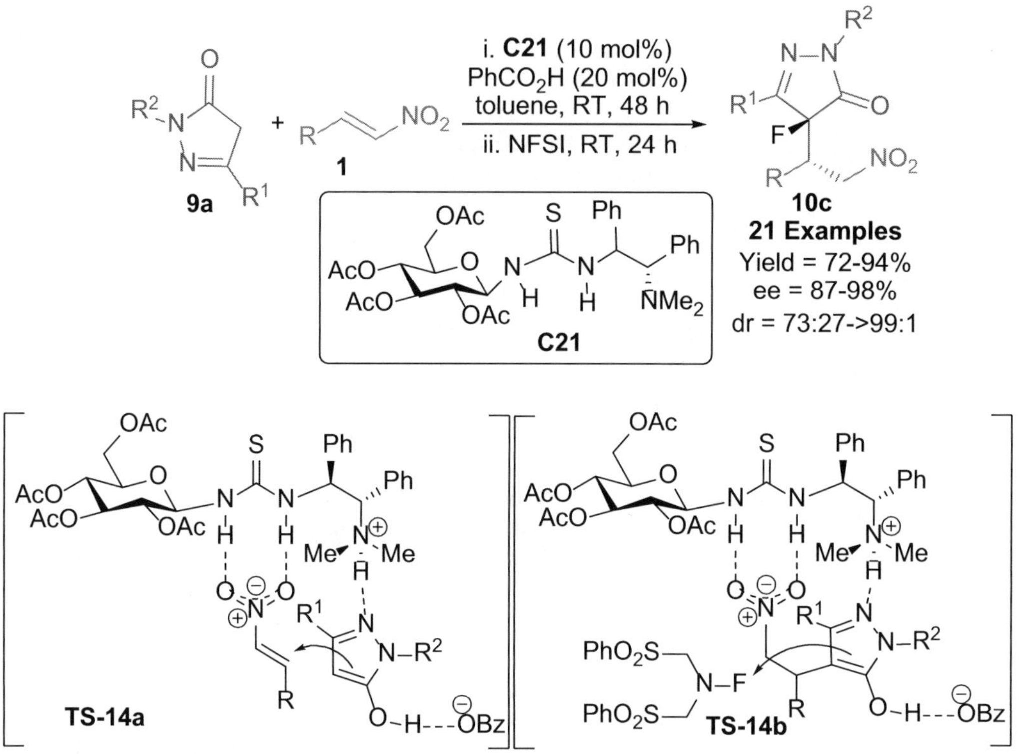

SCHEME 4.27 Sequential 1,4-addition/dearomative fluorination of pyrazolones on reaction with nitroalkenes in the presence of NFSI.

catalyst. Then, deprotonated pyrazolin-5-one adds to the nitroalkene from *Re*-face to give the product **10d** with *R*-configuration (**TS-15**, Scheme 4.28).

Yuan et al. disclosed the first bifunctional thiourea ***ent*-C1** catalyzed diastereo and enantioselective Michael addition of pyrazolin-5-ones **9b** to nitroalkenes **1** (Scheme 4.29).[29] The corresponding enantioenriched, multiple-substituted pyrazolin-5-ones **10e** bearing contiguous tertiary and quaternary stereocenters were obtained in high yields and enantioselectivities.

Recently, cinchona alkaloid-derived benzoyl thiourea organocatalysts were employed for the asymmetric Michael addition of pyrazolin-5-ones **9c** to nitroalkenes **1** (Scheme 4.30).[30] *R* or *S* enantiomer of the adduct **10f** or **10g** was formed with high yields and excellent enantioselectivities depending on the choice of the H-bonding catalyst (**C22** or **C23**).

Shibasaki et al. reported the Michael addition of α,β-unsaturated γ-butyrolactam **11a** to nitroalkenes **1**, and the corresponding vinylogous Michael adducts **12a** were obtained with high diastereoselectivity and excellent enantioselectivity up to 93%–99% in the presence of a nickel catalyst **C13** (Scheme 4.31).[31] The reaction proceeded smoothly in 1,4-dioxane at 50°C with 2.5 mol% of catalyst loading. Nitrodiene was also employed for the reaction, and the corresponding 1,4-adduct was obtained in 83% yield with 99% ee.

The Mukherjee group developed an organocatalytic vinylogous Michael reaction of α,β-unsaturated γ-butyrolactam **11a** to nitroalkenes **1** in the presence of a quinidine catalyst **C24** to give the Michael adduct **12b** (Scheme 4.32).[32] In the proposed mechanism, while the Brønsted acidic hydroxyl group activates the nitroalkene **1** through hydrogen bonding, the Brønsted basic tertiary amine provides nucleophilic activation. The vicinal tertiary amine and hydroxyl group in

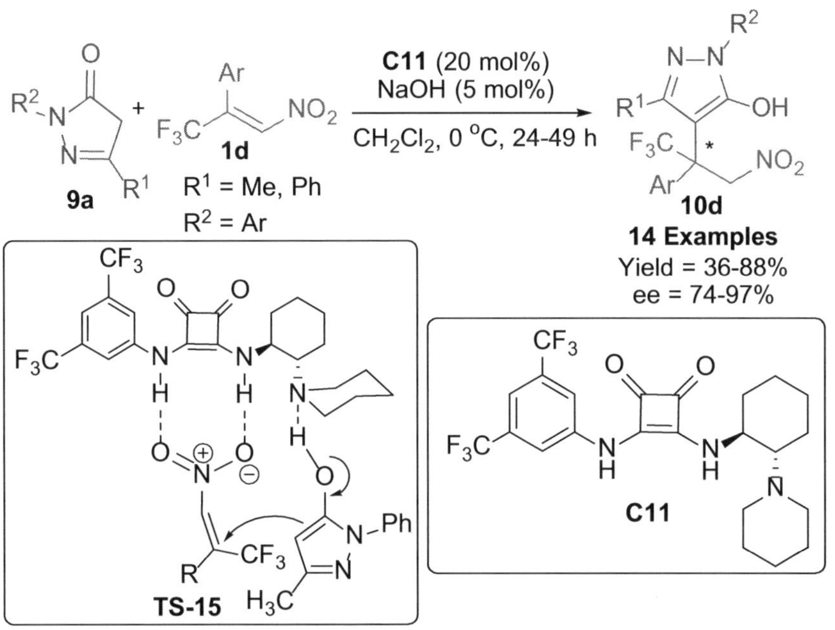

SCHEME 4.28 Asymmetric Michael addition of pyrazolin-5-ones to β-CF_3-β-disubstituted nitroalkenes in the presence of a chiral tertiary amine squaramide catalyst.

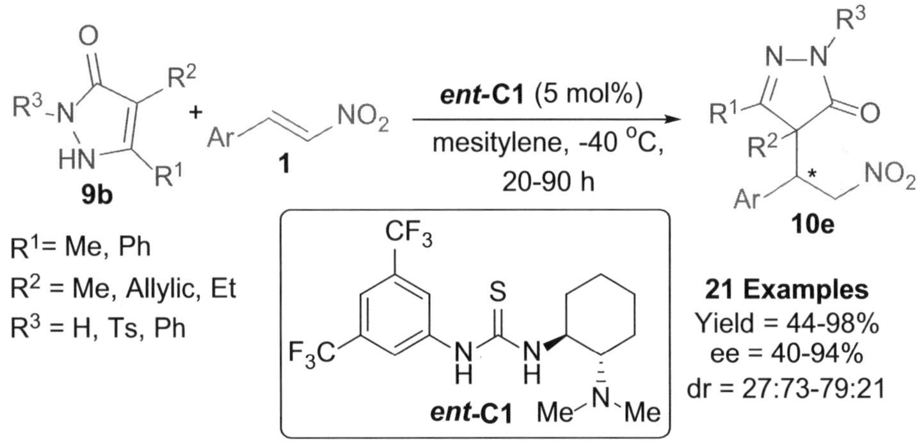

SCHEME 4.29 Michael addition of 4-substituted-pyrazolin-5-ones to nitroalkenes in the presence of a bifunctional thiourea catalyst.

the catalyst **C24** appeared to control the stereochemistry of the Michael adduct **12b** (**TS-16** and **TS-17**, Scheme 4.32).

Shen et al. reported an asymmetric Michael addition of pyrazolone acetates **13** to nitroalkenes **1** by employing a squaramide catalyst **C11** to give the corresponding adducts **14** in good-to-excellent yields and enantioselectivities (Scheme 4.33).[33] The reaction involves the initial formation of enolate of pyrazolone acetate, which adds to the activated nitroalkene from the *Re*-face to give the *R*-configured product (**TS-18**, Scheme 4.33).

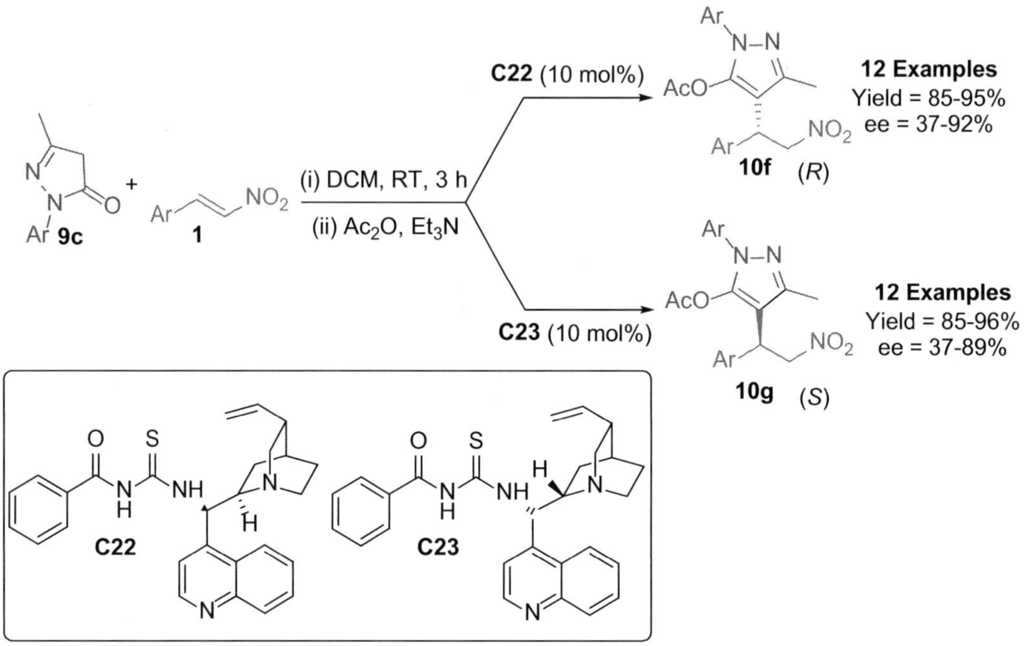

SCHEME 4.30 Cinchona alkaloid-derived benzoyl thiourea organocatalysts for Michael addition of pyrazolin-5-ones to nitroalkenes.

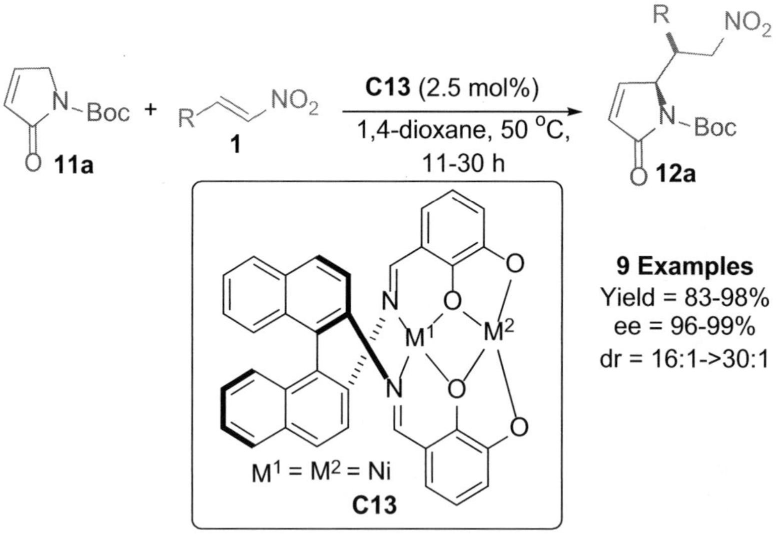

SCHEME 4.31 Michael addition of α,β-unsaturated γ-butyrolactam to nitroalkenes under dinuclear nickel catalysis.

Pedro and co-workers reported the synthesis of diverse alkenylpyrazolone adducts **15** containing a tetrasubstituted stereocenter by a dihydroquinine 2,5-diphenyl-4,6-pyrimidinediyl diether catalyst **C25** mediated enantioselective addition of 4-substituted pyrazolones **9e** to isatin-derived nitroalkenes **1e** with good regio and diastereoselectivity, as well as moderate enantioselectivity (Scheme 4.34).[34] The reaction includes a regioselective nucleophilic vinylic substitution where 3-alkylidene-2-oxindoles **15** were formed after elimination of the nitro group (**TS-19**).

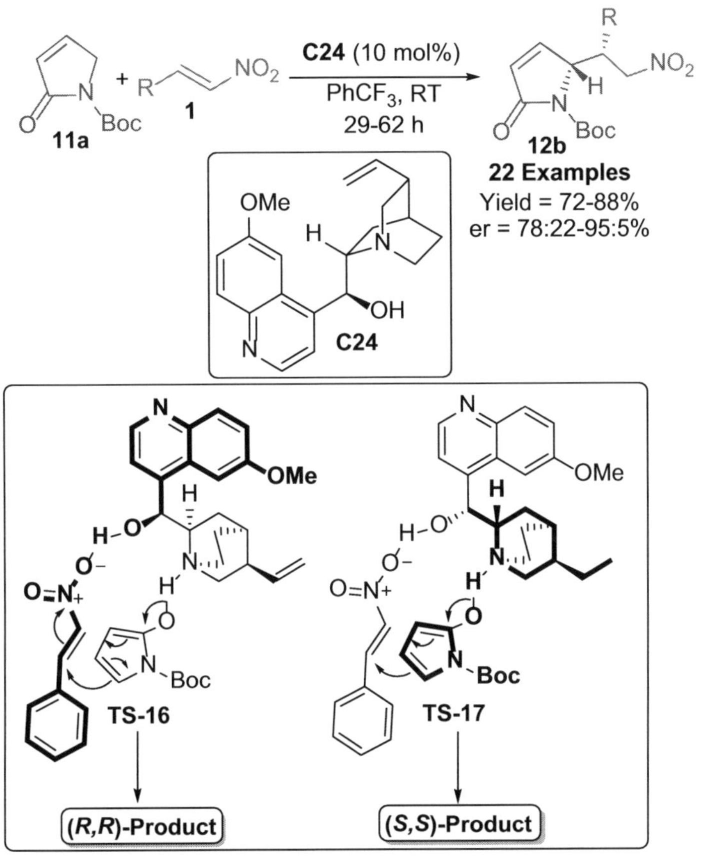

SCHEME 4.32 Direct vinylogous Michael reaction of α,β-unsaturated γ-butyrolactam to nitroalkenes in the presence of a quinidine catalyst.

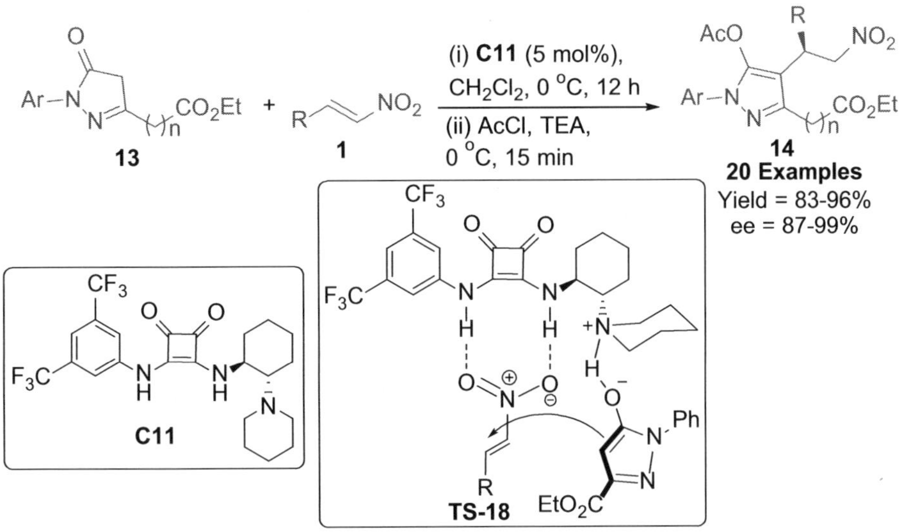

SCHEME 4.33 Asymmetric Michael addition of pyrazolone acetates to nitroalkenes catalyzed by a squaramide catalyst.

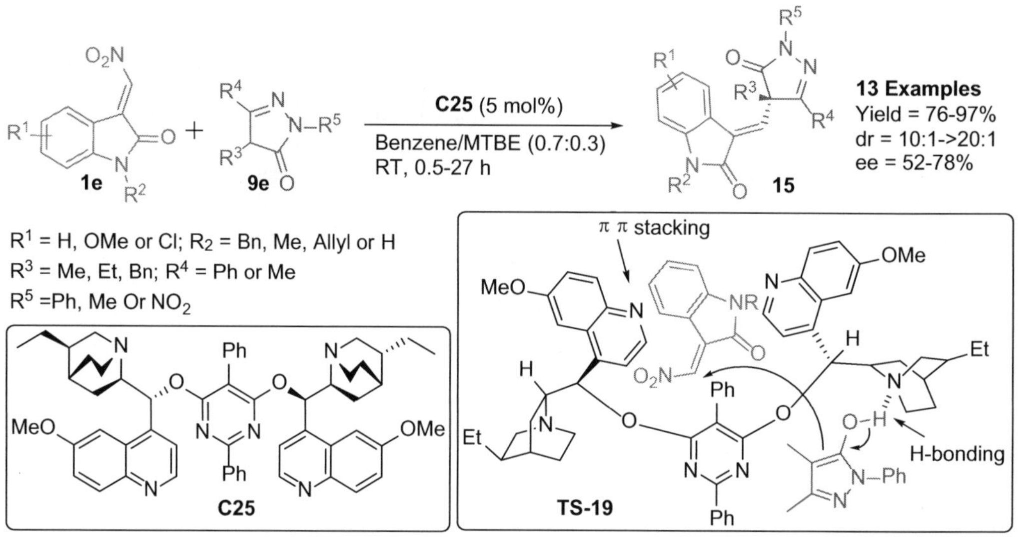

SCHEME 4.34 Regio, diastereo and enantioselective synthesis of diverse alkenylpyrazolone adducts.

4.4 CONCLUSIONS

Asymmetric Michael addition of various iminoesters, lactones and lactams to nitroalkenes proceeds well in the presence of suitable chiral organo- and metal-ligand catalysts. High yields and stereoselectivities are observed in these reactions, which produce nitroalkanes often bearing a heterocyclic moiety at the β-position. Another unique feature is the vinylogous nature of the Michael addition in the case of furanones and pyrrolidones.

REFERENCES

1. Li, Q.; Ding, C.-H.; Hou, X.-L.; Dai, L.-X. *Org. Lett.* **2010**, *12*, 1080.
2. Guerrero-Corella, A.; Esteban, F.; Iniesta, M.; Martin-Somer, A.; Parra, M.; Diaz-Tendero, S.; Fraile, A.; Aleman, J. *Angew. Chem. Int. Ed.* **2018**, *57*, 5350.
3. Imae, K.; Konno, T.; Ogata, K.; Fukuzawa, S. *Org. Lett.* **2012**, *14*, 4410.
4. Fallan, C.; Lam, H. W. *Chem. - Eur. J.* **2012**, *18*, 11214.
5. Koizumi, A.; Matsuda, Y.; Haraguchi, R.; Fukuzawa, S. *Tetrahedron: Asymmetry* **2017**, *28*, 428.
6. Meng, W.-T.; Zheng, Y.; Nie, J.; Xiong, H.-Y.; Ma, J.-A. *J. Org. Chem.* **2013**, *78*, 559.
7. Terada, M.; Ando, K. *Org. Lett.* **2011**, *13*, 2026.
8. Aleman, J.; Milelli, A.; Cabrera, S.; Reyes, E.; Jørgensen, K. A. *Chem. - Eur. J.* **2008**, *14*, 10958.
9. Liu, X.; Song, H.; Chen, Q.; Li, W.; Yin, W.; Kai, M.; Wang, R. *Eur. J. Org. Chem.* **2012**, *2012*, 6647.
10. Sekikawa, T.; Kitaguchi, T.; Kitaura, H.; Minami, T.; Hatanaka, Y. *Org. Lett.* **2016**, *18*, 646.
11. Trost, B. M.; Hitce, J. *J. Am. Chem. Soc.* **2009**, *131*, 4572.
12. Yang, M.; Chen, C.; Yi, X.; Li, Y.; Wu, X.; Li, Q.; Ban, S. *Org. Biomol. Chem.* **2019**, *17*, 2883.
13. Ding, M.; Zhou, F.; Liu, Y.-L.; Wang, C.-H.; Zhao, X.-L.; Zhou, J. *Chem. Sci.* **2011**, *2*, 2035.
14. Yang, W.; Wang, J.; Du, D.-M. *Tetrahedron: Asymmetry* **2012**, *23*, 972.
15. Rodríguez-Ferrer, P.; Naharro, D.; Maestro, A.; Andrés, J. M.; Pedrosa, R. *Eur. J. Org. Chem.* **2019**, 2019, 6539.
16. Wang, C.; Yang, X.; Enders, D. *Chem. –Eur. J.* **2012**, *18*, 4832.
17. Kato, Y.; Furutachi, M.; Chen, Z.; Mitsunuma, H.; Matsunaga, S.; Shibasaki, M. *J. Am. Chem. Soc.* **2009**, *131*, 9168.
18. Chen, Q.; Wang, G.; Jiang, X.; Xu, Z.; Lin, L.; Wang, R. *Org. Lett.* **2014**, *16*, 1394.
19. Noole, A.; Järving, I.; Werner, F.; Lopp, M.; Malkov, A.; Kanger, T. *Org. Lett.* **2012**, *14*, 4922.
20. Ding, R.; Wolf, C. *Org. Lett.* **2018**, *20*, 892.

21. Retini, M.; Bergonzini, G.; Melchiorre, P. *Chem. Commun.* **2012**, *48*, 3336.
22. Cui, B.-D.; Han, W.-Y.; Wu, Z.-J.; Zhang, X.-M.; Yuan, W.-C. *J. Org. Chem.* **2013**, *78*, 8833.
23. Dou, X.; Zhou, B.; Yao, W.; Zhong, F.; Jiang, C.; Lu, Y. *Org. Lett.* **2013**, *15*, 4920.
24. Zhan, G.; Shi, M.-L.; Lin, W.-J.; Ouyang, Q.; Du, W.; Chen, Y.-C. *Chem. - Eur. J.* **2017**, *23*, 6286.
25. Phelan, J. P.; Ellman, J. A. *Adv. Synth. Catal.* **2016**, *358*, 1713.
26. Li, J.-H.; Du, D.-M. *Org. Biomol. Chem.* **2013**, *11*, 6215.
27. Li, F.; Sun, L.; Teng, Y.; Yu, P.; Zhao, J. C.-G.; Ma, J.-A. *Chem. - Eur. J.* **2012**, *18*, 14255.
28. Lai, X.; Zha, G.; Liu, W.; Xu, Y.; Sun, P.; Xia, T.; Shen, Y. *Synlett* **2016**, *27*, 1983.
29. Liao, Y.-H.; Chen, W.-B.; Wu, Z.-J.; Du, X.-L.; Cun, L.-F.; Zhang, X.-M.; Yuan, W.-C. *Adv. Synth. Catal.* **2010**, *352*, 827.
30. Wang, Z.; Ban, S.; Yang, M.; Li, Q. *ChemistrySelect* **2017**, *2*, 3419.
31. Shepherd, N. E.; Tanabe, H.; Xu, Y.; Matsunaga, S.; Shibasaki, M. *J. Am. Chem. Soc.* **2010**, *132*, 3666.
32. Choudhury, A. R.; Mukherjee, S. *Org. Biomol. Chem.* **2012**, *10*, 7313.
33. Xu, Y.; Sun, P.; Song, Q.; Lai, X.; Liu, W.; Xia, T.; Huang, Y.; Shen, Y. *Eur. J. Org. Chem.* **2017**, *2017*, 2998.
34. Vila, C.; Dharmaraj, N. R.; Faubel, A.; Blay, G.; Cardona, M. L.; Muñoz, M. C.; Pedro, J. R. *Eur. J. Org. Chem.* **2019**, *2019*, 3040.

5 Catalytic Asymmetric Friedel–Crafts Reactions of Nitroalkenes

5.1 INTRODUCTION

Catalytic asymmetric Michael additions of various carbonyl compounds to nitroalkenes have been described in the previous chapters. However, nitroalkenes also participate in other reactions such as Friedel–Crafts reaction, which in the present context is a conjugate addition of aryl groups to nitroalkenes, which are covered in this chapter. The asymmetric Friedel–Crafts alkylation of indoles, pyrroles and electron-rich benzenoid aromatic compounds with nitroalkenes affords the corresponding enantioenriched indoles, pyrroles, etc. The indole and pyrrole rings can be subjected to further diastereoselective reduction to afford optically active derivatives, which are precursors to pharmaceuticals and natural products. A wide variety of metal-ligand complexes, such as Lewis acids and organocatalysts such as thioureas, squaramides and phosphoric acids as Brønsted acids have been employed as catalysts in Friedel–Crafts reactions.

5.2 FRIEDEL–CRAFTS REACTION

The asymmetric Friedel–Crafts alkylation of indoles with nitroalkenes has gained substantial importance owing to the synthetic utility and versatility of chiral indole scaffolds in the frameworks of diverse biologically active indole alkaloids. Over the last two decades, substantial contributions have been made for this reaction by the development of several efficient catalytic systems. For instance, bifunctional hydrogen-bond donor organocatalysts and metal-based catalysts provided dual activation of the nitro group and the indole NH moiety. In 2008, the Jørgensen group reviewed asymmetric Friedel–Crafts alkylation catalyzed by copper.[1] Later, Dalpozzo and co-workers published two review articles in 2010[2] and 2015[3] and highlighted the asymmetric functionalization of indoles, where few of the asymmetric Friedel–Crafts reactions of indole with nitroalkenes were discussed.[4–18] Thus, to avoid overlap, here we try to focus more on the uncovered literature.

In the asymmetric Friedel–Crafts alkylation of indole **2** with nitroalkenes **1**, the complex of diphenylamine-linked bis(imidazoline) ligand **L1** and $Zn(OTf)_2$ proved to be a good catalyst system and provided the product with excellent enantioselectivity. The adducts **3** could be transformed to useful heterocycles without loss of enantiomeric purity (Scheme 5.1).[19] In the same year, a diphenylamine-bis(oxazoline) ligand **L2** with an oxazoline ring with *trans*-diphenyl substitution was immobilized onto Fréchet-type dendrimers and used as a catalyst system in the asymmetric Friedel–Crafts alkylation of indole derivatives **2** with nitroalkenes **1** by Liu and Du.[20] The products **3** were obtained in moderate-to-high yields and excellent enantioselectivities (Scheme 5.1). The catalyst was reused and provided high yields and enantioselectivities even after four to five reaction cycles.

Wang et al. developed an efficient tridentate Schiff base ligand **L3**-$Zn(OTf)_2$ catalyst system for the asymmetric Friedel–Crafts alkylation of indoles **4** with nitroalkenes **1** (Scheme 5.2).[21] It was postulated that the catalyst acts in a bifunctional fashion wherein the hydroxyl group of the amino alcohol forms a hydrogen bond with indole. The indole moiety then attacks on the *Re*-face of the nitroalkene **1**, which is activated by zinc, as shown in the TS (**TS-1**, Scheme 5.2).

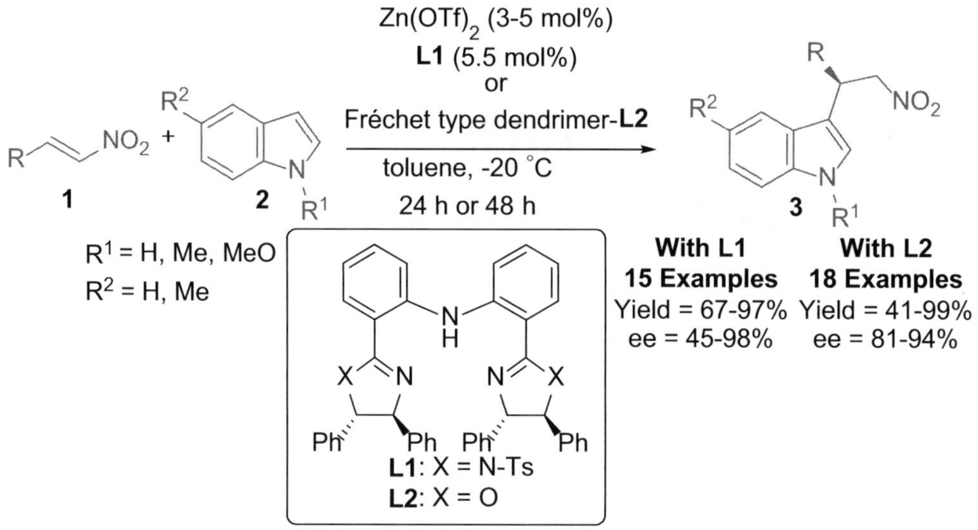

SCHEME 5.1 Asymmetric Friedel–Crafts alkylation of indoles with nitroalkenes in the presence of diphenylamine-linked bis(oxazoline) ligand.

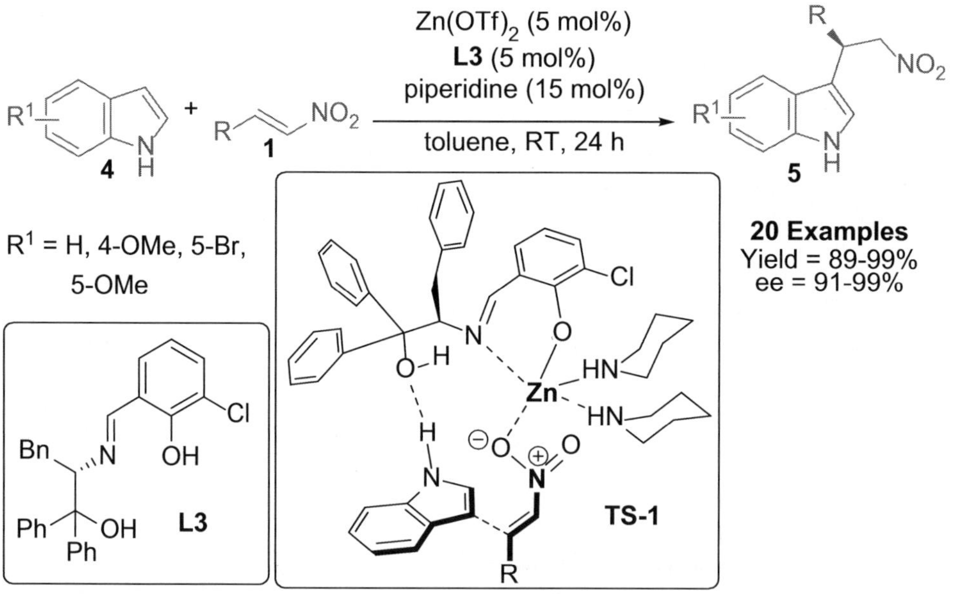

SCHEME 5.2 Enantioselective Friedel–Crafts alkylation of indoles with nitroalkenes in the presence of tridentate Schiff base-Zn$(OTf)_2$ catalyst system.

Recently, Ramanathan and co-workers reported an asymmetric Friedel–Crafts alkylation of indoles **4a** with nitroalkenes **1** mediated by a chiral bicyclic skeleton-tethered bipyridine **L4**-Zn$(OTf)_2$ complex (Scheme 5.3).[22] It was observed that the reaction afforded the alkylated indole products **5a** with good yields (up to 94%) and enantioselectivities (up to 91%). Here, the authors envisaged that the catalytic complex enhanced the electrophilicity of nitroalkene, which further facilitated its reaction with the indole nucleophile.

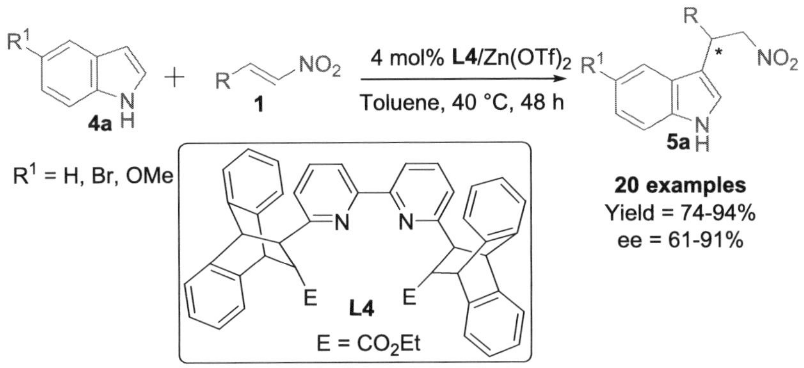

SCHEME 5.3 Asymmetric Friedel–Crafts alkylation of indoles with nitroalkenes mediated by a bicyclic bipyridine-$Zn(OTf)_2$ complex.

Oh et al. developed a catalyst system **L5**-$Cu(OTf)_2.C_6H_6$ that efficiently generated complementary enantioselectivities for the stereoselective Friedel–Crafts alkylation of indole **4a** with nitroalkenes **1** under both homogeneous as well as heterogeneous reaction conditions (Scheme 5.4).[23] Under homogeneous conditions, the catalyst was treated in the presence of triethylamine (TEA), whereas under heterogeneous conditions, a solid support was used to immobilize the catalyst to obtain the desired products **5** with excellent yields and enantioselectivities.

Wan et al. described the asymmetric Friedel–Crafts alkylation of indoles **4a** with nitroalkenes **1** by employing a $Cu(OTf)_2$/bis(sulfonamide)-diamine **L6** catalyst system.[24] It was observed that The sterically hindered aromatic nitroalkenes **1** preferably with *ortho*-substitution afforded the corresponding alkylated products **5** in 65%–99% yields with an enantiomeric excess of up to 97% (Scheme 5.5). The catalyst system acts in a bifunctional manner wherein the nitroalkene **1** was activated by chelating to Cu(II) through π-π interaction between its substituent with the phenyl group of the ligand **L6**. Subsequently, indolic hydrogen forms a weak hydrogen bond with Ts-protected

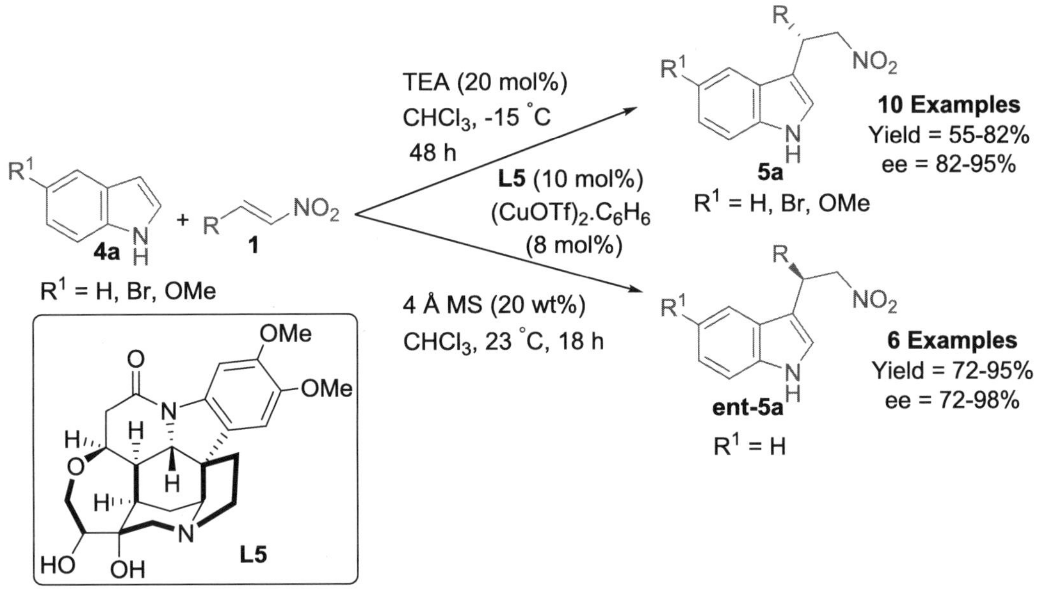

SCHEME 5.4 Asymmetric Friedel–Crafts alkylation of indoles with nitroalkenes in the presence of $Cu(OTf)_2.C_6H_6$ and brucine-derived ligand catalyst system.

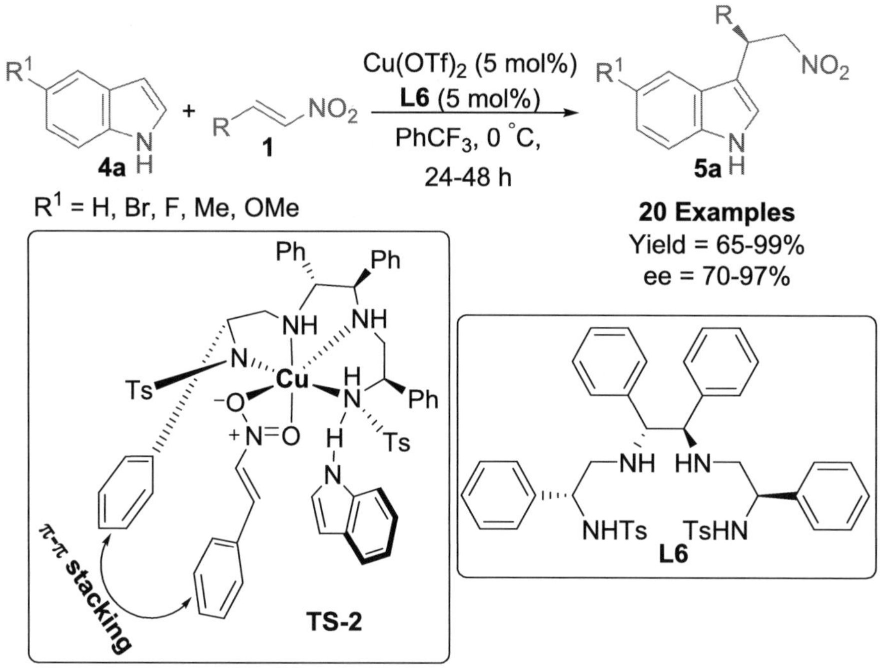

SCHEME 5.5 Asymmetric Friedel–Crafts alkylation of indoles with nitroalkenes using a $Cu(OTf)_2$/bis(sulfonamide)-diamine catalyst system.

amine of the ligand **L6**, and the bulky tosyl group directs the *Si*-face attack of indole to nitroalkene to generate the product **5** with '*R*' configuration (**TS-2**, Scheme 5.5).

Carmona et al. reported an uncommon effect of temperature and catalyst loading in the asymmetric Friedel–Crafts reaction of indoles **6** with nitroalkenes **1** in the presence of Rh-complex **C1** as a catalyst (Scheme 5.6).[25] They also conducted spectroscopic, kinetic and theoretical studies to disclose the mechanism of this Friedel–Crafts reaction and the relationship between the reaction conditions and enantioselectivity. At a fixed temperature, the ee changes with the reaction time. At temperatures

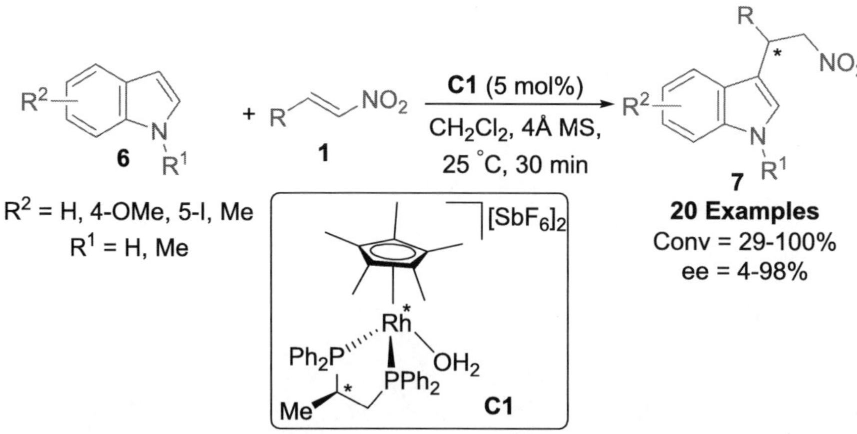

SCHEME 5.6 Asymmetric Friedel–Crafts reaction of indoles with nitroalkenes in the presence of chiral Rh-catalyst.

equal to or greater than 263 K, the concentration of the *S*-isomer increases with conversion through uncatalyzed prototropic path and involves dissociation of the *aci*-nitro compound. Conversely, at temperatures below 253 K, the relative amount of the *S*-isomer decreases with conversion along with an increase in *R*-isomer through Rh-catalyzed prototropic path. In this protocol, the same catalyst provides either enantiomer as the dominant product with excellent optical purity simply by changing the reaction temperature. Further, it was evident that increasing the catalyst loading favors the formation of the *R*-isomer over the *S*-isomer of the Friedel–Crafts adducts **7** and vice versa.

Saluzzo et al. described the Friedel–Crafts alkylation of indoles **4a** with nitroalkenes **1** in the presence of an organocatalyst **C2** derived from isomannide and isoidide to give the corresponding alkylated products **5a** in moderate yields and enantioselectivities (Scheme 5.7).[26]

Du et al. reported a catalytic asymmetric Friedel–Crafts alkylation of indoles **2** with various nitroalkenes **1** employing tridentate bis(oxazoline) **L7**-Zn$(OTf)_2$ complex as a catalyst to afford nitroalkylated indoles **3** in excellent yields with high enantioselectivities of up to 98% ee (Scheme 5.8).[27] The catalyst system activates the nitroalkene **1** through coordination of the nitro group to the Lewis acid center. The NH group between the two phenyl groups in the ligand **L7** served as a donor for the NH-π interaction that directed the *Si*-face attack of indole to the nitroalkene **1** to generate the product exclusively with *R* configuration. High enantioselectivities were observed even for N-protected indoles **2** (**TS-3**, Scheme 5.8).

Carmona et al. reported a Friedel–Crafts reaction between *trans*-β-nitroalkenes **1** and indoles **6a** mediated by the Rh-aqua complex **C1**, which involved uncommon metal-nitroalkene, metal-aci-nitro and free *aci*-nitro intermediates (Scheme 5.9).[28] In the postulated mechanism, indole **6a** is added to the activated nitroalkene **1** which is also the enantioselectivity-determining step to form the *aci*-nitro complex. This *aci*-nitro complex **8** reacts with nitroalkene **1**, eliminates the *aci*-nitro ligand and rearranges to the Friedel–Crafts adduct **7a** with the regeneration of catalyst **C1**.

Zhang reported enantioselective Friedel–Crafts reaction of indoles **9** with nitroalkenes **1** catalyzed by an air-stable, well-defined Cu/Eu/Cu heterotrimetallic complex **L8a**, which is based on a salen-type ligand **L8**.[29] The corresponding products **10** were obtained in excellent yields and enantioselectivities (Scheme 5.10). The possible mechanism involves a unique cooperative triple activation of the substrate, wherein one of the square planar Cu(II) centers initially bind to the indole and then the central unsaturated Eu(III) site coordinate to the nitroalkene **1**, which further interacts with another Cu(II) center by the nucleophilic oxygen atom and allows the substrate to undergo alkylation.

The enantioselective Friedel–Crafts fluoroalkylation of indoles **4** by employing chiral phosphoric acid **C3** has been described by Lin and Xiao (Scheme 5.11).[30] Chiral phosphoric acid **C3** assists the

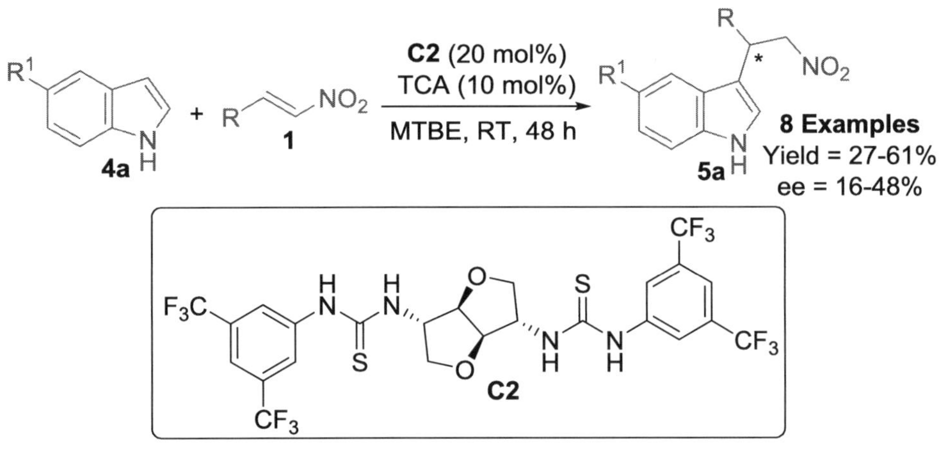

SCHEME 5.7 Asymmetric Friedel–Crafts alkylation of indoles with nitroalkenes in the presence of organocatalyst derived from isomannide.

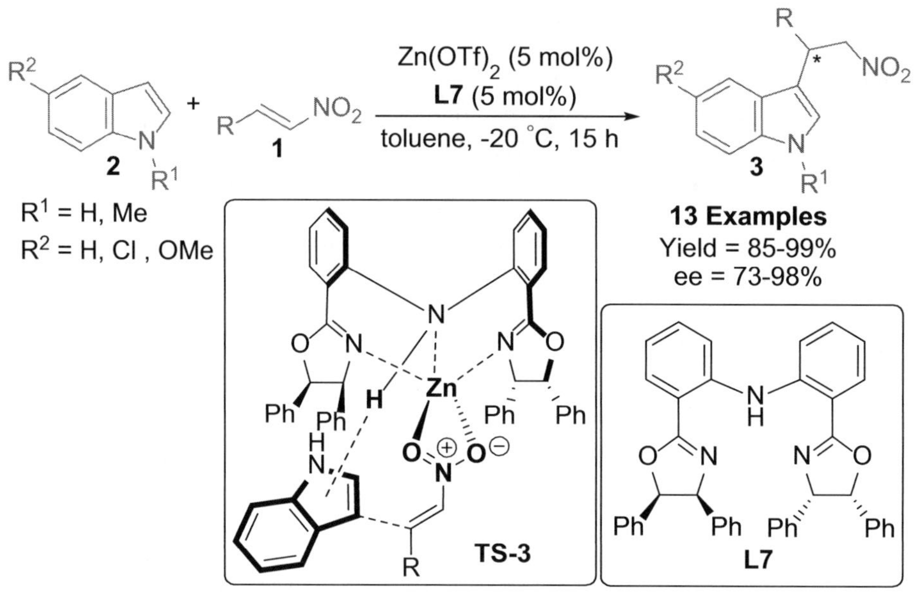

SCHEME 5.8 Asymmetric Friedel-Crafts alkylation of indoles with nitroalkenes catalyzed by Zn(II)-tridentate bis(oxazoline) ligand system.

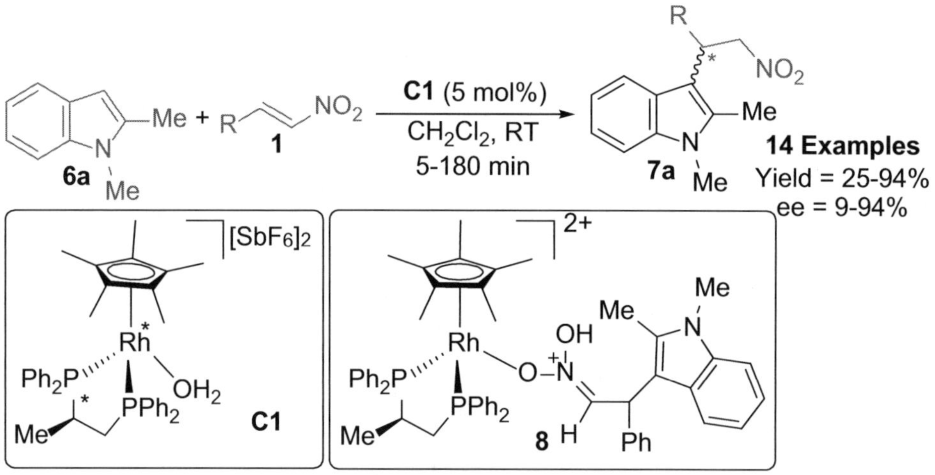

SCHEME 5.9 Asymmetric Friedel–Crafts alkylation of indoles with nitroalkenes catalyzed by Rh-aqua complex catalyst system.

reaction as a bifunctional catalyst and activates the nucleophile as well as the electrophile by hydrogen bonding. It was found that the absence of a hydrogen atom on the nitrogen atom or the presence of a methyl group at the 2-position of indole **4** considerably decreases the enantioselectivity as the N–H moiety is involved in hydrogen bonding.

The Kass group developed a charged thiourea catalyst for the Friedel–Crafts alkylation of β-nitrostyrenes **1** with indoles **4** (Scheme 5.12).[31] N-Alkylpyridinium center and 2-indanol containing thiourea catalyst **C4** exhibited excellent reactivity compared to the existing thiourea catalysts even in the absence of additional hydrogen bond donors.

Song et al. reported an asymmetric Friedel–Crafts alkylation of indoles **2** with nitroalkenes **1** in the presence of 5 mol% of cationic chiral NCN pincer Pt(II) aquo complex **C5** along with

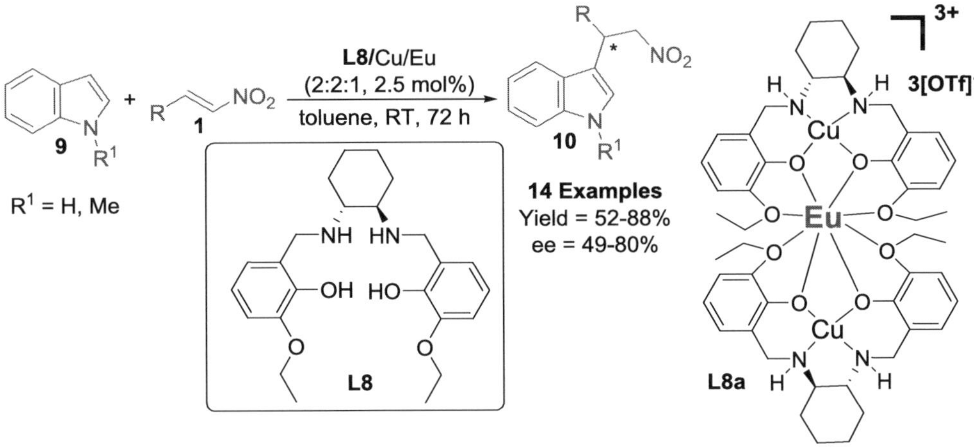

SCHEME 5.10 Asymmetric Friedel–Crafts alkylation of indoles with nitroalkenes catalyzed by trinuclear Cu/Eu/Cu complex.

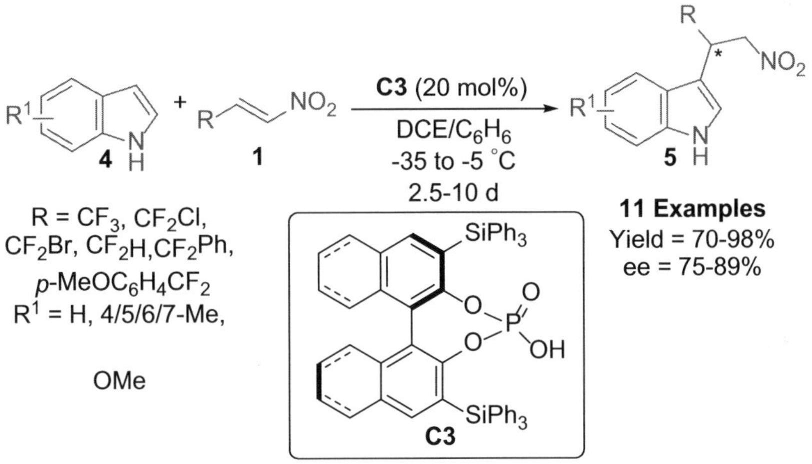

SCHEME 5.11 Enantioselective Friedel–Crafts fluoroalkylation of indoles using chiral phosphoric acid.

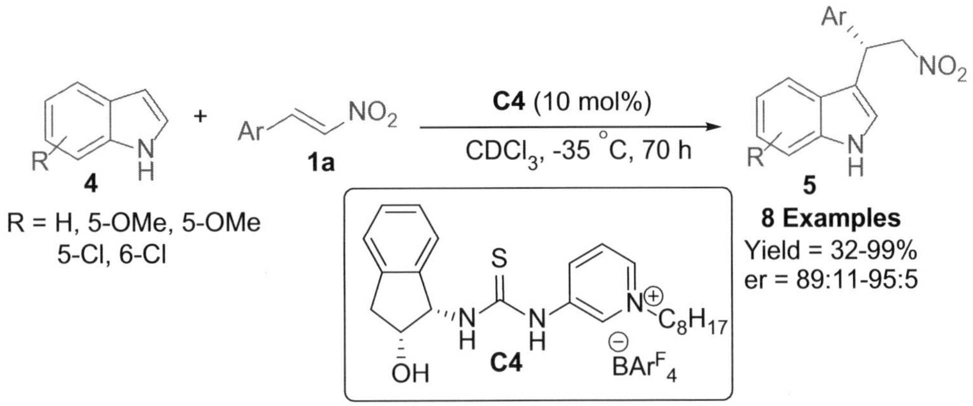

SCHEME 5.12 Asymmetric Friedel–Crafts alkylation of indoles with nitroalkenes catalyzed by charge-activated thiourea organocatalyst.

bis-(imidazolinyl)phenyl (Phebim) ligand. The products, nitroalkylated indoles **3**, were obtained in high yields with good enantioselectivities (up to 83% ee, Scheme 5.13).[32]

The asymmetric Friedel–Crafts alkylation of indole **4b** with nitroalkenes **1** using oxazoline–imidazoline **L9**-Zn$(OTf)_2$ complex was reported by Islam et al.[33] The catalyst activates nitroalkenes **1** and orients indoles **4b** effectively to afford the products **5b** in high yields and excellent enantioselectivities (Scheme 5.14). In the postulated mechanism, nitroalkene was activated by chelating to **L9**–Zn(II) forming a four-membered intermediate. In the subsequent step, indole **4b** adds to the nitroalkene from the *Si*-face to afford Friedel–Crafts adduct. This on proton-transfer and fragmentation results in the formation of the final product **5b** with the regeneration of the catalyst.

The Ricci group developed a catalytic asymmetric Friedel–Crafts alkylation of indoles **11** with nitroalkenes **1** to provide optically pure 2-indolyl-1-nitro derivatives **12** with high yields and enantioselectivities by employing simple thiourea-based organocatalyst **C6** (Scheme 5.15).[34] Here, the catalyst acts in a bifunctional fashion wherein the hydrogen atoms of the thiourea motif activate the nitroalkene **1** and the free alcoholic group of the catalyst **C6** forms a weak H-bonding interaction

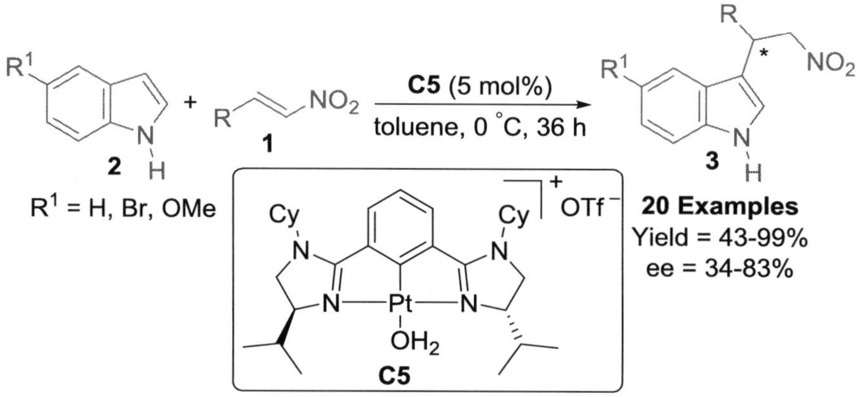

SCHEME 5.13 Asymmetric Friedel–Crafts alkylation between indoles and nitroalkenes catalyzed by cationic chiral NCN pincer Pt(II) aquo complex.

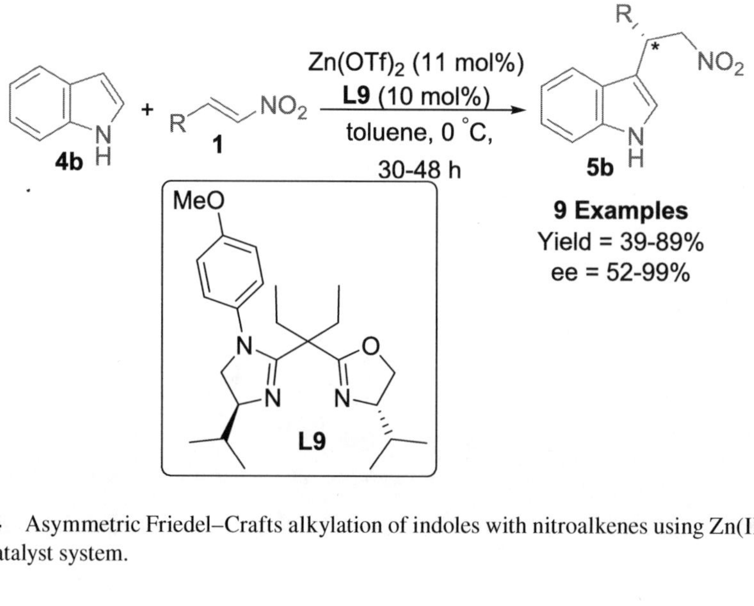

SCHEME 5.14 Asymmetric Friedel–Crafts alkylation of indoles with nitroalkenes using Zn(II)–oxazoline–imidazoline catalyst system.

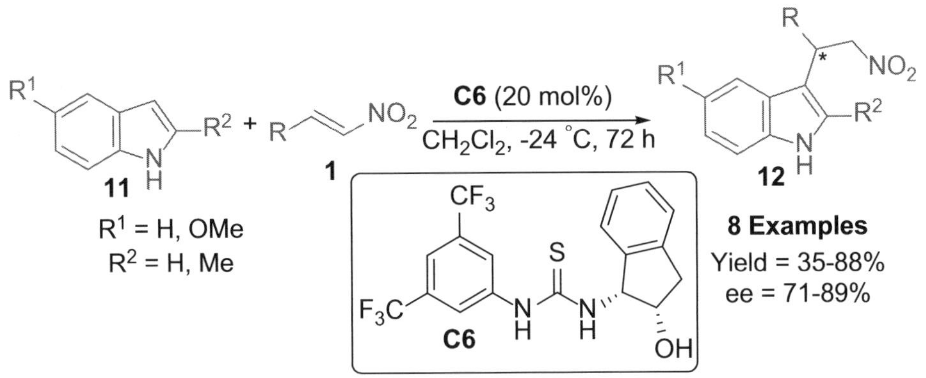

SCHEME 5.15 Asymmetric Friedel–Crafts alkylation of indoles with nitroalkenes using thiourea-based organocatalyst.

with the indolic proton, allowing the incoming nucleophile to attack on the *Si*-face of the nitroalkene **1** to yield the desired product **12**.

The catalytic asymmetric Friedel–Crafts alkylation reaction of indole and its derivatives **4** with nitroalkenes **1** employing Zn(II)-bisoxazoline complex **C7** was described by Zhou et al. (Scheme 5.16).[35] The corresponding nitroalkylated indoles **5** were obtained in high yields and enantioselectivities.

According to the proposed mechanism, Zn(II) activates the nitroalkene **1** by forming an intermediate **13**, and then the activated nitroalkene undergoes nucleophilic addition with indole to afford the alkylated adduct **14**. Subsequently, hydrogen transfer and dissociation afforded the product **5** with the regeneration of Zn(II)-bisoxazoline catalyst **C7** (Scheme 5.17).

In another report, the You group demonstrated Friedel–Crafts alkylation of indoles **2** with nitroalkenes **1** using biphenanthryl bis(oxazoline) **L10** ligand and a catalytic amount of $Zn(OTf)_2$ to give the products **3** in high yields and modest-to-good enantioselectivities (Scheme 5.18).[36] In this method, enantioselectivity was highly impacted by the steric hindrance between the substrates.

Pan et al. developed a catalytic asymmetric Friedel–Crafts alkylation of *N*-methylindole **2** with a variety of nitroalkenes **1** employing bifunctional abietic-acid-derived thiourea **L11**/$Zn(OTf)_2$ complex as a catalyst to give adducts **3** with high yields and good enantioselectivities. Nitroalkenes bearing alkyl groups reacted to give the products with low yields and low ee (Scheme 5.19).[16]

The Singh group described the synthesis of nitroalkylated indoles **5a** by the enantioselective Friedel–Crafts reaction of indoles **4a** with nitroalkenes **1** in the presence of $Cu(OTf)_2$ and

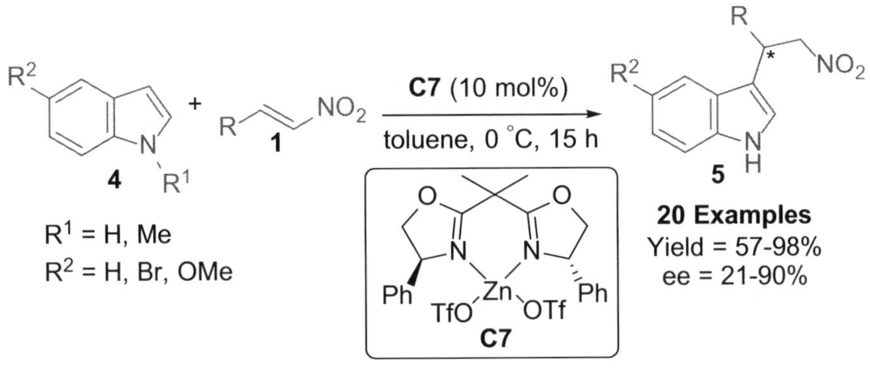

SCHEME 5.16 Asymmetric Friedel–Crafts alkylation of indoles with nitroalkenes catalyzed by Zn-(II)-bisoxazoline complex.

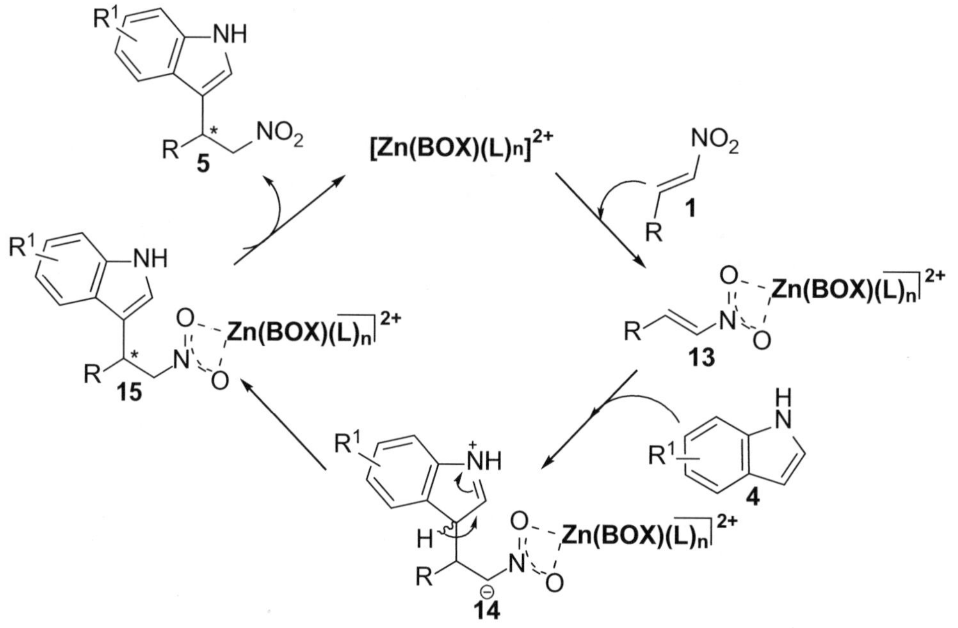

SCHEME 5.17 Proposed mechanism for the asymmetric Friedel–Crafts alkylation reaction of indoles with nitroalkenes in the presence of Zn(II)-bisoxazoline complex.

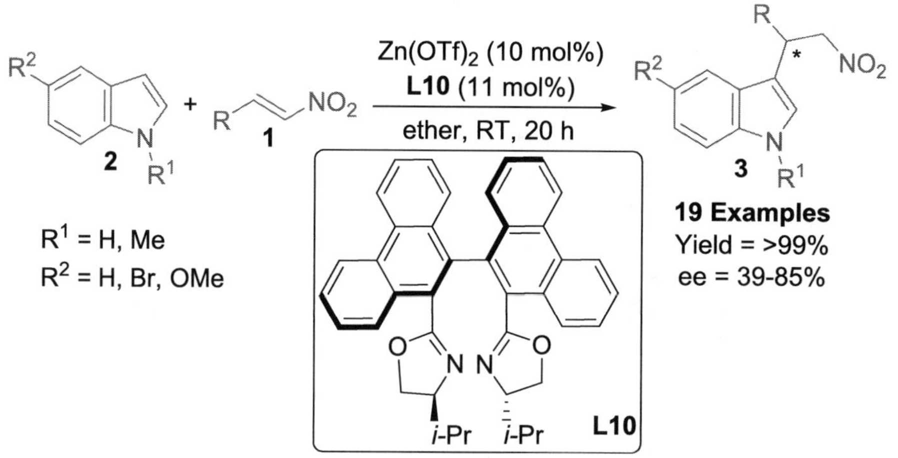

SCHEME 5.18 Asymmetric Friedel–Crafts alkylation of indoles with nitroalkenes using $Zn(OTf)_2$ and biphenanthryl bis(oxazoline) catalyst system.

the C_2-symmetric BOX ligand **L12** (Scheme 5.20). The reaction furnished various nitroalkylated indoles **5a** in moderate-to-excellent yields and high enantioselectivities (up to 86% ee).[37]

Gao et al. reported an asymmetric Friedel–Crafts alkylation of indoles **4** with β-trifluoromethylated nitroalkenes **16** mediated by a $Ni(ClO_4)_2$-bisoxazoline **L13** complex for the synthesis of indoles **17** bearing all-carbon quaternary chiral centers.[38] This methodology suggested an efficient approach to synthesize benzylic trifluoromethyl substituted chiral compounds **17** (Scheme 5.21).

The Meggers group reported an asymmetric Friedel–Crafts alkylation of indoles **4** with α-bromo nitroalkenes **18** in the presence of metal-templated H-bond-mediated catalyst **C8**, which gave the

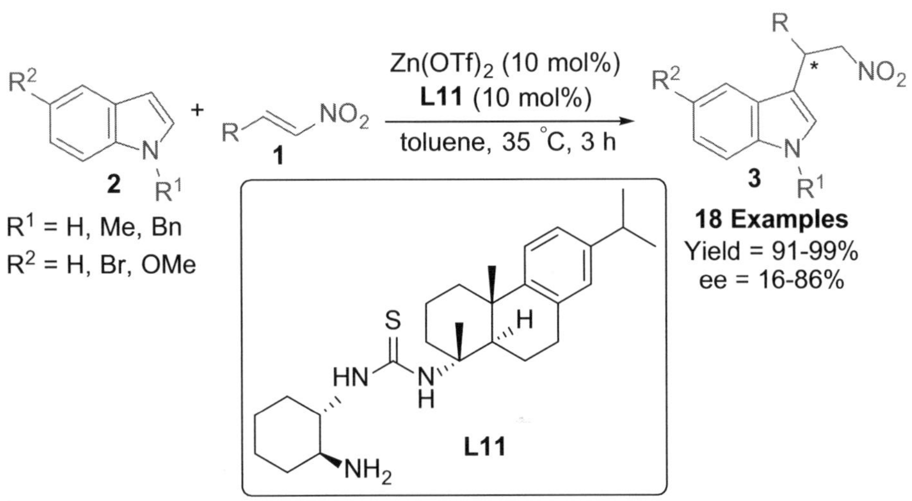

SCHEME 5.19 Asymmetric Friedel–Crafts alkylation of *N*-methylindole with nitroalkenes in the presence of $Zn(OTf)_2$ and bifunctional abietic acid-derived thiourea.

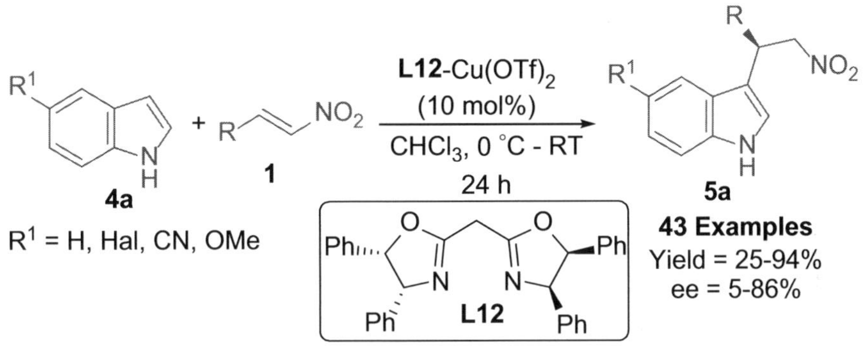

SCHEME 5.20 Asymmetric Friedel–Crafts reaction of indoles with nitroalkenes in the presence of $Cu(OTf)_2$–bisoxazoline complex.

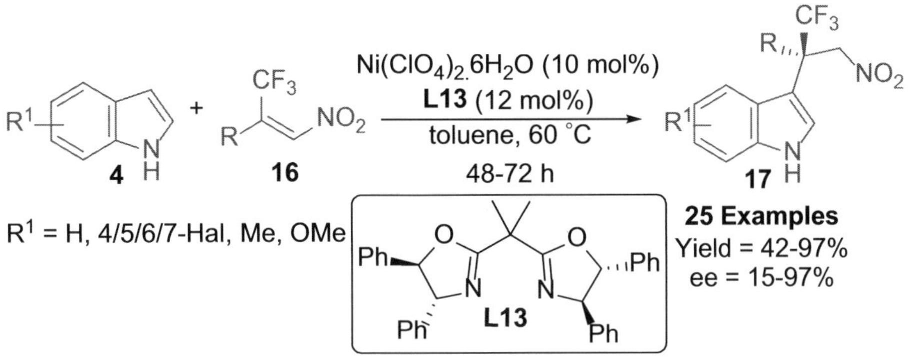

SCHEME 5.21 Asymmetric Friedel–Crafts alkylation of indoles with β-trifluoromethyl-β-disubstituted nitroalkenes.

alkylated products **19** in excellent yields and enantioselectivities (Scheme 5.22).[39] The bromo substituent on nitroalkene **18** plays a vital role in asymmetric induction and serves as a site for functional group transformation.

The Du group reported a highly enantioselective Friedel–Crafts alkylation of indoles **2** with 3-nitro-2H-chromenes **20a** employing diphenylamine-linked bis(oxazoline) **L14**-Zn(II) complex (Scheme 5.23).[40] An electron-withdrawing substituent such as chlorine in the 5-position of indole **2** provided the product **21** with decreased yield as well as selectivity. In this reaction, the catalyst follows a bifunctional pathway where the NH-π interaction directs the indole **2** to approach from the backside, whereas the Lewis acid (Zn cation) activates the chromene molecule **20a** by coordinating with oxygen atoms of the nitro group. Meanwhile, hydrogen atom orients toward the backside to eliminate the steric repulsion between the aryl group and the incoming indole (**TS-4**, Scheme 5.23).

The Jia group described an asymmetric Friedel–Crafts alkylation of indoles **4** with β,β-disubstituted nitroalkenes **22** in the presence of a nickel(II) perchlorate–bisoxazoline **L15** complex.[41] This reaction allowed the formation of product **23** with all-carbon quaternary stereocenters attached to the C3 position of the indole (Scheme 5.24). The nitroalkenes possessing electron-withdrawing substituents gave better yields and enantioselectivities than those having electron-donating groups.

Arai et al. demonstrated an enantioselective Friedel–Crafts reaction of 2-vinylindoles **24** with nitroalkenes **1** to yield chiral indoles **25** by employing chiral bis(imidazolidine)pyridine (PyBidine) **L16**-$Ni(OTf)_2$ complex (Scheme 5.25).[42] Here, the mild Lewis acidity of the PyBidine **L16**-$Ni(OTf)_2$ catalyst is responsible for this transformation. Moreover, the X-ray structure of the product revealed the *S-trans* configuration for C2–C3 double bond, which generally does not participate in [4+2] cycloaddition.

The same group described an imidazoline–aminophenol **L17**-Cu complex catalyzed asymmetric Friedel–Crafts/protonation of indoles or pyrroles **4/26** with nitroalkenes **27** to give the substituted β-heteroaryl-nitro adducts **28/29** preferably in an *anti*-selective manner. This allowed a diastereoselective construction of the chiral acyclic β-indolylalkylamines (Scheme 5.26).[13] Here, indole **4**

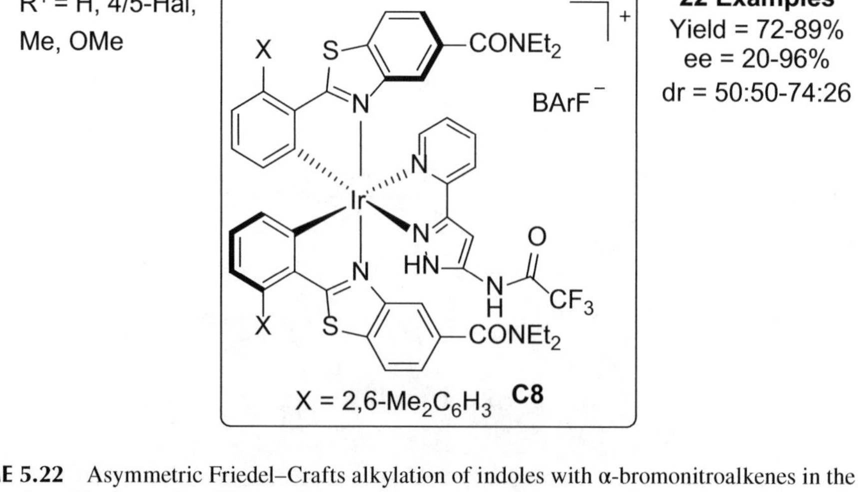

SCHEME 5.22 Asymmetric Friedel–Crafts alkylation of indoles with α-bromonitroalkenes in the presence of iridium-based hydrogen-bonding catalyst.

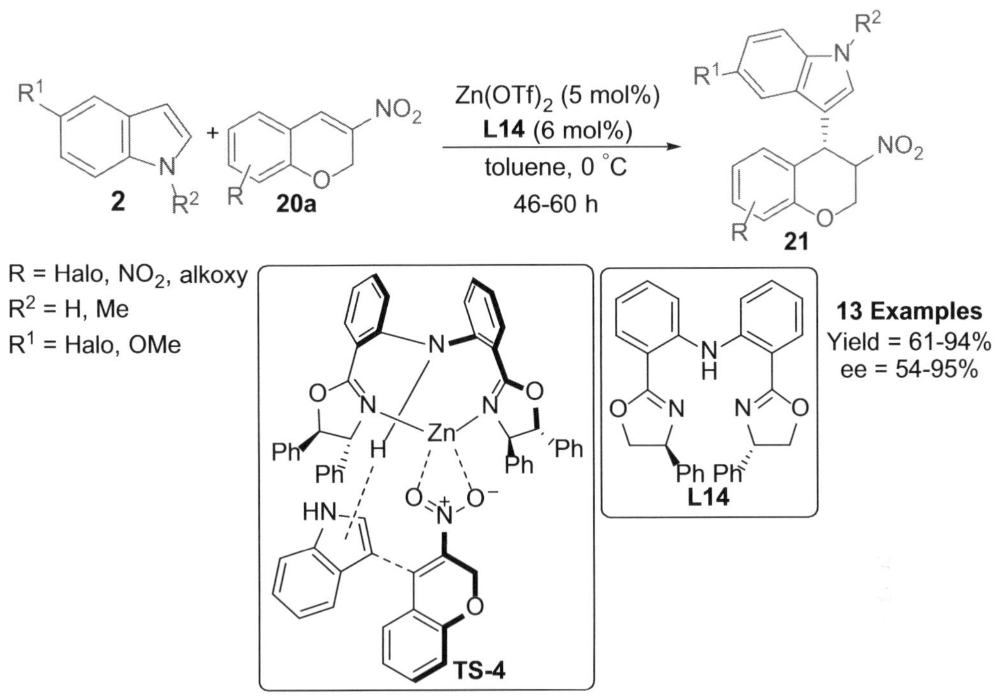

SCHEME 5.23 Asymmetric Friedel–Crafts alkylation between indoles and 3-nitro-2H chromenes using diphenylamine-linked bis(oxazoline)-Zn(II) complex.

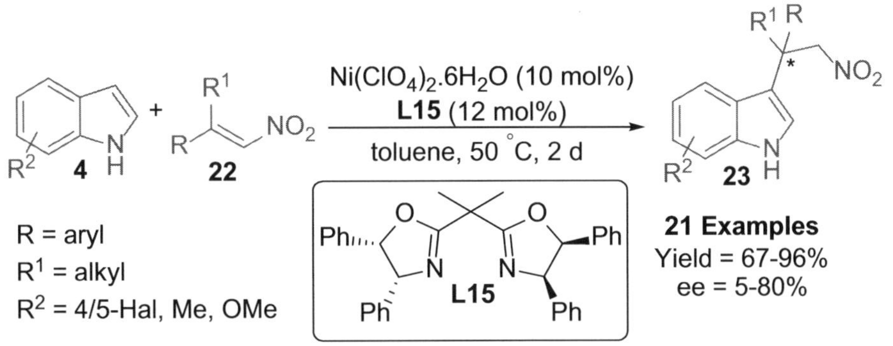

SCHEME 5.24 Asymmetric Friedel–Crafts alkylation of indoles with β,β-disubstituted nitroalkenes in the presence of a nickel(II) perchlorate–bisoxazoline complex.

attacks from the *Re*-face of the activated nitroalkenes to give the Cu-nitronate intermediate. This is followed by the reaction of acidic proton adjacent to the nitrogen atom of indole with nitronate to afford the *anti*-adduct **28** (**TS-5**, Scheme 5.26).

Takenaka et al. described the Friedel–Crafts alkylation of 4,7-dihydroindoles **30** to nitroalkenes **1** to afford β-nitro-indol-2-yl derivatives **31** with good yields and excellent enantioselectivities by employing helically chiral 2-aminopyridinium ion **C9** as hydrogen bond donor catalyst, which can be readily prepared from 1-azahelicene *N*-oxide (Scheme 5.27).[43] Further, the utility of the process was demonstrated by the stereoselective synthesis of monoamine oxidase (MAO) inhibitor in two steps with 74% overall yield and retention of configuration.

The You group developed an asymmetric Friedel–Crafts reaction of 4,7-dihydroindoles **30** with nitroalkenes **1** by employing chiral phosphoric acid **C10** as the catalyst (Scheme 5.28).[44] Very low

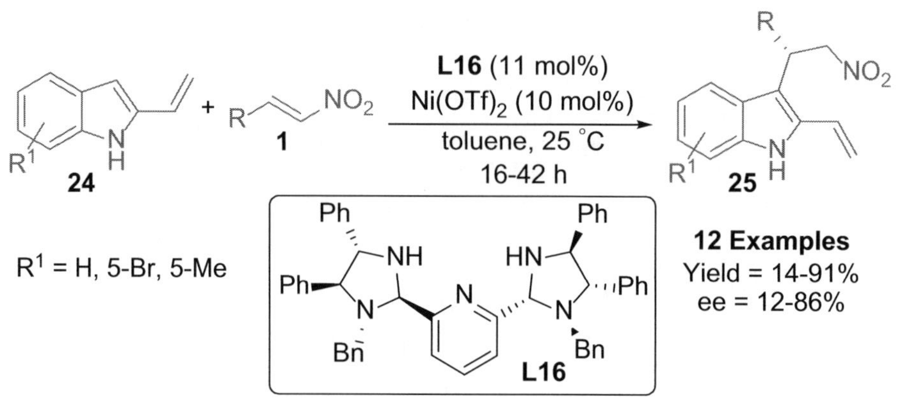

SCHEME 5.25 Asymmetric Friedel–Crafts reaction between 2-vinylindoles and nitroalkenes catalyzed by chiral bis(imidazolidine)pyridine-Ni(OTf)$_2$ complex.

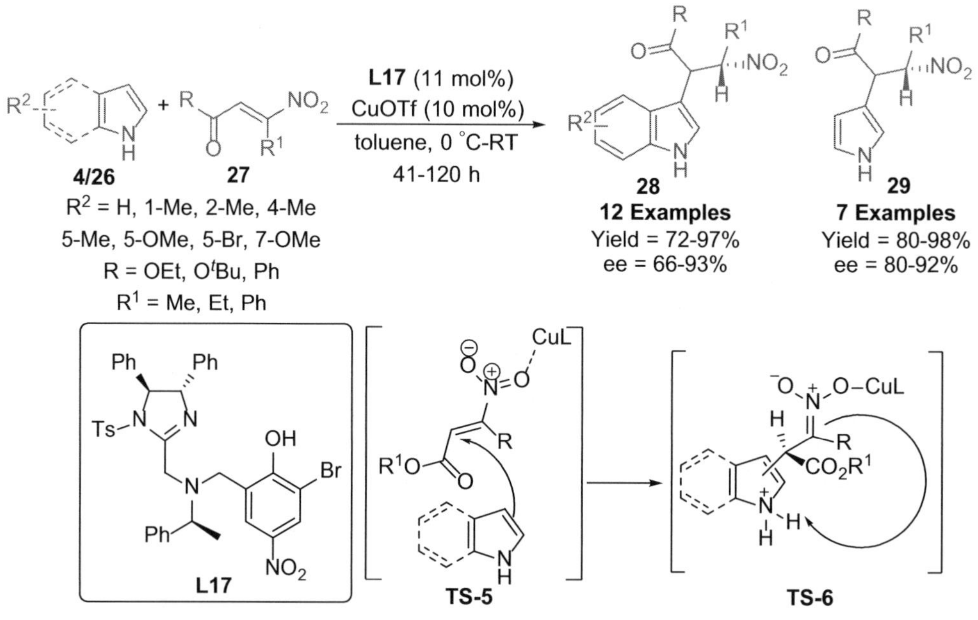

SCHEME 5.26 Asymmetric Friedel–Crafts alkylation of indoles and pyrroles with nitroalkenes catalyzed by the imidazoline–aminophenol–Cu complex.

catalyst (0.5 mol%) loading was achieved with slow addition of the nitroalkene using a syringe pump to afford the products **31** in good yields and enantioselectivity. The adducts thus obtained were then subjected to oxidation to yield the corresponding 2-substituted indole derivatives with retention of configuration. In this reaction, the chiral phosphoric acid **C10** serves as a bifunctional catalyst, wherein the acidic proton and the P=O moiety of the catalyst **C10** form hydrogen bonding with the nitroalkene **1** and dihydroindole **30**, respectively, which plays a vital role in achieving high ee values (**TS-7**, Scheme 5.28).

The Jia group employed Ni(ClO$_4$)$_2$–bisoxazoline **L13** complex as a catalyst for the Friedel–Crafts alkylations of 4,7-dihydroindoles **30** with β-CF$_3$-β-disubstituted nitroalkenes **16** to give alkylated dihydroindoles **32** bearing trifluoromethylated all-carbon quaternary chiral centers in high yields and high enantioselectivities (Scheme 5.29).[45] The obtained Friedel–Crafts alkylated products

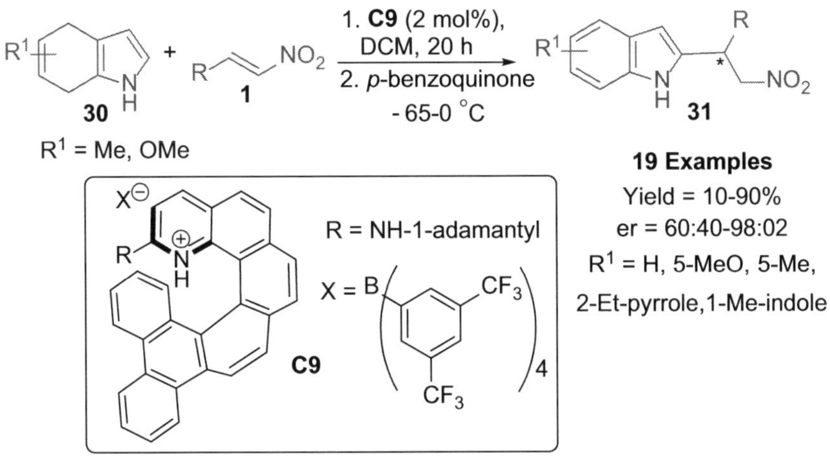

SCHEME 5.27 Asymmetric Friedel–Crafts alkylation of 4,7-dihydroindoles to nitroalkenes using helically chiral 2-aminopyridinium ion.

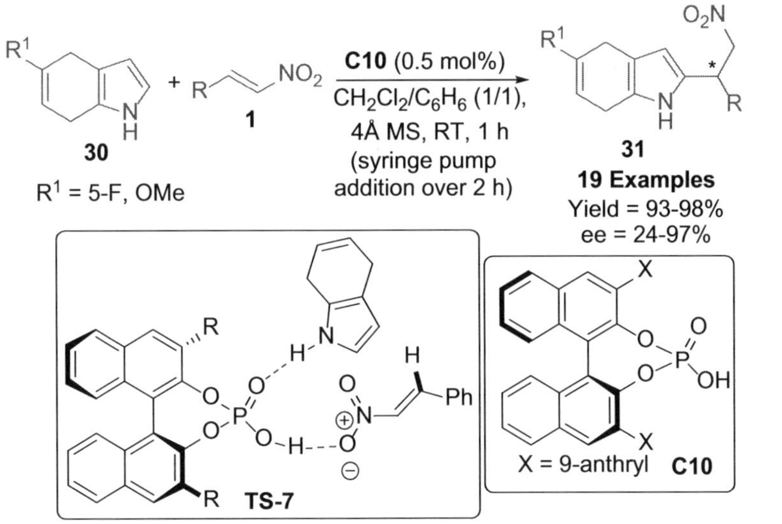

SCHEME 5.28 Asymmetric Friedel–Crafts reaction of 4,7-dihydroindoles with nitroalkenes catalyzed by chiral phosphoric acid.

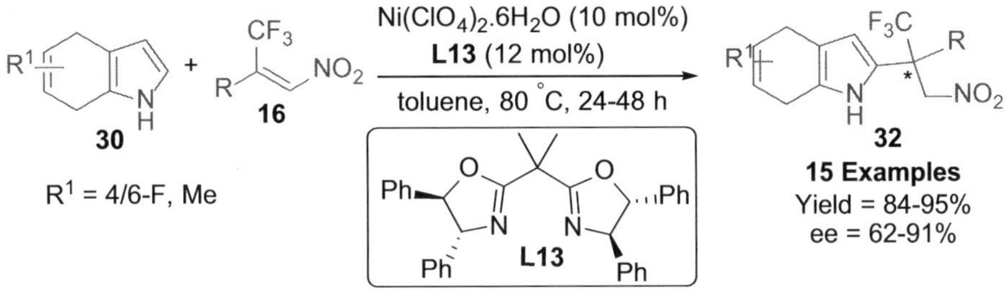

SCHEME 5.29 Asymmetric Friedel–Crafts reaction of 4,7-dihydroindoles with β-CF_3-β-disubstituted nitroalkenes in the presence of $Ni(ClO_4)_2$-bisoxazoline catalyst system.

were subjected to oxidation using 2,3-dichloro-5,6-dicyano-1,4-benzoquinone (DDQ) to afford corresponding 2-alkylated indoles without any loss of enantioselectivity.

The Arai group described a catalytic asymmetric Friedel–Crafts reaction of indoles **4**/pyrroles **26** with various nitroalkenes **1** by employing imidazoline–aminophenol ligand **L18**-CuOTf complex.[46] The introduction of nitro-functionality on the phenol ring of the ligand **L18** enhances the catalyst activity and selectivity (Scheme 5.30).

The enantioselective asymmetric Friedel–Crafts alkylation of hydroxyindoles **35** with β-nitrostyrenes **1** was reported by the Pedro group using Rawal's squaramide **C11** as the catalyst (Scheme 5.31).[47] Using this protocol, C-4, C-5 and C-7 functionalization was accomplished successfully with high yields and enantioselectivity using 5-hydroxy, 4-hydroxy and 6-hyroxyindole substrates, respectively. The absolute configuration of the product was assigned '*S*' by single-crystal X-ray analysis. Bifunctional squaramide catalyst **C11** plays a significant role in activating the substrates and *ortho*-selectivity of the Friedel–Crafts alkylation (**TS-8**, Scheme 5.31).

The Kiliç group developed an efficient bisphenol A-derived chiral tridentate Schiff-base ligand **L19**-Cu catalyst for the asymmetric Friedel–Crafts alkylation of pyrrole **26** with nitroalkenes **1** (Scheme 5.32).[48] The synthetic utility of the reaction was illustrated by the oxidative cleavage of the pyrrole ring to give 3-nitro-2-arylpropanamides **34** in enantiomerically pure form.

Trost and Muller demonstrated the use of dinuclear zinc bis-prophenol complex **C12** in Friedel–Crafts reaction of unprotected pyrroles **37** with various nitroalkenes **1** to synthesize both mono and disubstituted pyrroles **38** in a simple two-step sequence (Scheme 5.33).[49] In the proposed mechanism, the nitroalkene coordinates with the catalyst **C12** and undergoes the alkylation, meanwhile the proton exchange occurs to release the product **38** with the regeneration of the catalyst **C12**.

Asymmetric Friedel–Crafts alkylation of pyrroles **37** with nitroalkenes **1** was performed by the Zhang group employing a heterotrimetallic Pd-Sm-Pd catalyst along with a chiral ligand **L20** to afford the highly enantioenriched 2-substituted and 2,5-disubstituted pyrroles **38** (Scheme 5.34).[50] Further, it was proposed that during the catalytic reaction, this trinuclear complex serves as a bifunctional catalyst to activate both the substrates with Pd(II) and Sm(III) active sites, similar to a dinuclear Zn catalysis.

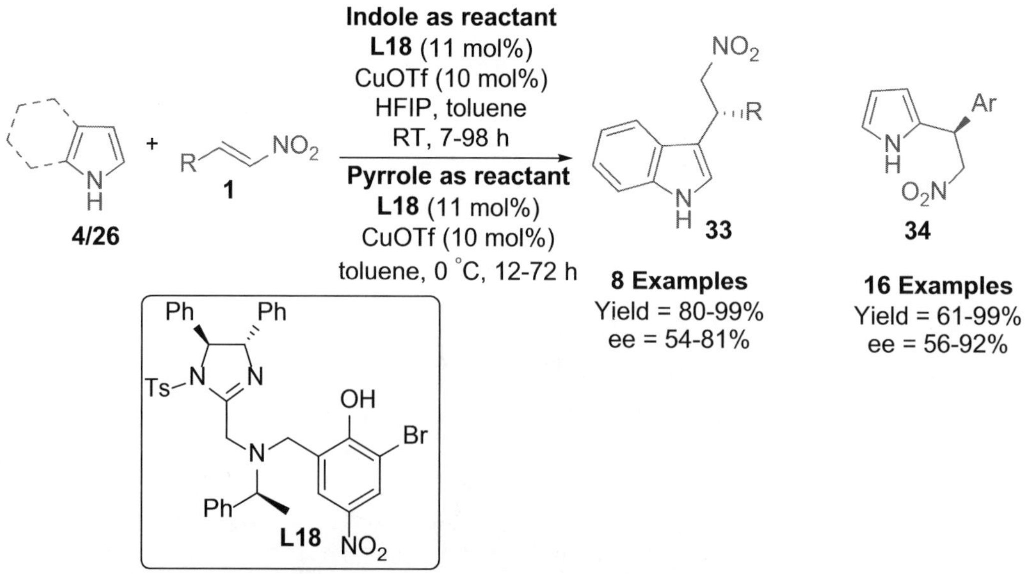

SCHEME 5.30 Asymmetric Friedel–Crafts reaction of indoles and pyrroles with various nitroalkenes in the presence of imidazoline–aminophenol–Cu complex.

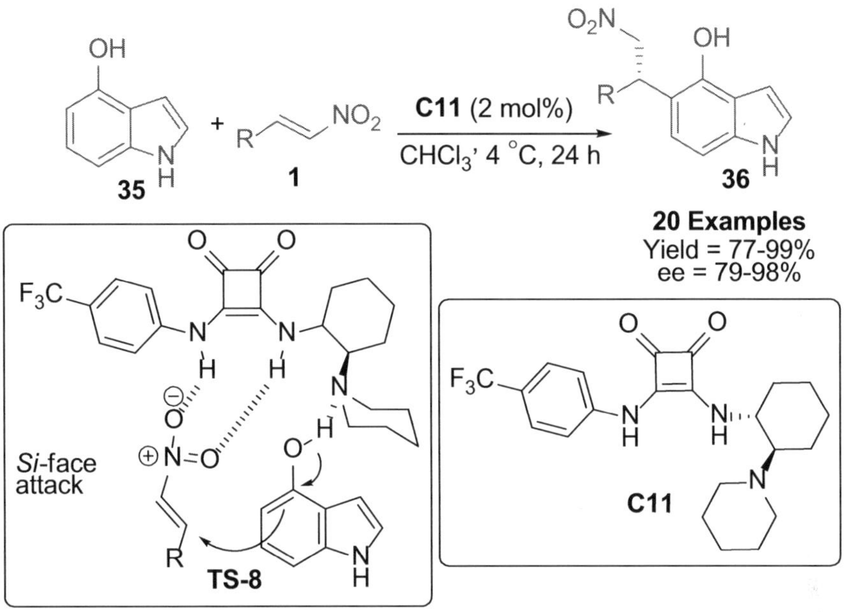

SCHEME 5.31 Rawal's squaramide-catalyzed enantioselective asymmetric Friedel–Crafts alkylation of hydroxyindoles with nitrostyrenes.

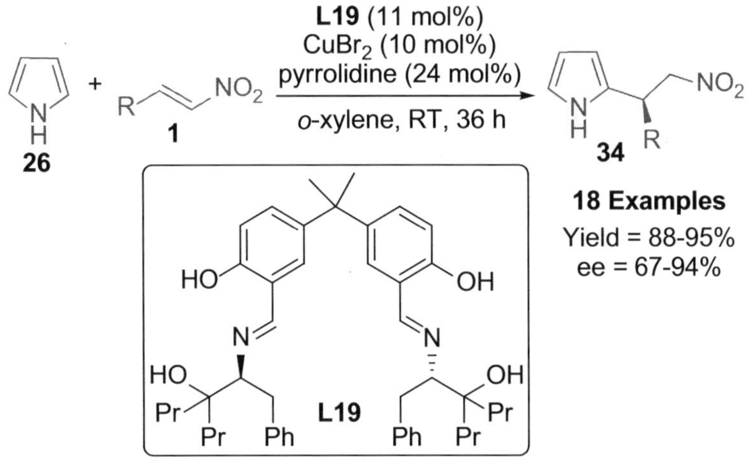

SCHEME 5.32 Asymmetric Friedel–Crafts alkylation of pyrrole with nitroalkenes catalyzed by Cu complex of bisphenol A-derived Schiff base.

The Wang group reported an efficient catalytic system, i.e., tridentate Schiff base-copper complex **C13** in an enantioselective Friedel–Crafts alkylation of pyrroles **37** with nitroalkenes **1** to afford substituted pyrroles **38** in high yields and enantioselectivities (Scheme 5.35).[51] The synthetic utility of this methodology was illustrated by the concise synthesis of nicotine analogs.

The asymmetric Friedel–Crafts alkylation of 2-methoxyfuran **39** with nitroalkenes **1** was reported by Du et al. using diphenylamine tethered bis(oxazoline) **L21**-$Zn(OTf)_2$ complex as a catalyst (Scheme 5.36).[52] Most aromatic and heteroaromatic groups, such as naphthalene, furan, and thiophene-substituted nitroalkenes, also participated well to afford the corresponding products **40** with good enantioselectivities. In the possible mechanism, it was presumed that the catalyst works in a bifunctional manner. Zinc(II) serves as a Lewis acid to activate the nitroalkene **1**, whereas

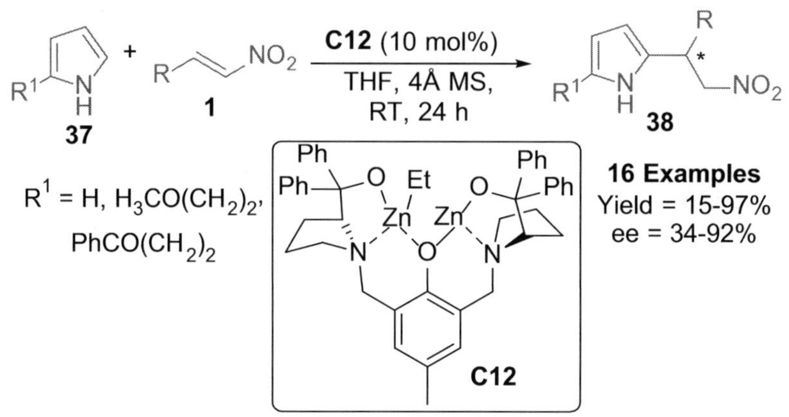

SCHEME 5.33 Asymmetric Friedel–Crafts alkylation between pyrrole and nitroalkenes catalyzed by dinuclear zinc bis-prophenol complex.

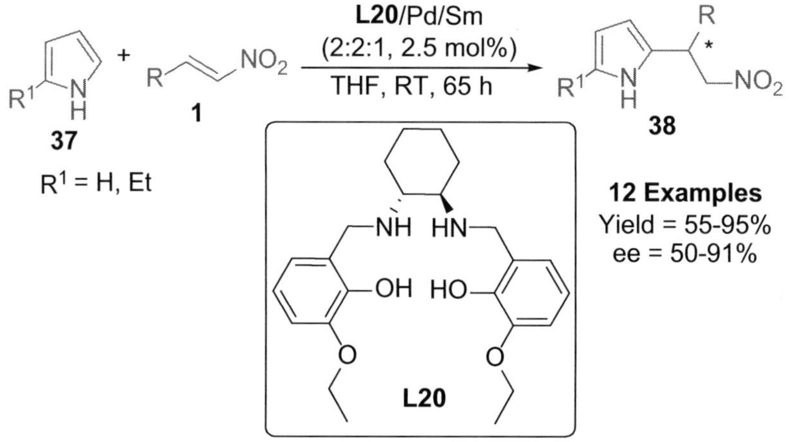

SCHEME 5.34 Asymmetric Friedel–Crafts alkylation of pyrrole with nitroalkenes catalyzed by heterotrimetallic Pd-Sm-Pd complex.

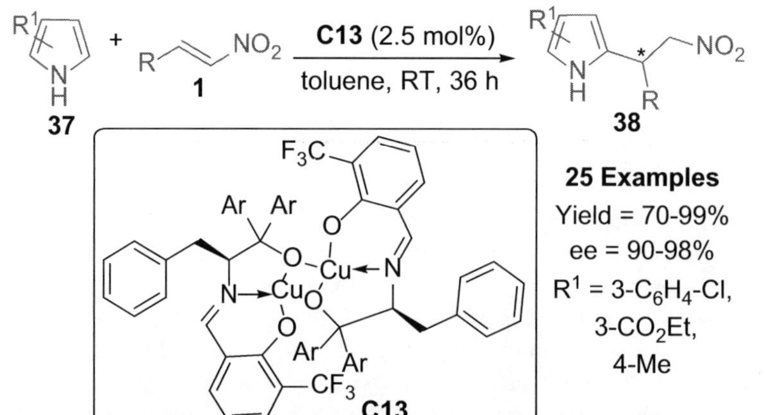

SCHEME 5.35 Asymmetric Friedel–Crafts alkylation of pyrrole with nitroalkenes catalyzed by tridentate Schiff base-copper complex.

the NH group works as a hydrogen bond donor via H-bonding interaction to direct the attack of 2-methoxyfuran **39** from the *Si*-face (**TS-9**, Scheme 5.36). The methoxyfuran motif in the product was transformed to the corresponding carboxylic acid via oxidative fragmentation without loss in stereoselectivity.

The Zeng group described an asymmetric Friedel–Crafts reaction of 3,5-dimethoxyphenol **41** with nitroalkenes **1** using 5 mol% of quinine-derived thiourea **C14** as a catalyst to afford products **42** with excellent yields and enantioselectivities (Scheme 5.37).[53] Here, the catalyst **C14** offers simultaneous activation of both the substrates via H-bonding interactions. This led to the increase

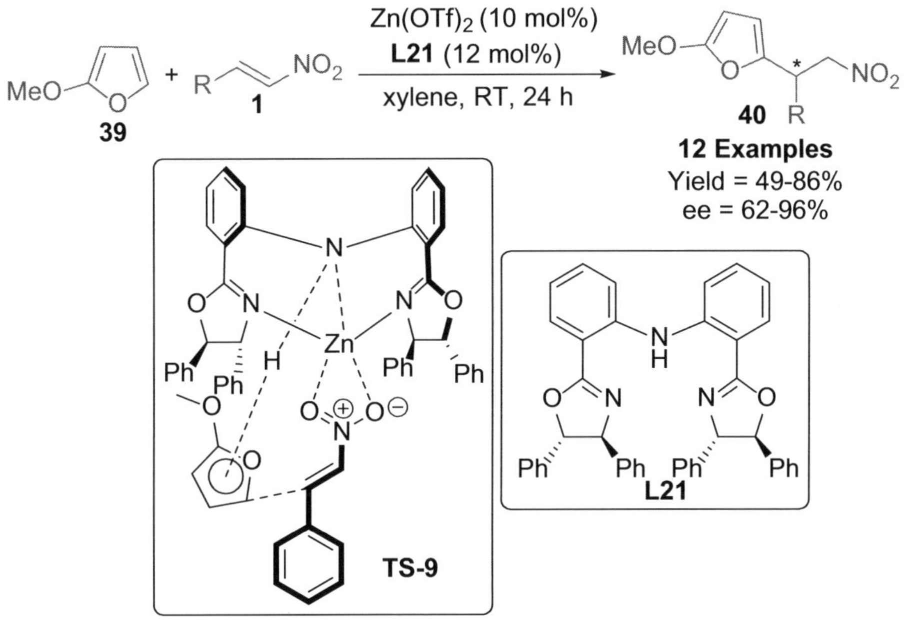

SCHEME 5.36 Asymmetric Friedel–Crafts alkylation of 2-methoxyfuran with nitroalkenes catalyzed by diphenylamine tethered bis(oxazoline)-$Zn(OTf)_2$ complex.

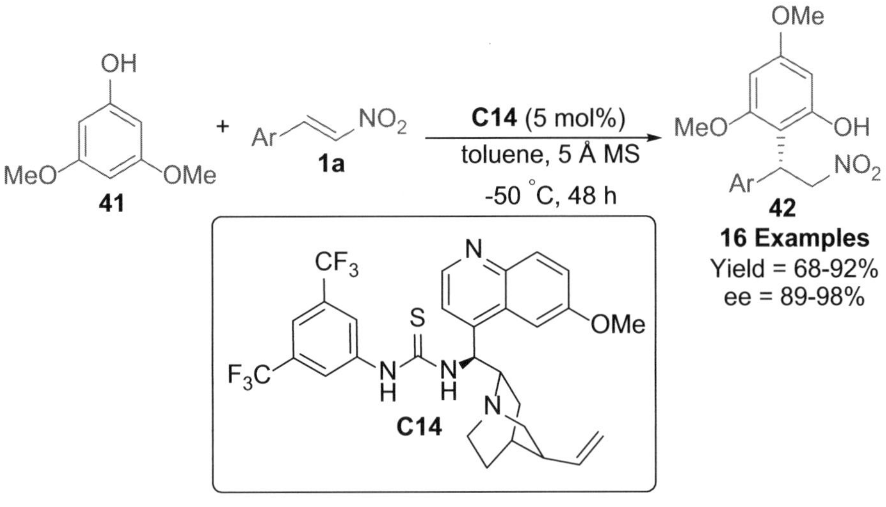

SCHEME 5.37 Asymmetric Friedel–Crafts reaction of 3,5-dimethoxyphenol with nitroalkenes using quinine-derived thiourea catalyst.

in the electrophilicity of nitroalkene **1**. On the other hand, the nucleophilicity of the catalyst **C14** was increased by the interaction of the nitro group with the tertiary amine of catalyst as well as the –OH group of the phenyl ring, thus facilitating chiral induction in the Friedel–Crafts reaction. The *R* configuration of the product **42** was confirmed by derivatization.

Carmona et al. utilized the chiral fragments Cp/Rh(*R*)-prophos **C15** to activate nitroalkenes **1** for the Friedel–Crafts alkylation of 1,3,5-trimethoxybenzene **43**. Quantitative and regioselective conversion to the monoalkylated adduct **44** was observed in this reaction (Scheme 5.38).[54] Further, the catalyst/adduct complexes formed by the reaction of **C15** with 1,3,5-trimethoxybenzene **43** was detected spectroscopically.

Sohtome et al. developed chiral C3-linked guanidine/bisthiourea **C16** catalyzed *ortho*-selective alkylation of phenols **45** with nitroalkenes **1** (Scheme 5.39).[55] The catalyst **C16** effectively promotes nucleophilic addition at the C6 position of phenol derivatives **45** selectively to afford the corresponding product **46** with 82%–94% ee. The kinetic studies using Eyring plots provided evidence that differences in the activation entropies play a vital role in the stereodiscrimination observed in the Friedel–Crafts reaction.

Tanaka and co-workers reported the first asymmetric Friedel–Crafts alkylation of *N,N*-dialkylanilines **47** with nitroalkenes **1b** mediated by a homochiral organometallic framework R-Cu-MOF **C17** (Scheme 5.40).[56] The corresponding products **48** were obtained in high yields (up to 96%) and excellent enantioselectivities (up to 98%) via the *Si*-face addition of dialkylaniline **47** to nitroalkene **1b** (**TS-10**). The authors also reported addition of pyrrole and indole under similar conditions with 64%–87% yield and 78%–90% ee.

Du and co-workers reported the stereoselective Friedel–Crafts reaction of indoles **2** with nitrodienes **49** and 2-propargyloxy-β-nitrostyrenes **51** mediated by diphenylamine-linked bis(oxazoline) **L22**–$Zn(OTf)_2$ complexes as catalysts (Scheme 5.41).[57] The alkylation products **52** of indoles **2** with 2-propargyloxy-β-nitrostyrenes **51** could be further transformed to access the chiral isoxazolobenzoxepane, which are medicinally significant compounds.

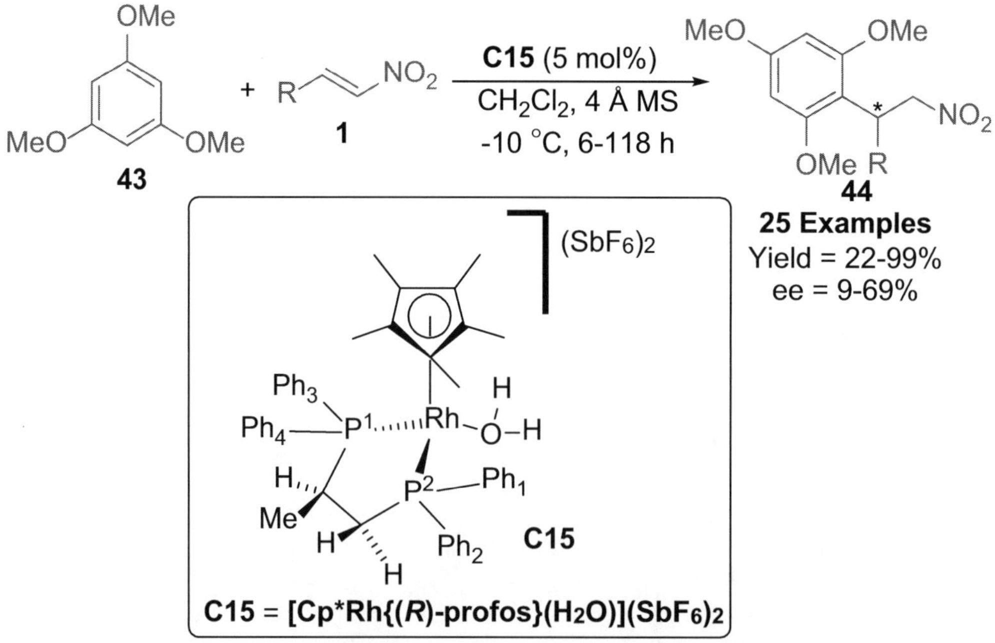

SCHEME 5.38 Asymmetric Friedel–Crafts alkylation of 1,3,5-trimethoxybenzene with nitroalkenes using Cp/Rh(*R*)-prophos catalyst.

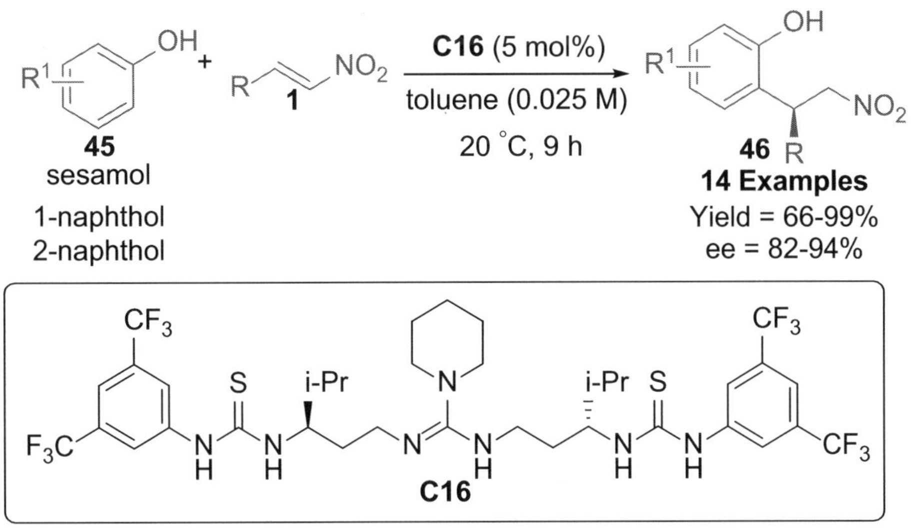

SCHEME 5.39 Asymmetric Friedel–Crafts alkylation of phenols with nitroalkenes in the presence of guanidine/bisthiourea organocatalyst.

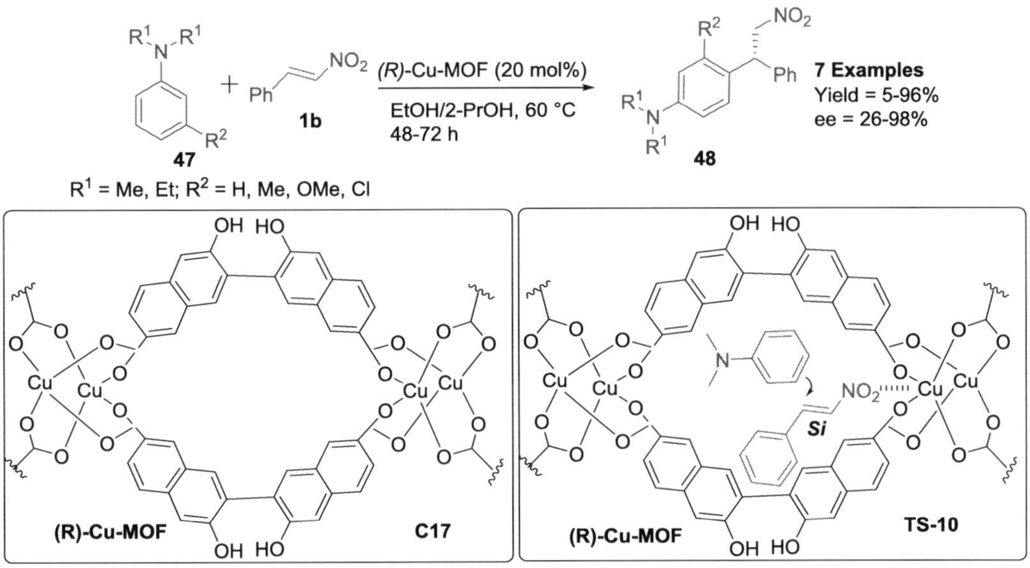

SCHEME 5.40 Asymmetric Friedel–Crafts alkylation of *N,N*-dialkylanilines with nitroalkenes mediated by a chiral organometallic framework.

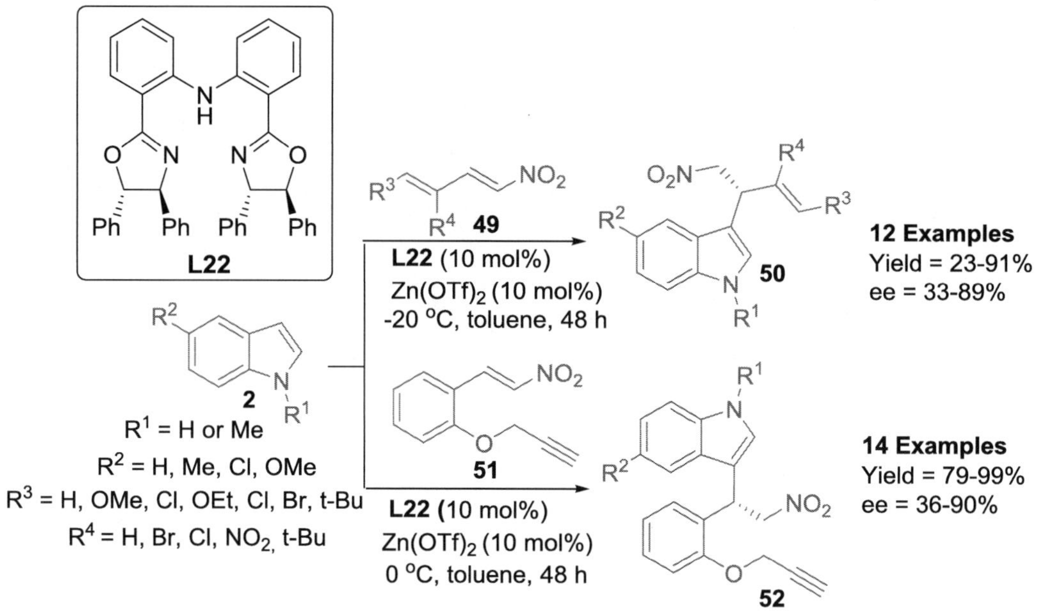

SCHEME 5.41 Asymmetric Friedel–Crafts reaction of indoles with nitrodienes and 2-propargyloxy-β-nitrostyrenes.

5.3 CONCLUSIONS

Nitroalkenes undergo catalytic asymmetric Friedel–Crafts reaction with aromatic groups such as indole, pyrrole and other electron-rich arenes and heteroarenes in the presence of various chiral Lewis and Brønsted acid catalysts. The enantioenriched β-aryl nitroalkanes, which are formed in high yield and selectivity in most cases, are valuable precursors to drugs and bioactive natural products.

REFERENCES

1. Poulsen, T. B.; Jørgensen, K. A. *Chem. Rev.* **2008**, *108*, 2903.
2. Bartoli, G.; Bencivenni, G.; Dalpozzo, R. *Chem. Soc. Rev.* **2010**, *39*, 4449.
3. Dalpozzo, R. *Chem. Soc. Rev.* **2015**, *44*, 742.
4. Yuan, Z.-L.; Lei, Z.-Y.; Shi, M. *Tetrahedron: Asymmetry* **2008**, *19*, 1339.
5. Zhuang, W.; Hazell, R. G.; Jørgensen, K. A. *Org. Biomol. Chem.* **2005**, *3*, 2566.
6. Fleming, E. M.; McCabe, T.; Connon, S. J. *Tetrahedron Lett.* **2006**, *47*, 7037.
7. Sui, Y.; Liu, L.; Zhao, J.-L.; Wang, D.; Chen, Y.-J. *Tetrahedron* **2007**, *63*, 5173.
8. Arai, T.; Yokoyama, N. *Angew. Chem. Int. Ed.* **2008**, *47*, 4989.
9. Ganesh, M.; Seidel, D. *J. Am. Chem. Soc.* **2008**, *130*, 16464.
10. Itoh, J.; Fuchibe, K.; Akiyama, T. *Angew. Chem. Int. Ed.* **2008**, *47*, 4016.
11. Liu, H.; Lu, S.-F.; Xu, J.; Du, D.-M. *Chem. - Asian J.* **2008**, *3*, 1111.
12. McKeon, S. C.; Müller-Bunz, H.; Guiry, P. J. *Eur. J. Org. Chem.* **2009**, *2009*, 4833.
13. Arai, T.; Awata, A.; Wasai, M.; Yokoyama, N.; Masu, H. *J. Org. Chem.* **2011**, *76*, 5450.
14. Hirata, T.; Yamanaka, M. *Chem. - Asian J.* **2011**, *6*, 510.
15. Marqués-López, E.; Alcaine, A.; Tejero, T.; Herrera, R. P. *Eur. J. Org. Chem.* **2011**, *2011*, 3700.
16. Huang, W.; Wang, H.; Huang, G.; Wu, Y.; Pan, Y. *Eur. J. Org. Chem.* **2012**, 2012, 5839.
17. Arai, T.; Yamamoto, Y.; Awata, A.; Kamiya, K.; Ishibashi, M.; Arai, M. A. *Angew. Chem. Int. Ed.* **2013**, *52*, 2486.
18. Chen, L.-A.; Tang, X.; Xi, J.; Xu, W.; Gong, L.; Meggers, E. *Angew. Chem. Int. Ed.* **2013**, *52*, 14021.

19. Liu, H.; Du, D.-M. *Adv. Synth. Catal.* **2010**, *352*, 1113.
20. Liu, H.; Du, D.-M. *Eur. J. Org. Chem.* **2010**, 2010, 2121.
21. Guo, F.; Lai, G.; Xiong, S.; Wang, S.; Wang, Z. *Chem. - Eur. J.* **2010**, *16*, 6438.
22. Venkatanna, K.; Yeswanth Kumar, S.; Karthick, M.; Padmanaban, R.; Ramaraj Ramanathan, C. *Org. Biomol. Chem.* **2019**, *17*, 4077.
23. Kim, H. Y.; Kim, S.; Oh, K. *Angew. Chem. Int. Ed.* **2010**, *49*, 4476.
24. Wu, J.; Li, X.; Wu, F.; Wan, B. *Org. Lett.* **2011**, *13*, 4834.
25. Méndez, I.; Rodríguez, R.; Polo, V.; Passarelli, V.; Lahoz, F. J.; García-Orduña, P.; Carmona, D. *Chem. - Eur. J.* **2016**, *22*, 11064.
26. Chen, L.-Y.; Guillarme, S.; Saluzzo, C. *ARKIVOC* **2013**, *iii*, 227.
27. Lu, S.-F.; Du, D.-M.; Xu, J. *Org. Lett.* **2006**, *8*, 2115.
28. Carmona, D.; Méndez, I.; Rodríguez, R.; Lahoz, F. J.; García-Orduña, P.; Oro, L. A. *Organometallics* **2014**, *33*, 443.
29. Zhang, G. *Inorg. Chem. Commun.* **2014**, *40*, 1.
30. Lin, J.-H.; Xiao, J.-C. *Eur. J. Org. Chem.* **2011**, 2011, 4536.
31. Fan, Y.; Kass, S. R. *J. Org. Chem.* **2017**, *82*, 13288.
32. Hao, X.-Q.; Xu, Y.-X.; Yang, M.-J.; Wang, L.; Niu, J.-L.; Gong, J.-F.; Song, M.-P. *Organometallics* **2012**, *31*, 835.
33. Islam, M. S.; Al Majid, A. M. A.; Al-Othman, Z. A.; Barakat, A. *Tetrahedron: Asymmetry* **2014**, *25*, 245.
34. Herrera, R. P.; Sgarzani, V.; Bernardi, L.; Ricci, A. *Angew. Chem. Int. Ed.* **2005**, *44*, 6576.
35. Jia, Y.-X.; Zhu, S.-F.; Yang, Y.; Zhou, Q.-L. *J. Org. Chem.* **2006**, *71*, 75.
36. Lin, S.; You, T. *Tetrahedron* **2009**, *65*, 1010.
37. Singh, P. K.; Bisai, A.; Singh, V. K. *Tetrahedron Lett.* **2007**, *48*, 1127.
38. Gao, J.-R.; Wu, H.; Xiang, B.; Yu, W.-B.; Han, L.; Jia, Y.-X. *J. Am. Chem. Soc.* **2013**, *135*, 2983.
39. Huang, K.; Ma, Q.; Shen, X.; Gong, L.; Meggers, E. *Asian J. Org. Chem.* **2016**, *5*, 1198.
40. Jia, Y.; Yang, W.; Du, D.-M. *Org. Biomol. Chem.* **2012**, *10*, 4739.
41. Wu, H.; Sheng, W.-J.; Chen, B.; Liu, R.-R.; Gao, J.-R.; Jia, Y.-X. *Synlett* **2015**, *26*, 2817.
42. Arai, T.; Tsuchida, A.; Miyazaki, T.; Awata, A. *Org. Lett.* **2017**, *19*, 758.
43. Takenaka, N.; Chen, J.; Captain, B.; Sarangthem, R. S.; Chandrakumar, A. *J. Am. Chem. Soc.* **2010**, *132*, 4536.
44. Sheng, Y.-F.; Li, G.-Q.; Kang, Q.; Zhang, A.-J.; You, S.-L. *Chem. - Eur. J.* **2009**, *15*, 3351.
45. Wu, H.; Liu, R.-R.; Shen, C.; Zhang, M.-D.; Gao, J.; Jia, Y.-X. *Org. Chem. Front.* **2015**, *2*, 124.
46. Yokoyama, N.; Arai, T. *Chem. Commun.* **2009**, 2009, 3285.
47. Vila, C.; Rostoll-Berenguer, J.; Sanchez-Garcia, R.; Blay, G.; Fernandez, I.; Munoz, M. C.; Pedro, J. R. *J. Org. Chem.* **2018**, *83*, 6397.
48. Özdemir, H. S.; Şahin, E.; Çakici, M.; Kiliç, H. *Tetrahedron* **2015**, *71*, 2882.
49. Trost, B. M.; Müller, C. *J. Am. Chem. Soc.* **2008**, *130*, 2438.
50. Zhang, G. *Org. Biomol. Chem.* **2012**, *10*, 2534.
51. Guo, F.; Chang, D.; Lai, G.; Zhu, T.; Xiong, S.; Wang, S.; Wang, Z. *Chem. - Eur. J.* **2011**, *17*, 11127.
52. Liu, H.; Xu, J.; Du, D.-M. *Org. Lett.* **2007**, *9*, 4725.
53. Han, X.; Ye, C.; Chen, F.; Chen, Q.; Wang, Y.; Zeng, X. *Org. Biomol. Chem.* **2017**, *15*, 3401.
54. Carmona, D.; Lamata, M. P.; Sánchez, A.; Viguri, F.; Oro, L. A. *Tetrahedron: Asymmetry* **2011**, *22*, 893.
55. Sohtome, Y.; Shin, B.; Horitsugi, N.; Takagi, R.; Noguchi, K.; Nagasawa, K. *Angew. Chem. Int. Ed.* **2010**, *49*, 7299.
56. Tanaka, K.; Sakuragi, K.; Ozaki, H.; Takada, Y. *Chem. Commun.* **2018**, *54*, 6328.
57. Peng, J.; Du, D.-M. *Eur. J. Org. Chem.* **2012**, *2012*, 4042.

6 Catalytic Asymmetric Michael Addition of Miscellaneous Carbon-centered Nucleophiles to Nitroalkenes

6.1 INTRODUCTION

Having discussed the catalytic asymmetric Michael addition of various enolizable carbonyl compounds and electron-rich aromatic compounds to nitroalkenes in the previous chapters, similar addition of other carbon-centered nucleophiles is discussed in this chapter. In particular, nitroalkanes, cyanides, malononitrile, boronic acids and organozinc reagents have been successfully added to nitroalkenes in the presence of various chiral catalysts, including organo- and metal-ligand catalysts. Interestingly, unlike in the previous cases, various metal-ligand catalysts have been extensively employed in the addition of the above-mentioned carbon-centered nucleophiles to achieve high yields and stereoselectivities, as described in the following sections.

6.2 ADDITION OF NITROALKANES

Among the various substrates for Michael addition, nitro compounds are of great importance primarily due to the strong electron-withdrawing nature of the nitro group. The nitro group is also amenable for transformation to a host of diverse functional groups. Nitroalkenes **1a** function as excellent Michael acceptors and the anions of nitroalkanes **2** (nitronates) as important Michael donors. The conjugate addition of nitroalkanes **2** to nitroalkenes **1a** is a very useful reaction as the products obtained, i.e., 1,3-dinitro compounds **3**, are important starting materials for various complex molecules. However, controlling the addition of nitronate to nitroalkene is sometimes a difficult task. The initial product formed is the nitronate, which is sufficiently reactive to add to another molecule of nitroalkene **1**, thus resulting in a mixture of oligomers. In view of this, few groups have been successful in overcoming the above-said difficulties to afford the desired products in high yields and excellent enantioselectivities.

The Feng group reported a highly efficient catalytic asymmetric conjugate addition of nitroalkanes **2a** to nitroalkenes **1a** using a novel chiral $La(OTf)_3$/*N*, *N′*-dioxide **L1** complex. This process provided various 1,3-dinitro compounds **3** bearing two chiral centers in high diastereoselectivity and enantioselectivity (up to 97% ee) under mild conditions (Scheme 6.1).[1]

The chiral DMAP-thiourea hybrid **C1** catalyzed Michael addition of nitroalkanes **2a** to nitroalkenes **1a** proceeds with excellent asymmetric induction (91%–95% ee).[2] Remarkably, the asymmetric induction increases with decreasing catalyst loading, and the optimal amount of catalyst is approximately 2 mol% (Scheme 6.2).

Maruoka and co-workers achieved an efficient, asymmetric formal Michael addition of nitroalkanes to nitroalkenes **1b** by employing chiral quaternary ammonium bifluoride **C2** as the catalyst.[3] Here, silyl nitronates **2b** serve as nitroalkene counterparts (Scheme 6.3). This reaction expanded the use of these organonitro compounds to provide enantiopure 1,3-dinitro compounds **3b**, which are of pharmaceutical importance.

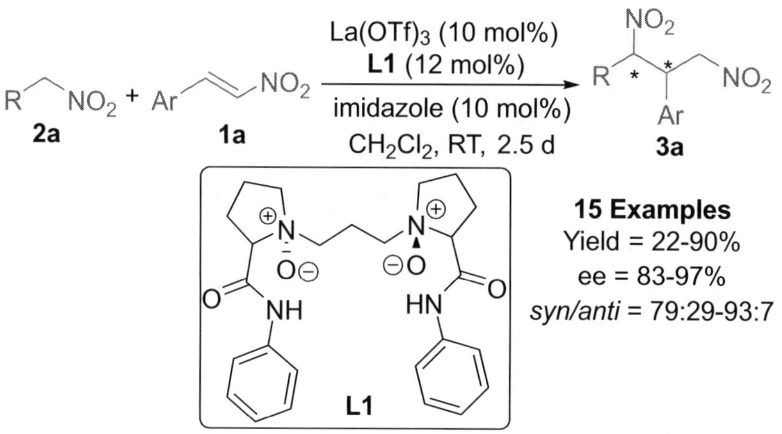

SCHEME 6.1 Asymmetric conjugate addition of nitroalkanes to nitroalkenes using chiral $La(OTf)_3$/*N,N′*-dioxide complex.

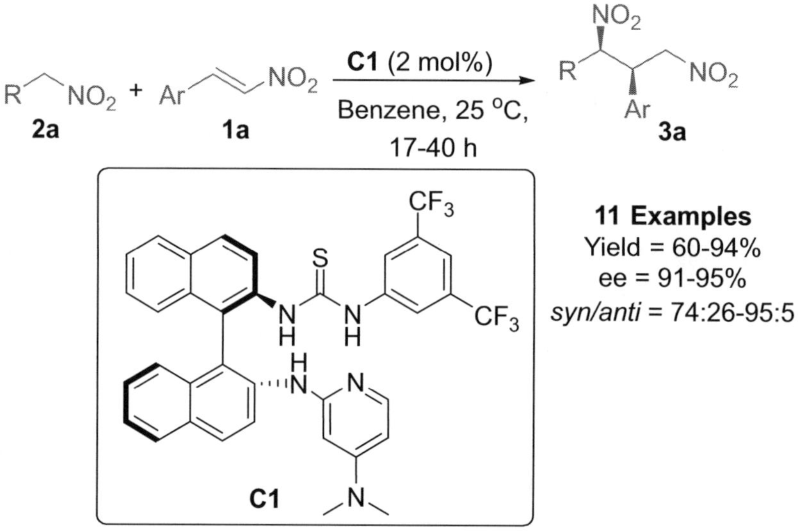

SCHEME 6.2 Michael addition of nitroalkanes to nitroalkenes catalyzed by chiral DMAP-thiourea hybrid.

Lu et al. have developed an enantioselective synthesis of optically pure 1,3-dinitro compounds **3c** via the Michael addition of nitroalkanes **2c** to nitroalkenes **1b** with good enantio and diastereoselectivities (Scheme 6.4).[4] Two mole percent of the catalyst was sufficient for the asymmetric induction without compromising on the reaction rate, and a wide variety of nitroalkanes and nitroalkenes were well-tolerated in yielding the products in high enantioselectivities. In the plausible mechanism involving the cinchona alkaloid **C3** as the catalyst, the chiral basic moiety first deprotonates nitroalkane **2c**. Meanwhile, catalyst **C3** simultaneously activates the Michael donor and the acceptor through double hydrogen bonding (**TS-1**). After Michael addition of the nitronate to the nitroalkene **1b** (**TS-2**), the chiral base is released to afford 1,3-dinitro compounds **3c** (Scheme 6.4).

Du and co-workers developed a squaramide **C4** catalyzed enantioselective Michael addition of nitroalkanes **2a** to nitroalkenes **1b**. This catalytic system with a 2 mol% catalyst loading afforded the corresponding Michael adducts **3b** with good yields and high diastereo and enantioselectivities of up to 97% ee (Scheme 6.5).[5]

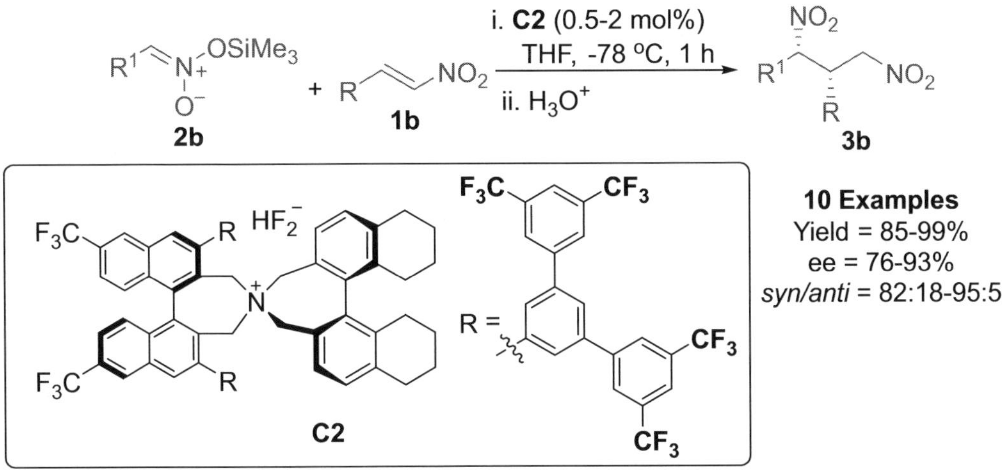

SCHEME 6.3 Michael addition of nitroalkanes to nitroalkenes using chiral quaternary ammonium bifluoride as a catalyst.

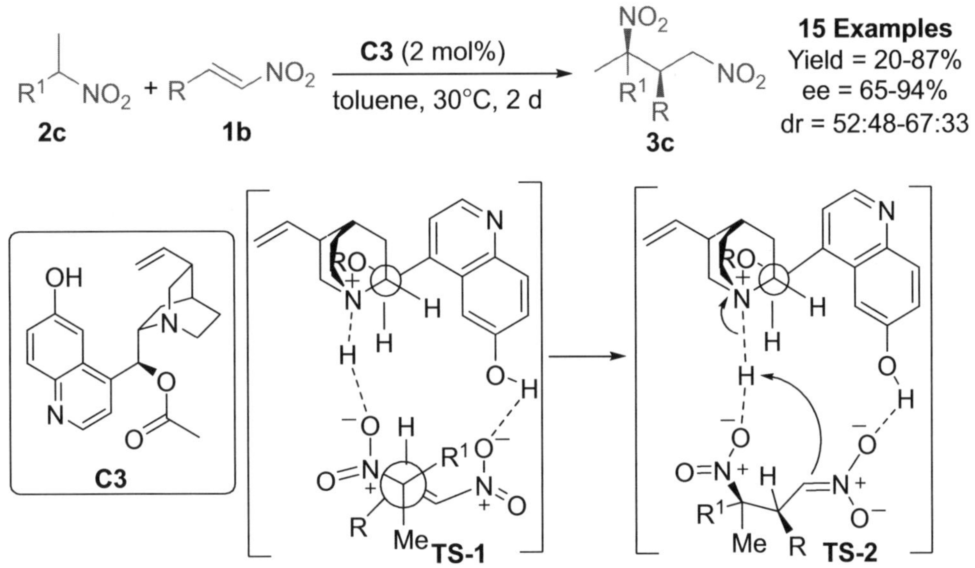

SCHEME 6.4 Michael addition of nitroalkanes to nitroalkenes using cinchona alkaloid as a catalyst.

Bis(oxazoline) ligand **L2** was found to promote the Zn(II)-catalyzed stereoselective addition of nitroalkanes **2d** to various nitroalkenes **1b**.[6] This new procedure allowed the synthesis of 1,3-dinitroalkanes **3d** with high diastereo and enantioselectivities. Ten mole percent of the ligand in the presence of 25 mol% of dialkylzinc reagent and 80 mol% of $Ti(O^{i}Pr)_4$ gave Michael adducts **3d** in good yields and enantioselectivities (Scheme 6.6).

The Wang group described an efficient direct Michael addition of nitroalkanes **2d** to nitroalkenes **1b** employing chiral bifunctional amine-thiourea catalyst **C5**.[7] This catalytic system performed well for a various substrates and provided various 1,3-dinitro compounds **3d** with high diastereo and enantioselectivities (Scheme 6.7). Multiple H-bonding donors play a vital role in accelerating the reaction, as well as improving the diastereo and enantioselectivities of the products **3d**.

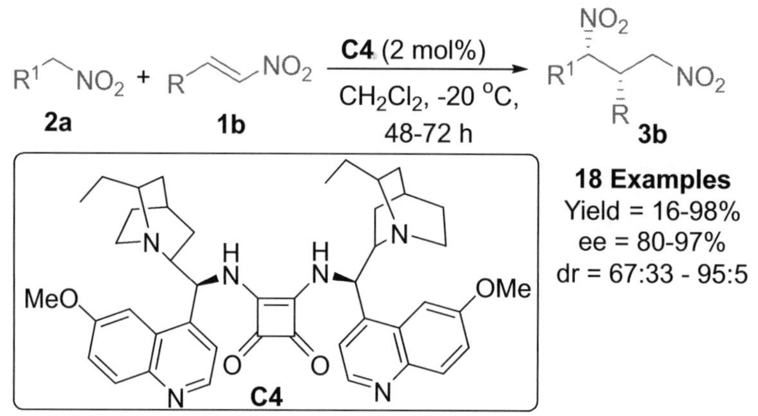

SCHEME 6.5 Squaramide-catalyzed Michael addition of nitroalkanes to nitroalkenes.

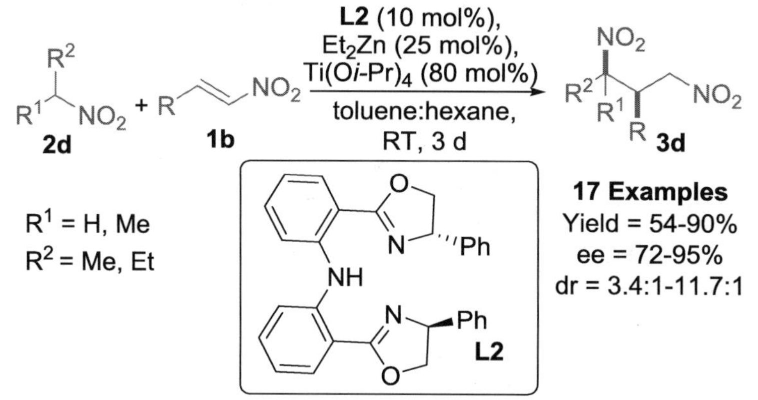

SCHEME 6.6 Michael addition of nitroalkanes to nitroalkenes in the presence of Zn(II)/bis(oxazoline) catalyst system.

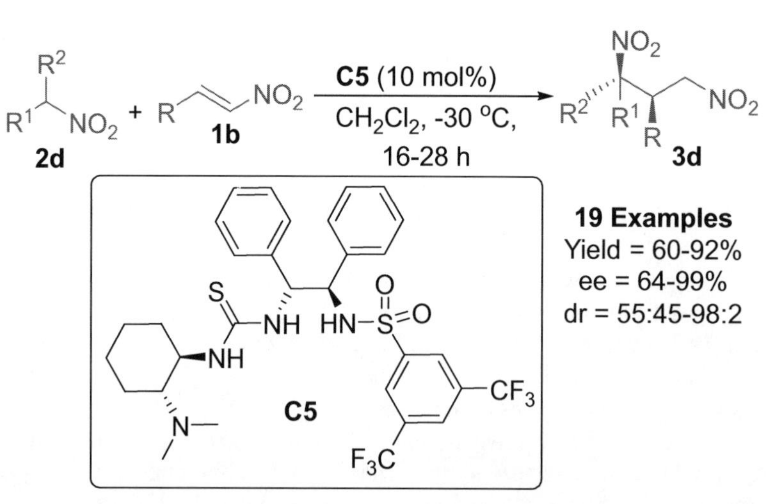

SCHEME 6.7 Michael addition of nitroalkanes to nitroalkenes using chiral bifunctional amine-thiourea catalyst.

Lu and co-workers developed an asymmetric Michael addition of α-fluoro-α-nitroalkanes **2e** to nitroalkenes **1b** in the presence of amino acid-incorporated multifunctional catalyst **C6** to afford the Michael adducts **3e** in good yields with high selectivities (Scheme 6.8).[8] This method allowed access to the fluorinated amines with quaternary stereogenic center and tetrahydropyrimidines in an optically enriched form. It appeared that the bifunctional activation of the substrates was necessary for the observed stereoselectivity. However, fluorinated nitroalkanes **2e** with a simple alkyl substituent were not suitable for the reaction.

The same group developed an asymmetric Michael addition of α-fluoro-α-nitroesters **2f** to nitroalkenes **1b** employing an amino acid-incorporated multifunctional catalyst **C6**. This was the first report that described the synthesis of α-fluoro-α-aminoester **3f** with a quaternary α-carbon with high yields and selectivities (Scheme 6.9).[9]

The Tanyeli group described the application of a rigid and sterically crowded unit attached to the quinine-derived squaramide moiety **C7** for the Michael addition of 1-nitropropane **2** (R = Et) to various nitroalkenes **1a**. The corresponding Michael adducts **3g** were obtained in high yields and high enantioselectivity. The reaction was carried out at 0°C with 2 mol% of the catalyst (Scheme 6.10).[10]

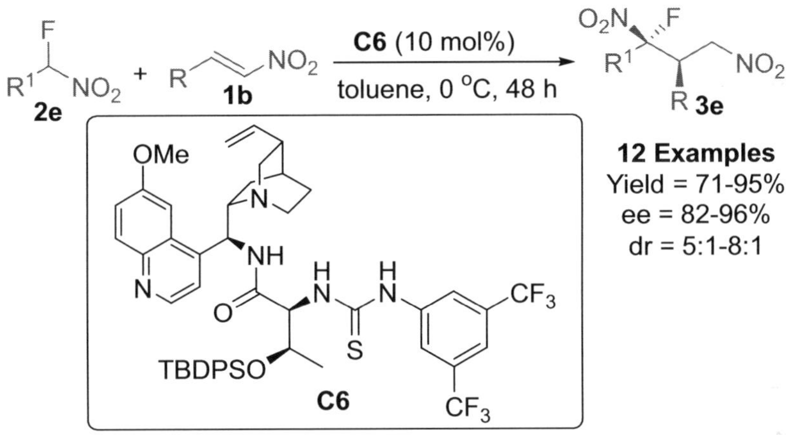

SCHEME 6.8 Michael addition of α-fluoro-α-nitroalkanes to nitroalkenes in the presence of an amino acid-incorporated multifunctional catalyst.

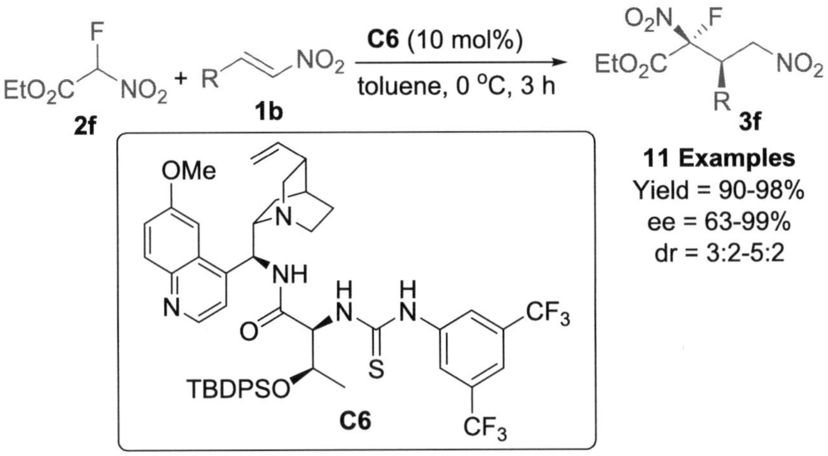

SCHEME 6.9 Michael addition of α-fluoro-α-nitroesters to nitroalkenes in the presence of an amino acid-incorporated multifunctional catalyst.

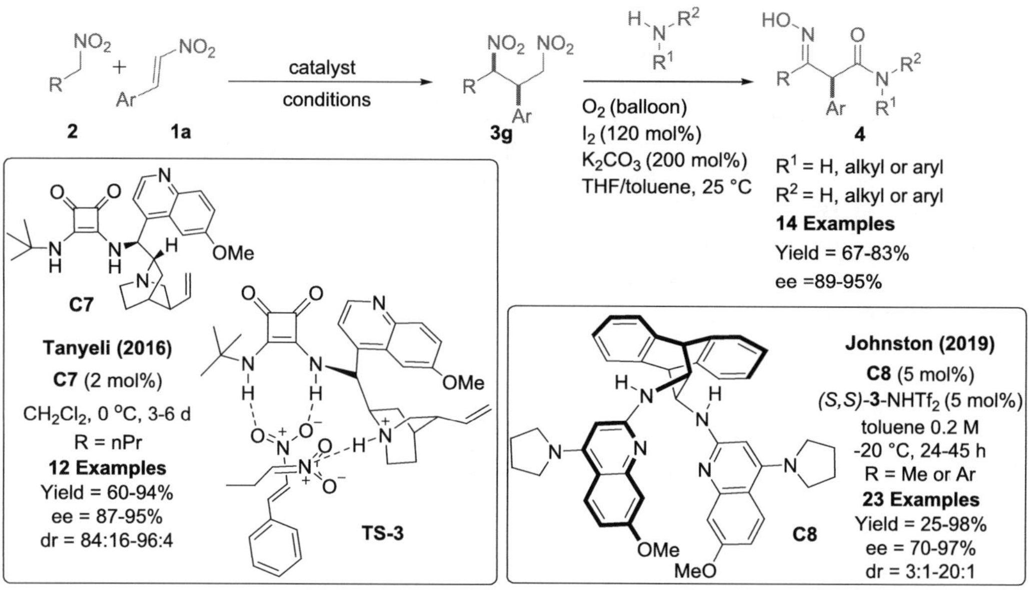

SCHEME 6.10 Michael addition of nitroalkanes to nitroalkenes in the presence of quinine-squaramide and bis(amidine)-based bifunctional organocatalysts.

In the possible mechanism, it was assumed that the bifunctional organocatalyst activates both the substrates. In the transition state **TS-3** (Scheme 6.10), the deprotonation of 1-nitropropane **2** (R = Et) occurs by a H-bonding interaction with the quinine unit of the catalyst **C7**, whereas the squaramide moiety activates nitroalkene **1a** through a double H-bonding interaction. The anion of 1-nitropropane **2** (R = Et) then attacks the activated nitroalkene from *Si*-face to give the desired product **3g**.

The Johnston group reported a stereoselective synthesis of β-amino amides **4** where for the first time nitroalkanes **2** and nitroalkenes **1a** were used in achieving a double umpolung protocol (Scheme 6.10).[11] Using a conformationally restricted bis(amidine)-based catalyst **C8**, a significant level of stereoselectivities were found in the obtained products **3g** and their derivatives **4**. This synthetic strategy is an example of an unusual double umpolung involving an inverted ketene, where both C–C and C–N bonds are formed with uncommon polarities.

6.3 ADDITION OF CYANIDES

The synthetic versatility of both the nitro and the nitrile groups in C–C bond-forming reactions coupled with their ready availability made them significant functionalities in organic synthesis. However, the reaction is complicated by the propensity of the resulting β-nitronitriles to undergo elimination of either hydrogen cyanide or nitrous acid to form nitroalkenes. The latter would then polymerize via either an anionic or a radical-anion mechanism.

In view of the aforementioned shortcomings, the use of preformed and structurally well-defined metal (salen) complexes allows a significant enhancement in the activity as exhibited by metal (salen) complexes in the enantioselective addition of trimethylsilyl cyanide **5** to nitroalkenes **1b**. Loading of 2 mol% of catalyst afforded products **6a** in good yields and excellent enantioselectivities at 0°C. According to the literature survey, complexes **C9b** are vanadium-based catalysts utilized in the present context, and the catalysts **C9a** are the titanium-based ones which were as active as that of vanadium-based catalysts. Three mole percent of the catalyst affords the products with moderate-to-good yields and enantioselectivities at 0°C (Scheme 6.11).[12] In this reaction, the nitro group serves as a bidentate ligand and assists in bridging the two metal ions. Moreover, the stepped

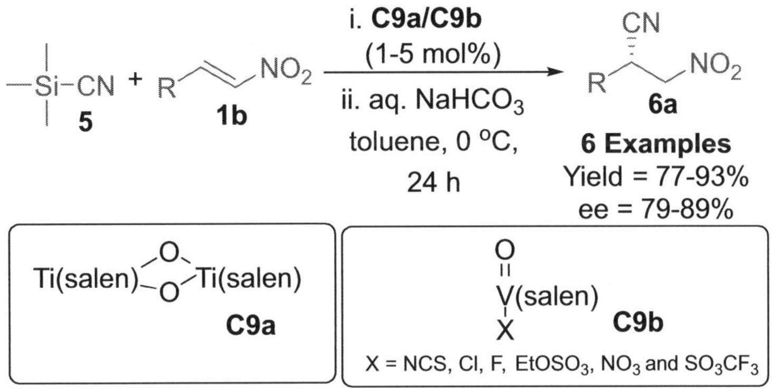

SCHEME 6.11 Enantioselective addition of trimethylsilyl cyanide to nitroalkenes in the presence of metal (salen) complexes.

conformation of the silane ligand **C9a/C9b** allows the reaction of cyanide on the less hindered *Si*-face of the coordinated nitroalkene.

The cyanosilylation of nitroalkenes **1c** was achieved by employing thiourea/tetraalkylammonium bifunctional catalyst **C10** that binds both the substrate and the reagent (Scheme 6.12).[13] While the catalyst activates the nitroalkene **1c** by hydrogen bonding in step 1, the conjugate addition of cyanide **5** affords the product **6b**. In fact, the bifunctional catalyst **C10** incorporates the "cyanide" nucleophile, whereas Me_3SiCN **5** acts as a silylating agent for the nitronate intermediate as well as a source of cyanide.

Kondo and co-workers reported an asymmetric conjugate addition of dithioacetonitrile **7** to nitroalkenes **1b** mediated by chiral bis(imidazoline)–Pd complex **C11** (Scheme 6.13).[14] The corresponding Michael adducts **6c** were formed in excellent yields (up to 99%) as well as high enantioselectivities (up to 98%), which were further converted into γ-lactams and (R)-rolipram.

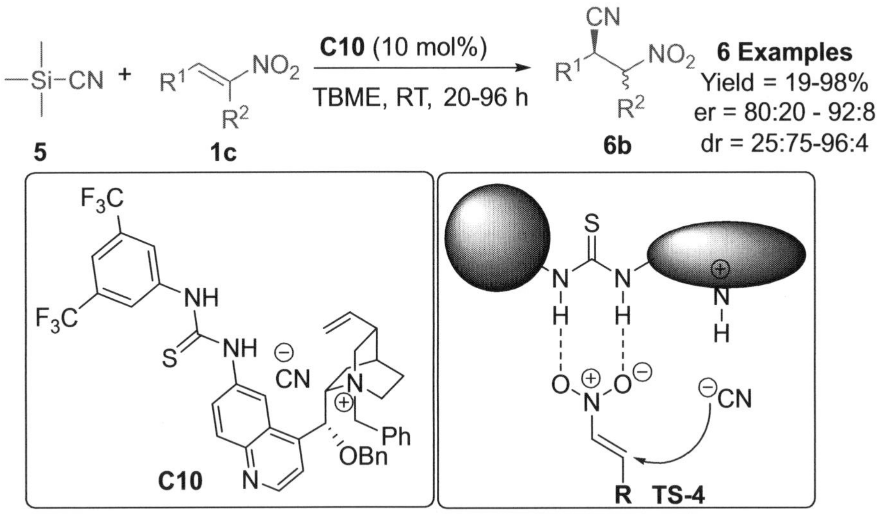

SCHEME 6.12 Cyanosilylation of nitroalkenes in the presence of thiourea/tetraalkylammonium bifunctional catalyst.

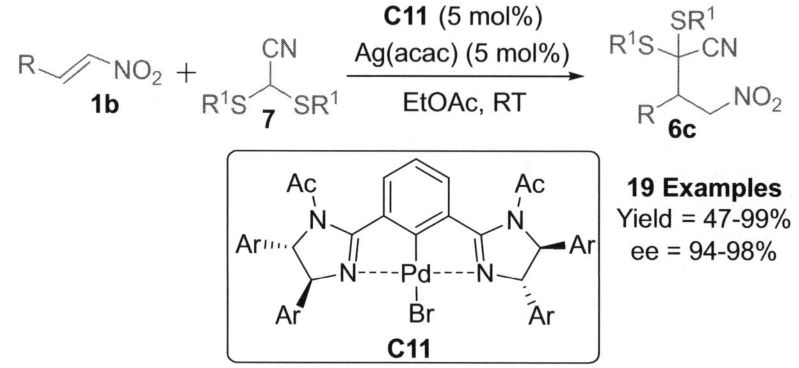

SCHEME 6.13 Asymmetric 1,4-addition of dithioacetonitrile to nitroalkenes mediated by chiral ligand-Pd complex.

6.4 ADDITION OF BORONIC ACIDS

Feringa and co-workers employed chiral monodentate phosphoramidite ligands in combination with rhodium for an enantioselective 1,4-addition of boronic acids **8** to nitroalkene **9** (Scheme 6.14).[15] The ligand combination of two different phosphoramidites (**L2** and **L3**) results in the simultaneous formation of two homocomplexes, Rh(**L2**)$_2$ and Rh(**L3**)$_2$, and heterocomplex Rh(**L2**–**L3**). ^{31}P NMR studies revealed that among these homo and heterocomplexes, homo-rhodium-phosphoramidite complexes are less active and less selective, whereas the corresponding heterocomplexes are more selective and active catalysts.

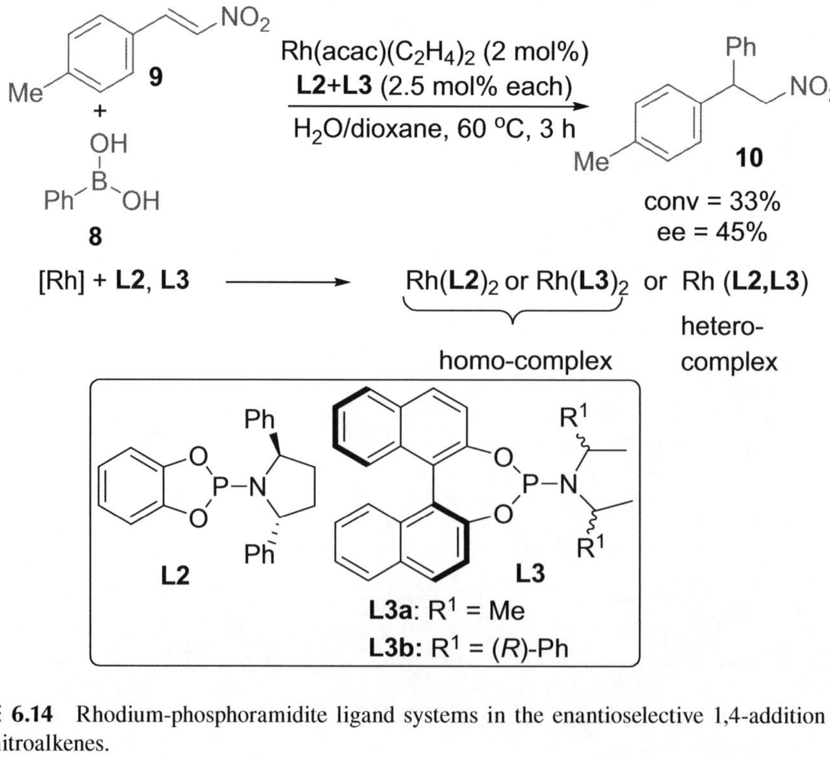

SCHEME 6.14 Rhodium-phosphoramidite ligand systems in the enantioselective 1,4-addition of boronic acids to nitroalkenes.

Palladium–diphosphine-catalyzed asymmetric conjugate addition of aryl boronic acids **8a** to 2-nitroacrylamide derivatives **11** was developed by the Gutnov group.[16] The reaction can be performed without an inert atmosphere using catalyst **C12** as low as 0.05–0.25 mol% to obtain enantiomerically enriched 2-aryl-3-nitropropionamides **12** with 73%–89% ee. Additionally, temperature did not have any major influence on the enantioselectivity but significantly accelerated the rate of the reaction (Scheme 6.15).

Xue et al. reported an efficient rhodium/olefin-sulfoxide **L4** catalyzed asymmetric conjugate addition of organoboronic acids **8b** to a range of nitroalkenes **1b** to obtain corresponding products **10b** in moderate-to-high yields and good enantioselectivities (Scheme 6.16).[17] The electronic properties of the substituents on the substrates did not significantly affect the stereoselectivity of the reaction. Although the steric hindrance of the substituents on the aromatic ring had a significant effect on the activity, it had little effect on the enantioselectivity. The stereochemical outcome of the Rh-complex catalyzed 1,4-addition can be attributed to the π-π stacking between the phenyl ring of the ligand and the metalated phenyl ring of phenylboronic acid in the transition state

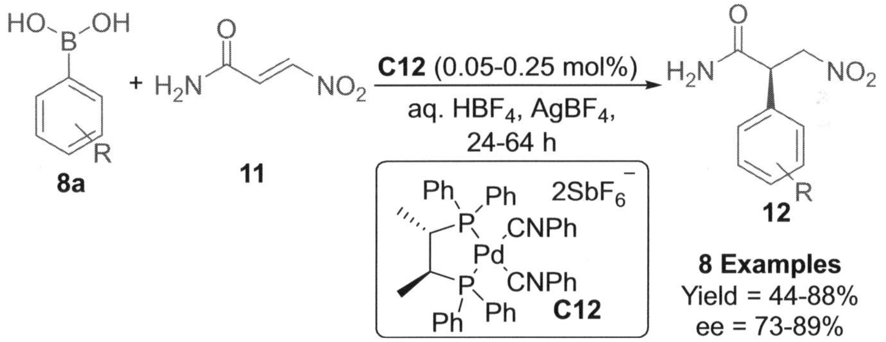

SCHEME 6.15 Palladium–diphosphine-catalyzed asymmetric conjugate addition of aryl boronic acids to 2-nitroacrylamide derivative.

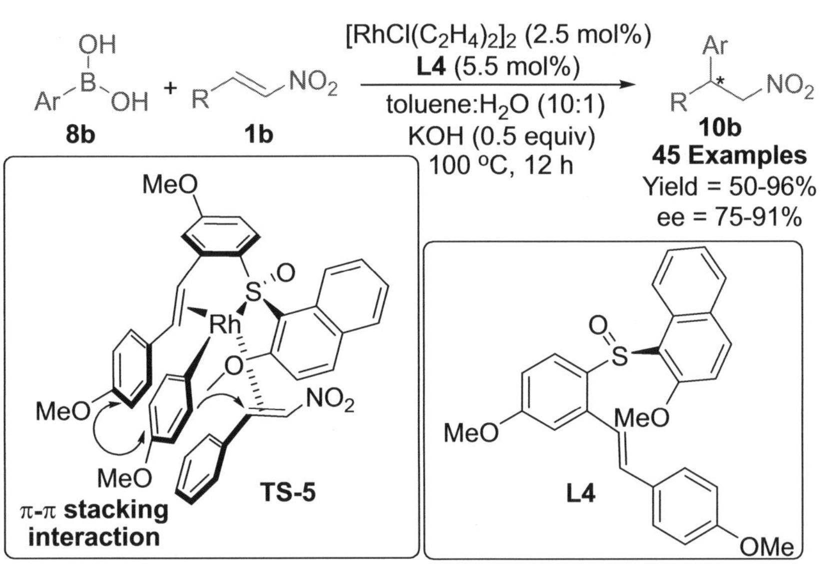

SCHEME 6.16 Rhodium/olefin-sulfoxide-catalyzed asymmetric conjugate addition of organoboronic acids to nitroalkenes.

TS-5 (Scheme 6.16). Owing to the steric repulsion between the 2-methoxy-1-naphthyl group of the ligand **L4** and nitro group of the nitroalkene **1b**, Rh complex recognizes the alkene moiety of the nitroalkenes **1b.** Hence, coordination of the nitroalkene **1b** from the less hindered side would be more favorable, leading to the '*S*' isomer.

Monodentate phosphoramidites **L5/L6** have been used for the first time as chiral ligands in the Rh-catalyzed enantioselective conjugate addition of phenylboronic acid **8** to *p*-methyl styrene-derived nitroalkene **9** to afford the product **10** with 100% conversion and 44% ee (Scheme 6.17).[18]

Lin et al. employed chiral bicyclo[3.3.0] diene **L7**, which was identified as a superior ligand in the rhodium-diene catalyzed asymmetric conjugate addition of organoboronic acids **8b** to nitroalkenes **1b** (Scheme 6.18).[19] The reaction proceeded with high enantioselectivity to give a wide range of synthetically useful enantioenriched β,β-disubstituted nitroalkanes **10b**. The authors used boronic acids in combination with KHF_2 as reactive organoborons, a promising reagent alternative to organotrifluoroborates in transition metal-catalyzed reactions. The absolute configuration of the product **10b** thus formed was determined to be '*R*' by X-ray crystallography.

In 2000, Hayashi and co-workers for the first time reported the rhodium-catalyzed asymmetric 1,4-addition of aryl boronic acids **8b** to 1-nitrocyclohexene **13** (Scheme 6.19).[20] The reaction of 1-nitrocyclohexene **13** proceeded with high diastereoselectivity in the presence of *S*-BINAP **L8** yielding thermodynamically less stable *cis*-isomer **14** preferentially. It was also observed that the quantity of boronic acid, the temperature and the solvent influenced the yield and selectivity of the products. The highest *cis*-selectivity and the highest enantioselectivity up to 99.3% ee were observed when the reaction was carried out at 80°C with DMA/H_2O as a solvent medium (Scheme 6.19).

Lang et al. developed an efficient rhodium-catalyzed conjugate addition of aryl boronic acids **8b** to nitroalkenes **1b** under mild conditions to afford products **10b** in excellent yields (up to 99%) and enantioselectivities up to 98% by employing *tert*-butanesulfinylphosphine **L9** as a ligand (Scheme 6.20).[21] The substituents on the aromatic ring had a considerable effect on the enantioselectivity, but little effect on the activity. Electron-donating substituents were good for this transformation, whereas electron-withdrawing substituents adversely affected the enantioselectivity. In the proposed mechanism, the initial coordination of the nitroalkene **1b** *trans* to the phosphine group was favored by at least 6.0 kcal mol^{-1} compared to the nitroalkene **1b** *cis* to the phosphine group.

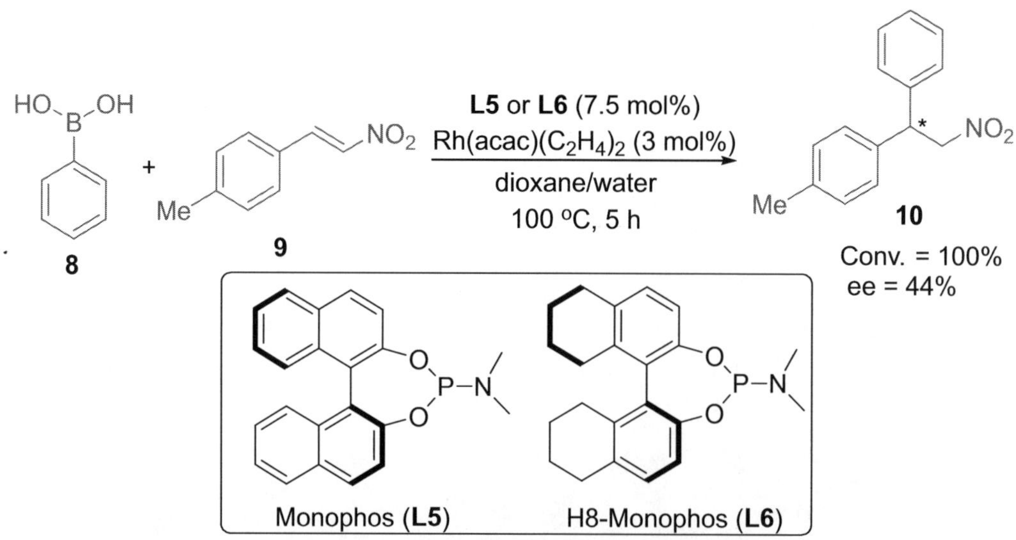

SCHEME 6.17 Rhodium-monodentate phosphoramidite catalyst systems in asymmetric conjugate addition of boronic acid to nitroalkenes.

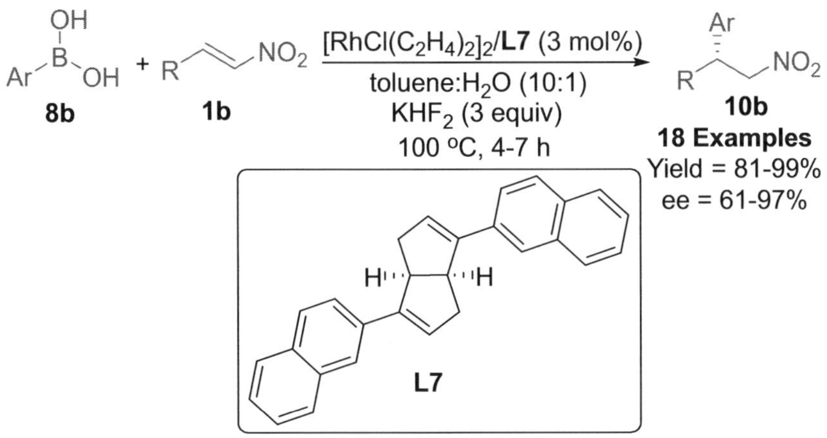

SCHEME 6.18 Rhodium-chiral bicyclo[3.3.0]diene catalyst system in the asymmetric conjugate addition of organoboronic acids to nitroalkenes.

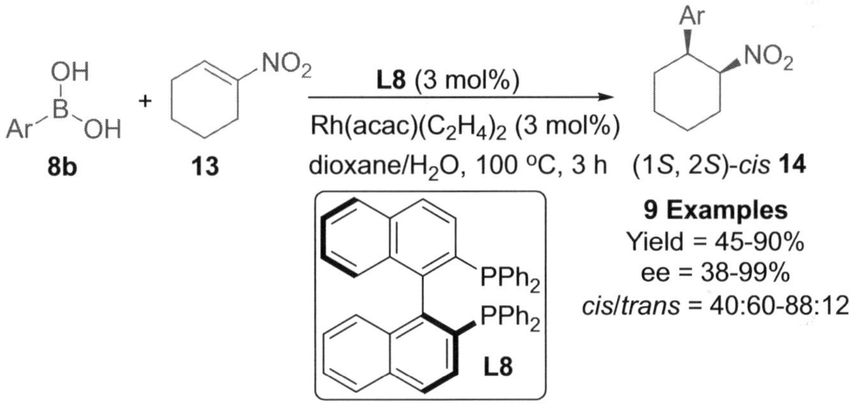

SCHEME 6.19 Rhodium-*S*-BINAP-catalyzed asymmetric 1,4-addition of aryl boronic acids to 1-nitrocyclohexene.

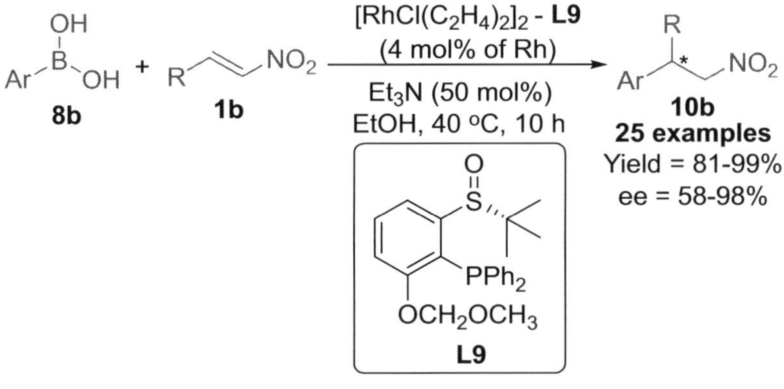

SCHEME 6.20 Rhodium-*tert*-butanesulfinylphosphine catalyst system in conjugate addition of aryl boronic acids to nitroalkenes.

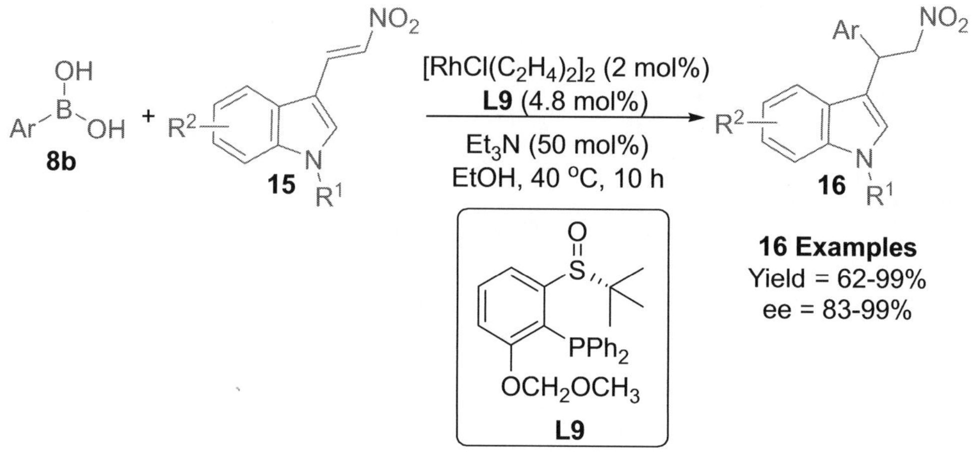

SCHEME 6.21 Rhodium-(*tert*-butylsulfinyl)phosphane catalyst system in the asymmetric 1,4-addition of aryl boronic acids to indolyl-nitroalkenes.

Further DFT calculations showed that the TS leading to the '*S*' product is about 5.0 kcal mol^{-1} higher in energy than that of the '*R*' product, which is in agreement with the experimental observation.

Liao group developed rhodium-catalyzed asymmetric 1,4-addition of aryl boronic acids **8b** to indolyl-nitroalkenes **15** to obtain optically pure α-aryl-3-indolylnitroethanes **16** by employing (*tert*-butylsulfinyl)phosphane **L9** as a ligand in excellent yields (up to 99% yield) and enantioselectivities (up to 99% ee).[22] The steric hindrance of the substituents on the aryl boronic acids affected the yield, and only trace amounts of products were obtained in the case of *o*-substituted boronic acid and 1-naphthaleneboronic acid (Scheme 6.21). Notably, lower catalyst loading led to the failure of the reaction, which suggested that the indolylnitroalkene was less reactive than other nitroalkenes.

6.5 ADDITION OF MALONONITRILE

Malononitrile is analogous to a 1,3-dicarbonyl compound, and the nitrile group is a versatile functionality for many synthetic transformations. The nucleophilic reactivity of malononitrile in organocatalytic asymmetric Michael reactions is relatively less explored due to its high reactivity and its inability to participate in two-point binding with the catalyst. However, there are a few reports describing the asymmetric Michael addition of malononitriles **17** to nitroalkenes **1**.

For the first time, Guo et al. described the enantioselective Michael addition of malononitrile **17** to nitroalkenes **1b** employing amine-thiourea organocatalyst **C13**, which afforded the products **18a** in high yields and moderate-to-high enantioselectivities (Scheme 6.22).[23] Electron-donating or neutral aryl group at the *para*-position of an aromatic ring or heterocyclic group on the nitroalkene **1b** favored the reaction.

Du and co-workers described an organocatalytic enantioselective tandem Michael addition–cyclization of malononitrile **17** with nitroalkenes **19** for the synthesis of chiral 2-amino-4H-chromene-3-carbonitriles **20** in chloroform at 60°C (Scheme 6.23).[24] Ten mole percent of catalyst **C14** was sufficient to afford the chromene derivatives **20** with good yields and enantioselectivities of up to 91% ee.

Zhao et al. demonstrated the enantioselective Michael reaction of malononitrile **17** with nitroalkenes **21**.[25] This reaction was catalyzed by a cinchona alkaloid-derived thiourea catalyst **C14** to afford products **22** with a quaternary chiral center at high yields with varying enantioselectivity, and the protocol is of significant interest as it allows the synthesis of dihydropyrrole derivatives (Scheme 6.24).

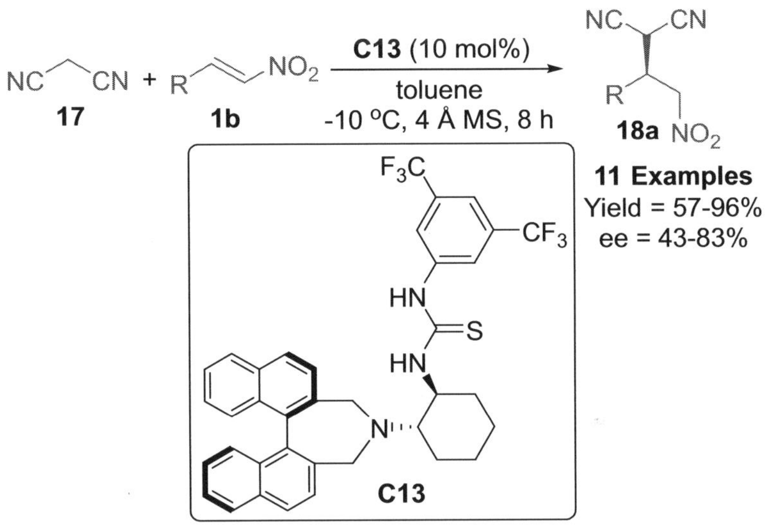

SCHEME 6.22 Michael addition of malononitrile to nitroalkenes using amine-thiourea organocatalyst.

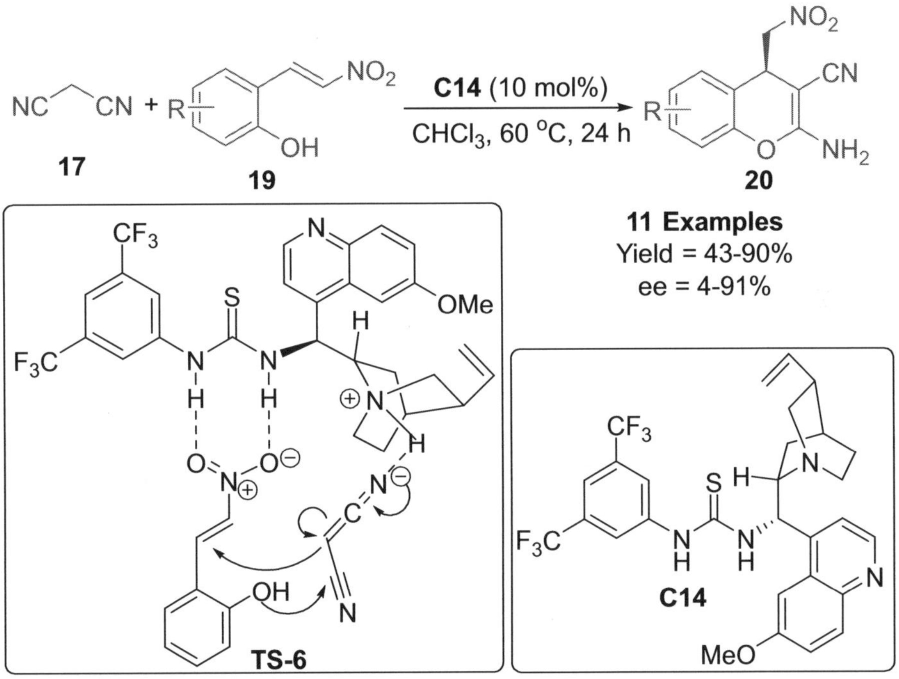

SCHEME 6.23 Synthesis of chiral 2-amino-4H-chromene-3-carbonitriles by organocatalytic enantioselective tandem Michael addition–cyclization of malononitrile with nitroalkenes.

A highly enantioselective vinylogous addition of γ-enolizable-α,β-unsaturated masked carbonyl compounds **23** to nitroalkenes **1b** catalyzed by commercially available $[DHQD]_2PYR$ **C15** as cinchona alkaloid has been described (Scheme 6.25).[26] α-Tetralone was masked as malononitrile owing to its higher value of Hammett substituent constant for α,α-dicyanovinyl group $[–CH=C(CN)_2]$ compared to nitromethyl group. Thus, various electron-deficient vinyl malononitriles were screened for

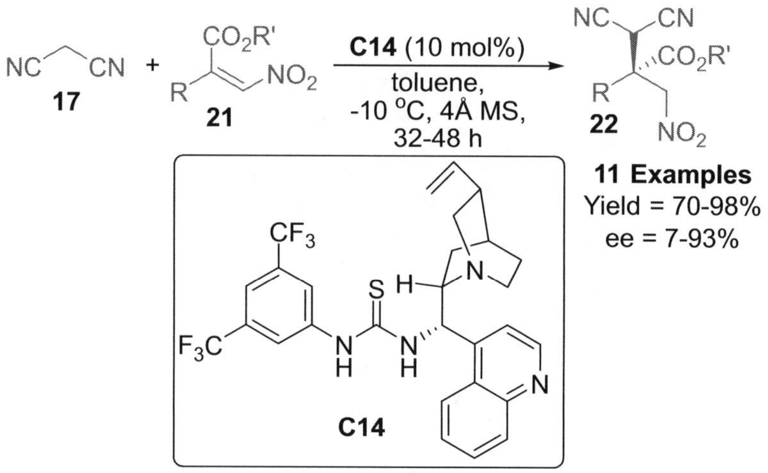

SCHEME 6.24 Cinchona alkaloid-derived thiourea catalyzed enantioselective Michael addition of malononitrile to nitroalkenes.

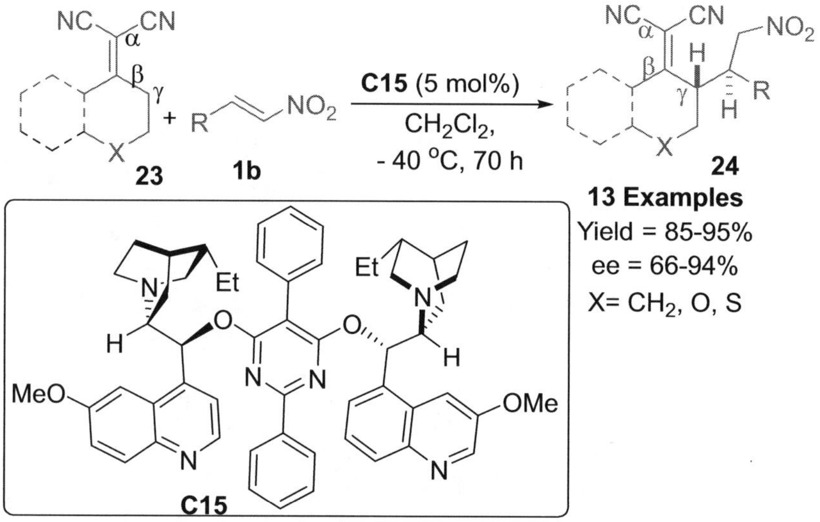

SCHEME 6.25 Enantioselective vinylogous addition of γ-enolizable-α,β-unsaturated masked carbonyl compounds to nitroalkenes.

the reaction with a range of aromatic nitroalkenes **1b** catalyzed by chiral tertiary amine as a base to form γ-selective Michael adducts **24** with high regio, diastereo and enantioselectivities.

6.6 ADDITION OF ORGANO-ZINC REAGENTS

Feringa's group demonstrated the utility of chiral copper-bis-(1-phenylethyl)amine-derived phosphoramidite ligands **L10/L11** for the enantioselective conjugate addition of diethylzinc to nitroalkenes **1b** (Scheme 6.26).[27] In this one-pot, multisubstrate procedure, nine aromatic β-nitroalkene derivatives **1** were used and a total of 18 Michael adducts **25a** were analyzed in a single chiral gas chromatography run without any overlap in the peaks. In contrast to aromatic substituents, aliphatic nitroalkenes also provided good yields of the products with high enantioselectivities using either **L10** or **L11** phosphoramidite ligand.

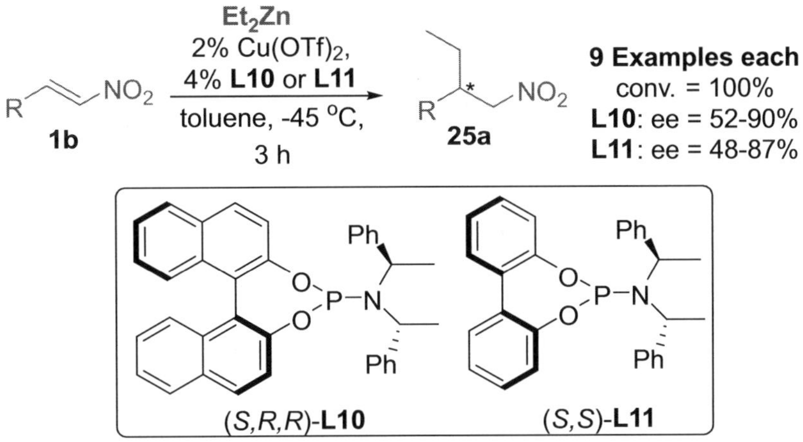

SCHEME 6.26 Copper-phosphoramidite as catalyst systems in the conjugate addition of diethylzinc to nitroalkenes.

The Hoveyda group reported an efficient method for the catalytic asymmetric addition of alkylzinc reagents to small-, medium-, and large-ring nitroalkenes **1d** promoted by amino acid-derived chiral phosphine **L12** (Scheme 6.27).[28] The reaction provided *syn*-stereoisomer of five- and six-membered rings **26** in good yields and enantioselectivities of up to 96%. The treatment of *syn*-isomers with DBU provided corresponding *anti*-isomers without a lowering of enantiomeric excess. For medium- and large-ring nitroalkenes, the reaction exclusively afforded the ketone derivatives **27**.

The asymmetric conjugate addition of simple aliphatic dialkylzinc reagents to acyclic nitropropene acetal **1e** was demonstrated by the Feringa group. The reaction proceeded with yields

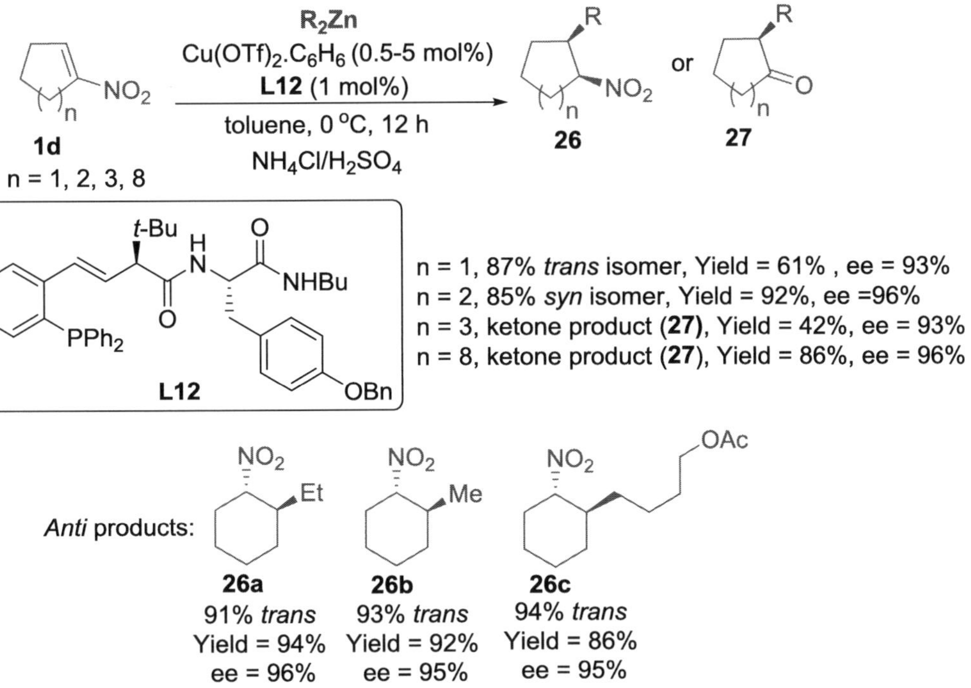

SCHEME 6.27 Catalytic asymmetric conjugate addition of alkylzinc reagents to small-, medium- and large-ring nitroalkenes.

up to 58%–78% and enantioselectivities in the range of 88%–98% in the presence of copper-phosphoramidite (*S, R, R*)-**L10** as the catalyst (Scheme 6.28).[29] The reaction was conducted on 1 g scale affording corresponding Michael adduct **25b** with methyl stereogenic center. This was converted to N-Boc-β-amino acid via intermediate oxidation of amino-acetal under acidic conditions, which serve as intermediates in the total synthesis of cryptophycins. Further, β-aminoalcohol **29**, a starting material in the synthesis of β-methyl carbapenem antibiotics, was obtained by deprotection of the acetal followed by a reduction of the intermediate aldehyde.

1,1′-Bi-2-naphthol (BINOL)-based thioether ligands **L13** and **L14** were developed by Kang's group for the enantioselective conjugate addition of dialkylzinc reagents to nitroalkene **9a** to afford the corresponding adduct **25c** with high yield and excellent enantioselectivity (Scheme 6.29).[30] Copper-based *tert*-butyl-thio ligand **L14** gave better enantioselectivity compared to methylthio ligand **L13**.

An efficient and highly enantioselective synthesis of β,β′-arylalkyl and β,β′-dialkylnitroalkanes **25d** was reported by Hoveyda group in their Cu-catalyzed asymmetric conjugate addition of alkylzinc reagents to acyclic nitroalkenes **1b**, promoted by dipeptide phosphine **L15** as an optimal catalyst (Scheme 6.30).[31] The substrate scope was extended to aliphatic nitroalkenes **1b** and alkylzinc reagents other than Et_2Zn. Less reactive dimethylzinc also yielded the desired product. The reaction tolerated electron-withdrawing and electron-donating aryl substituents and sterically bulky *ortho*-substituents. The reaction scope was extended to the synthesis of various optically active amines via palladium-mediated hydrogenation whose selectivities were in the range 87%–94% *ee*.

The conjugate addition of diethylzinc to various aromatic and heteroaromatic nitroalkenes **1b** employing chiral 5,5′, 6,6′-tetramethylbiphenol-based monodentate phosphoramidite ligand (*S, R, R*)-**L16**-$Cu(OTf)_2$ was demonstrated by the Ojima group (Scheme 6.31).[32] The substituents on the aryl ring with a variety of electronic and steric patterns influenced the enantioselectivity of the reaction. The electron-withdrawing *para*-substituents of *trans*-β-nitrostyrene such as fluoro and trifluoromethyl decrease the selectivity to 91% and 77%, respectively. The *ortho*-substituents such as -OMe or fluoro- provided selectivities lower than 75% with an exception of the trifluoromethyl group. The latter afforded the product **25a** with 88% ee as a result of the bulky CF_3 group at the *ortho* position, which affects the co-planarity of the substrate and cancels the electron-withdrawing effect of the CF_3 group.

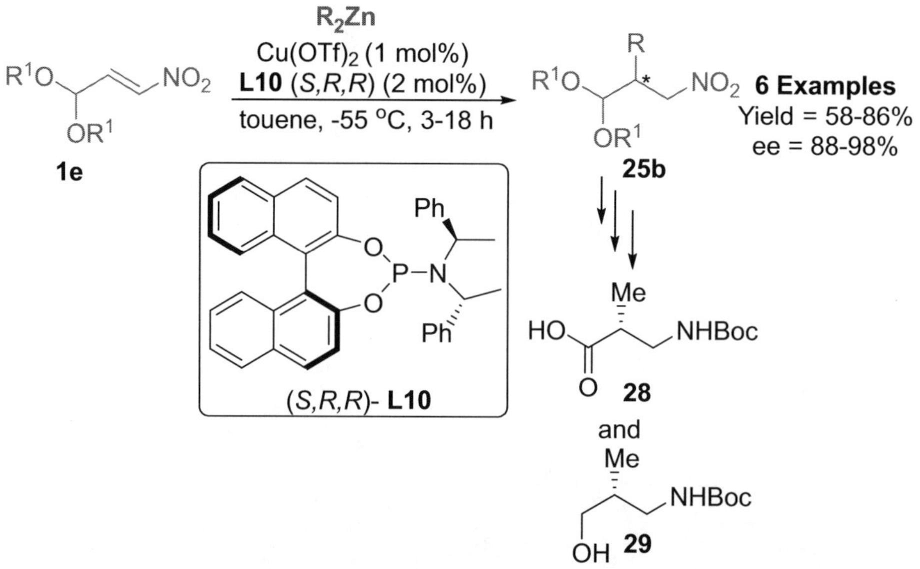

SCHEME 6.28 Asymmetric conjugate addition of aliphatic dialkylzinc to acyclic nitropropene acetal.

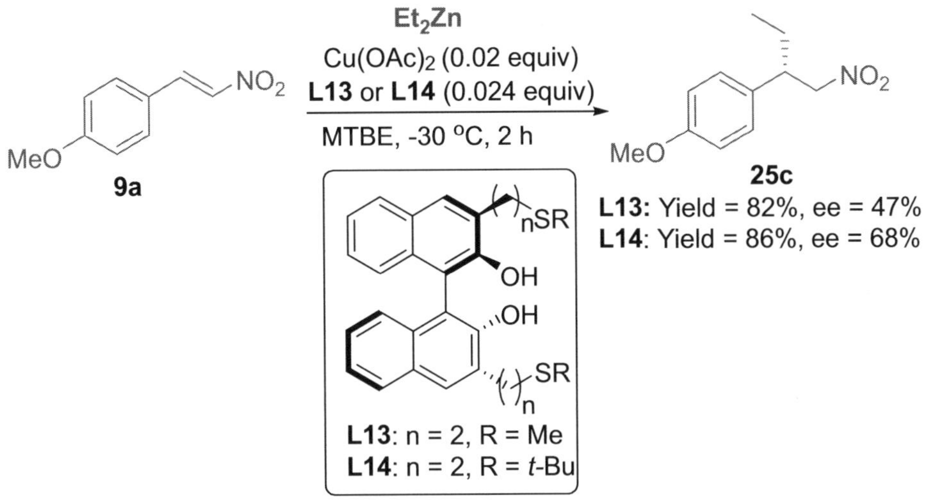

SCHEME 6.29 Enantioselective conjugate addition of dialkylzinc reagents to nitroalkene in the presence of copper acetate and BINOL-based thioether ligands.

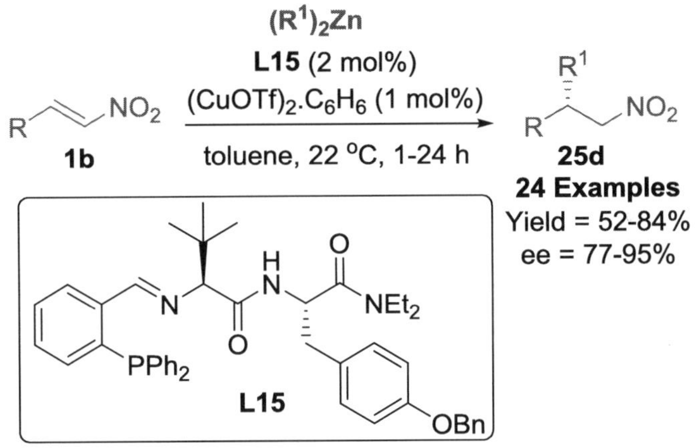

SCHEME 6.30 Cu/dipeptide phosphine catalyst system in the asymmetric conjugate addition of alkylzinc reagents to acyclic nitroalkenes.

Achiral and chiral conjugate addition of organozinc reagents to Michael acceptors such as nitroalkenes was achieved by employing copper (I)-derived BINOL-phosphoramidite **L17**-complex, which provided the Michael adducts **25a** with enantioselectivities in the range of 16%–87% (Scheme 6.32).[33] The reaction was even extended to 3-alkyl substituted-3-nitroacrylates and applied in the synthesis of Boc-protected homoaminoacid **30**.

The Hoveyda group reported the catalytic conjugate addition of dialkylzinc reagents with longer chains or a heteroatom functionality to β,β-disubstituted nitroalkenes **1f** in the presence of chiral peptide-based ligand **L15**.[34] The resulting Michael products **25e** were obtained in high yields and excellent enantioselectivities (Scheme 6.33). Because of the presence of a quaternary carbon stereogenic center, the products thus obtained serve as precursors for optically active amines, nitriles, aldoximes and even carboxylic acids **31**, with the latter obtained by Nef reaction using a reagent combination of sodium nitrite and DMSO.

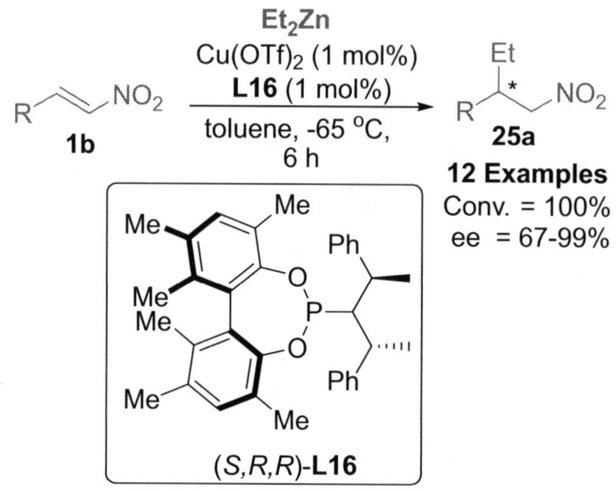

SCHEME 6.31 Conjugate addition of diethylzinc to nitroalkenes in the presence of tetramethylbiphenol-based monodentate phosphoramidite ligand and $Cu(OTf)_2$.

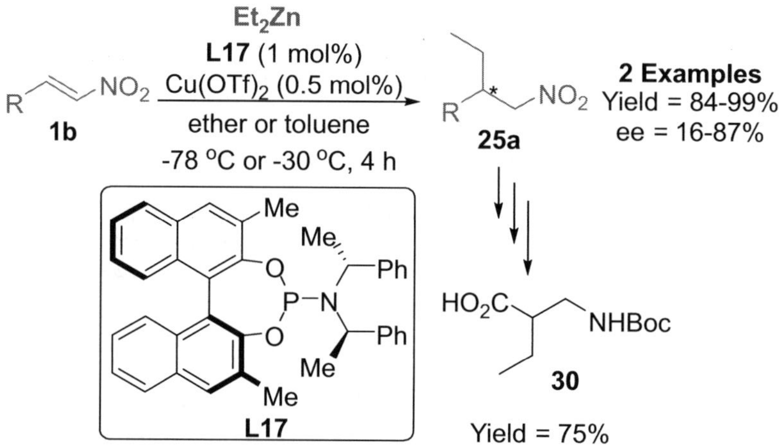

SCHEME 6.32 Conjugate addition of organozinc reagents to nitroalkenes using copper(I) derived BINOL-phosphoramidite catalyst.

A chiral ligand-controlled asymmetric conjugate addition of dialkylzinc mediated addition of acetylenes **32** to nitroalkenes **1g** was reported by the Tomioka group.[35] A set of cyclic and acyclic nitroalkenes **1g** were chosen as substrates in the conjugate addition of acetylenes **32** in the presence of dialkylzinc reagents and chiral controller amino alcohol **L18** as an additive to provide products **25f** in moderate-to-good yields with moderate diastereoselectivity and excellent enantioselectivity (Scheme 6.34). The radical scavenger galvinoxyl helped to improve the yield. The absolute configuration of the product **25f** was determined to be (*R*) by conversion to a known compound following Nef reaction and hydrogenolysis. The transition state model **TS-7** (Scheme 6.34) depicts Lewis acidic coordination of zinc metal in the bimetallic complex with the oxygen of a nitro group, followed by the attack of the alkynyl group on the double bond to complete the reaction.

Catalytic enantioselective approach to β,β-disubstituted nitroalkanes **25d** was promoted by copper-amidophosphane **L19** for the conjugate addition of dialkylzinc to several aromatic and aliphatic nitroalkenes **1b** (Scheme 6.35).[36] Moderate-to-good enantioselectivities were obtained using two approaches. In one of the approaches, the nitroalkene **1b** was added to the mixture of

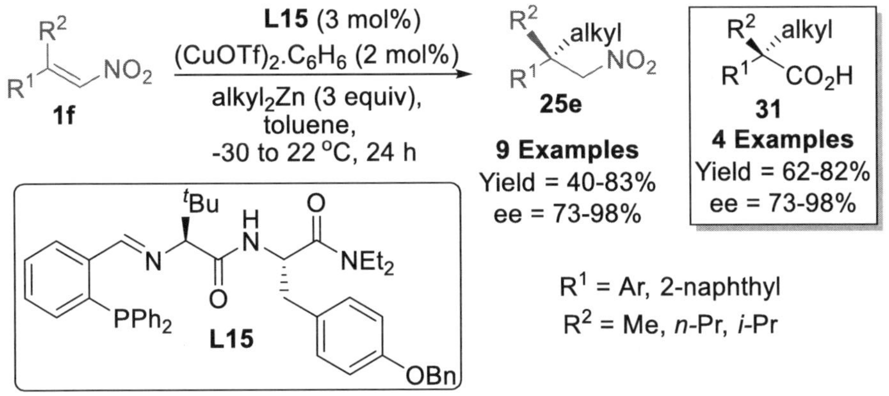

SCHEME 6.33 Cu-catalyzed conjugate addition of dialkylzinc reagents to β,β-disubstituted nitroalkenes in the presence of chiral peptide-based ligand.

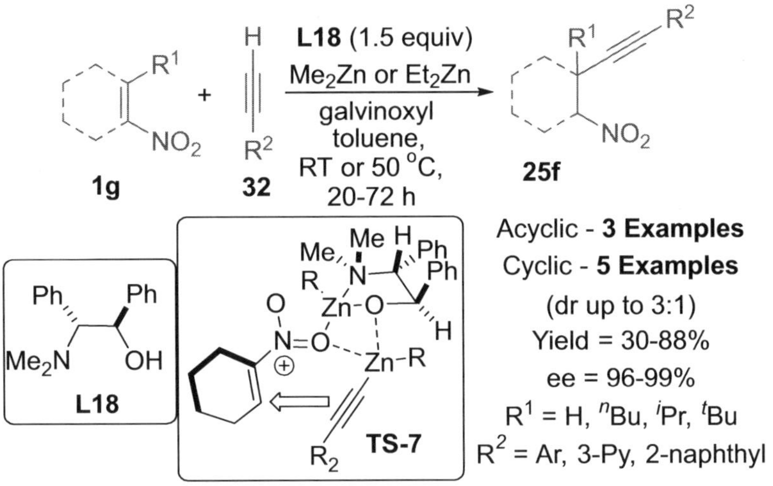

SCHEME 6.34 Asymmetric conjugate addition of arylalkynes to nitroalkenes mediated by dialkylzinc reagents and chiral amino alcohol.

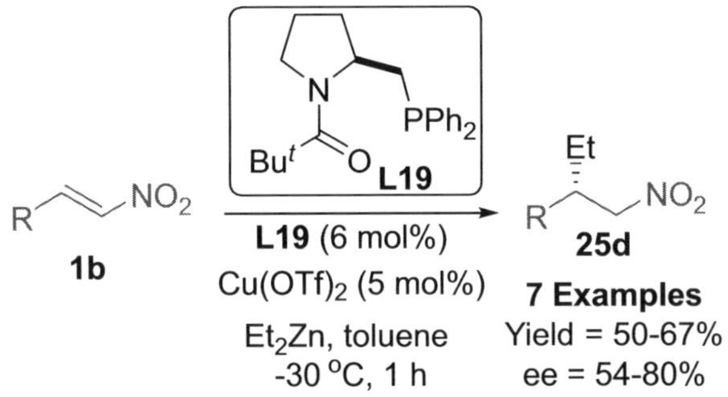

SCHEME 6.35 Copper-amidophosphane promoted conjugate addition of dialkylzinc to nitroalkenes.

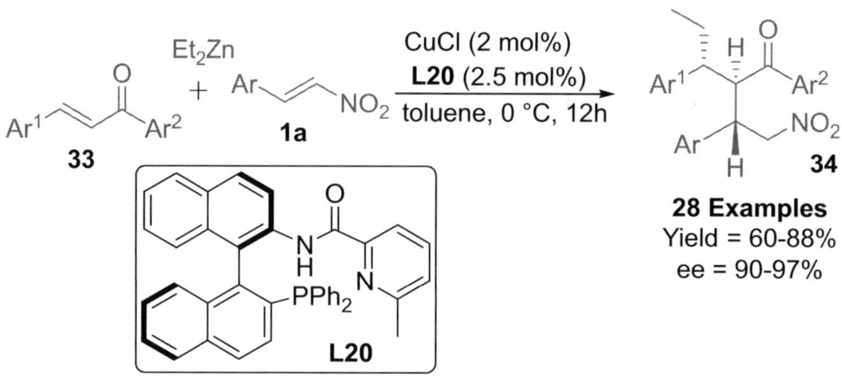

SCHEME 6.36 Asymmetric tandem double Michael addition of diethylzinc to unsaturated ketones followed by trapping with nitroalkenes mediated by a copper-based catalyst.

copper-amidophosphane **L19** and dialkylzinc and higher selectivity was achieved. In the second approach, dialkylzinc was added into a mixture of copper-amidophosphane **L19** and nitroalkene **1b**, which gave moderate selectivity.

Hu and co-workers reported asymmetric tandem double Michael addition reaction of diethylzinc with unsaturated ketones **33** mediated by copper, which was subsequently followed by the trapping with nitroalkenes **1a** using a chiral P,N-ligand **L20** (Scheme 6.36).[37] The corresponding adducts **34** possessing three contiguous chiral centers were formed in good yields (up to 88%) and high enantioselectivities (up to 97%).

6.7 CONCLUSIONS

Nitroalkenes undergo Michael addition of various nucleophiles such as nitroalkanes, cyanides, malononitrile, boronic acids and organozinc reagents in the presence of suitable chiral organo- and metal-ligand catalysts. Nitroalkanes bearing a diverse array of functional groups at the β-position of the nitro group, such as nitroalkyl, cyano, dicyanoalkyl, aryl and alkyl, are generated in the process with high chemical yield and good-to-excellent stereoselectivity in most cases.

REFERENCES

1. Yang, X.; Zhou, X.; Lin, L.; Chang, L.; Liu, X.; Feng, X. *Angew. Chem. Int. Ed.* **2008**, *47*, 7079.
2. Rabalakos, C.; Wulff, W. D. *J. Am. Chem. Soc.* **2008**, *130*, 13524.
3. Ooi, T.; Takada, S.; Doda, K.; Maruoka, K. *Angew. Chem. Int. Ed.* **2006**, *45*, 7606.
4. Deng, Y.-Q.; Zhang, Z.-W.; Feng, Y.-H.; Chan, A. S. C.; Lu, G. *Tetrahedron: Asymmetry* **2012**, *23*, 1647.
5. Yang, W.; Du, D.-M. *Chem. Commun.* **2011**, *47*, 12706.
6. Lu, S.-F.; Du, D.-M.; Xu, J.; Zhang, S.-W. *J. Am. Chem. Soc.* **2006**, *128*, 7418.
7. Dong, X.-Q.; Teng, H.-L.; Wang, C.-J. *Org. Lett.* **2009**, *11*, 1265.
8. Kwiatkowski, J.; Lu, Y. *Chem. Commun.* **2014**, *50*, 9313.
9. Kwiatkowski, J.; Lu, Y. *Org. Biomol. Chem.* **2015**, *13*, 2350.
10. Kanberoğlu, E.; Tanyeli, C. *Asian J. Org. Chem.* **2016**, *5*, 114.
11. Vishe, M.; Johnston, J. N. *Chem. Sci.* **2019**, *10*, 1138.
12. North, M.; Watson, J. M. *ChemCatChem* **2013**, *5*, 2405.
13. Bernal, P.; Fernández, R.; Lassaletta, J. M. *Chem. - Eur. J.* **2010**, *16*, 7714.
14. Nakamura, S.; Tokunaga, A.; Saito, H.; Kondo, M. *Chem. Commun.* **2019**, *55*, 5391.
15. Duursma, A.; Hoen, R.; Schuppan, J.; Hulst, R.; Minnaard, A. J.; Feringa, B. L. *Org. Lett.* **2003**, *5*, 3111.
16. Petri, A.; Seidelmann, O.; Eilitz, U.; Leßmann, F.; Reißmann, S.; Wendisch, V.; Gutnov, A. *Tetrahedron Lett.* **2014**, *55*, 267.
17. Xue, F.; Wang, D.; Li, X.; Wan, B. *J. Org. Chem.* **2012**, *77*, 3071.
18. Boiteau, J.-G.; Imbos, R.; Minnaard, A. J.; Feringa, B. L. *Org. Lett.* **2003**, *5*, 681.

19. Wang, Z.-Q.; Feng, C.-G.; Zhang, S.-S.; Xu, M.-H.; Lin, G.-Q. *Angew. Chem. Int. Ed.* **2010**, *49*, 5780.
20. Hayashi, T.; Senda, T.; Ogasawara, M. *J. Am. Chem. Soc.* **2000**, *122*, 10716.
21. Lang, F.; Chen, G.; Li, L.; Xing, J.; Han, F.; Cun, L.; Liao, J. *Chem. - Eur. J.* **2011**, *17*, 5242.
22. Xing, J.; Chen, G.; Cao, P.; Liao, J. *Eur. J. Org. Chem.* **2012**, *2012*, 1230.
23. Guo, H.-M.; Li, J.-G.; Qu, G.-R.; Zhang, X.-M.; Yuan, W.-C. *Chirality* **2011**, *23*, 514.
24. Gao, Y.; Yang, W.; Du, D.-M. *Tetrahedron: Asymmetry* **2012**, *23*, 339.
25. Chen, S.; Lou, Q.; Ding, Y.; Zhang, S.; Hu, W.; Zhao, J. *Adv. Synth. Catal.* **2015**, *357*, 2437.
26. Xue, D.; Chen, Y.-C.; Wang, Q.-W.; Cun, L.-F.; Zhu, J.; Deng, J.-G. *Org. Lett.* **2005**, *7*, 5293.
27. Duursma, A.; Minnaard, A. J.; Feringa, B. L. *Tetrahedron* **2002**, *58*, 5773.
28. Luchaco-Cullis, C. A.; Hoveyda, A. H. *J. Am. Chem. Soc.* **2002**, *124*, 8192.
29. Duursma, A.; Minnaard, A. J.; Feringa, B. L. *J. Am. Chem. Soc.* **2003**, *125*, 3700.
30. Kang, J.; Lee, J. H.; Lim, D. S. *Tetrahedron: Asymmetry* **2003**, *14*, 305.
31. Mampreian, D. M.; Hoveyda, A. H. *Org. Lett.* **2004**, *6*, 2829.
32. Choi, H.; Hua, Z.; Ojima, I. *Org. Lett.* **2004**, *6*, 2689.
33. Rimkus, A.; Sewald, N. *Synthesis **2004***, *2004*, 135.
34. Wu, J.; Mampreian, D. M.; Hoveyda, A. H. *J. Am. Chem. Soc.* **2005**, *127*, 4584.
35. Yamashita, M.; Yamada, K.-i.; Tomioka, K. *Org. Lett.* **2005**, *7*, 2369.
36. Valleix, F.; Nagai, K.; Soeta, T.; Kuriyama, M.; Yamada, K.-i.; Tomioka, K. *Tetrahedron* **2005**, *61*, 7420.
37. Wang, Q.; Li, S.; Hou, C.-J.; Chu, T.-T.; Hu, X.-P. *Tetrahedron* **2019**, *75*, 3943.

7 Catalytic Asymmetric Michael Addition of Heteroatom-centered Nucleophiles to Nitroalkenes

7.1 INTRODUCTION

Besides carbon-centered nucleophiles, a wide variety of heteroatom-centered nucleophiles also participate in the Michael addition to nitroalkenes in the presence of various chiral catalysts to afford the corresponding Michael adducts. These include P-, S-, N- and O-centered nucleophiles. Among P-centered nucleophiles, diaryl phosphites and phosphine oxides are the prominent ones though there is at least one report on the addition of dialkyl phosphite as well. Both organo- and metal-catalysts were employed for this purpose. Similarly, S-centered nucleophiles such as thiols and thioacetic acids have been added to nitroalkenes in the presence of various Brønsted base-Brønsted acid organocatalysts. As far as N-centered nucleophiles are concerned, hydrazones, hydrazides, imines, protected hydroxylamines, imides and even amines have been added to nitroalkenes in the presence of several organocatalysts. A couple of O-centered nucleophiles such as hydroperoxides and hydroxylamines have also been added to nitroalkenes in the presence of chiral organocatalysts.

7.2 ADDITION OF PHOSPHORUS-CENTERED NUCLEOPHILES

Rawal et al. showed that squaramide **C1** is a remarkably effective catalyst for the asymmetric Michael reaction of diphenyl phosphite (DPP) **2** to nitroalkenes **1** (Scheme 7.1).[1] The reaction provided an easy and highly enantioselective route to synthesize chiral β-nitrophosphonates **3**. Here the authors mentioned that both aryl- and alkyl-substituted nitroalkenes including sterically demanding substituents worked well under reported conditions.

The Zhao group described the Michael addition of DPP **2** to nitroalkenes **1** using the quinidine-thiourea organocatalyst **C2** for the enantioselective synthesis of β-nitrophosphonates **3** (Scheme 7.2).[2] Efficiency of the reaction was improved by adding molecular sieves (MS) to the reaction mixture

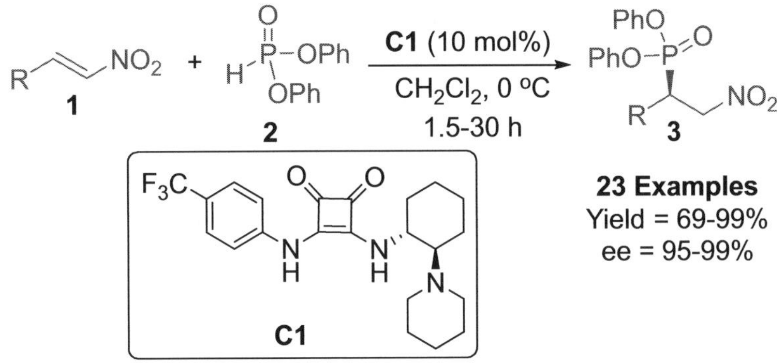

SCHEME 7.1 Michael addition of DPP to nitroalkenes in the presence of squaramide-based organocatalyst.

along with MS-treated DPP. MS played a key role in the reaction as it performed not only as a water and acid scavenger but also removed the impurities in DPP samples. It might also facilitate the equilibrium between the phosphonate and the phosphite tautomers of the DPP.

The Namboothiri group also demonstrated the formation of β-nitrophosphonates **3** via the Michael addition of dialkyl phosphites **2** to nitroalkenes **1a** using heterobimetallic (*S*)-(-)-aluminum lithium bis(binaphthoxide) **C3** (Scheme 7.3).[3] The bifunctional catalyst binds and activates both the substrates and orients them in such a manner that dialkyl phosphite participates in face-selective addition in a chiral environment. The corresponding Michael adducts **3** were formed in good yields from nitroalkenes with electron-donating substituents, whereas electron-donating substituents containing nitroalkenes afforded Michael adducts in moderate yields.

Herrera and co-workers reported bifunctional thiourea **C4** catalyzed enantioselective phospha-Michael addition reaction of DPP **2** to nitroalkenes **1**.[4] This methodology provided facile access to enantiomerically enriched β-nitrophosphonates **3** (Scheme 7.4). DFT calculations revealed the fact that the attack of the DPP **2** to the nitroalkene **1** occurs from the *Re*-face (**TS-1**, Scheme 7.4).

Quinine **C5** catalyzed enantioselective conjugate addition of DPP **2** to nitroalkenes **1** has been developed by Wang et al., which gave access to a wide range of enantiomerically enriched β-nitrophosphates **3** (Scheme 7.5).[5] The proposed mechanism involves the activation of nitroalkene **1** by the hydroxyl group of the catalyst **C5** (**TS-2,** Scheme 7.5). Meanwhile, the amino group of the

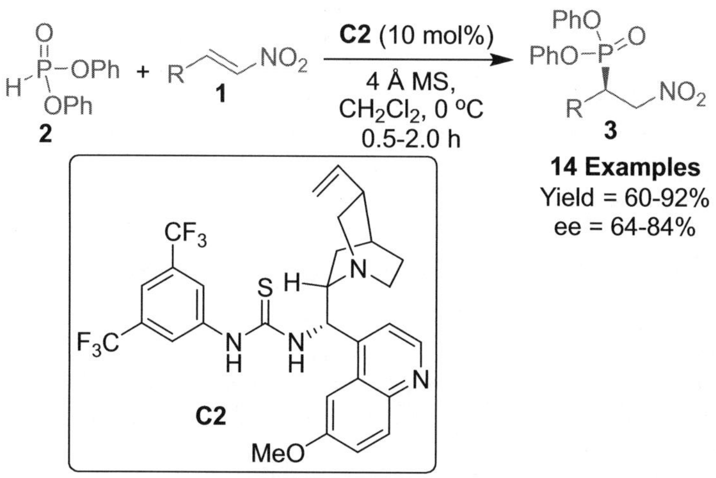

SCHEME 7.2 Michael addition of DPP to nitroalkenes in the presence of quinidine-thiourea organocatalyst.

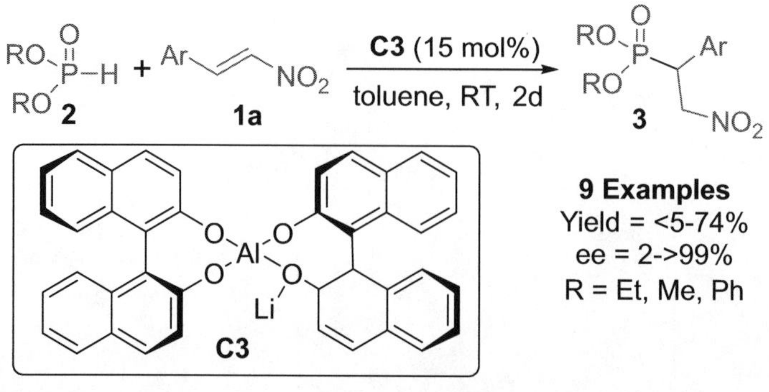

SCHEME 7.3 Michael addition of dialkyl phosphites to nitroalkenes using heterobimetallic (*S*)-(-)-aluminum lithium bis(binaphthoxide) catalyst.

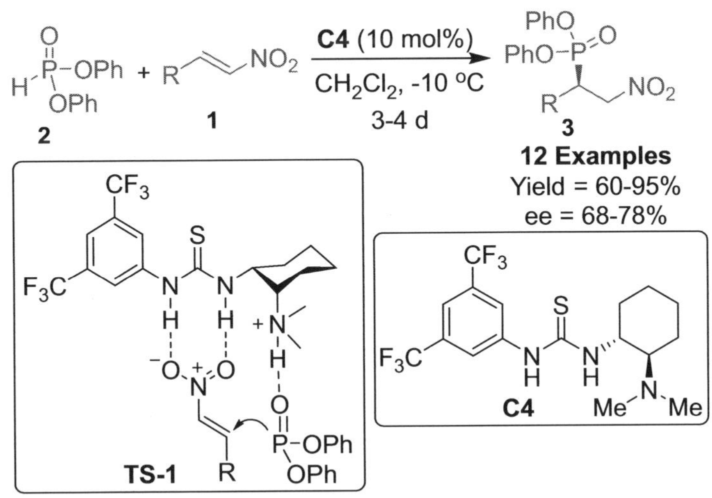

SCHEME 7.4 Bifunctional thiourea catalyzed phospha-Michael addition of DPP to nitroalkenes.

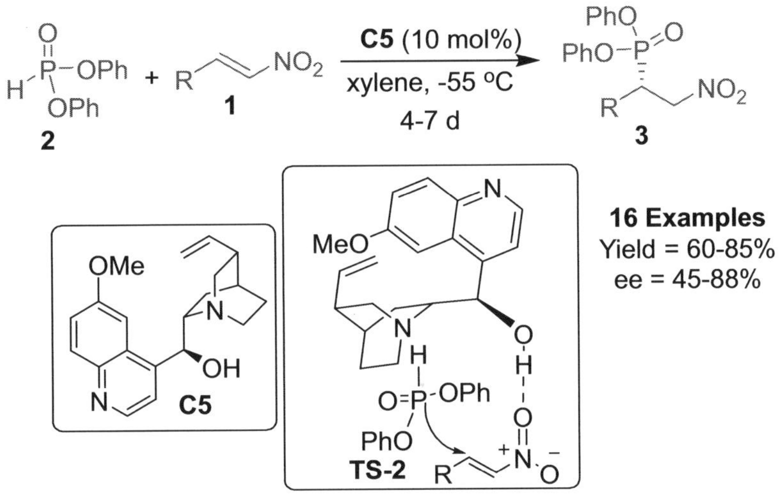

SCHEME 7.5 Conjugate addition of DPP to nitroalkenes catalyzed by quinine.

catalyst via the second H-bonding interaction activates and orients the phosphite for *Si*-face attack on nitroalkene to form the *S*-configured product **3**.

Terada et al. described a highly enantioselective 1,4-addition reaction of DPP **2** with nitroalkenes **1** catalyzed by a newly developed axially chiral guanidine **C6** (Scheme 7.6).[6] A wide variety of nitroalkenes **1**, bearing aromatic as well as aliphatic substituents, were suitable for the present enantioselective reaction. The method can be implemented to synthesize enantiopure α-amino phosphonates which are of biological significance.

Recently, Khan and co-workers demonstrated the Michael addition of diphenylphosphonates **2** to nitroalkenes **1** using a secondary amine bis-thiourea organocatalyst **C7** with MS as an additive (Scheme 7.7).[7] MS plays a significant role in reproducing high yields and enantioselectivity. The Michael acceptor was activated by hydrogen bonding interactions through nitrogen moieties in the catalyst **C7** (**TS-3**, Scheme 7.7).

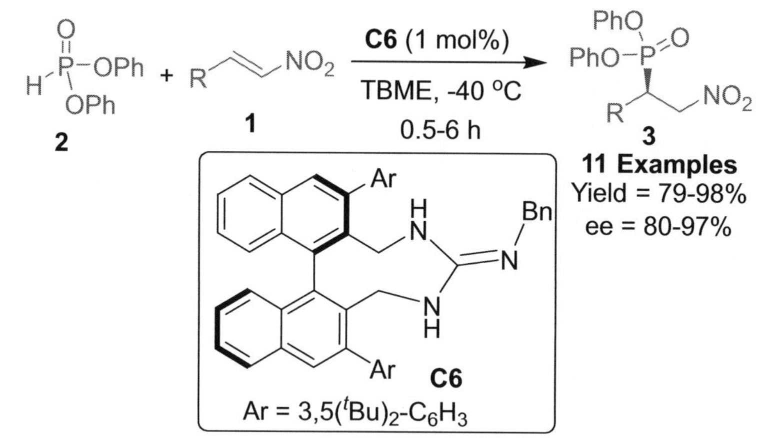

SCHEME 7.6 Enantioselective 1,4-addition reaction of DPP to nitroalkenes in the presence of axially chiral guanidine organocatalyst.

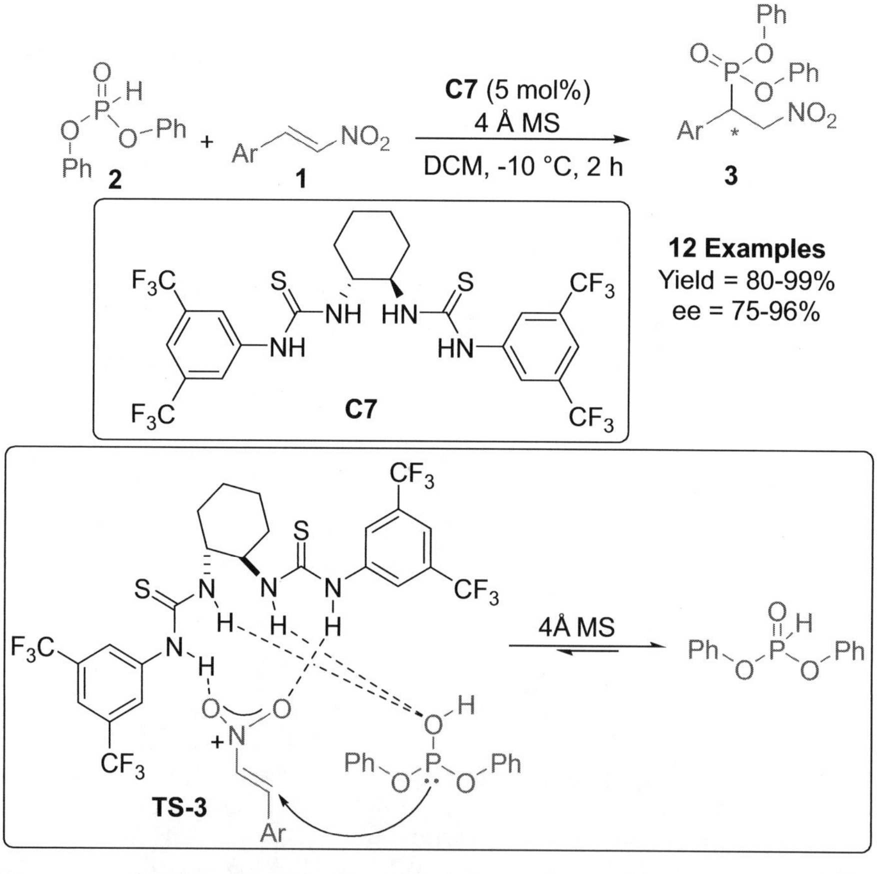

SCHEME 7.7 Secondary amine bis-thiourea catalyzed Michael addition of diphenylphosphonates to nitroalkenes.

The Tan group disclosed the catalytic asymmetric phospha-Michael reaction between diaryl phosphine oxides **4** and nitroalkenes **1** using bicyclic guanidine **C8** as a catalyst. The reaction provided chiral α-substituted β-phosphine oxides **5** in good yields and enantioselectivities (Scheme 7.8).[8]

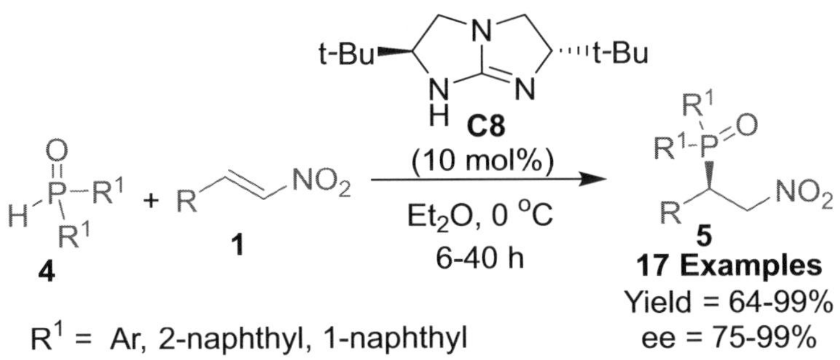

SCHEME 7.8 Bicyclic guanidine as organocatalyst in the asymmetric phospha-Michael reaction between diaryl phosphine oxides and nitroalkenes.

Ding et al. designed and synthesized novel P-stereogenic PCP pincer–Pd complex **C9** and demonstrated the potential utility as a chiral catalyst in the asymmetric addition of diarylphosphines **6** to nitroalkenes **1** (Scheme 7.9).[9] The reaction proceeded well to afford the products **7** with high yields and excellent enantioselectivities.

A highly stereoselective 1,4-addition of diarylphosphines **6** to nitroalkenes **1** catalyzed by a pincer–Pd complex **C10** was developed by Duan et al. to synthesize chiral phosphination products **8** with good-to-excellent enantioselectivities (Scheme 7.10).[10]

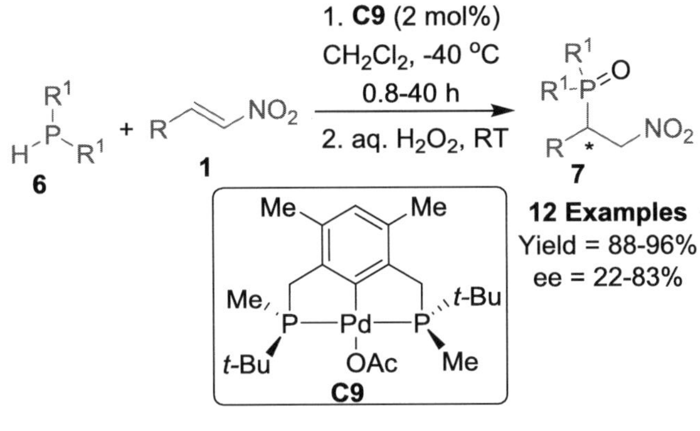

SCHEME 7.9 PCP pincer–Pd complex as a catalyst in the asymmetric addition of diarylphosphines to nitroalkenes.

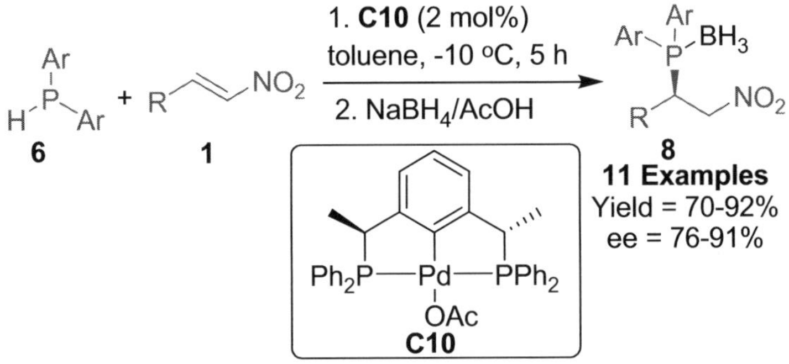

SCHEME 7.10 Bis(phosphine) pincer–Pd complex as a catalyst in asymmetric 1,4-addition of diarylphosphines to nitroalkenes.

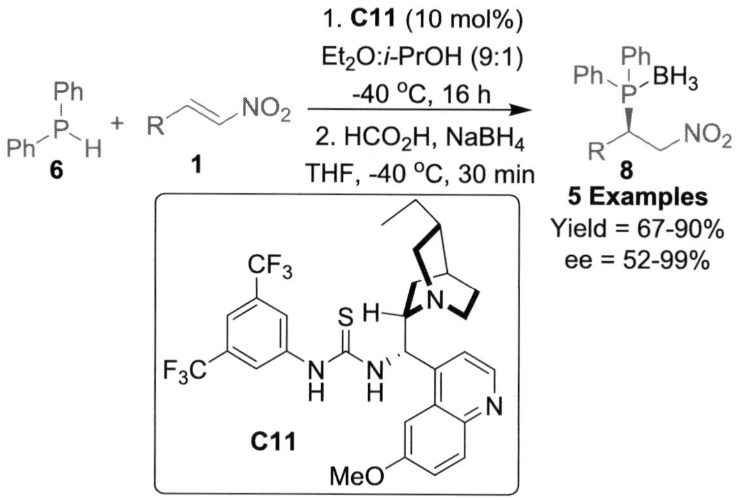

SCHEME 7.11 Bifunctional cinchona alkaloid catalyzed enantioselective addition of diphenylphosphine to nitroalkenes.

Melchiorre et al. described an enantioselective addition of diphenylphosphine **6** to various nitroalkenes **1** in the presence of a chiral base organocatalyst **C11**, affording optically active β-nitrophosphines **8**. The catalyst **C11** efficiently activates secondary phosphines toward asymmetric nucleophilic addition (Scheme 7.11).[11]

7.3 ADDITION OF SULPHUR-CENTERED NUCLEOPHILES

Enders and co-workers[12] recently reviewed some of the asymmetric sulfa-Michael additions of nitroalkenes,[12–24] and in this book we have attempted to emphasize more on the uncovered details of other articles.

The asymmetric sulfa-Michael addition of thioacetic acid **10** to α,β-disubstituted nitroalkenes **9** employing a chiral squaramide catalyst **C12** was developed by the Du group (Scheme 7.12).[25] Low catalyst loading of 0.2 mol% afforded the corresponding β-nitro sulfides **11** in high yields and diastereoselectivities with excellent enantioselectivities. Squaramide catalyst **C12** acts in a bifunctional manner, where it activates nitroalkene **9** via double H-bonding interactions. Meanwhile, thioacetic acid **10** which is deprotonated by the tertiary amine of the catalyst adds to the activated nitroalkene from the *Si*-face to afford the intermediate, which abstracts a proton from ammonium via the transition state **TS-4** to form the *anti*-product **11** (Scheme 7.12).

The Ellman group described an enantioselective addition of thioacetic acid **10** to various aromatic and aliphatic nitroalkenes **1** using N-sulfinyl urea **C13** as an organocatalyst (Scheme 7.13).[15] Nitroalkenes bearing electron-deficient *para*-substituents provided higher yield, whereas electron-rich derivatives provided higher enantioselectivities. The reaction was speculated to proceed through bifunctional organocatalysis, involving activation of nitroalkene **1** by urea via hydrogen bonding and deprotonation of thioacetic acid **10** by amine. Further, in another report, the same group showed that sulfinyl urea organocatalyst **C13** promotes the highly enantioselective addition of thioacetic acid **10** to aliphatic and aromatic β-substituted nitroalkenes **1**, as well as various cyclic nitroalkenes to introduce two chiral centers.[26]

The Yuan group reported an asymmetric sulfa-Michael addition of various S-nucleophiles to a range of α,β-disubstituted nitroalkenes **9** in the presence of a bifunctional organocatalyst **C14**.[27] The reaction proceeded well with quinidine derivative **C14** as a bifunctional organocatalyst, and the adducts **14** were obtained in high yields with high enantio and diastereoselectivities (Scheme 7.14). Here the catalyst acts in a bifunctional manner. The tertiary amine moiety of catalyst **C14** can

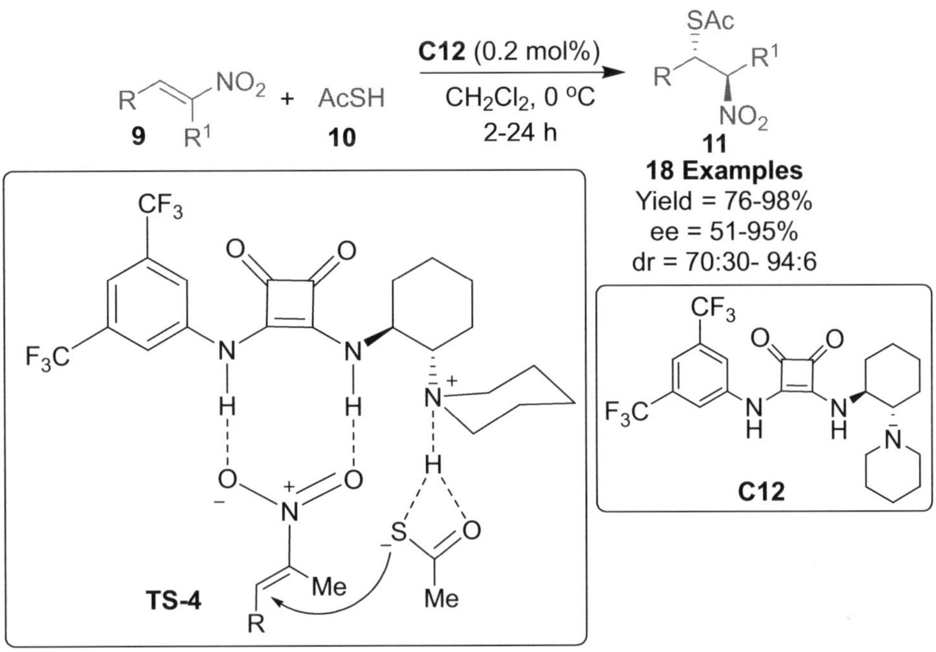

SCHEME 7.12 Asymmetric sulfa-Michael addition of thioacetic acid to α,β-disubstituted nitroalkenes in the presence of a squaramide-based catalyst.

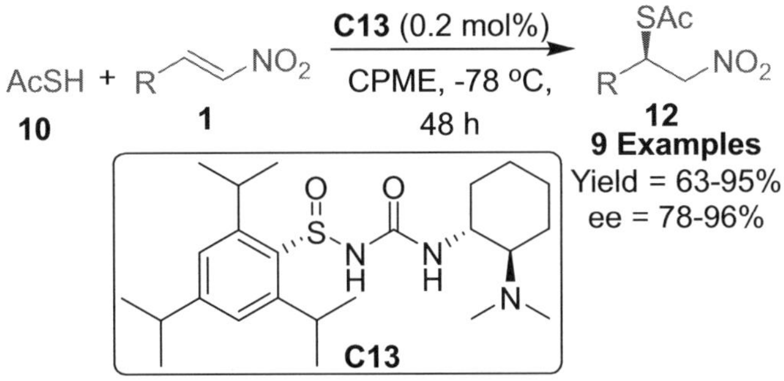

SCHEME 7.13 N-sulfinyl urea catalyzed enantioselective addition of thioacetic acid to nitroalkenes.

deprotonate the active hydrogen atom from the S-nucleophiles. Simultaneously, the OH group of catalyst **C14** activates the nitroalkenes **9** via hydrogen bonding to the oxygen atom of the nitro group. In the postulated mechanism, the thiol anion attacks the β-position of activated nitroalkenes **9** and, at the same time, the intermediate abstracts proton from ammonium and forms the *syn*-products **14** (**TS-5**, Scheme 7.14).

The Xiao group reported an organocatalytic asymmetric Michael reaction of thiols **13** to substituted nitroalkenes **15** using 9-thiourea cinchona alkaloid catalyst **C2** (Scheme 7.15). The reaction exhibited excellent enantioselectivities under low loading of catalyst **C2**.[16] The synergistic cooperative activation of the thiol **13** and nitroalkene **15** seemed necessary for the observed high enantioselectivity (**TS-6**, Scheme 7.15).

Peng and co-workers reported a catalytic asymmetric three-component sulfa-Michael/aldol cascade reaction involving ketones **17**, thiols **13** and nitroalkenes **1** as substrates. L-*tert*-leucine-derived multifunctional catalyst **C15** catalyzed the above reaction to afford γ-thio-β-nitro-α-hydroxyesters

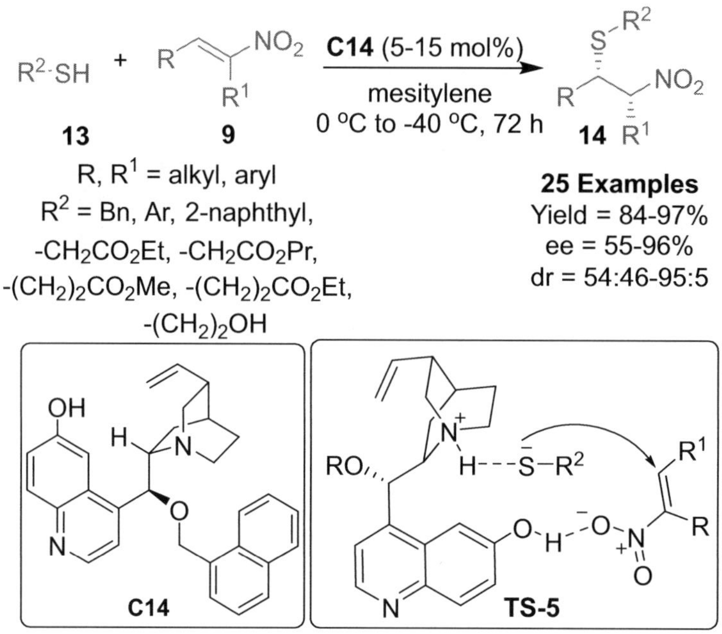

SCHEME 7.14 Asymmetric sulfa-Michael addition of various S-nucleophiles to a range of α,β-disubstituted nitroalkenes in the presence of demethylated quinidine organocatalyst.

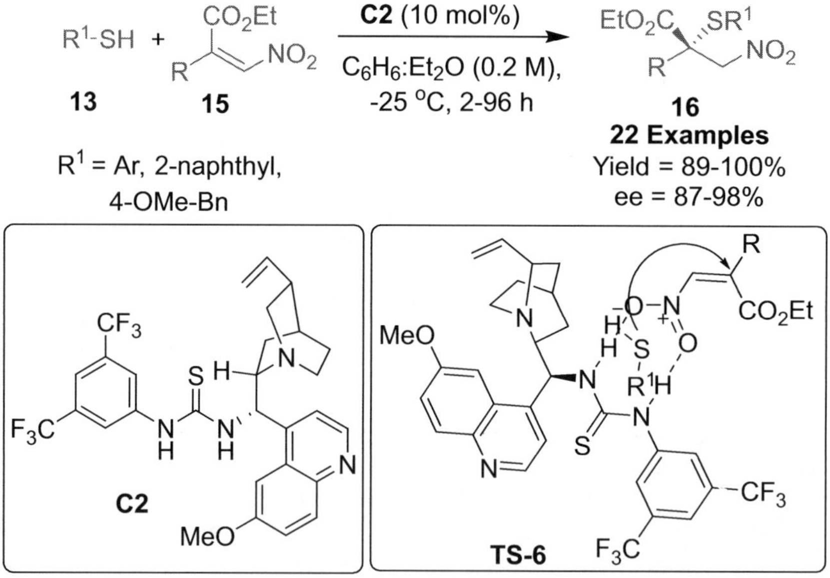

SCHEME 7.15 Asymmetric Michael reaction of thiols to substituted nitroalkenes in the presence of 9-thiourea cinchona alkaloid catalyst.

18 bearing three consecutive linear stereocenters with high yields and excellent enantio and diastereoselectivities (Scheme 7.16).[28] Nitroalkenes **1** with an electron-donating or electron-withdrawing substituent on their aryl ring afforded the corresponding products **18** in good yields with high enantio and diastereoselectivities. Less reactive aliphatic nitroalkenes **1** also gave the corresponding

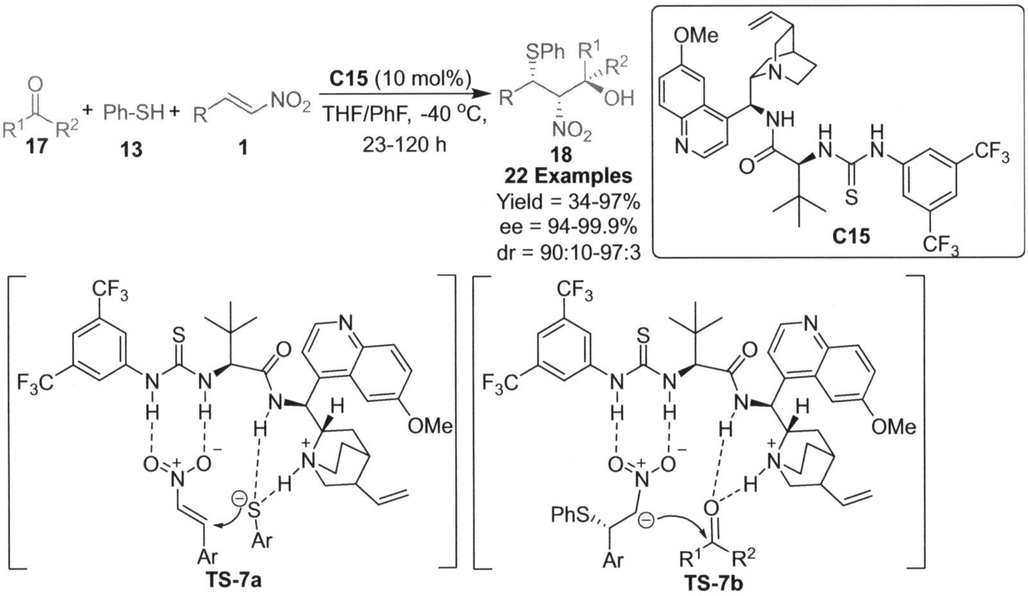

SCHEME 7.16 Asymmetric three-component sulfa-Michael/aldol cascade reaction of ketones, thiols and nitroalkenes using L-*tert*-leucine-derived organocatalyst.

products **18** in moderate yield with high diastereo and enantioselectivities. The proposed transition states (**TS-7a** and **TS-7b**) for the cascade reaction are shown in Scheme 7.16.

The same group reported the sulfa-Michael addition with aromatic thiols **13** as nucleophiles and β-nitroalkenes **19** using multifunctional thiourea catalyst **C16** (Scheme 7.17).[29] This Michael addition was extended to α,β- and β,β-substituted nitroalkenes **19** without loss of reactivity. Nitroalkenes **19** were activated by the thiourea protons of the catalyst **C16** through hydrogen bonding, whereas the quinuclidine moiety present in the catalyst **C16** deprotonates thiol **13** and generates thiolate anion, which forms hydrogen bonding with amide proton. '*S*'-configured sulfa-Michael adducts **20**

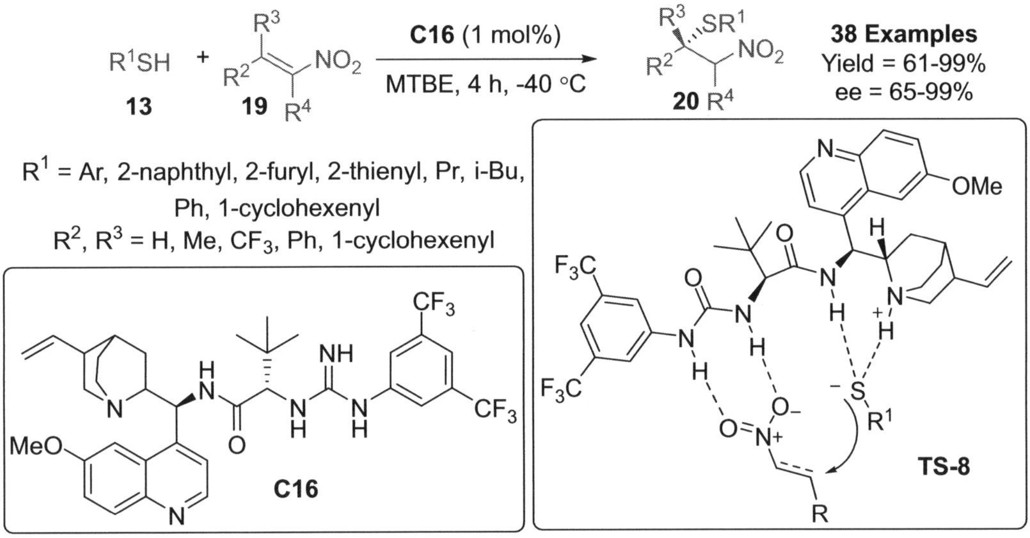

SCHEME 7.17 Sulfa-Michael addition of aromatic thiols to β-nitroalkenes using L-*tert*-leucine-derived organocatalyst.

were obtained by the *Si*-face attack on the nitrostyrene (**TS-8**, Scheme 7.17). Catalytic performance was maintained in scale-up experiments, and excellent results were observed with even 0.05 mol% of catalytic loading.

7.4 ADDITION OF NITROGEN-CENTERED NUCLEOPHILES

Monge et al. reported a diaza-ene reaction of formaldehyde *N-tert*-butyl hydrazone **21** with nitroalkenes **1**. It was catalyzed by an axially chiral bis-thiourea **C17** to afford the diazenes **22** in good-to-excellent yields and moderate enantioselectivities.[30] Further, the products **22** were transformed into β-nitronitriles **23** (Scheme 7.18).

Herrera and co-workers reported a thiourea catalyst **C4**-mediated *aza*-Michael addition of hydrazides **24** to nitroalkenes **1** (Scheme 7.19).[31] The accepted mechanism involves a bidentate activation of the nitro group by the thiourea moiety. The most acidic proton in N1 could be involved in assisting the *aza*-Michael addition through N–H bond coordination. The amine would facilitate the approach of the nucleophile to the *Si*-face of the nitroalkene **1** providing adducts **25** with *S* configuration (**TS-10**, Scheme 7.19).

Jørgensen and co-workers reported an enantioselective thiourea **C18** catalyzed *aza*-Michael addition of benzophenone imine **26** to nitroalkenes **1** (Scheme 7.20).[32] A wide range of aromatic and aliphatic nitroalkenes **1** were employed for the synthesis of protected optically active β-amino nitro compounds **27** in high yields and enantioselectivities.

Maruoka et al. reported an efficient catalytic asymmetric amination of nitroalkenes **1** in the presence of neutral phase-transfer conditions, i.e., in the presence of a chiral bifunctional tetraalkylammonium bromide **C19** in a water-rich biphasic solvent (Scheme 7.21).[33] Here, the chiral ammonium amide is generated from *tert*-butyl benzyloxycarbamate **28** and the catalyst **C19**. Stabilization of the ammonium amide via ionic interaction as well as hydrogen bond between the hydroxyl group in the catalyst and the oxygen atom of the amide moiety has been proposed. The observed absolute configuration was rationalized based on the approach of nitroalkene **1** from the topside to avoid the steric repulsion between the Boc group of amide and the phenyl group of nitrostyrene.

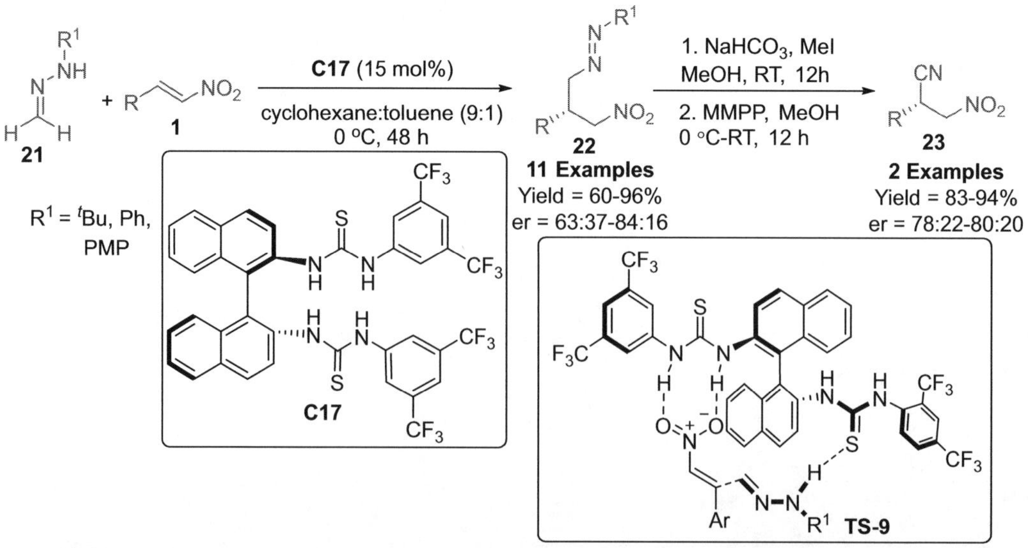

SCHEME 7.18 Axially chiral bis-thiourea catalyzed diaza-ene reaction of formaldehyde *N-tert*-butyl hydrazone with nitroalkenes.

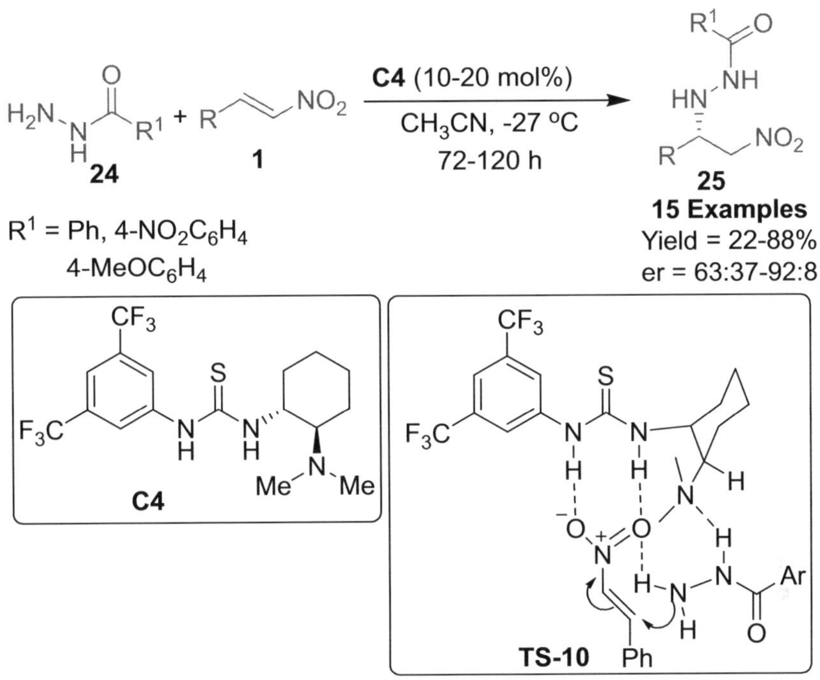

SCHEME 7.19 Thiourea-catalyzed *aza*-Michael addition reaction of hydrazides to nitroalkenes.

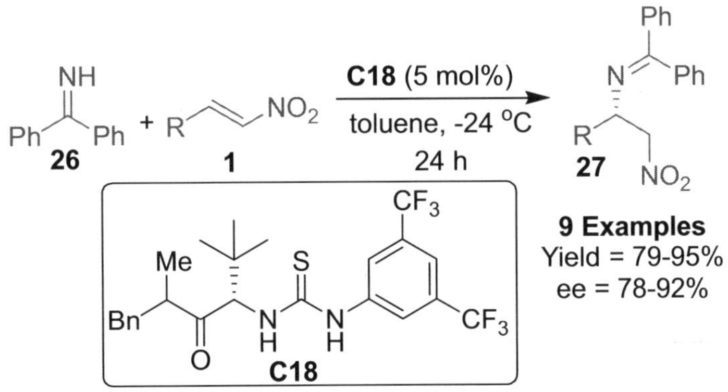

SCHEME 7.20 Thiourea-based organocatalyst in *aza*-Michael addition of benzophenone imine to nitroalkenes.

Recently, Wang et al. presented *tert*-leucine-derived squaramide phase-transfer catalysts **C20** and **C21** for asymmetric amination under water-rich conditions (Scheme 7.22).[34] 2-Aminonitroalkenes **29** are obtained in good yields (up to 96%) and enantioselectivities (up to 93%). Nitroalkene **1** was activated by H-bonding interactions with the catalyst **C20** or **C21**, and static electronic interaction of cationic ammonium center in the catalyst activates the *tert*-butyl benzyloxycarbamate **28** to attack from the *Si*-face (**TS-11**, Scheme 7.22).

The Wang group showed an application of 4-nitrophthalimide **30** as N-centered nucleophile for enantioselective *aza*-Michael addition to nitroalkenes **1** (Scheme 7.23).[35] The process has been catalyzed by chiral thiourea catalyst **C22** derived from cinchonine to give Michael adducts **31** with up to 87% ee.

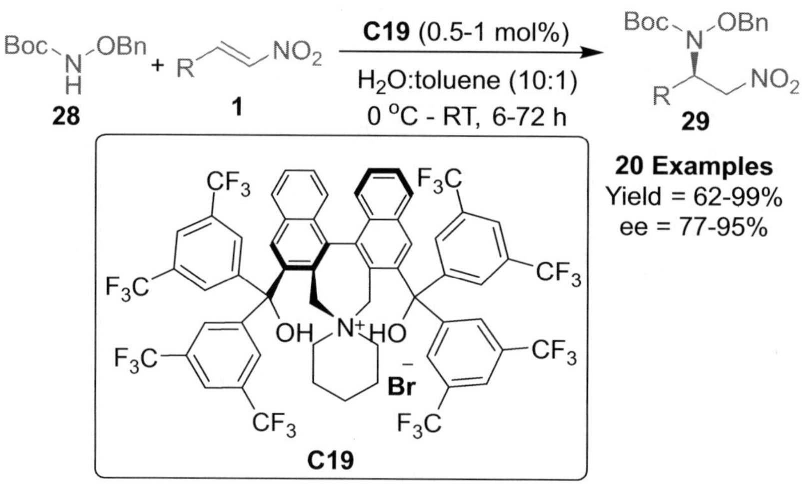

SCHEME 7.21 Asymmetric amination of nitroalkenes 1 in the presence of chiral bifunctional ammonium salt.

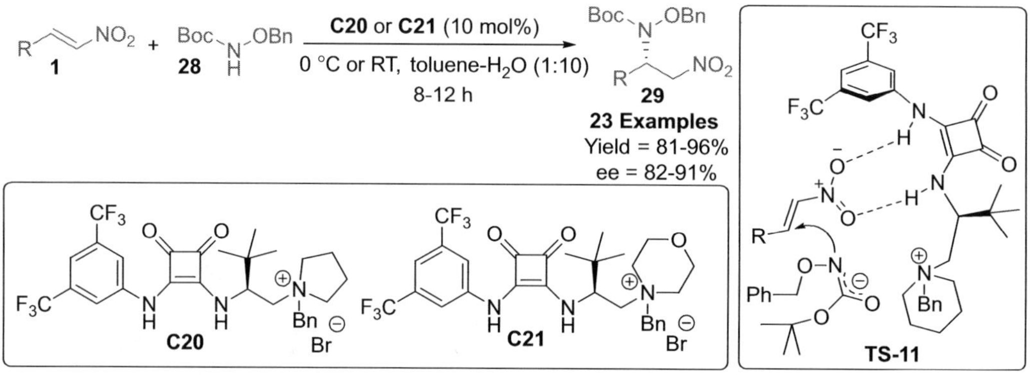

SCHEME 7.22 *tert*-Leucine-derived squaramide phase-transfer catalysts in the asymmetric amination of nitroalkenes.

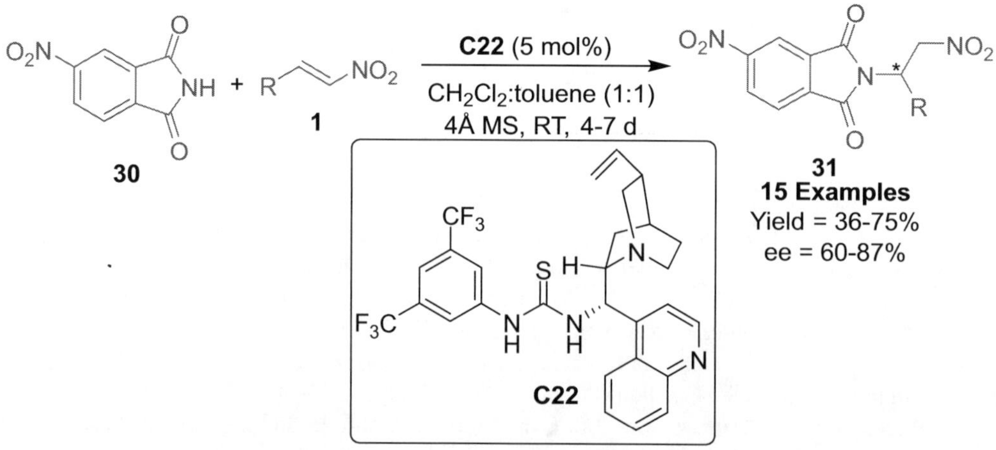

SCHEME 7.23 Enantioselective *aza*-Michael addition of 4-nitrophthalimide to nitroalkenes in the presence of chiral thiourea catalyst.

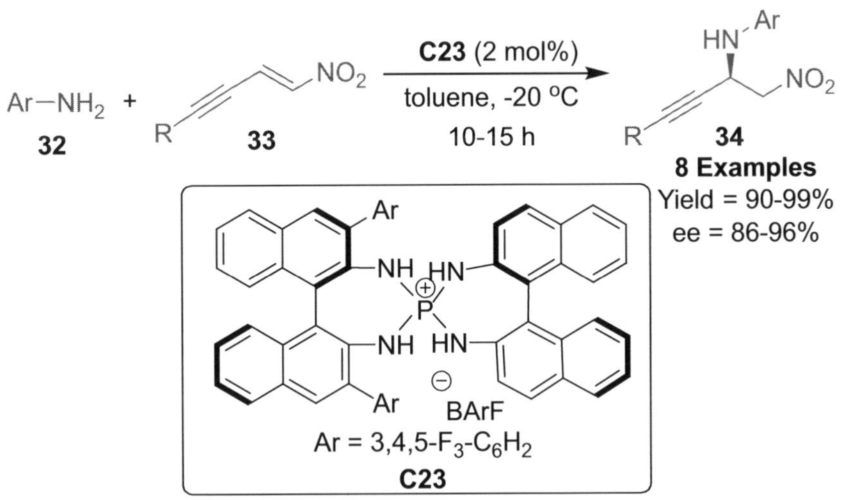

SCHEME 7.24 *aza*-Michael addition of conjugated nitroenynes catalyzed by chiral arylaminophosphonium barfates.

The Ooi group described an enantioselective *aza*-Michael addition of conjugated nitroenynes **33** using *P*-spiro chiral arylaminophosphonium barfate **C23** as a catalyst to afford β-amino homopropargylic nitro compounds **34** in good yields with excellent enantioselectivities (Scheme 7.24).[36] The absolute configuration of the product **34** was assigned to be '*R*' by single-crystal X-ray analysis of the compound.

7.5 ADDITION OF OXYGEN-CENTERED NUCLEOPHILES

The Lattanzi group described an asymmetric β-peroxidation of nitroalkenes **1** to afford optically active peroxides **36** in good yields and moderate-to-good enantioselectivities.[37] The reaction used a commercially available diaryl-2-pyrrolidinemethanol derivative **C24** as an organocatalyst and *tert*-butyl hydroperoxide **35** as the reactant (Scheme 7.25). In the proposed mechanism, the intermolecular hydrogen bonding between the OH group of the catalyst and the nitro group appears to favor the attack of *tert*-butyl peroxy anion to the *Si*-face of nitroalkene **1** (**TS-12**) affording the product **36** (Scheme 7.25).

The Deng group developed the enantioselective peroxidation of nitroalkenes **1** utilizing peroxide **37** as the reactant in the presence of a readily accessible acid-base bifunctional organocatalyst **C25** (Scheme 7.26).[38] Aromatic nitroalkenes **1** bearing both the *meta*- and *para*-substituted groups afforded corresponding optically active chiral peroxides **38** in good yields and enantioselectivities. The reaction is sensitive to the steric factor as reactions with bulky α-alkoxyl hydroperoxides failed to occur.

The Xiao group described intermolecular *oxa*-Michael addition of oximes **39** to β-CF_3-β-disubstituted nitroalkenes **40** by employing a chiral bifunctional cinchona alkaloid-based thiourea catalyst **C2**. Various structurally diverse CF_3-containing oxime ethers **41** were formed in excellent yields and enantioselectivities (Scheme 7.27).[39]

An intermolecular conjugate addition of oxime **39** to nitroalkenes **15** was developed by the Xiao group in the presence of a chiral catalyst **C26** to yield the corresponding products **42** with high stereoselectivities (Scheme 7.28).[40] In the proposed mechanism, the OH group of the catalyst **C26** activates the electrophile through H-bonding interaction, and the oxime **39** would attack the *Si*-face of the nitroalkene **15** to form the major stereoisomers (**TS-14**, Scheme 7.28).

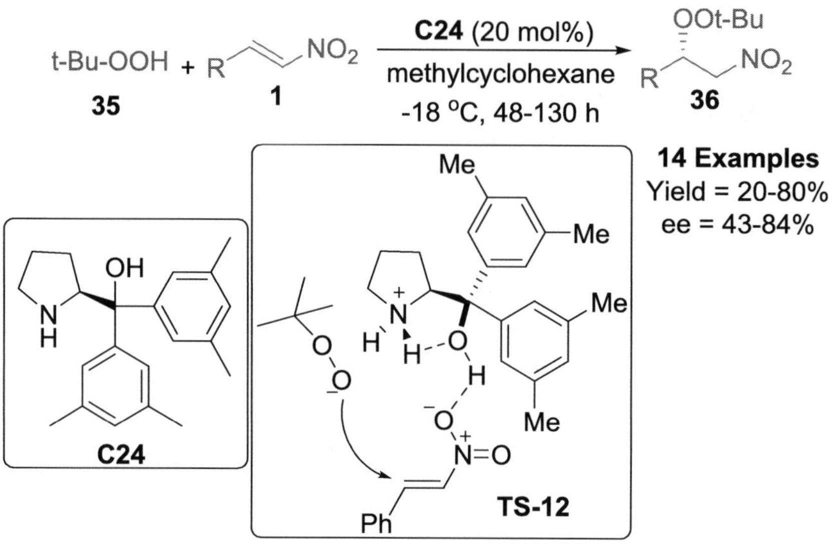

SCHEME 7.25 Asymmetric β-peroxidation of nitroalkenes catalyzed by diaryl-2-pyrrolidinemethanol derivative.

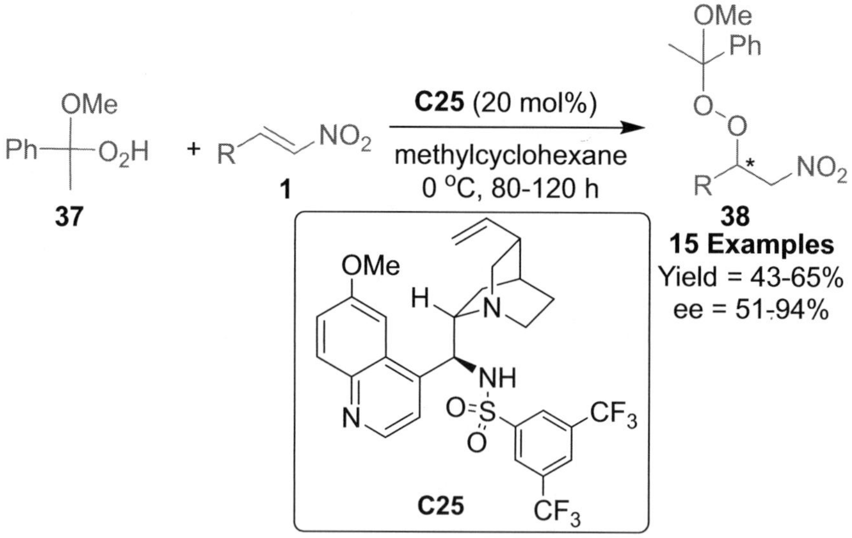

SCHEME 7.26 Enantioselective peroxidation of nitroalkenes catalyzed by 9-sulfonamide cinchona alkaloid.

The Jørgensen group reported the catalytic highly enantioselective conjugate hydroxylation of nitroalkenes **1** by employing oximes **43** as oxygen-centered nucleophile and bifunctional thiourea cinchona organocatalyst **C2** to afford optically active products **44** (Scheme 7.29).[41] Here the catalyst **C2** has the potential of activating the oxime by hydrogen bonding through tertiary nitrogen atom of the catalyst and by Lewis-acid activation of the nitroalkene **1** by the thiourea moiety of the catalyst. The absolute configuration of the product was determined to be *R*.

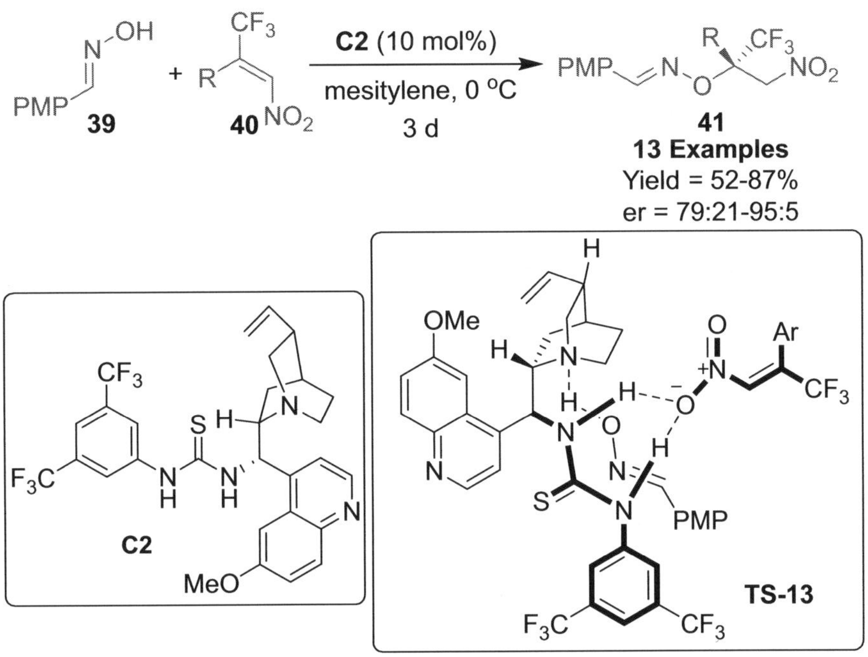

SCHEME 7.27 Asymmetric intermolecular *oxa*-Michael addition of oximes to β-CF_3-β-disubstituted nitroalkenes using bifunctional cinchona alkaloid-based thiourea catalyst.

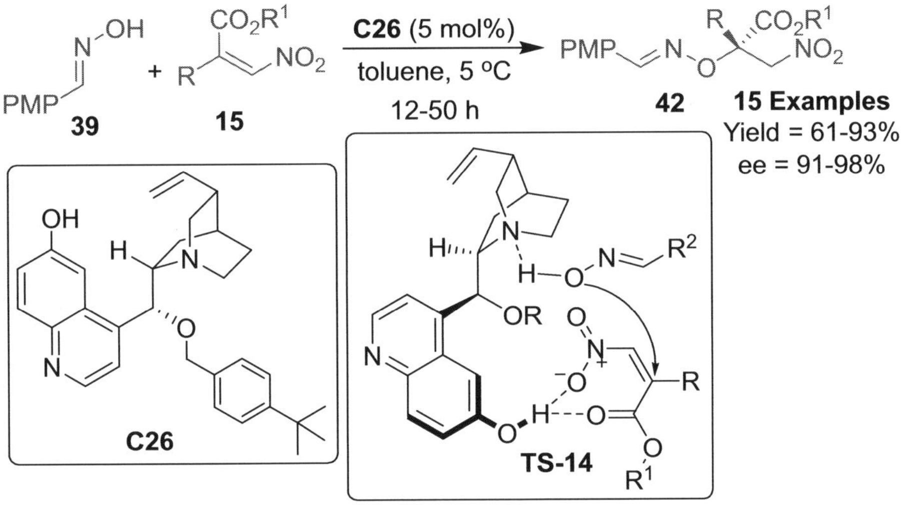

SCHEME 7.28 Enantioselective conjugate addition of oxime to nitroalkenes catalyzed by cinchona alkaloid-based thiourea.

7.6 ADDITION OF ARSENIC-CENTERED NUCLEOPHILES

Leung and co-workers reported asymmetric hydroarsination of nitroalkenes **1b** mediated by chiral Ni/Pd/Pt-based pincer complexes **C27** (Scheme 7.30).[42] The corresponding arsine adducts **46** were formed in good yields and enantioselectivities (up to 80% ee). The best result (80% ee) was obtained using the NiCl-complex as the catalyst. The stereochemistry of the hydroarsination product **46** was confirmed by X-ray crystallographic analysis of the Au(I) chloride complex.

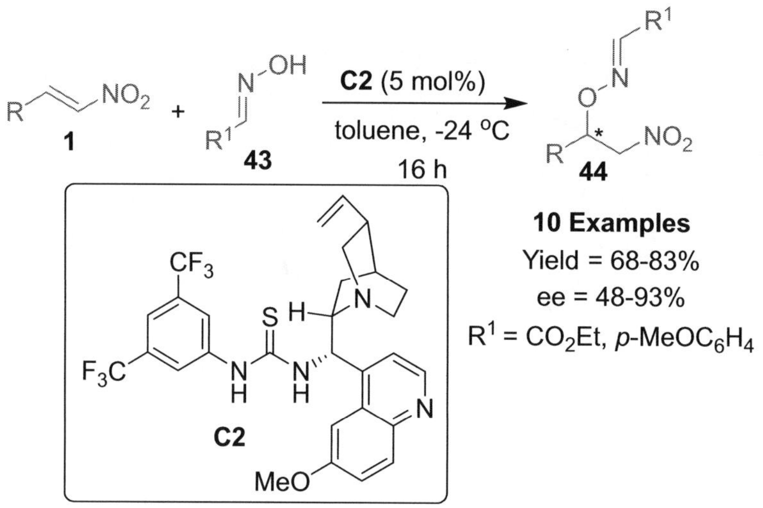

SCHEME 7.29 Enantioselective conjugate hydroxylation of nitroalkenes by employing oximes as oxygen-centered nucleophiles using bifunctional thiourea cinchona organocatalyst.

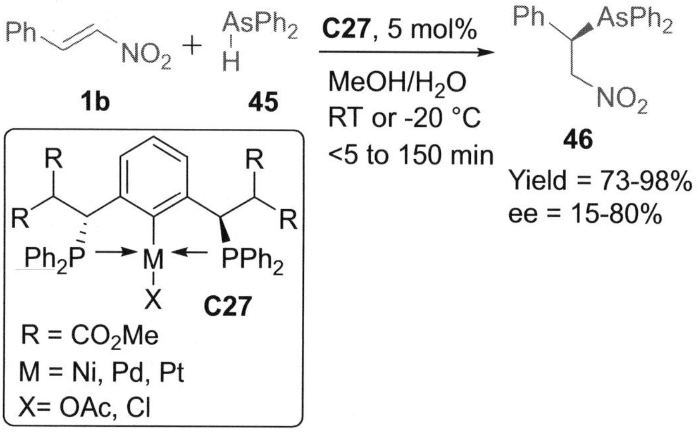

SCHEME 7.30 Stereoselective hydroarsination of nitroalkenes catalyzed by metal-based complexes.

7.7 CONCLUSIONS

Conjugated nitroalkenes are suitable acceptors for various heteroatom-centered nucleophiles despite the fact that the nucleophilic atoms, such as P, S, N and O, possess a wide range of electronegativities. The products were generated in high yields and selectivities in most cases by employing appropriate catalysts and conditions. Reports on the applications of the enantioenriched Michael adducts are expected in the foreseeable future.

REFERENCES

1. Zhu, Y.; Malerich, J. P.; Rawal, V. H. *Angew. Chem. Int. Ed.* **2010**, *49*, 153.
2. Abbaraju, S.; Bhanushali, M.; Zhao, C.-G. *Tetrahedron* **2011**, *67*, 7479.
3. Rai, V.; Namboothiri, I. N. N. *Tetrahedron: Asymmetry* **2008**, *19*, 2335.
4. Alcaine, A.; Marques-Lopez, E.; Merino, P.; Tejero, T.; Herrera, R. P. *Org. Biomol. Chem.* **2011**, *9*, 2777.
5. Wang, J.; Heikkinen, L. D.; Li, H.; Zu, L.; Jiang, W.; Xie, H.; Wang, W. *Adv. Synth. Catal.* **2007**, *349*, 1052.

6. Terada, M.; Ikehara, T.; Ube, H. *J. Am. Chem. Soc.* **2007**, *129*, 14112.
7. Nazish, M.; Jakhar, A.; Gupta, N.; Khan, N. H.; Kureshy, R. I. *Synlett* **2018**, *2018*, 1385.
8. Fu, X.; Jiang, Z.; Tan, C.-H. *Chem. Commun.* **2007**, *2007*, 5058.
9. Ding, B.; Zhang, Z.; Xu, Y.; Liu, Y.; Sugiya, M.; Imamoto, T.; Zhang, W. *Org. Lett.* **2013**, *15*, 5476.
10. Feng, J.-J.; Huang, M.; Lin, Z.-Q.; Duan, W.-L. *Adv. Synth. Catal.* **2012**, *354*, 3122.
11. Bartoli, G.; Bosco, M.; Carlone, A.; Locatelli, M.; Mazzanti, A.; Sambri, L.; Melchiorre, P. *Chem. Commun.* **2007**, *2007*, 722.
12. Chauhan, P.; Mahajan, S.; Enders, D. *Chem. Rev.* **2014**, *114*, 8807.
13. Dodda, R.; Goldman, J. J.; Mandal, T.; Zhao, C.-G.; Broker, G. A.; Tiekink, E. R. T. *Adv. Synth. Catal.* **2008**, *350*, 537.
14. Wang, J.; Xie, H.; Li, H.; Zu, L.; Wang, W. *Angew. Chem. Int. Ed.* **2008**, *47*, 4177.
15. Kimmel, K. L.; Robak, M. T.; Ellman, J. A. *J. Am. Chem. Soc.* **2009**, *131*, 8754.
16. Lu, H.-H.; Zhang, F.-G.; Meng, X.-G.; Duan, S.-W.; Xiao, W.-J. *Org. Lett.* **2009**, *11*, 3946.
17. Wang, X.-F.; Hua, Q.-L.; Cheng, Y.; An, X.-L.; Yang, Q.-Q.; Chen, J.-R.; Xiao, W.-J. *Angew. Chem. Int. Ed.* **2010**, *49*, 8379.
18. Yu, C.; Zhang, Y.; Song, A.; Ji, Y.; Wang, W. *Chem. - Eur. J.* **2011**, *17*, 770.
19. Du, Z.; Zhou, C.; Gao, Y.; Ren, Q.; Zhang, K.; Cheng, H.; Wang, W.; Wang, J. *Org. Biomol. Chem.* **2012**, *10*, 36.
20. Palacio, C.; Connon, S. J. *Chem. Commun.* **2012**, *48*, 2849.
21. Uraguchi, D.; Kinoshita, N.; Nakashima, D.; Ooi, T. *Chem. Sci.* **2012**, *3*, 3161.
22. Kowalczyk, R.; Nowak, A. E.; Skarżewski, J. *Tetrahedron: Asymmetry* **2013**, *24*, 505.
23. Wu, L.; Wang, Y.; Song, H.; Tang, L.; Zhou, Z.; Tang, C. *Adv. Synth. Catal.* **2013**, *355*, 1053.
24. Yang, W.; Yang, Y.; Du, D.-M. *Org. Lett.* **2013**, *15*, 1190.
25. Yang, W.; Du, D.-M. *Org. Biomol. Chem.* **2012**, *10*, 6876.
26. Kimmel, K. L.; Robak, M. T.; Thomas, S.; Lee, M.; Ellman, J. A. *Tetrahedron* **2012**, *68*, 2704.
27. Pei, Q.-L.; Han, W.-Y.; Wu, Z.-J.; Zhang, X.-M.; Yuan, W.-C. *Tetrahedron* **2013**, *69*, 5367.
28. Hou, W.; Wei, Q.; Peng, Y. *Adv. Synth. Catal.* **2016**, *358*, 1035.
29. Wei, Q.; Hou, W.; Liao, N.; Peng, Y. *Adv. Synth. Catal.* **2017**, *359*, 2364.
30. Monge, D.; Daza, S.; Bernal, P.; Fernandez, R.; Lassaletta, J. M. *Org. Biomol. Chem.* **2013**, *11*, 326.
31. Alcaine, A.; Marques-Lopez, E.; Herrera, R. P. *RSC Adv.* **2014**, *4*, 9856.
32. Lykke, L.; Monge, D.; Nielsen, M.; Jørgensen, K. A. *Chem. - Eur. J.* **2010**, *16*, 13330.
33. Wang, L.; Shirakawa, S.; Maruoka, K. *Angew. Chem. Int. Ed.* **2011**, *50*, 5327.
34. Zhu, J.; Cui, D.; Li, Y.; He, J.; Chen, W.; Wang, P. *Org. Biomol. Chem.* **2018**, *16*, 3012.
35. Ma, S.; Wu, L.; Liu, M.; Huang, Y.; Wang, Y. *Tetrahedron* **2013**, *69*, 2613.
36. Uraguchi, D.; Kinoshita, N.; Kizu, T.; Ooi, T. *Synlett* **2011**, *2011*, 1265.
37. Russo, A.; Lattanzi, A. *Adv. Synth. Catal* **2008**, *350*, 1991.
38. Lu, X.; Deng, L. *Org. Lett.* **2014**, *16*, 2358.
39. Liu, F.-L.; Chen, J.-R.; Feng, B.; Hu, X.-Q.; Ye, L.-H.; Lu, L.-Q.; Xiao, W.-J. *Org. Biomol. Chem.* **2014**, *12*, 1057.
40. Zhang, F.-G.; Yang, Q.-Q.; Xuan, J.; Lu, H.-H.; Duan, S.-W.; Chen, J.-R.; Xiao, W.-J. *Org. Lett.* **2010**, *12*, 5636.
41. Diner, P.; Nielsen, M.; Bertelsen, S.; Niess, B.; Jørgensen, K. A. *Chem. Commun.* **2007**, *2007*, 3646.
42. Tay, W. S.; Yang, X.-Y.; Li, Y.; Pullarkat, S. A.; Leung, P.-H. *Chem. Commun.* **2017**, *53*, 6307.

8 Catalytic Asymmetric Cycloadditions of Nitroalkenes

8.1 INTRODUCTION

Catalytic asymmetric Michael additions of various nucleophiles to nitroalkenes and Friedel–Crafts alkylation of electron-rich aromatic and heteroaromatic compounds with nitroalkenes have been described in the previous chapters. However, nitroalkenes also participate in cycloadditions, including Diels–Alder and 1,3-dipolar reactions, which are covered in this chapter. Various metal-ligand complexes and organocatalysts have been employed as catalysts in these cycloadditions.

In 1996, Denmark and Thorarensen reviewed tandem [4+2]/[3+2] cycloadditions of nitroalkenes.[1] Later, several other groups published reviews[2–4] on the asymmetric cycloadditions of nitroalkenes as a part of other topics.[5–40] Therefore, here, we highlight primarily the literature which is not covered in those reviews.

8.2 [3+2] CYCLOADDITIONS

The Zhou group described an asymmetric synthesis of polysubstituted pyrrolidines **3** bearing a trifluoromethylated quaternary chiral center via the CuI/Si-FOXAP **L1**-catalyzed 1,3-dipolar cycloaddition of azomethine ylides derived from imines **2a** with β-trifluoromethyl nitroalkenes **1a** (Scheme 8.1).[41] The reaction exhibited excellent diastereo and enantioselectivity.

The Najera group described the use of chiral phosphoramidites **L2** as ligands in the copper(II)-catalyzed 1,3-dipolar cycloaddition of azomethine ylides of **2b** with β-nitrostyrenes **1** at room temperature to afford the corresponding tetrasubstituted proline esters **4** mainly as *exo*-cycloadducts in a high enantiomeric ratio (Scheme 8.2).[42] The diastereoselective formation of *exo*-adducts was further supported by density functional theory (DFT) calculations. Moreover, the isolation of Michael adduct at lower temperature gives evidence that the reaction follows a stepwise mechanism.

1,3-Dipolar cycloadditions using α-imino γ-lactones **5** as azomethine ylide precursors with nitroalkenes **1** as dipolarophiles for the synthesis of constrained spiro-nitroprolinate cycloadducts **6** was achieved by Sansano et al. in moderate-to-good yields with excellent enantio and

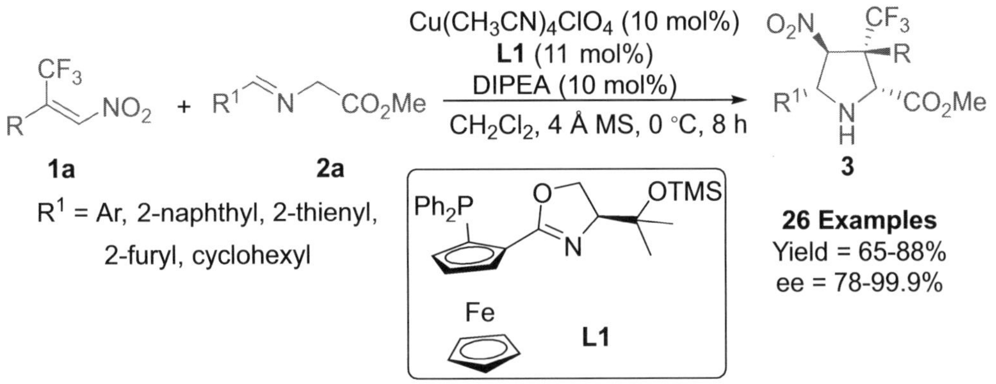

SCHEME 8.1 Cu^{I}/Si-FOXAP-Catalyzed 1,3-dipolar cycloaddition of azomethine ylides with β-CF_3-nitroalkenes.

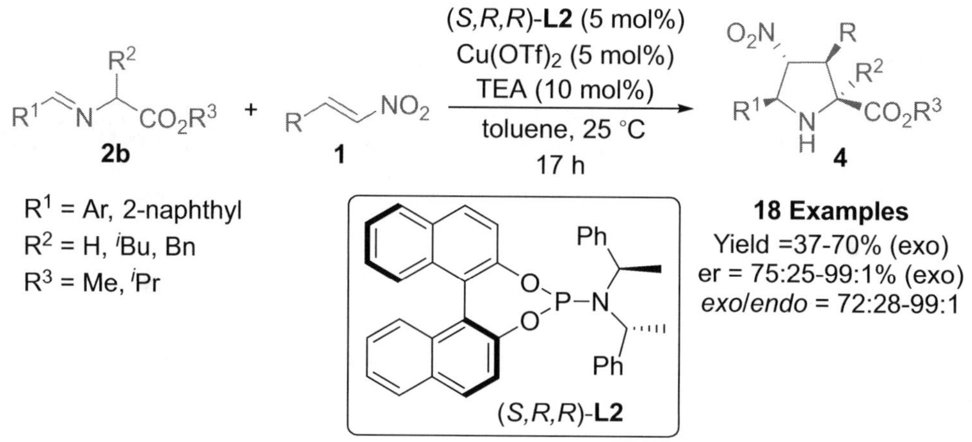

SCHEME 8.2 1,3-Dipolar cycloaddition of azomethine ylides with β-nitrostyrenes in the presence of chiral phosphoramidite-$Cu(OTf)_2$ catalyst system.

diastereoselectivities (Scheme 8.3).[43] Here, (*R*, *R*)-Me-DuPhos and AgF forms a complex that serves as a bifunctional catalyst wherein fluoride behaves as a base and participates in the reaction. DFT calculations revealed the absolute configuration and other stereochemical parameters of the product.

Highly substituted chiral pyrrolidines **7** were synthesized by the organocatalytic stereoselective 1,3-dipolar cycloaddition of azomethine ylides from **2c** with nitroalkenes **1** with high yields, moderate enantioselectivities and very high diastereoselectivities (Scheme 8.4).[21]

The Hou group reported the Cu-P, N-ferrocene catalyzed 1,3-dipolar cycloaddition of nitroalkenes to iminoester. The reaction proceeded with high stereo (*endo/exo*) and enantioselectivity. *Exo* and *endo*-selectivity was achieved by controlling the electronic properties of the ligand.[9] The assignment of absolute configuration as 2*S*, 3*S*, 4*R*, 5*R* for *exo*-**8a** and as 2*S*, 3*R*, 4*S*, 5*R* for *endo*-**8b** was made by X-ray diffraction analysis. The 2*S*, 5*R* configuration for both the products suggests that the nitroalkene **1** adds from the *Si*-face of the Cu-bound iminoester **2a**. The *Re*-face of the Cu-bound iminoester complex is shielded by the isopropyl group. However, because the *Si*-face is not shielded, the attack of nitroalkene **1** could occur at the *Si*-face of the Cu-bound iminoester anion to provide the experimentally observed product. The transition states of *exo* and *endo*-adducts are **TS-1A** and **TS-1B**, respectively (Scheme 8.5).

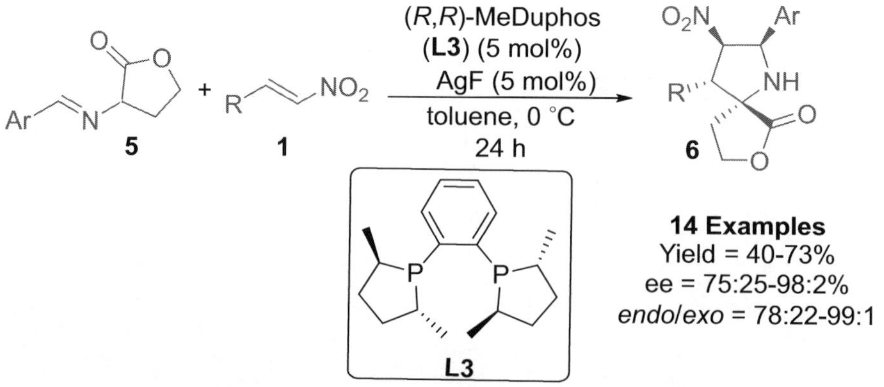

SCHEME 8.3 (R,R)-Me-DuPhos·AgF-Catalyzed 1,3-dipolar cycloaddition of azomethine ylides generated from α-imino-γ-lactones with nitroalkenes.

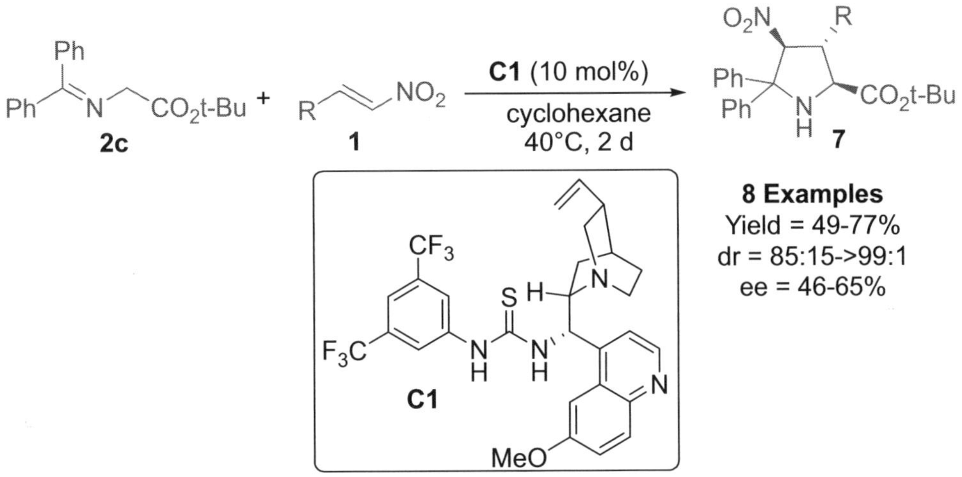

SCHEME 8.4 Chiral thiourea-catalyzed stereoselective 1,3-dipolar cycloaddition of azomethine ylides with nitroalkenes.

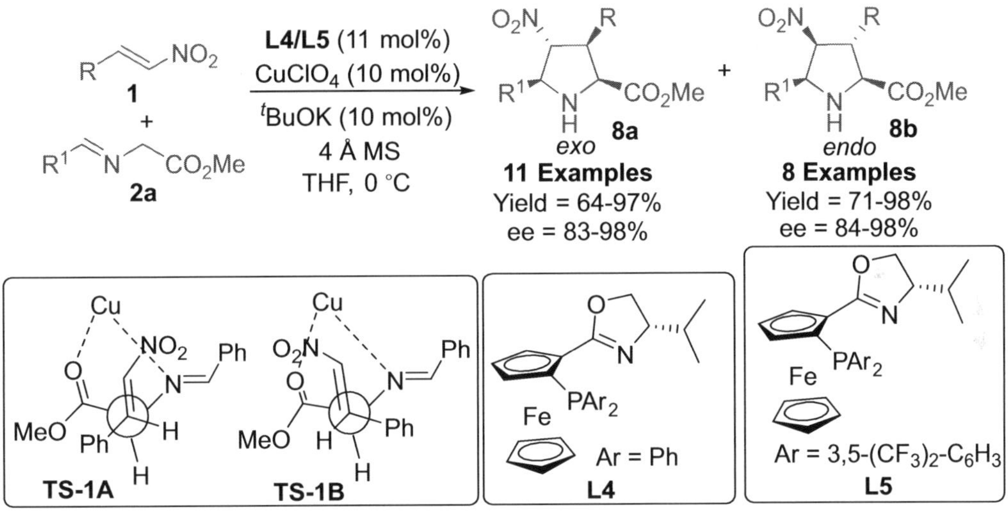

SCHEME 8.5 1,3-Dipolar cycloaddition of nitroalkenes to iminoester in the presence of Cu-P, N-ferrocene catalyst system.

Later, the same group synthesized 3-(fluoromethyl)-4-nitroproline derivatives **9a-b** via cycloaddition of azomethine ylides of **2a** with fluoromethyl-substituted nitroalkenes **1b** in the presence of copper(I) perchlorate and a chiral Walphos ligand **L6** (Scheme 8.6).[37]

Arai et al. reported a catalytic asymmetric *exo′*-selective [3+2] cycloaddition of iminoesters **2a** and nitroalkenes **1** using a Ligand **L7**/Ni(OAc)$_2$ catalytic system (Scheme 8.7).[26] The authors explained that the plausible mechanism involves an initial nickel-catalyzed Michael addition of the iminoester **2a** to the nitroalkene **1**. The interaction between the nitro functionality and the nickel center appears to direct the *anti*-selective nucleophilic addition at the C2 position of the nitroalkene **1**, as shown in **TS-2** (Scheme 8.7). After the *anti*-selective Michael addition, the nickel atom spontaneously flips to the nitronate for opening the strained cyclic intermediate. It was also mentioned that, prior to the subsequent Mannich reaction, rotation of the C–N single bond occurs to give the *exo′* isomer via the most stable transition state **TS-2**, which was further supported through DFT calculations.

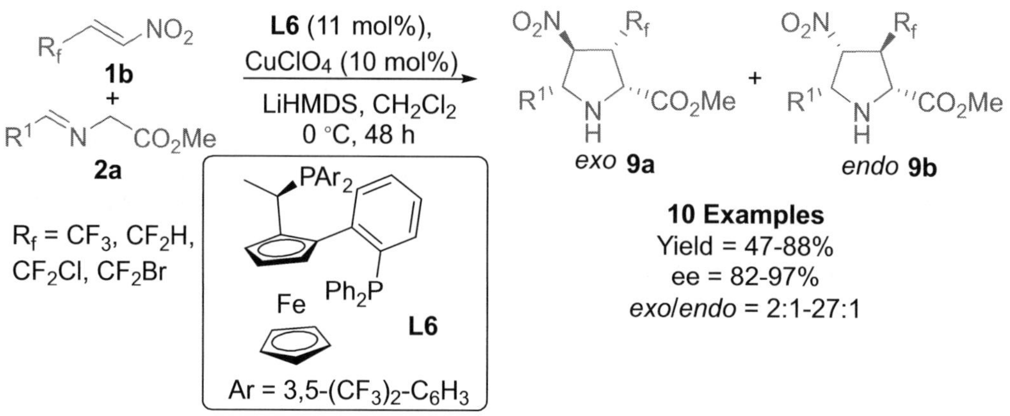

SCHEME 8.6 Cycloaddition of fluoromethyl-substituted nitroalkenes with azomethine ylides using copper(I) perchlorate and chiral Walphos ligand.

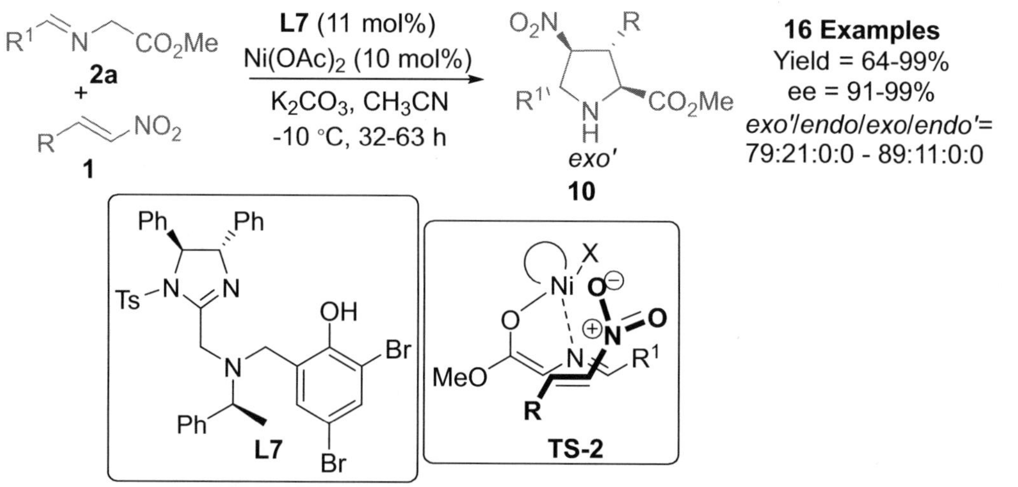

SCHEME 8.7 Asymmetric *exo*-selective reaction of iminoesters and nitroalkenes in the presence of imidazoline-aminophenol/Ni(OAc)$_2$ catalytic system.

Padilla et al. reported an asymmetric Cu(I)-catalyzed [3+2] cycloaddition of *N*-(2-pyridylmethyl) imines **11** with nitroalkenes (Scheme 8.8).[44] In this reaction, Cu(CH$_3$CN)$_4$PF$_6$/bisoxazoline **L8** was used as a chiral catalyst system, and the corresponding products **12a-b** were obtained in excellent enantioselectivities and moderate-to-good *exo*-selectivities.

Arai et al. employed bis(imidazolidine)pyridine-Cu complex for the highly *endo*-selective [3+2] cycloaddition of iminoesters **2d** to nitroalkenes **1d** resulting in the selective formation of the (2*S*, 3*R*, 4*S*, 5*S*)-pyrrolidines (Scheme 8.9).[25] The authors explained that the Cu catalyst formed a Cu-bound iminoester complex, and, consequently, the access of the nitroalkene **1d** to the activated iminoester in a sterically restricted environment provided the product **13a-b** in an *endo*-selective manner.

Recently, Xiao and co-workers designed phosphoramidite-thioether ligands **L10a-b** for highly enantioselective copper-catalyzed diastereodivergent 1,3-dipolar addition of azomethine ylides generated from **2a** to nitroalkenes **1** (Scheme 8.10).[45] Both *endo*- and *exo*-pyrrolidines **14a-b** are produced in high diastereo and enantioselectivity, and this diastero switching was accomplished because of the conformation and chiral environment of the catalyst attained by minor ligand modification.

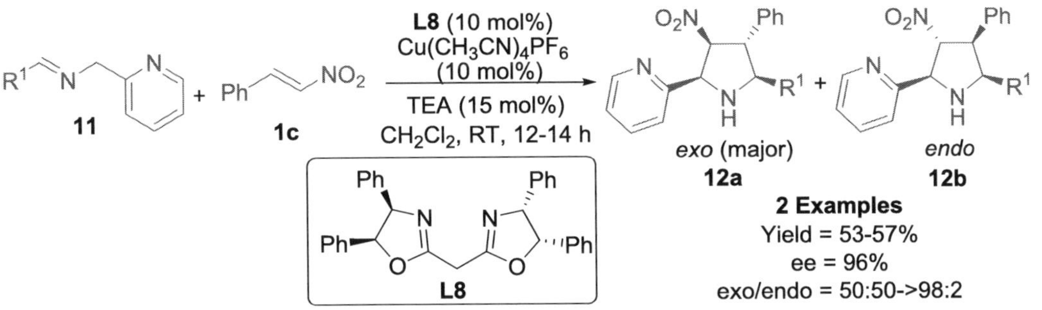

SCHEME 8.8 Asymmetric Cu(I)-bisoxazoline catalyzed [3+2] cycloaddition of *N*-(2-pyridylmethyl) imines with nitroalkenes.

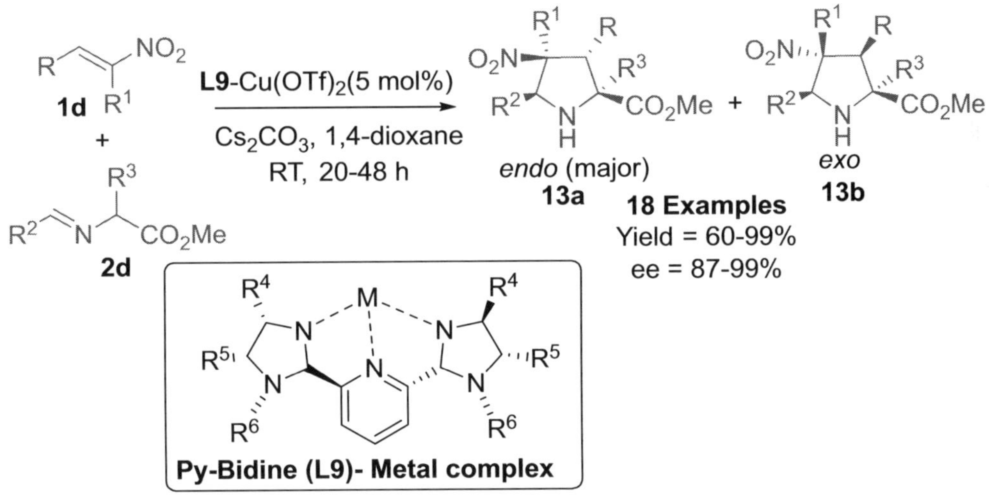

SCHEME 8.9 Bis(imidazolidine)pyridine-Cu complex as catalyst for *endo*-selective [3+2] cycloaddition of iminoesters with nitroalkenes.

The absolute configuration of the *endo*- and *exo*-pyrrolidines was determined by single-crystal X-ray crystallography as (2*R*, 3*S*, 4*R*, 5*R*) and (2*R*, 3*R*, 4*S*, 5*R*), respectively.

The same group disclosed the umpolung reactivity of imines in [3+2] cycloaddition with nitroalkenes. The regio-reversed asymmetric reaction works in the presence of Cu catalyst and chiral phosphoramidite ligand **L10a** to furnish substituted pyrrolidines **15** in high yield and enantioselectivity (Scheme 8.11).[46] This mechanism of the cycloaddition was explained based on the control experiments. α-Aryl group present in the iminoester **2d** stabilizes the *aza*-allyl anion of azomethine ylide intermediate, which promotes the umpolung reactivity of iminoester **2d**. N-metalated azomethine ylide attacks the nitroalkene **1d** from the *Si*-face. Moreover, the steric hindrance created between the imine group and the phenyl group of nitroalkene **1d** leads to a chiral *exo*-product (**TS-3**, Scheme 8.11).

Oxindole-based azomethine ylides were used as substrates with 3-nitro-2H-chromenes **1e** for the synthesis of substituted polycyclic spirooxindole-chromane adducts **18e** via asymmetric [3+2] cycloaddition (Scheme 8.12).[47] Bifunctional phosphoric acid catalyst **C2** activates both nitroalkene **1e** and dipoles derived from **16** and **17** simultaneously to afford respective *endo*-fused cycloadduct **18e**. The synthetic potential of the cycloaddition strategy was proved by scale-up experiments and synthetic transformations of the product without loss of diastereoselectivity.

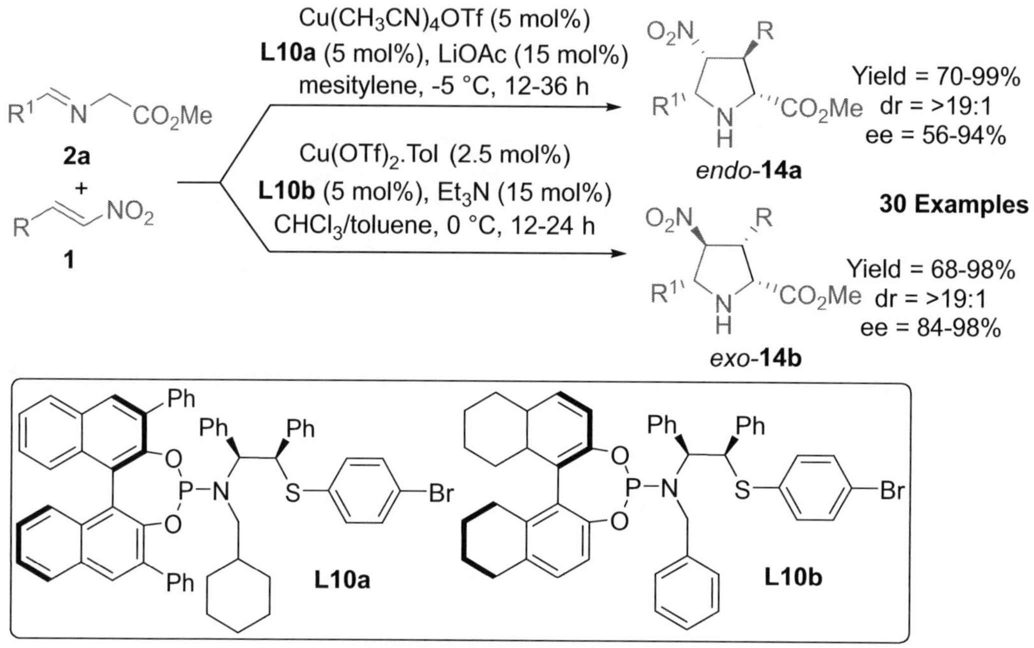

SCHEME 8.10 Phosphoramidite-thioether-Cu complexes as catalysts for 1,3-dipolar cycloaddition of azomethine ylides with nitroalkanes.

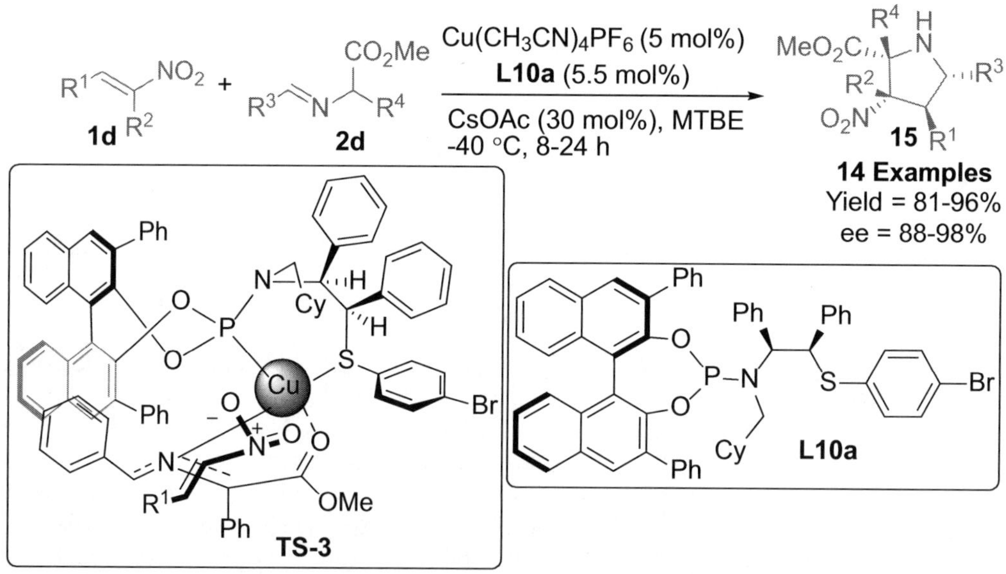

SCHEME 8.11 Copper-catalyzed regio-reversed [3+2] cycloadditions of iminoesters with nitroalkenes.

Wang and co-workers reported an asymmetric [3+2] cycloaddition of isatin-derived trifluoromethylamines **19** with nitroalkenes **1** by employing a squaramide catalyst **C3** to afford substituted 5′-CF_3-spiro-[pyrrolidin-3,2′-oxindoles] **20** in good yields and excellent enantioselectivities under mild conditions (Scheme 8.13).[48] In the transition state model, ketimine **19** forms hydrogen bonding with tertiary –NH of the catalyst **C3** and, at the same time, squaramide –NH of catalyst **C3** activates nitroalkene **1**. At this juncture, the catalyst draws the two substrates spatially close

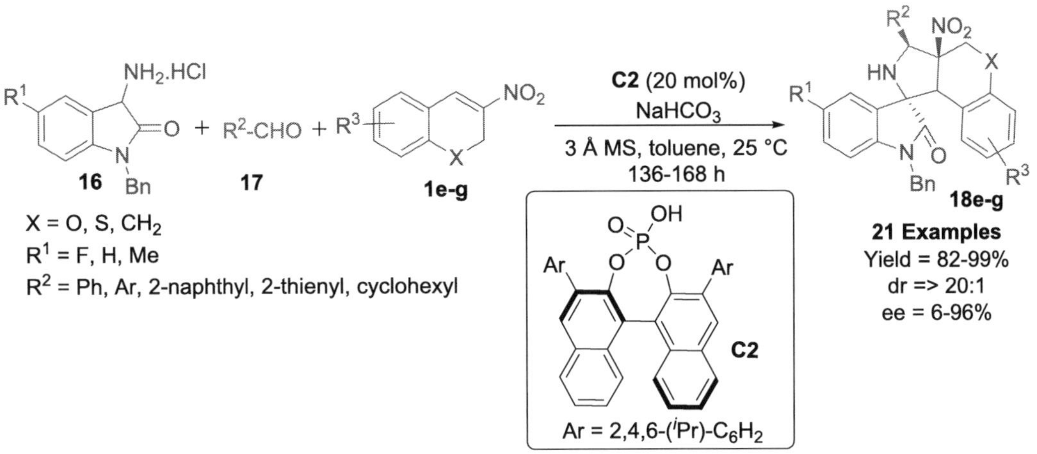

SCHEME 8.12 Asymmetric [3+2] cycloaddition of oxindole-based azomethine ylides with 3-nitro-2H-chromenes in the presence of a bifunctional phosphoric acid catalyst.

enough for two *Re*-face attacks to occur and to form the product **20** with definite configuration (**TS-4**, Scheme 8.13).

The Alemán group presented an organocatalytic 1,3-dipolar cycloaddition strategy for the synthesis of tetrasubstituted pyrrolidines **22** from azomethine ylides **21** derived from salicylaldehyde and nitroalkenes **1h** (Scheme 8.14).[49] Reactivity of the imine was retained by the *ortho*-hydroxyl group through intramolecular hydrogen bonding and allows the reaction to proceed successfully in the absence of an activating group with a high enantiomeric ratio and *exo/endo*-selectivity.

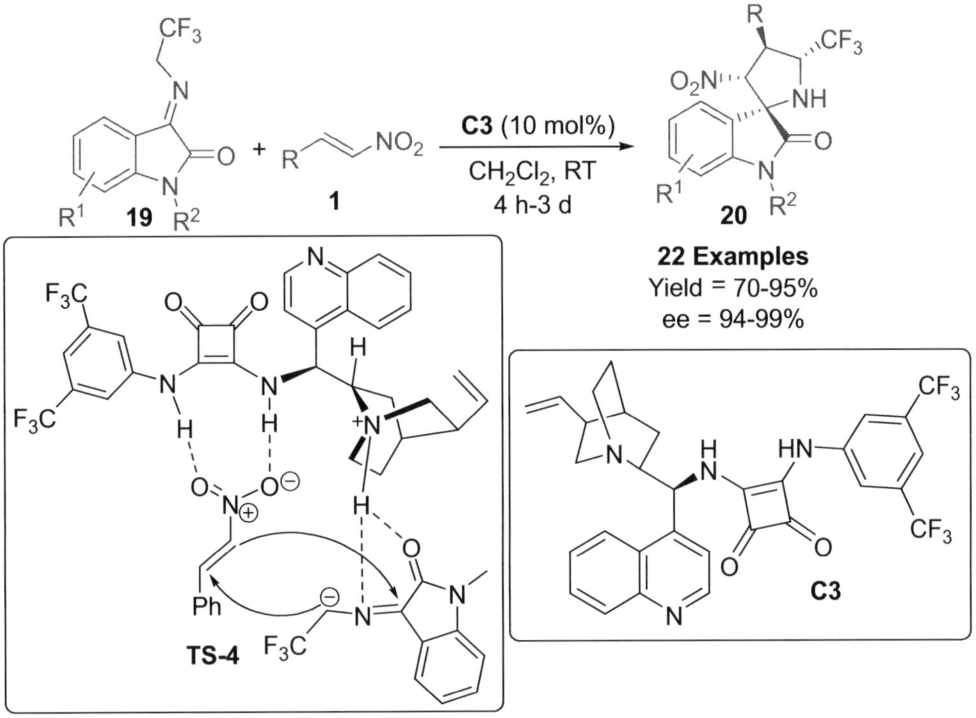

SCHEME 8.13 Asymmetric [3+2] cycloaddition of isatin derived trifluoromethylamines with nitroalkenes by employing a squaramide catalyst.

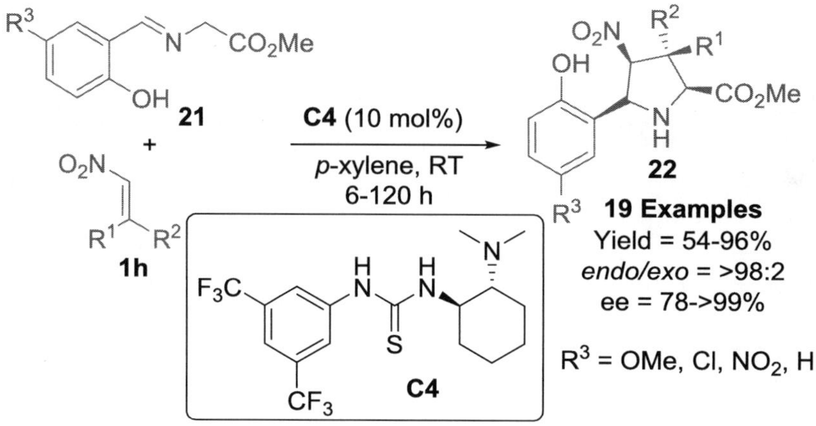

SCHEME 8.14 Thiourea-organocatalyzed 1,3-dipolar cycloadditions of salicylaldehyde-derived azomethine ylides with nitroalkenes.

Intermolecular hydrogen bonding between the bifunctional catalyst **C4** and nitroalkene also helps to enhance the reactivity.

The Oh group reported a stereodivergent catalytic asymmetric conjugate addition of glycine ketimines **2c** and glycine iminoesters **2a** with nitroalkenes **1** using various chiral catalyst systems derived from a multidentate amino alcohol **L11** (Scheme 8.15).[36] The authors also demonstrated the stepwise [3+2] cycloaddition of *N*-metalated azomethine ylides for the stereoselective synthesis of *exo*-**24** and *endo*-**26** cyclic products from the respective *syn*-**23** and *anti*-**25** conjugate adducts in a one-pot tandem manner.

The Deng group conducted the stereoselective reaction of cyclic azomethine ylides **27** with nitroalkenes **1**, which afforded diastereodivergent Michael adducts (*syn* **28a** or *anti* **28b**) using N, O-ligand **L12**/$Cu(OAc)_2$ and N, P-ligand **L5**/$Cu(OAc)_2$ catalytic systems (Scheme 8.16).[50] Both *syn*- and *anti*-adducts were transformed into the corresponding 1,7-diazaspiro[4.4]nonane derivatives by $NaBH_4$ reduction followed by cyclization.

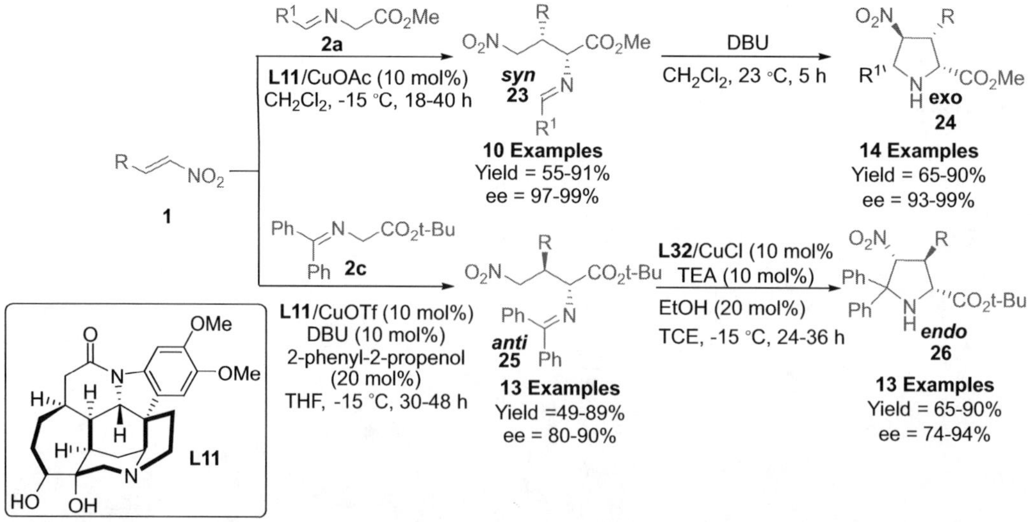

SCHEME 8.15 Asymmetric conjugate addition of glycine derived (ket)imines to nitroalkenes and stepwise [3+2] cycloaddition of N-metalated azomethine ylides.

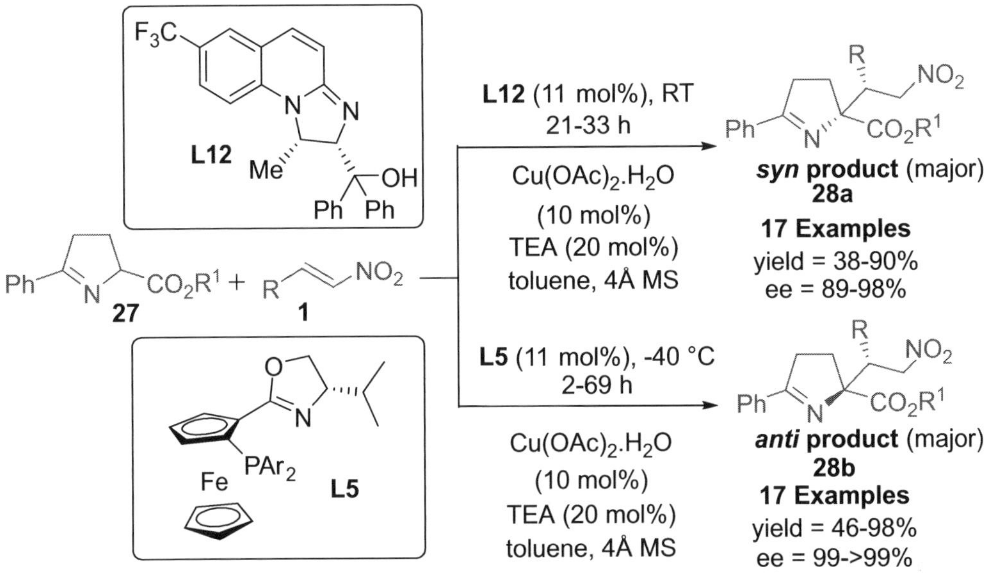

SCHEME 8.16 Reaction of cyclic azomethine ylides with nitroalkenes in the presence of N, O-ligand/$Cu(OAc)_2$ and N, P-ligand/$Cu(OAc)_2$ catalytic systems.

The Xu group established a catalytic enantioselective route to biologically important spiro[pyrrolidin-3,2′-oxindole] scaffold **31** with four contiguous stereogenic centers, including one quaternary stereocenter with good selectivity using a bifunctional squaramide catalyst **C5** (Scheme 8.17).[51] In the possible mechanism, the reaction may be initiated by a base-catalyzed 1,3-proton shift to form an intermediate, which subsequently undergoes synergistic activation with nitroalkene **1** in the presence of catalyst **C5** to give another intermediate responsible for cycloaddition reaction to afford the product **31** (**TS-5**, Scheme 8.17).

The Chen group reported an organocatalytic one-pot, three-component reaction of aldehydes **17a**, diethylamino malonate **32** and nitroalkenes **1** (Scheme 8.18).[16] According to the experimental evidence, a concerted mechanism, rather than a stepwise Michael addition-cyclization, has been proposed for the 1,3-dipolar cycloaddition of in-situ generated azomethine ylides and nitroalkenes **1**. In the plausible mechanism, the double hydrogen bonding between the thiourea moiety and nitrostyrene is facilitated by the steric hindrance of the bulky 2,5-diarylpyrrole on the catalyst **C7**. Subsequent *endo*-attack on the in-situ generated azomethines ylide from the *Re*-face of nitroalkene would afford the product **34** (**TS-6**, Scheme 8.18). The bifunctional urea-tertiary amine organocatalyst **C6** selectively furnished Michael addition product **33** (Scheme 8.18).

The asymmetric palladium-catalyzed [3+2] cycloaddition of trimethylenemethane (TMM) **35a** with nitroalkenes **1** was described by Trost et al. (Scheme 8.19).[52] The functionalization of these cycloadducts **36** proceeds with excellent diastereoselectivity, and thus allowed access to several important synthetic intermediates, such as cyclopentyl amines, cyclopentenones and cyclopentanes bearing tetrasubstituted stereocenters.

The same group also established highly enantioselective palladium-catalyzed cycloaddition of TMM **35a** with β,β-disubstituted nitroalkenes **1h** (Scheme 8.20).[53] The reaction allows both unsubstituted and substituted TMM donors. The cyano-substituted TMM donor **35b**, in particular, allows the formation of highly substituted cyclopentane products **37b**. Unusual reversal of asymmetric induction observed in this case is dependent on the structure of the TMM intermediate.

The Hou group developed a palladium-catalyzed diastereo and enantioselective [3+2]-cycloaddition of 2-vinyl-2-methyloxirane **38** and nitroalkenes **1c** to afford 2,3-*trans*-3,4-*cis*-

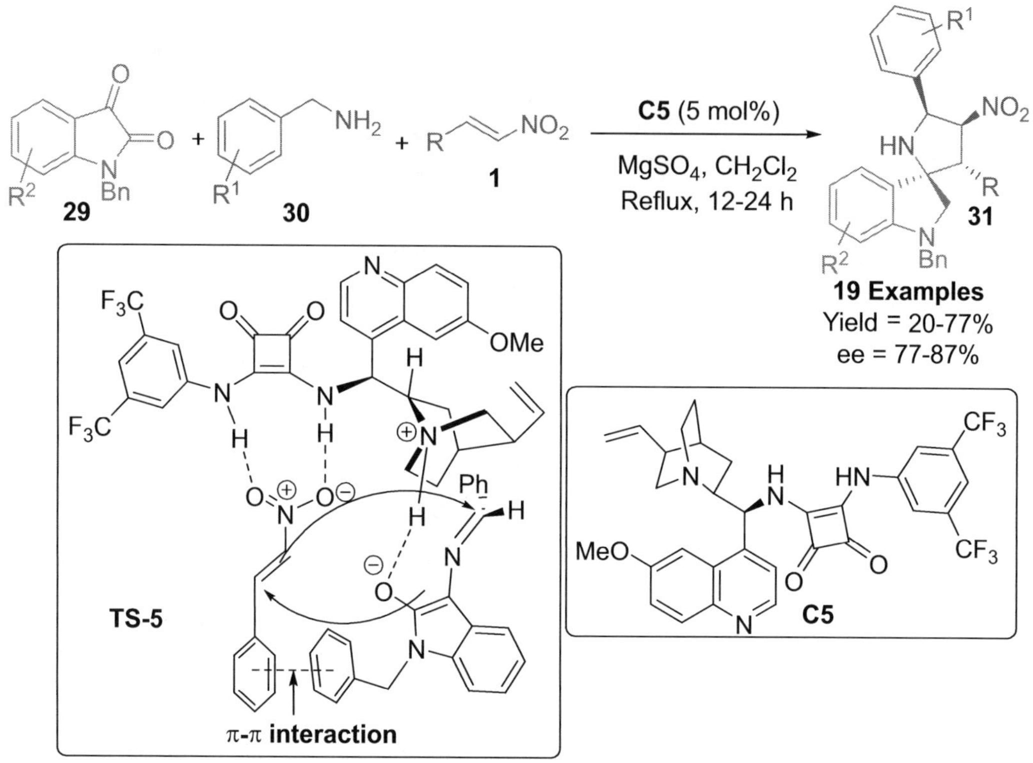

SCHEME 8.17 Synthesis of spiro[pyrrolidin-3,2′-oxindole] scaffolds by the reaction of isatins, benzyl amines and nitroalkenes in the presence of bifunctional squaramide.

trisubstituted tetrahydrofurans **39** in high yields with good enantio and diastereoselectivities in the presence of 1,1′-*P*, *N*-ferrocene ligands **L14** (Scheme 8.21).[54] Here, the authors also observed reversal of enantioselectivity when –H and $-CH_3$ were used as the substituent at the benzylic position of the ligand, respectively. The same authors reported the synthesis of multifunctionalized tetrahydrofurans **39** using Pd-catalyzed asymmetric cycloaddition of vinyl epoxides **38** to 1,1,2-trisubstituted alkenes **1c** (Scheme 8.21).[55] Transformations of the obtained products **39** further added to the usability of the protocol. Further, the process helped by showing an easier way to employ trisubstituted alkenes with a single activator in metal-catalyzed [3+2] cycloaddition reactions.

Umemiya and Hayashi reported an enantioselective synthesis of nitrocyclopentene derivatives **42** in good yields via a domino Michael–Henry reaction of succinaldehyde **40** and nitroalkenes **1** in the presence of diphenylprolinol silyl ether catalyst **C8a** (Scheme 8.22).[56] The reaction yielded thermodynamically unstable *cis* isomers exclusively. The reaction mechanism involves the formation of an enamine by the reaction of catalyst **C8a** and succinaldehyde **40.** This enamine reacts with nitroalkene **1** in a Michael fashion to give an iminium ion (**TS-7**, Scheme 8.22), which undergoes intramolecular Henry reaction followed by hydrolysis to provide the corresponding products **41.**

The Liu group reported an asymmetric [3+2] cycloaddition of vinyl cyclopropane appended on indanedione **43** with nitroalkenes **1** by employing palladium(0) with a chiral bis(*tert*-amine) ligand **L16** to afford the corresponding spiro-cyclopentane derivatives **44** with excellent yields and enantioselectivities (Scheme 8.23).[57] During the reaction, vinylcyclopropane complexes with the catalyst to form a zwitterionic (π-allyl)-Pd intermediate, which adds to the nitroalkene **1** to form spiro[cyclopentane-1,2′-indene]-1′, 3′-dione adduct **44** (**TS-8**, Scheme 8.23). The stereoselectivity of this reaction is attributed to the sterically hindered substituents on the N atom of the ligand.

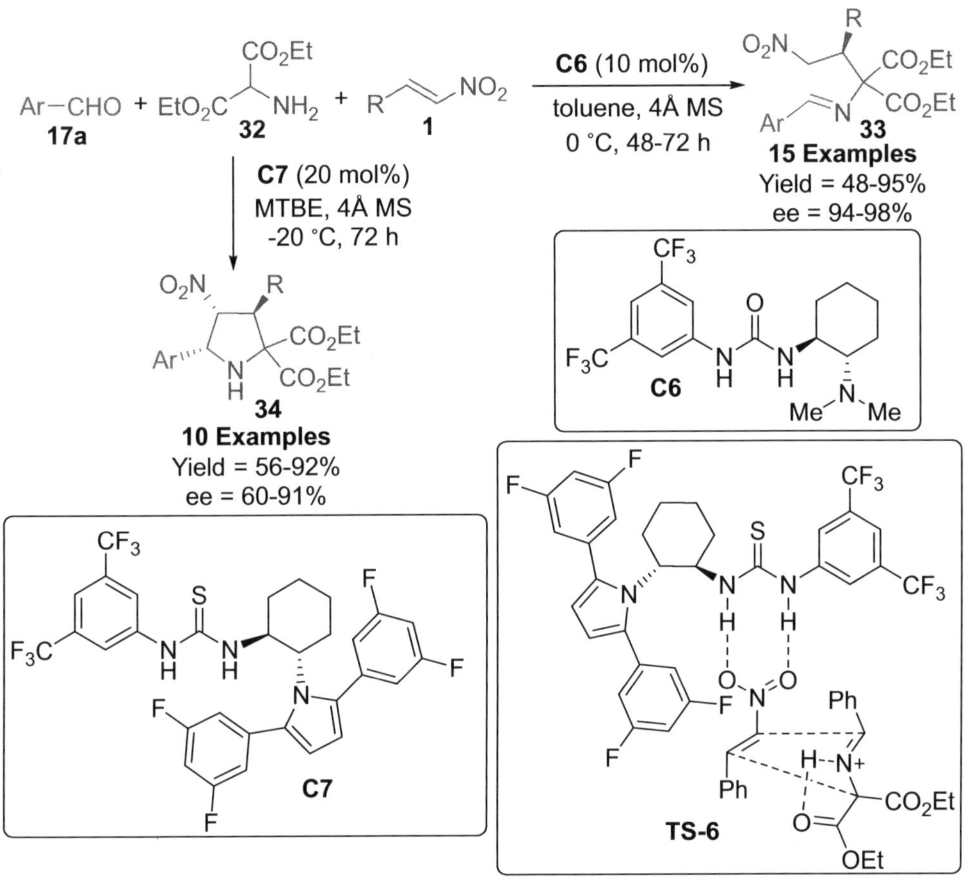

SCHEME 8.18 1,3-Dipolar cycloaddition of in-situ formed azomethine ylides and nitroalkenes catalyzed by thiourea-based organocatalyst.

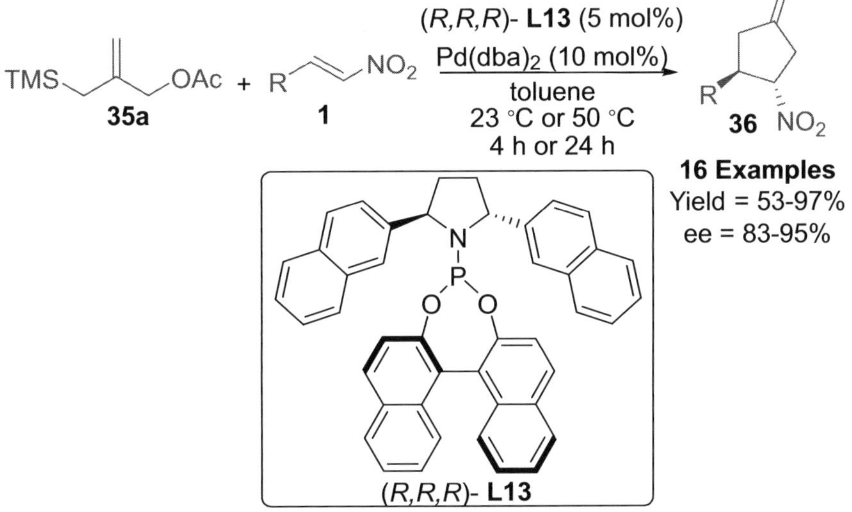

SCHEME 8.19 Asymmetric palladium-catalyzed [3+2] cycloaddition of trimethylenemethane (TMM) with nitroalkenes.

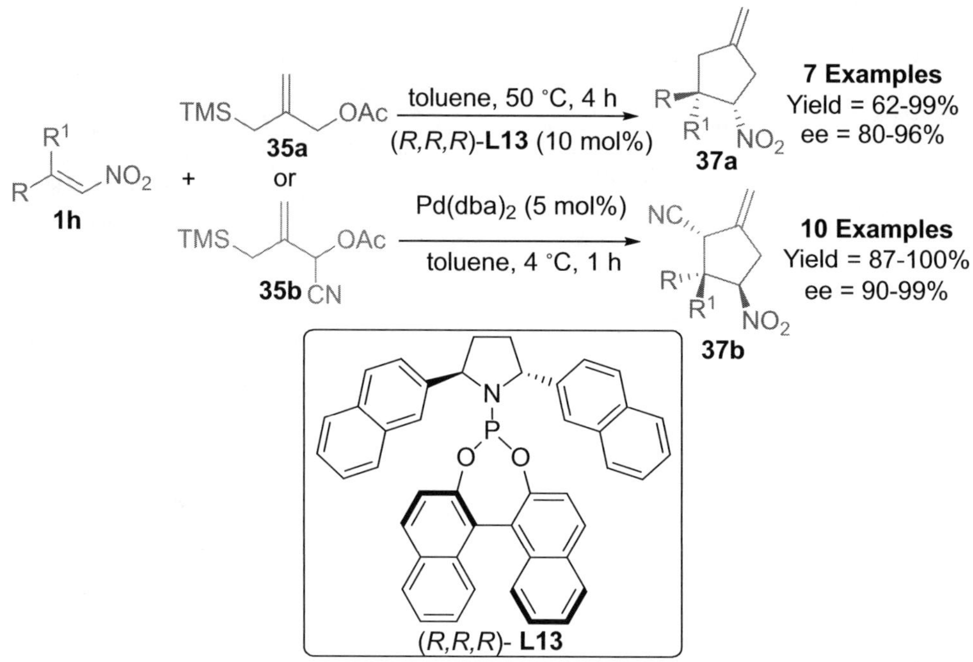

SCHEME 8.20 Asymmetric palladium-catalyzed [3+2] cycloaddition of trimethylenemethane (TMM) with β,β-disubstituted nitroalkenes.

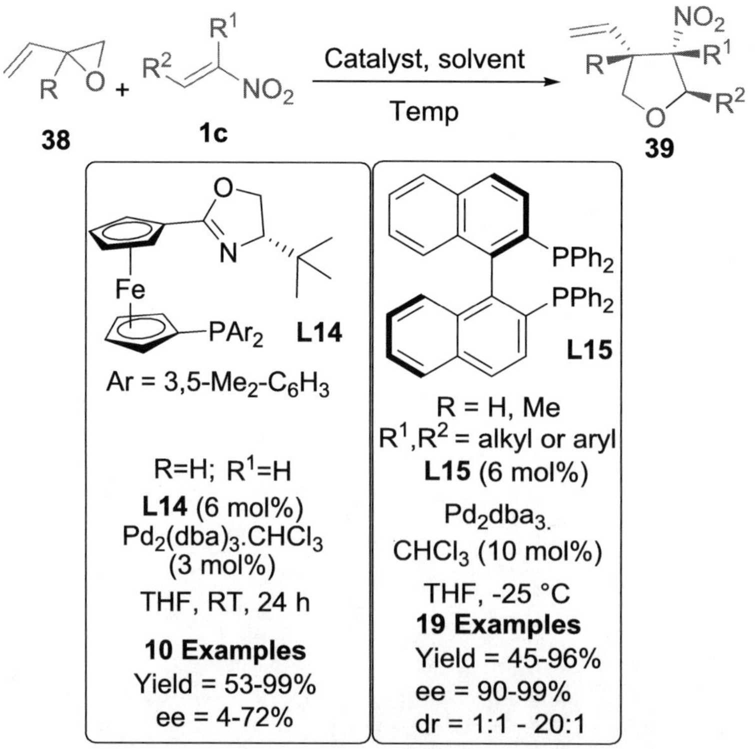

SCHEME 8.21 Palladium-catalyzed enantioselective [3+2]-cycloaddition of vinyl epoxides with substituted nitroalkenes.

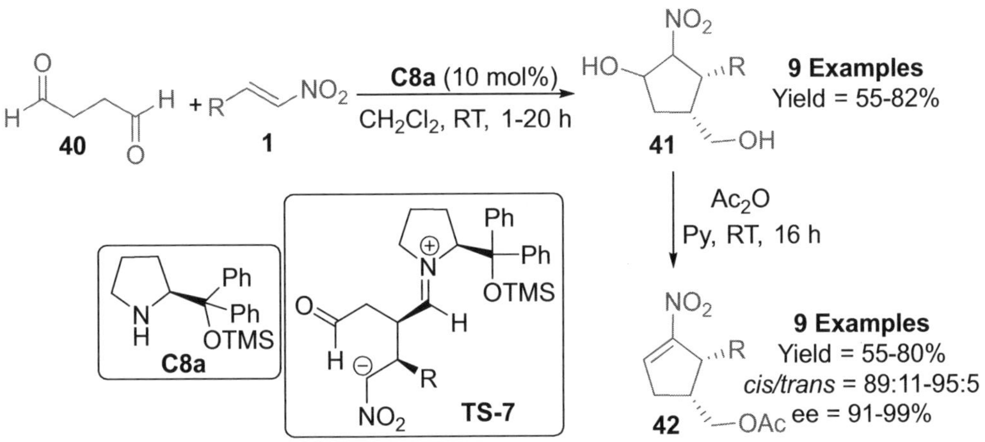

SCHEME 8.22 Domino Michael/Henry reaction of succinaldehyde with nitroalkenes in the presence of diphenylprolinol silyl ether catalyst.

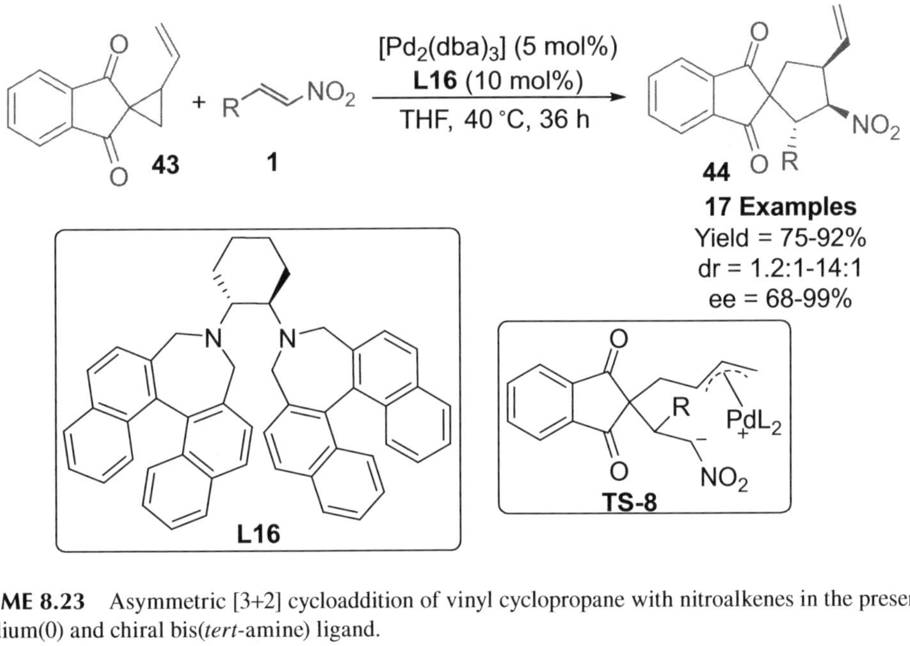

SCHEME 8.23 Asymmetric [3+2] cycloaddition of vinyl cyclopropane with nitroalkenes in the presence of palladium(0) and chiral bis(*tert*-amine) ligand.

Hong et al. described an enantioselective [3+2] cycloaddition of 4-hydroxybut-2-enal **45** with nitroalkenes **1** to form cyclopentanecarbaldehydes **46a-b** with four contiguous stereocenters with high yields and excellent enantioselectivities in the presence of a catalyst **C9** through Michael–Henry reaction (Scheme 8.24).[58] The possible mechanism of the reaction involves the activation of enal **45** by the catalyst **C9** to form an iminium intermediate, which on isomerization transforms into dienamine intermediate. This then adds to the nitroalkene **1** to form an adduct **47a**, which on tautomerization followed by Henry reaction provides the diastereomeric products **46a-b** (Scheme 8.24).

The Yuan group reported an asymmetric dearomative [3+2] cycloaddition of 2-nitrobenzofurans **1i** with 3-isothiocyanato oxindoles **48** to produce spirooxindole derivatives **49** under $Zn(OTf)_2$/chiral bis(oxazoline) **L17** catalytic system with excellent stereoselectivity (Scheme 8.25).[59] This synthetic strategy gives easy access to diversely substituted, pharmaceutically relevant

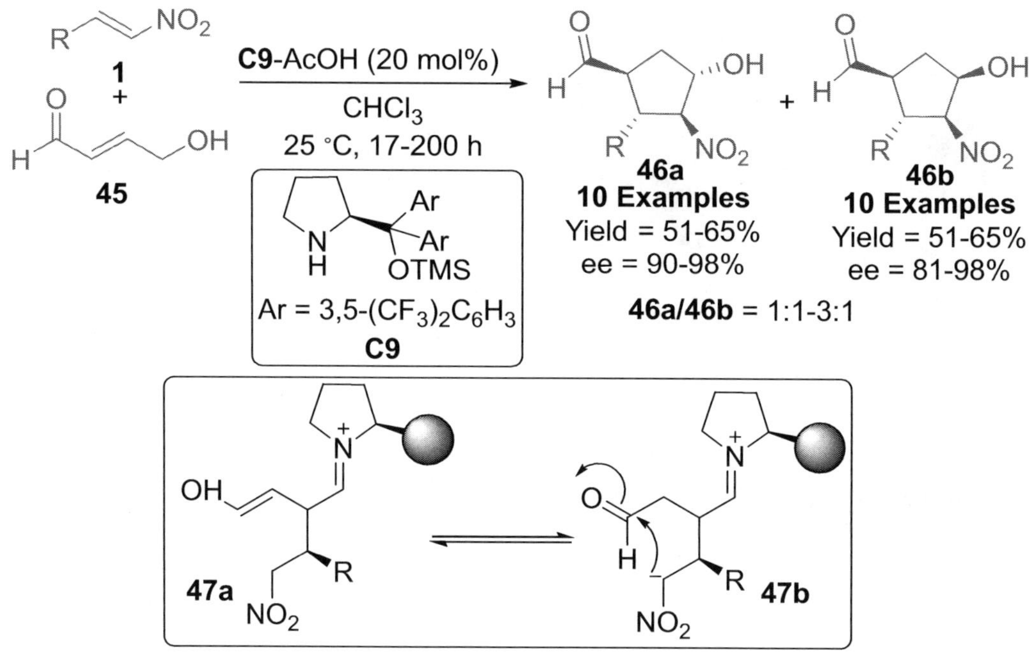

SCHEME 8.24 [3+2] Cycloaddition of 4-hydroxybut-2-enal with nitroalkenes in the presence of prolinol silyl ether catalyst.

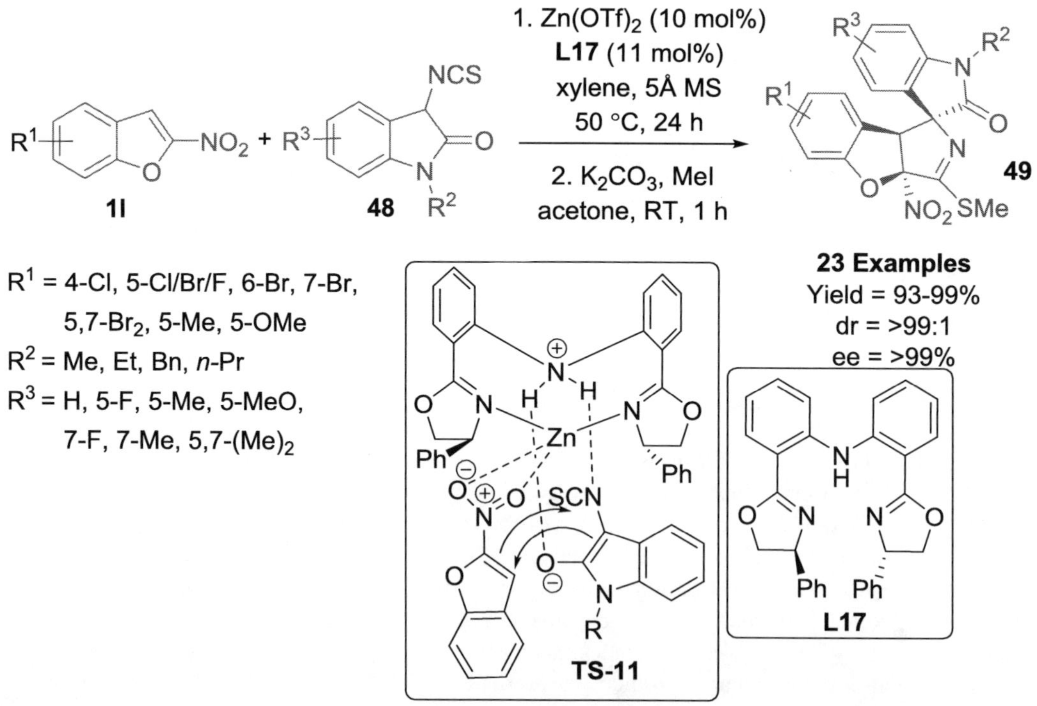

SCHEME 8.25 Asymmetric dearomative [3+2] cycloaddition of 2-nitrobenzofurans with 3-isothiocyanato oxindoles in the presence of $Zn(OTf)_2$/chiral bis(oxazoline) catalyst system.

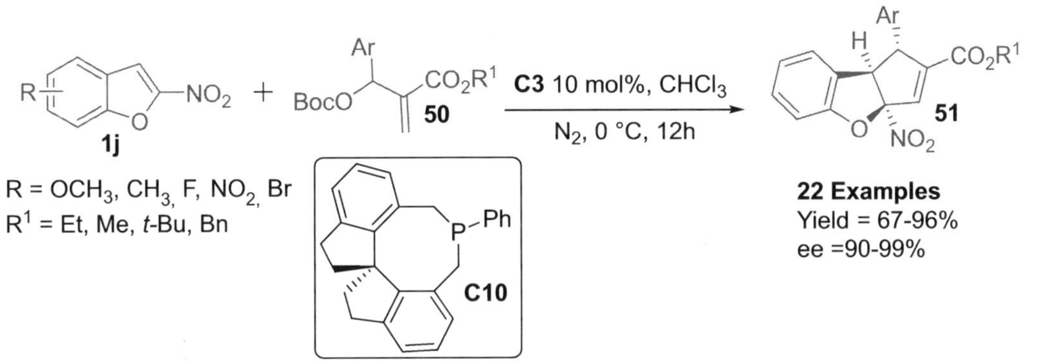

SCHEME 8.26 Stereoselective synthesis of cyclopentabenzofurans from 2-nitrobenzofurans and Morita–Baylis-Hillman carbonates.

motif 2,3-dihydrobenzofuran containing spirooxindole moiety **49** with three contiguous stereocenters. The Zn(OTf)$_2$/**L17** complex acts as a bifunctional catalyst, the acidic part of the complex activates nitrobenzofuran **1i** through Zn–NO_2 interaction and isothiocyanate group was activated by the base part (NH) of the complex. The attack of the activated isothiocyanate occurs from the *Re*-face of the nitrobenzofuran **1i** followed by intramolecular annulation at the C2 position (nitrobenzofuran), which yields the spirooxindole skeleton. The proposed transition state (**TS-11**) for the [3+2] cycloaddition is shown in Scheme 8.25. The synthetic potential of the reaction was also established by scale-up experiments and different synthetic transformations of the product spirooxindole **49**.

Guo and co-workers reported the stereoselective synthesis of a series of cyclopentabenzofurans by a phosphine **C10**-catalyzed dearomative [3+2] cycloaddition of 2-nitrobenzofurans **1j** with Morita–Baylis–Hillman carbonates **50** (Scheme 8.26). The resulting products **51** containing three contiguous centers were formed in good yields (up to 96%) and excellent enantioselectivities (up to 99%).[60]

8.3 [4+2] CYCLOADDITIONS

The Vicario group described the difference in reactivity of nitroalkenes leading to products **53** exclusively with β-selectivity in excellent yields and enantioselectivities (Scheme 8.27).[61] They demonstrated that unconjugated dienals **52** react under trienamine activation and can be used as alternative substrates for reactions, in which the fully conjugated α,β-unsaturated aldehydes failed to participate. The effect of breaking the conjugation translates into a better ability to condense with the chiral secondary amine catalyst **C8b**. This leads to the same vinylogous trienamine intermediate, which reacts subsequently in a very efficient manner. The Diels–Alder reaction between the unconjugated dienal **52** and nitroalkenes **1d**, the former acting as the electron-rich diene, leads to the formation of cyclohexenes **53**.

A catalytic asymmetric version of the Diels–Alder reaction of anthracenes **54** has been developed by employing a chiral, bifunctional aminocatalyst **C11** at remarkably low loadings using nitroalkenes **1h** as dienophiles by the Jørgensen group (Scheme 8.28).[62] The anthracene-derived [4+2] cycloadducts **55** were synthesized in excellent yields and enantioselectivities. The experimental observations were supported by DFT calculations.

The Deng group reported the cycloaddition of 2-pyrones **56** with aliphatic nitroalkenes **1** catalyzed by a bifunctional cinchona alkaloid-derived catalyst **C12** bearing a bulky TIPS-ether at the 9-position.[63] This reaction results in the [2.2.2] bicyclic adducts **57a**, **57b** and **57c** in high yields with excellent enantio and diastereoselectivities (Scheme 8.29). Carbon isotope effects measured by ^{13}C NMR spectroscopic analysis indicate that the reaction follows a stepwise mechanism.

The periselective Diels–Alder reaction of 5-substituted pentamethyl cyclopentadienes **58** and nitroethylene **1k** has been reported by Takenaka and co-workers using helical-chiral H-bond donor

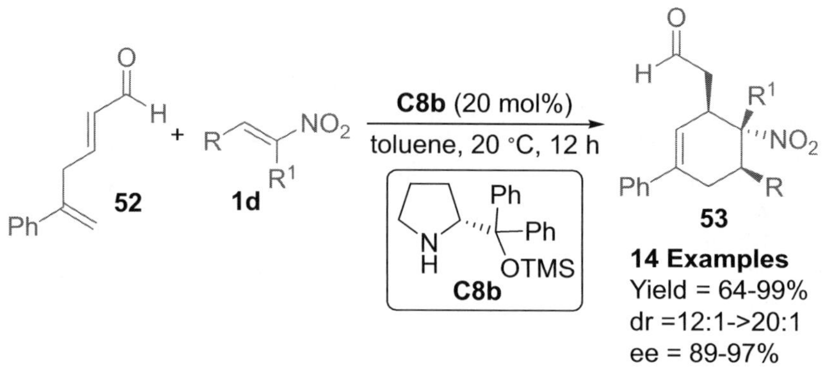

SCHEME 8.27 Diels–Alder reaction between unconjugated dienals and nitroalkenes under trienamine activation.

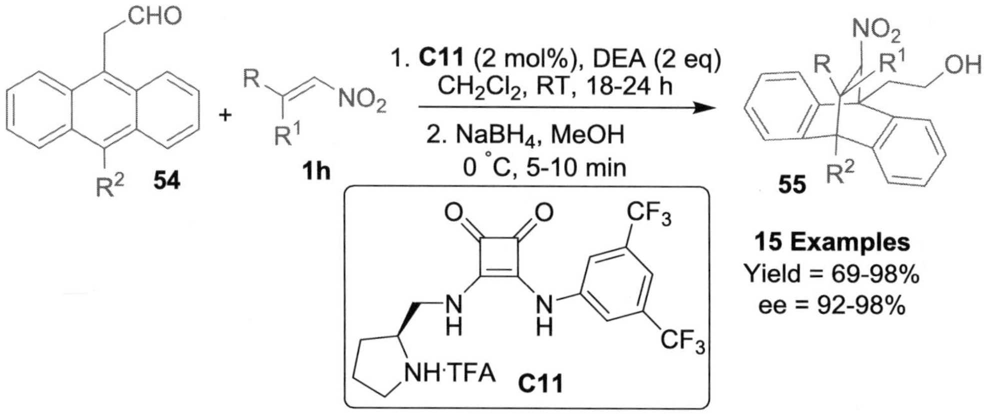

SCHEME 8.28 Asymmetric Diels–Alder reaction of anthracenes with nitroalkenes in the presence of bifunctional amine-squaramide catalyst.

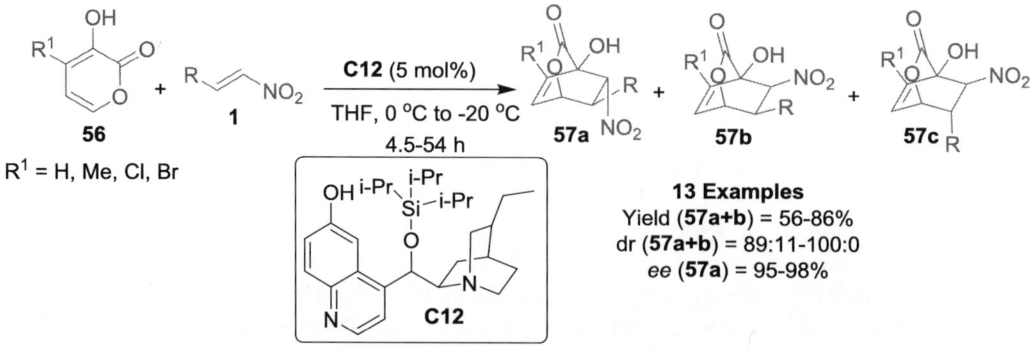

SCHEME 8.29 Asymmetric [4+2] cycloaddition of 2-pyrones with aliphatic nitroalkenes catalyzed by bifunctional cinchona alkaloid-derived catalyst.

catalyst. In the recent report, they have described that helical-chiral double H-bond donor catalysts **C13/C14** promote asymmetric catalytic Diels–Alder reaction by LUMO-lowering catalysis in an enantio and periselective manner (Scheme 8.30).[64,65]

The Sun group reported an organocatalytic formal [4+2] cycloaddition reaction between β-aryl or alkyl nitroalkenes **1** with α′-ethoxycarbonyl cyclopentenones **60** providing synthetically useful bicyclo[2.2.1]heptane-1-carboxylates **61** in high yields with excellent enantioselectivities (Scheme 8.31).[66]

Chen et al. described the Diels–Alder reaction of 2,4-dienals **62** and nitroalkenes **1** by employing trienamine catalysis to afford densely substituted chiral cyclohexene derivatives **63** in high diastereo and enantioselectivities (Scheme 8.32).[67] The strategy of raising the HOMO energy through the introduction of appropriate substituents in the skeleton of 2,4-dienals **62** proved to be successful as the unsubstituted 2,4-hexadienal and 2,4-heptadienal were inactive under the same catalytic conditions.

The Xiao group reported an enantioselective [4+2] cycloaddition of 3-nitro-2H-chromenes **1e** with 1-benzyl-2-vinyl-1H-indoles **64** using chiral $Zn(OTf)_2$/bis(oxazoline) **L18** complex (Scheme 8.33).[68] Here, the reaction follows a bifunctional mode of activation, where zinc(II) acts as a Lewis acid to activate the nitro group of 3-nitro-2H-chromene **1e**, whereas the NH group of the ligand **L18** serves

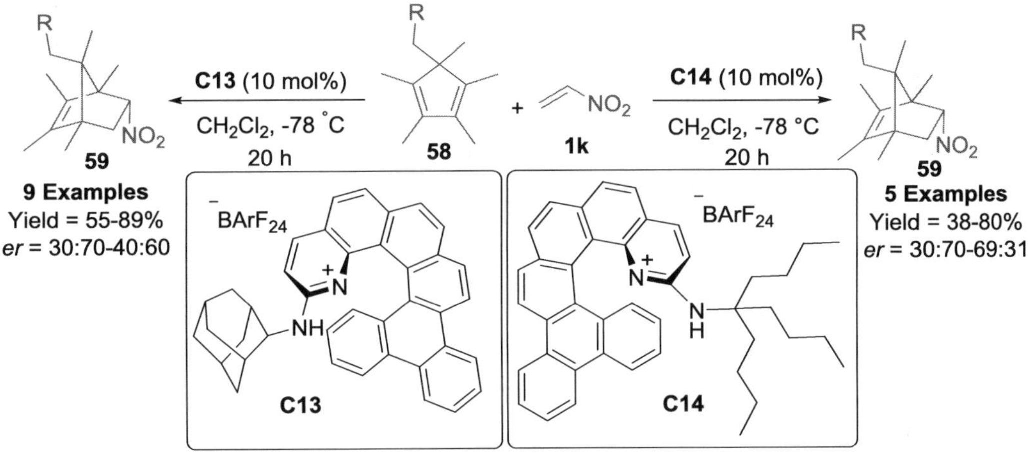

SCHEME 8.30 Diels–Alder reaction of 5-substituted pentamethyl cyclopentadienes and nitroethylene using helically chiral H-bond donor catalysts.

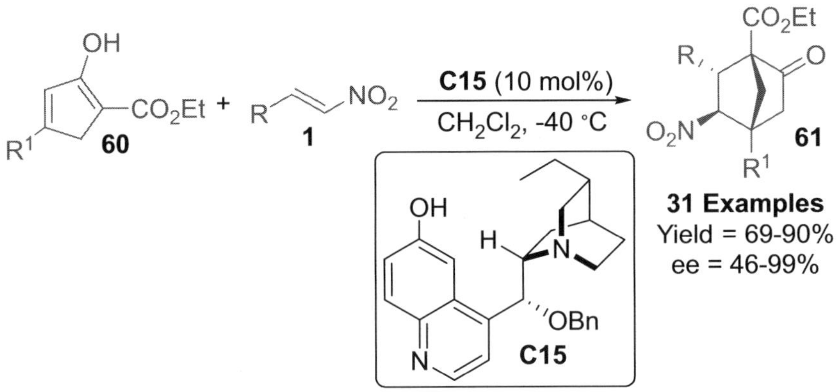

SCHEME 8.31 [4+2] Cycloaddition between β-aryl or alkyl nitroalkenes with α′-ethoxycarbonyl cyclopentenones.

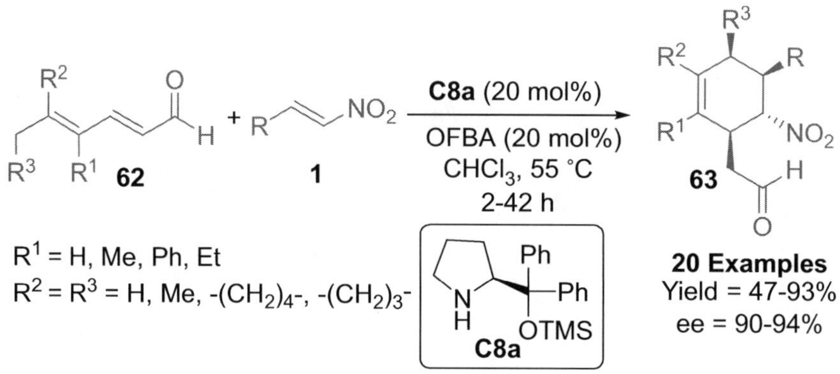

SCHEME 8.32 Diels–Alder reaction of 2,4-dienals and nitroalkenes using diphenylprolinol trimethylsilyl ether catalyst.

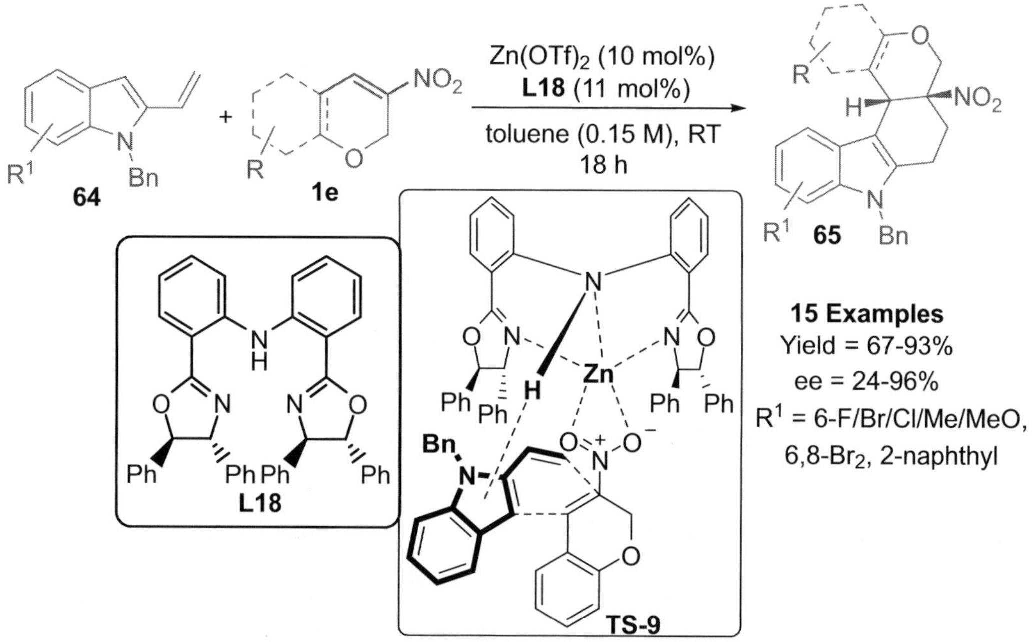

SCHEME 8.33 Asymmetric [4+2] cycloaddition between 3-nitro-2H-chromenes and 1-benzyl-2-vinyl-1H-indoles catalyzed by a chiral $Zn(OTf)_2$/bis(oxazoline) complex.

as a donor for the 2-vinylindole interaction. This kind of *endo*-selective cycloaddition delivers the product **65** with (7*S*, 8*R*) configuration.

Guo and co-workers reported the first catalytic asymmetric [4+2] cycloaddition of 3-vinylindole **66** and nitroolefin **1**, where 1-nitrohydrocarbazoles **67** were produced in good yields (up to 70%) and remarkable enantioselectivities (up to 93%) (Scheme 8.34).[69] It was worth noting that all the products were obtained as a single diastereomer. The report revealed that the chiral carbonyl–thiourea combination in the organocatalyst **C16** assisting an asymmetric organic reaction could be an interesting attribute for efficient catalyst designing.

Xu et al. reported the activation of α,β-unsaturated aldehydes **68** using Brønsted base-Brønsted acid catalysts **C17a-b** and demonstrated the [4+2] cyclization with nitroalkenes **1** to furnish cyclohexenes **69a-b** with four contiguous stereocenters (Scheme 8.35).[70] Bifunctional Brønsted base-Brønsted acid

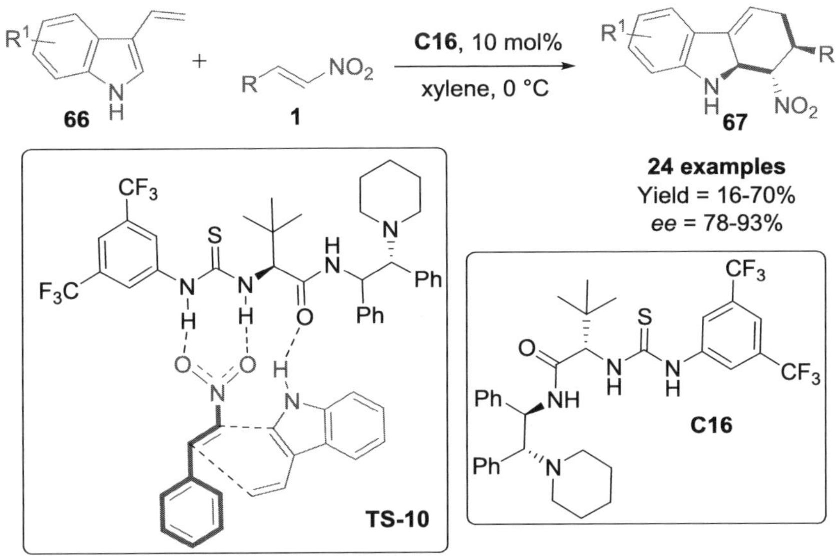

SCHEME 8.34 Asymmetric [4+2] cycloaddition of 3-vinylindoles and nitroalkenes mediated by a chiral carbonyl–thiourea catalyst.

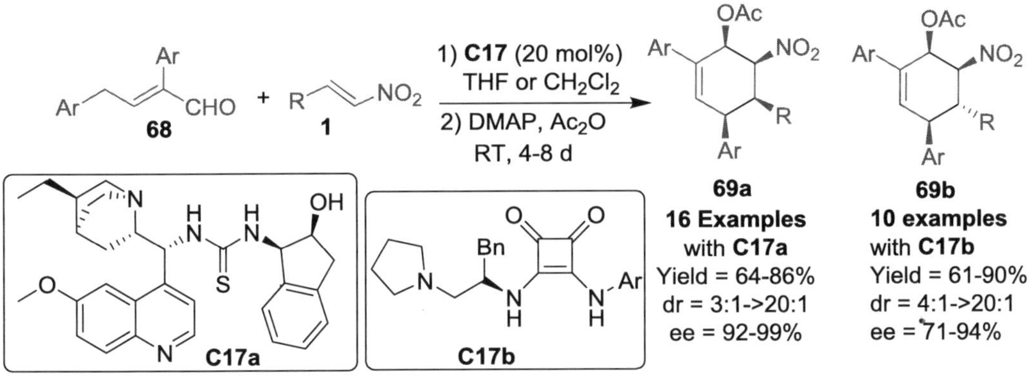

SCHEME 8.35 [4+2] Cyclization of α,β-unsaturated aldehydes with nitroalkenes using a bifunctional Brønsted base catalyst.

catalysts **C17a-b** facilitates the in-situ generation of chiral dienolate complexes from α,β-unsaturated aldehydes, which undergo all-carbon [4+2] annulation. They have also demonstrated the diastereodivergent synthesis of cyclohexenes using (pseudo)enantiomers of the catalyst **C17**.

The Zeitler group reported an asymmetric, multicomponent reaction of nitroalkenes **1** with aldehydes **70** in a three-step, one-pot strategy to afford dihydropyranones **72** (Scheme 8.36).[71] In this reaction, nitroalkene serves as latent 1,2-bielectrophile and thereby generates active enones in situ. The initially formed Michael adduct in the presence of catalysts **C18** and **C19** further undergoes hetero-Diels–Alder reaction with α,β-unsaturated aldehyde **71** under NHC C**20** catalysis to give chiral dihydropyranones **72** in low-to-high yields and enantioselectivities (Scheme 8.36).

The Rovis group described an asymmetric [4+2] cycloaddition of 1-azadienes **73** with nitroalkenes **1** to afford piperidine derivatives **74** by employing Zn catalyst (Scheme 8.37).[72] Mechanistic evidence suggests that the reaction proceeds via a stepwise mechanism. *Ortho*- substitution of the

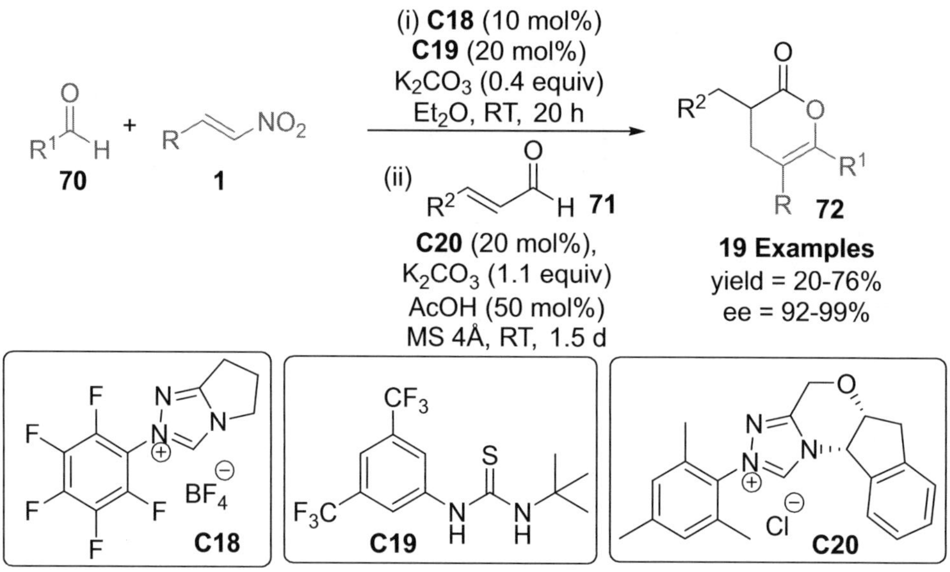

SCHEME 8.36 N-Heterocyclic carbene-mediated multicatalytic protocol for the synthesis of dihydropyranones in a sequential three-component one-pot reaction.

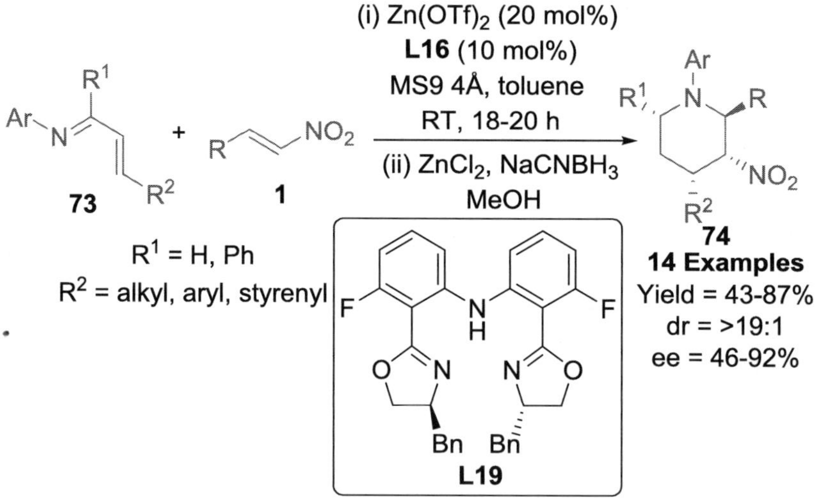

SCHEME 8.37 Zn-catalyzed asymmetric [4+2] cycloaddition of 1-azadienes with nitroalkenes.

BOPA ligand **L19** disfavors the undesired coordination of 1-azadienes **73** to Zn, and the use of this ligand played a key role in achieving high enantioselectivity.

Tamura cycloaddition was demonstrated by employing nitroalkenes **1l** as electrophilic candidates and homophthalic anhydride **75** as nucleophile using bifunctional squaramide-based catalyst **C21** (Scheme 8.38).[73] The cycloaddition provided fused bicyclic ketone diastereomers **76a** and **76b** containing three stereogenic centers with poor-to-moderate diastereocontrol and moderate-to-excellent enantiocontrol. The epimerization of kinetic diastereomer to the thermodynamic diastereomer in methanol occurs in the absence of a catalyst.

Yuan et al. described an asymmetric [4+2] dearomative annulation of 3-nitroindoles **1l** with Nazarov reagents **77** by employing bifunctional thiourea organocatalyst **C22** to produce

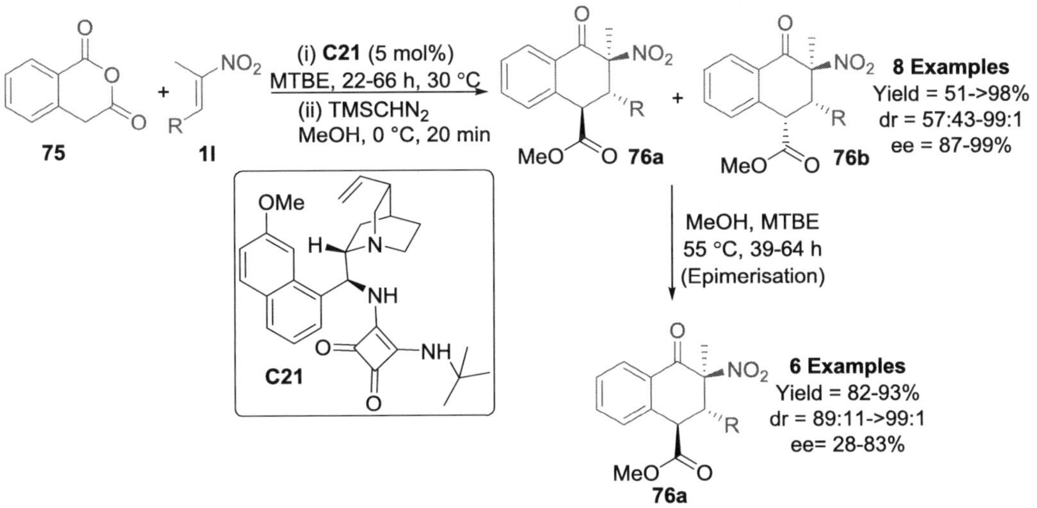

SCHEME 8.38 Tamura cycloaddition between nitroalkenes and homophthalic anhydride using a bifunctional squaramide-based catalyst.

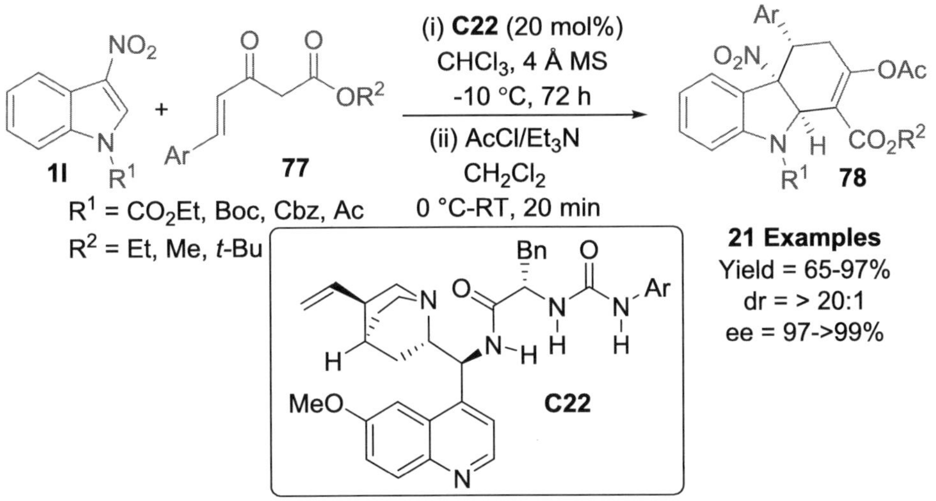

SCHEME 8.39 Asymmetric [4+2] dearomative annulation of 3-nitroindoles with Nazarov reagents using bifunctional thiourea organocatalyst.

hydrocarbazole skeletons with three continuous stereocenters (Scheme 8.39).[74] Multiple hydrogen-bonding ability of the thiourea catalyst **C22** plays a pivotal role in controlling the reactivity (up to 97% yield) and stereoselectivity (>20:1 dr and >99% ee). The synthetic potential of this protocol was established by the transformation of hydrocarbazole skeleton to polyheterocyclic compounds.

8.4 OTHER CYCLOADDITIONS

The Jørgensen group reported a bifunctional activation of α,β-unsaturated aldehydes **79** and nitroalkenes **1** by amino- and H-bonding catalysis to provide cyclobutanes **80** with four contiguous stereocenters (Scheme 8.40).[75] The products were obtained with complete enantio and diastereocontrol by a formal [2+2]-cycloaddition. The squaramide-based bifunctional amine catalyst **C23** was utilized for this transformation. The first step in the formal [2+2]-cycloaddition involves the C–C

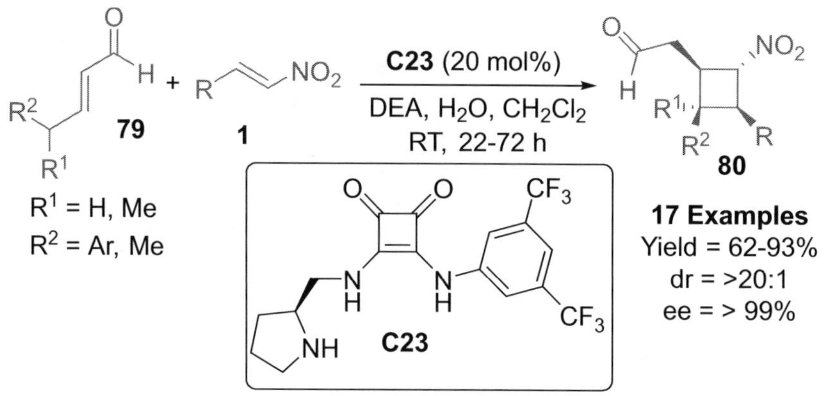

SCHEME 8.40 Asymmetric [2+2]-cycloaddition between α,β-unsaturated aldehydes and nitroalkenes catalyzed by bifunctional squaramide-based aminocatalyst.

bond formation between the nitroalkene and dienamine through their β- and γ-carbons, respectively. In the second step, the β-carbon of the iminium ion forms C–C bond with the carbon adjacent to the nitro group resulting in an intermediate. This step possesses a higher energy barrier (17.6 kcal mol^{-1}) that likely results from the weakening of the H-bonding interactions present in the intermediate. This mechanism was consistent with the experimentally observed reactivity and selectivity.

Vicario et al. utilized two different organocatalysts, i.e., a chiral secondary amine catalyst **C8a** and achiral thiourea catalyst **C24** for a diastereo and enantioselective synthesis of substituted cyclobutanes **82** by reacting enolizable α,β-unsaturated aldehydes **81** and α-hydroxymethylstyrenes **1m** (Scheme 8.41).[76] In this reaction, enal **81** gets activated by the amine catalyst **C8a**, whereas nitroalkene **1m** is activated by thiourea catalyst **C24** through H-bonding. Then both engage in a cascade Michael/Michael reaction to afford the product **82** with excellent enantioselectivity (**TS-11**, Scheme 8.41).

Recently, Jørgenson and co-workers demonstrated the formation of polycyclic compounds by catalytic enantioselective hetero-[6+2] cycloaddition of substituted pyrrole-2-carbaldehydes **83** with nitroalkenes **1** in the presence of pyrrolidine-based catalyst **C25** (Scheme 8.42).[77] The reaction

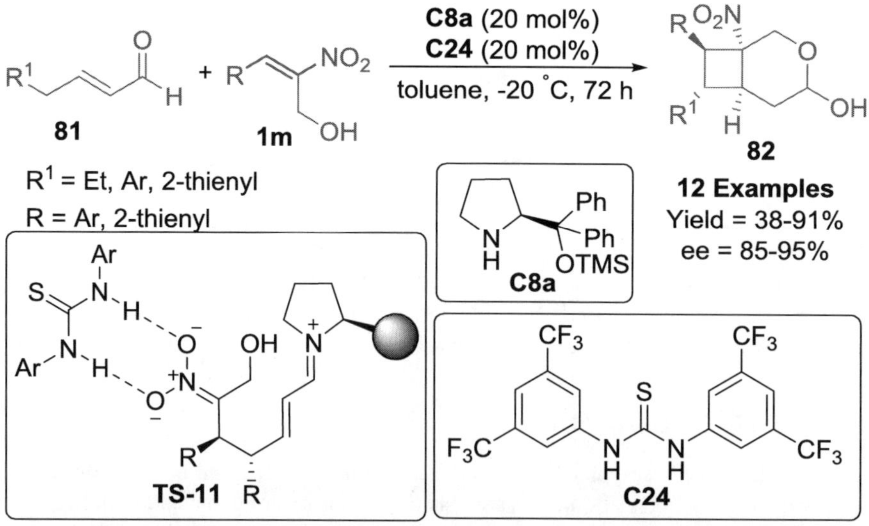

SCHEME 8.41 Enantioselective formal [2+2] cycloaddition of enals with nitroalkenes under cooperative dienamine/hydrogen-bonding catalysis.

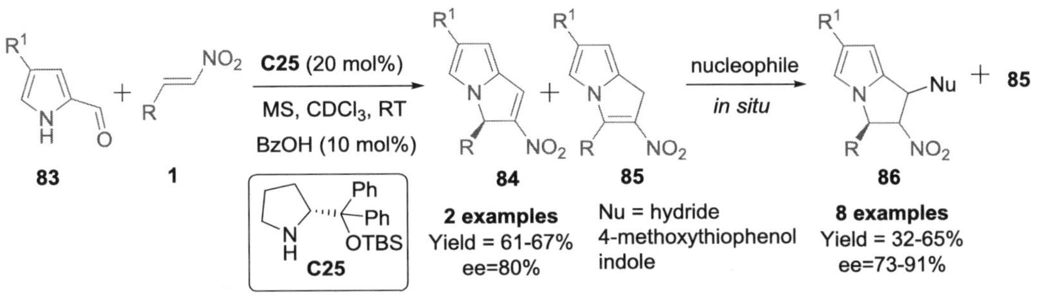

SCHEME 8.42 Hetero-[6+2] cycloaddition of pyrrole-2-carbaldehydes with nitroalkenes mediated by pyrrolidine-based catalyst.

afforded pyrrolizidine scaffolds that underwent further cyclization to obtain the corresponding products **86** in moderate yields (up to 67%) but good to excellent enantioselectivities (up to 91%). Mechanistic studies using NMR investigations revealed the formation of an iminium ion as an intermediate followed by the deprotonation of the heteroatomic N–H, which leads to two isomeric forms of the reactive hetero-6π-intermediate, where only one isomeric form reacts with the olefin for cycloaddition. In addition, the authors performed intensive theoretical studies and concluded that slight modification in catalyst structure can further enhance the enantioselectivity.

8.5 CONCLUSIONS

Nitroalkenes participate in catalytic asymmetric cycloadditions as two carbon components in the presence of various Lewis and Brønsted acid catalysts. Most reported cycloadditions include azomethine ylide (1,3-dipolar) and Diels–Alder-type cycloadditions, though sporadic examples of formal [2+2] and higher-order cycloadditions are also reported.

REFERENCES

1. Denmark, S. E.; Thorarensen, A. *Chem. Rev.* **1996**, *96*, 137.
2. Gothelf, K. V.; Jørgensen, K. A. *Chem. Rev.* **1998**, *98*, 863.
3. Moyano, A.; Rios, R. *Chem. Rev.* **2011**, *111*, 4703.
4. Hashimoto, T.; Maruoka, K. *Chem. Rev.* **2015**, *115*, 5366.
5. Ayerbe, M.; Arrieta, A.; Cossío, F. P.; Linden, A. *J. Org. Chem.* **1998**, *63*, 1795.
6. Vivanco, S.; Lecea, B.; Arrieta, A.; Prieto, P.; Morao, I.; Linden, A.; Cossío, F. P. *J. Am. Chem. Soc.* **2000**, *122*, 6078.
7. Wittkopp, A.; Schreiner, P. R. *Chem. - Eur. J.* **2003**, *9*, 407.
8. Hoashi, Y.; Yabuta, T.; Yuan, P.; Miyabe, H.; Takemoto, Y. *Tetrahedron* **2006**, *62*, 365.
9. Yan, X.-X.; Peng, Q.; Zhang, Y.; Zhang, K.; Hong, W.; Hou, X.-L.; Wu, Y.-D. *Angew. Chem. Int. Ed.* **2006**, *45*, 1979.
10. Li, H.; Wang, J.; Xie, H.; Zu, L.; Jiang, W.; Duesler, E. N.; Wang, W. *Org. Lett.* **2007**, *9*, 965.
11. Deng, X.; Mani, N. S. *J. Org. Chem.* **2008**, *73*, 2412.
12. Du, W.; Liu, Y.-K.; Yue, L.; Chen, Y.-C. *Synlett* **2008**, *2008*, 2997.
13. Enders, D.; Hüttl, M. R. M.; Raabe, G.; Bats, J. W. *Adv. Synth. Catal.* **2008**, *350*, 267.
14. Enders, D.; Wang, C.; Bats, J. W. *Angew. Chem. Int. Ed.* **2008**, *47*, 7539.
15. Guo, C.; Xue, M.-X.; Zhu, M.-K.; Gong, L.-Z. *Angew. Chem. Int. Ed.* **2008**, *47*, 3414.
16. Liu, Y.-K.; Liu, H.; Du, W.; Yue, L.; Chen, Y.-C. *Chem. - Eur. J.* **2008**, *14*, 9873.
17. Lu, M.; Zhu, D.; Lu, Y.; Hou, Y.; Tan, B.; Zhong, G. *Angew. Chem. Int. Ed.* **2008**, *47*, 10187.
18. Tan, B.; Shi, Z.; Chua, P. J.; Zhong, G. *Org. Lett.* **2008**, *10*, 3425.
19. Xie, J.; Yoshida, K.; Takasu, K.; Takemoto, Y. *Tetrahedron Lett.* **2008**, *49*, 6910.
20. Xu, D.-Q.; Wang, Y.-F.; Luo, S.-P.; Zhang, S.; Zhong, A.-G.; Chen, H.; Xu, Z.-Y. *Adv. Synth. Catal.* **2008**, *350*, 2610.

21. Xue, M.-X.; Zhang, X.-M.; Gong, L.-Z. *Synlett* **2008**, *2008*, 691.
22. Wang, Y.; Han, R.-G.; Zhao, Y.-L.; Yang, S.; Xu, P.-F.; Dixon, D. J. *Angew. Chem. Int. Ed.* **2009**, *48*, 9834.
23. Wang, Y.-F.; Zhang, W.; Luo, S.-P.; Li, B.-L.; Xia, A.-B.; Zhong, A.-G.; Xu, D.-Q. *Chem. - Asian J.* **2009**, *4*, 1834.
24. Zhang, F.-L.; Xu, A.-W.; Gong, Y.-F.; Wei, M.-H.; Yang, X.-L. *Chem. - Eur. J.* **2009**, *15*, 6815.
25. Arai, T.; Mishiro, A.; Yokoyama, N.; Suzuki, K.; Sato, H. *J. Am. Chem. Soc.* **2010**, *132*, 5338.
26. Arai, T.; Yokoyama, N.; Mishiro, A.; Sato, H. *Angew. Chem. Int. Ed.* **2010**, *49*, 7895.
27. Ding, D.; Zhao, C.-G.; Guo, Q.; Arman, H. *Tetrahedron* **2010**, *66*, 4423.
28. Enders, D.; Krüll, R.; Bettray, W. *Synthesis* **2010**, *2010*, 567.
29. Jiang, K.; Jia, Z.-J.; Chen, S.; Wu, L.; Chen, Y.-C. *Chem. - Eur. J.* **2010**, *16*, 2852.
30. Li, N.; Song, J.; Tu, X.-F.; Liu, B.; Chen, X.-H.; Gong, L.-Z. *Org. Biomol. Chem.* **2010**, *8*, 2016.
31. Rueping, M.; Kuenkel, A.; Fröhlich, R. *Chem. - Eur. J.* **2010**, *16*, 4173.
32. Tan, B.; Lu, Y.; Zeng, X.; Chua, P. J.; Zhong, G. *Org. Lett.* **2010**, *12*, 2682.
33. Wang, X.-F.; Chen, J.-R.; Cao, Y.-J.; Cheng, H.-G.; Xiao, W.-J. *Org. Lett.* **2010**, *12*, 1140.
34. Wang, Y.; Yu, D.-F.; Liu, Y.-Z.; Wei, H.; Luo, Y.-C.; Dixon, D. J.; Xu, P.-F. *Chem. - Eur. J.* **2010**, *16*, 3922.
35. Xie, J.-W.; Fan, L.-P.; Su, H.; Li, X.-S.; Xu, D.-C. *Org. Biomol. Chem.* **2010**, *8*, 2117.
36. Kim, H. Y.; Li, J.-Y.; Kim, S.; Oh, K. *J. Am. Chem. Soc.* **2011**, *133*, 20750.
37. Li, Q.; Ding, C.-H.; Li, X.-H.; Weissensteiner, W.; Hou, X.-L. *Synthesis* **2012**, *2012*, 265.
38. Narayan, R.; Bauer, J. O.; Strohmann, C.; Antonchick, A. P.; Waldmann, H. *Angew. Chem. Int. Ed.* **2013**, *52*, 12892.
39. Awata, A.; Arai, T. *Angew. Chem. Int. Ed.* **2014**, *53*, 10462.
40. Lykke, L.; Carlsen, B. D.; Rambo, R. S.; Jørgensen, K. A. *J. Am. Chem. Soc.* **2014**, *136*, 11296.
41. Tang, L.-W.; Zhao, B.-J.; Dai, L.; Zhang, M.; Zhou, Z.-M. *Chem. - Asian J.* **2016**, *11*, 2470.
42. Castelló, L. M.; Nájera, C.; Sansano, J. M.; Larrañaga, O.; Cózar, A. C., Fernando, P. *Org. Lett.* **2013**, *15*, 2902.
43. Cayuelas, A.; Ortiz, R.; Nájera, C.; Sansano, J. M.; Larrañaga, O.; de Cózar, A.; Cossío, F. P. *Org. Lett.* **2016**, *18*, 2926.
44. Padilla, S.; Tejero, R.; Adrio, J.; Carretero, J. C. *Org. Lett.* **2010**, *12*, 5608.
45. Feng, B.; Chen, J.-R.; Yang, Y.-F.; Lu, B.; Xiao, W.-J. *Chem. - Eur. J.* **2018**, *24*, 1714.
46. Feng, B.; Lu, L.-Q.; Chen, J.-R.; Feng, G.; He, B.-Q.; Lu, B.; Xiao, W.-J. *Angew. Chem. Int. Ed.* **2018**, *57*, 5888.
47. Wu, S.; Zhu, G.; Wei, S.; Chen, H.; Qu, J.; Wang, B. *Org. Biomol. Chem.* **2018**, *16*, 807.
48. Sun, Q.; Li, X.; Su, J.; Zhao, L.; Ma, M.; Zhu, Y.; Zhao, Y.; Zhu, R.; Yan, W.; Wang, K.; Wang, R. *Adv. Synth. Catal.* **2015**, *357*, 3187.
49. Esteban, F.; Cieslik, W.; Arpa, E. M.; Guerrero-Corella, A.; Diaz-Tendero, S.; Perles, J.; Fernandez-Salas, J. A.; Fraile, A.; Aleman, J. *ACS Catal.* **2018**, *8*, 1884.
50. Li, C.-Y.; Yang, W.-L.; Luo, X.; Deng, W.-P. *Chem. - Eur. J.* **2015**, *21*, 19048.
51. Tian, L.; Hu, X.-Q.; Li, Y.-H.; Xu, P.-F. *Chem. Commun.* **2013**, *49*, 7213.
52. Trost, B. M.; Bringley, D. A.; Seng, P. S. *Org. Lett.* **2012**, *14*, 234.
53. Trost, B. M.; Bringley, D. A.; O'Keefe, B. M. *Org. Lett.* **2013**, *15*, 5630.
54. Wu, W.-Q.; Ding, C.-H.; Hou, X.-L. *Synlett* **2012**, *2012*, 1035.
55. Du, J.; Jiang, Y.-J.; Suo, J.-J.; Wu, W.-Q.; Liu, X.-Y.; Chen, D.; Ding, C.-H.; Wei, Y.; Hou, X.-L. *Chem. Commun.* **2018**, *54*, 13143.
56. Umemiya, S.; Hayashi, Y. *Eur. J. Org. Chem.* **2015**, *2015*, 4320.
57. Wei, F.; Ren, C.-L.; Wang, D.; Liu, L. *Chem. - Eur. J.* **2015**, *21*, 2335.
58. Hong, B.-C.; Chen, P.-Y.; Kotame, P.; Lu, P.-Y.; Lee, G.-H.; Liao, J.-H. *Chem. Commun.* **2012**, *48*, 7790.
59. Zhao, J.-Q.; Zhou, X.-J.; Zhou, Y.; Xu, X.-Y.; Zhang, X.-M.; Yuan, W.-C. *Org. Lett.* **2018**, *20*, 909.
60. Yang, X.-H.; Li, J.-P.; Wang, D.-C.; Xie, M.-S.; Qu, G.-R.; Guo, H.-M. *Chem. Commun.* **2019**, *55*, 9144.
61. Prieto, L.; Talavera, G.; Uria, U.; Reyes, E.; Vicario, J. L.; Carrillo, L. *Chem. - Eur. J.* **2014**, *20*, 2145.
62. Jiang, H.; Rodríguez-Escrich, C.; Johansen, T. K.; Davis, R. L.; Jørgensen, K. A. *Angew. Chem. Int. Ed.* **2012**, *51*, 10271.
63. Bartelson, K. J.; Singh, R. P.; Foxman, B. M.; Deng, L. *Chem. Sci.* **2011**, *2*, 1940.
64. Narcis, M. J.; Sprague, D. J.; Captain, B.; Takenaka, N. *Org. Biomol. Chem.* **2012**, *10*, 9134.
65. Peng, Z.; Narcis, M. J.; Takenaka, N. *Molecules* **2013**, *18*, 9982.
66. Fu, J.-G.; Shan, Y.-F.; Sun, W.-B.; Lin, G.-Q.; Sun, B.-F. *Org. Biomol. Chem.* **2016**, *14*, 5229.
67. Jia, Z.-J.; Zhou, Q.; Zhou, Q.-Q.; Chen, P.-Q.; Chen, Y.-C. *Angew. Chem. Int. Ed.* **2011**, *50*, 8638.

68. Tan, F.; Xiao, C.; Cheng, H.-G.; Wu, W.; Ding, K.-R.; Xiao, W.-J. *Chem. - Asian J.* **2012**, *7*, 493.
69. Yang, X.; Zhou, Y.-H.; Yang, H.; Wang, S.-S.; Ouyang, Q.; Luo, Q.-L.; Guo, Q.-X. *Org. Lett.* **2019**, *21*, 1161.
70. Xie, J.-K.; Wang, Y.; Lin, J.-B.; Ren, X.-R.; Xu, P.-F. *Chem. - Eur. J.* **2017**, *23*, 6752.
71. Fuchs, P. J. W.; Zeitler, K. *Org. Lett.* **2017**, *19*, 6076.
72. Chu, J. C. K.; Dalton, D. M.; Rovis, T. *J. Am. Chem. Soc.* **2015**, *137*, 4445.
73. Manoni, F.; Farid, U.; Trujillo, C.; Connon, S. J. *Org. Biomol. Chem.* **2017**, *15*, 1463.
74. Yue, D.-F.; Zhao, J.-Q.; Chen, X.-Z.; Zhou, Y.; Zhang, X.-M.; Xu, X.-Y.; Yuan, W.-C. *Org. Lett.* **2017**, *19*, 4508.
75. Albrecht, Ł.; Dickmeiss, G.; Acosta, F. C.; Rodríguez-Escrich, C.; Davis, R. L.; Jørgensen, K. A. *J. Am. Chem. Soc.* **2012**, *134*, 2543.
76. Talavera, G.; Reyes, E.; Vicario, J. L.; Carrillo, L. *Angew. Chem. Int. Ed.* **2012**, *51*, 4104.
77. Bertuzzi, G.; Thøgersen, M. K.; Giardinetti, M.; Vidal-Albalat, A.; Simon, A.; Houk, K. N.; Jørgensen, K. A. *J. Am. Chem. Soc.* **2019**, *141*, 3288.

9 Catalytic Asymmetric Reduction of Nitroalkenes

9.1 INTRODUCTION

Catalytic asymmetric Michael reactions, Friedel–Crafts reactions and cycloadditions of nitroalkenes have been described in the previous chapters. This chapter features the catalytic asymmetric and bioreduction of nitroalkenes. The catalytic asymmetric conjugate reduction of β,β-disubstituted nitroalkenes is one of the most practical ways for synthesizing optically active β,β-disubstituted nitroalkanes. Asymmetric catalytic methods such as transition metal catalysis, organocatalysis and biocatalysis have been realized for the enantioselective reduction of nitroalkenes.[1–24] Diverse metal-ligand complexes along with various hydrogen/hydride sources dominated catalytic asymmetric reduction, though organocatalysts such as thiourea along with Hantzsch ester as a hydride source also catalyzed the reduction of nitroalkenes. Baker's yeast and several specific enzymes have been employed for the bioreduction of nitroalkenes.

9.2 METAL CATALYZED REDUCTION

Czekelius and Carreira reported a copper-catalyzed asymmetric reduction of substituted nitroalkenes **1** to access enantioenriched β,β-disubstituted nitroalkanes **2** by employing just 0.1 mol% of the catalyst (Scheme 9.1).[1] The additional significance of this work is that halides can inhibit this kind of catalytic process that uses copper salts. The copper-phosphane complex of (*S*, *R*)-Josiphos **L1a** was found to be suitable for the enantioselective reduction of β,β-disubstituted nitroalkenes **1**. The asymmetric reduction was performed in the presence of phenyl silane, which serves as a reducing agent with substoichiometric amounts of polymethylhydrosiloxane (PMHS). The substrate scope of nitroalkenes was extended to aliphatic, protected and even unprotected alcohol functionalities. The products β,β-disubstituted nitroalkanes **2** can also be generated using *(S)-p*-tol-BINAP catalyst and provides access to optically active amines.

Conjugate reduction of an isomeric mixture of nitroalkenes **1a** and **3** was also achieved using a combination of CuO^{t}Bu and Josiphos **L1b** (Scheme 9.2).[2] In the presence of a quaternary ammonium base, these isomers can be interconverted and, at the same time, reduced to corresponding nitroalkanes **2a** in a highly enantioselective manner. However, the use of a base led to a decrease in the yield because of the formation of *p*-acetyl-α-methyl styrene by the elimination of nitrous acid.

Later, highly enantioselective conjugate reduction of β,β-disubstituted nitroalkenes **1** was reported by the Carreira group using a new catalyst system comprising copper difluoride and chiral Josiphos

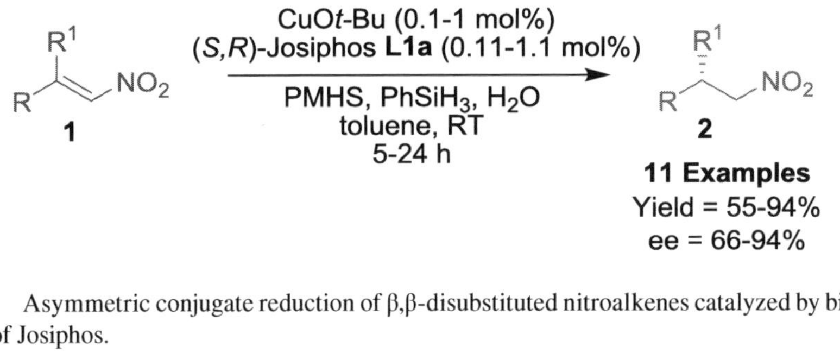

SCHEME 9.1 Asymmetric conjugate reduction of β,β-disubstituted nitroalkenes catalyzed by bisphosphane-Cu complex of Josiphos.

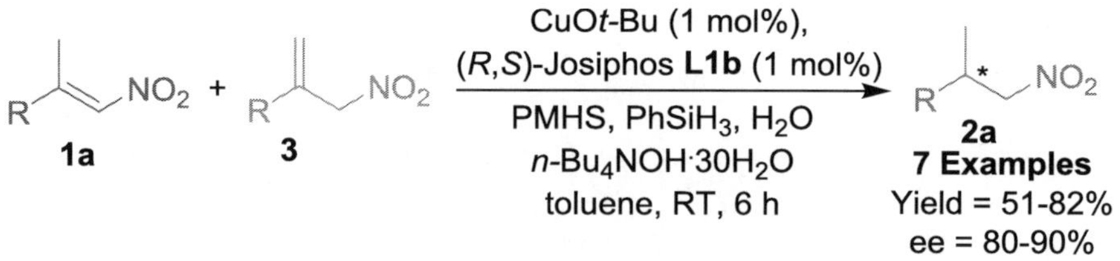

SCHEME 9.2 Conjugate reduction of an isomeric mixture of nitroalkenes using a combination of CuO-t-Bu and Josiphos.

L1a as a ligand (Scheme 9.3).[3] This protocol involves the utilization of bench-stable copper(II) salt to generate catalytically active species in situ by the addition of nitromethane, which helps the initial catalyst activation and facilitates the reaction. The scope was extended to electron-rich nitroalkenes as substrates in the presence of nitromethane or 2-phenyl-1-nitropropane as an additive to produce catalytically active species for effective reduction.

Asymmetric transfer hydrogenation of β,β-disubstituted nitroalkenes **1a** by employing a chiral diamine **L2**–rhodium complex along with a hydrogen source, i.e., HCO_2Na–HCO_2H was reported by Deng and co-workers (Scheme 9.4).[4] The reaction was performed in water to afford the saturated products **2a** in high yields and excellent enantioselectivities. Most substrates were reduced within 10h. In this reaction, the metal catalyst and the pH of the solution play a vital role in determining

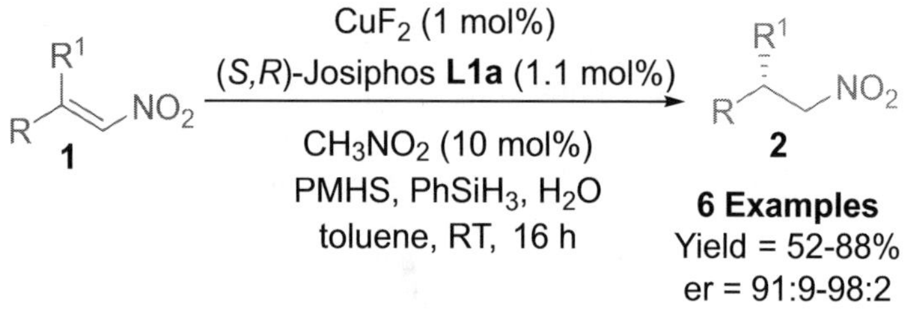

SCHEME 9.3 Asymmetric conjugate reduction of β,β-disubstituted nitroalkenes catalyzed by copper difluoride and chiral Josiphos ligand.

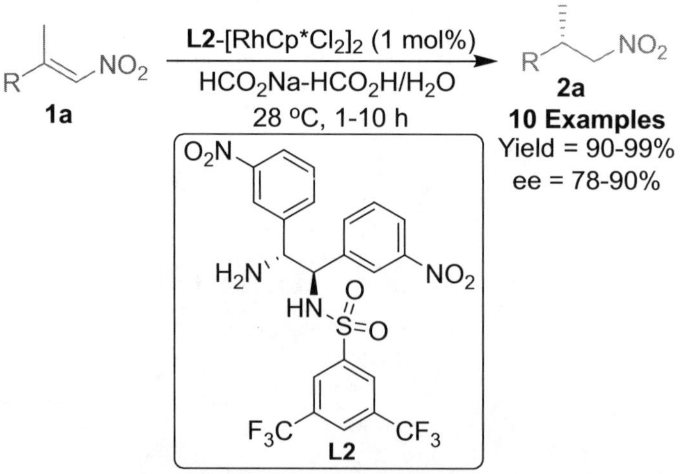

SCHEME 9.4 Asymmetric conjugate reduction of β,β-disubstituted nitroalkenes catalyzed by chiral diamine-rhodium complex in water.

chemoselectivity as well as reactivity. Further, substituents on both phenyl and the sulfonyl groups of TsDPEN **L2** have significant effects on the enantioselectivity. A stepwise conjugate addition mechanism of rhodium hydride to nitroalkenes **1a** was proposed based on the deuterium-labeling experiment and reduction of α-methyl-substituted nitrostyrene.

Carreira and co-workers reported a mild and convenient process for the selective reduction of β,β-disubstituted nitroalkenes **1a** where formic acid was employed as an economical, benign and readily accessible reductant.[5] The reaction is conducted in water at low pH and in open air to give corresponding adducts **2a** in relatively good yields and selectivity (Scheme 9.5).

Zhang and co-workers designed chiral bisphosphine-thiourea ligand **L4** based on the concept of combining metal catalysis along with organocatalysis and employed the ligand **L4** in combination with Rh catalyst for the asymmetric hydrogenation of challenging β,β-disubstituted nitroalkenes **2**. Here, it was observed that excellent results were obtained even with low (0.25 mol %) catalyst loading (Scheme 9.6).[6]

Later the same group developed an efficient method for synthesizing chiral β-amino nitroalkanes **2b** by Ni-catalyzed asymmetric hydrogenation of β-acylamino nitroalkenes **1b** using H_2 as the reductant (Scheme 9.7).[7] $Ni(OAc)_2$ and (*S*)-Binapine ligand **L5** along with additive Bu_4NI under

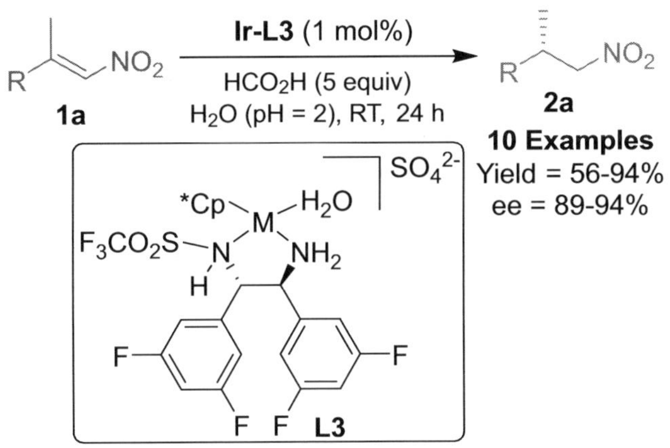

SCHEME 9.5 Asymmetric conjugate reduction of β,β-disubstituted nitroalkenes utilizing formic acid as a reductant in the presence of iridium catalyst.

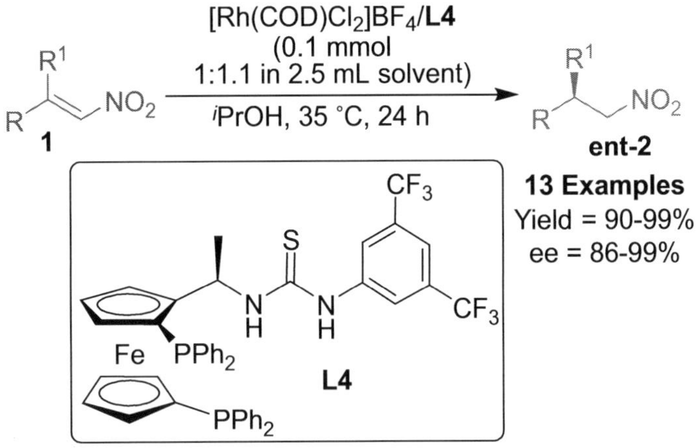

SCHEME 9.6 Asymmetric conjugate reduction of β,β-disubstituted nitroalkenes in the presence of Rh-bisphosphine-thiourea catalyst system.

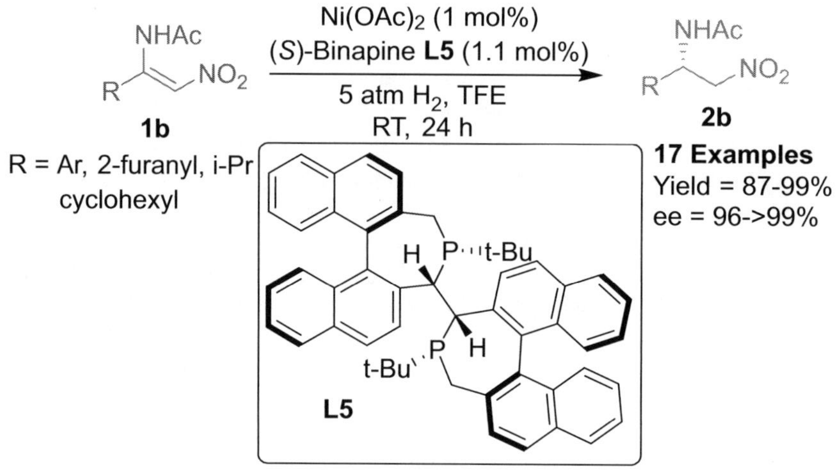

SCHEME 9.7 Ni-catalyzed asymmetric hydrogenation of β-acylamino nitroalkenes using H_2 as the reductant.

50 atm of H_2 at 50°C in MeOH provided β-amino nitroalkanes **2b** in high yields with excellent enantioselectivities.

The Meggers group described chiral-at-metal iridium(III) complex **C1** catalyzed enantioselective reduction of β,β′-disubstituted nitroalkenes **1**. The reaction mediated by Hantzsch ester **3** and the corresponding reduced products **2** were formed in good enantiomeric excess (Scheme 9.8).[8] In the possible reaction pathway, the 5-amino-3-(2-pyridyl)-1H-pyrazole ligand coordinated to the metal (Ir) activates the nitroalkene **1** by forming hydrogen bonds. At the same time, the hydroxymethyl motif on a benzoxazole ligand acts as a hydrogen bond acceptor for the nucleophile and facilitates the hydride donor ability of Hantzsch ester **3**.

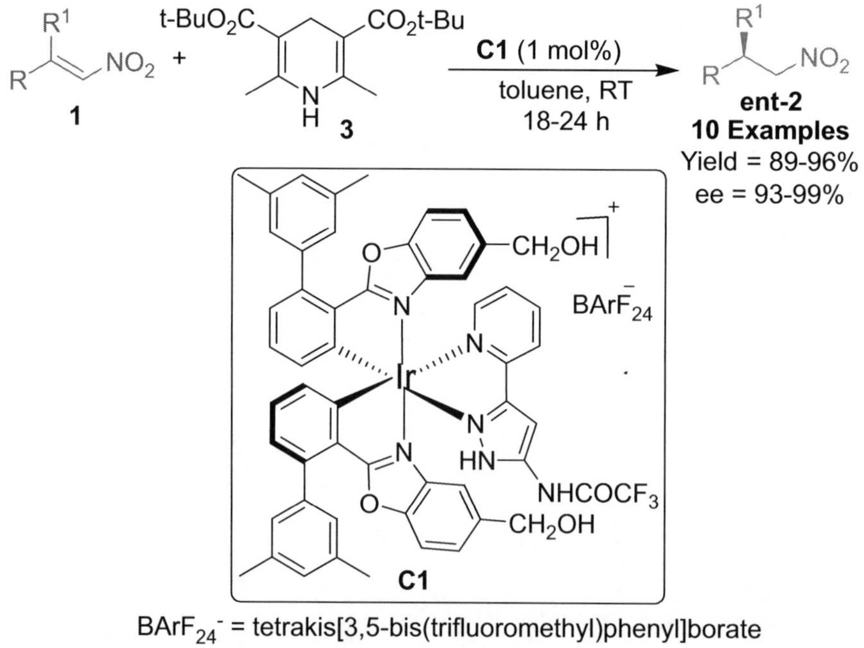

SCHEME 9.8 Chiral-at-metal iridium(III) complex for the catalytic asymmetric reduction of β,β′-disubstituted nitroalkenes.

The Zhang group employed a combination of rhodium and bis(phosphine)-thiourea **L6** for the asymmetric hydrogenation of β-amino nitroalkenes **1** (Scheme 9.9).[9] Here, the reaction progresses through the synergistic activation pathway by cooperative metal catalysis and organocatalysis. Further, the electronic properties of the substituents didn't play any role in yield as well as enantioselectivity. Even the less reactive isopropyl- and propyl-substituted nitroalkenes worked well with excellent enantioselectivity under these conditions. In this reaction, the catalyst–ligand–substrate ratio was maintained at 1:1.1:100 or 4:4.4:100, the conversion rate was >95% and turnover number was up to 1000.

The Zhang group described the asymmetric hydrogenation of β-aryl-β-alkyl disubstituted nitroalkenes **1** to corresponding nitroalkanes **2** with high enantioselectivity.[10] The complex $[Rh(nbd)_2]SbF_6$ [nbd=2,5-norbornadiene] and Josiphos ligand **L7** gave the best enantioselectivity (Scheme 9.10). Irrespective of the electronic nature of the substituents, nitroalkenes **1** with *ortho*-substitution suffered low conversion. Nitroalkenes **1** with halogen substitution, except iodine and naphthyl substitution, delivered the corresponding products **2** in good yields and enantioselectivity.

The same group in another report demonstrated an Rh/Duanphos **L8** catalyst system for the asymmetric hydrogenation of β-β′-disubstituted nitroalkenes **1** and its isomeric mixture into enantiomerically pure chiral β-nitroalkanes **2** (Scheme 9.11).[11] In this transformation, the catalyst suffers lag in the conversion rate when the substituent is at the *ortho*-position of the aryl ring of

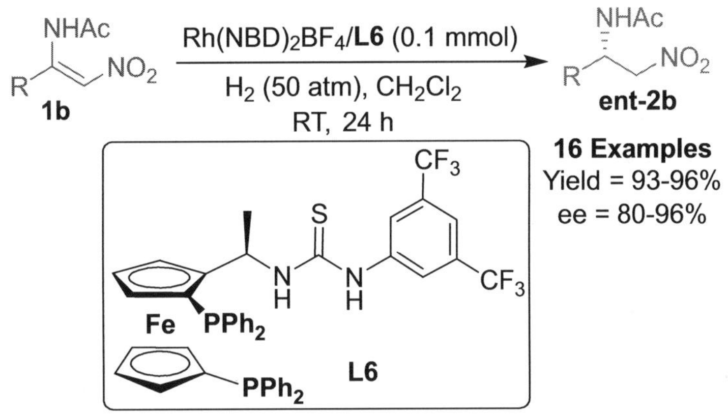

SCHEME 9.9 Rhodium-bis(phosphine)-thiourea catalyst system for the asymmetric hydrogenation of β-amino nitroalkenes.

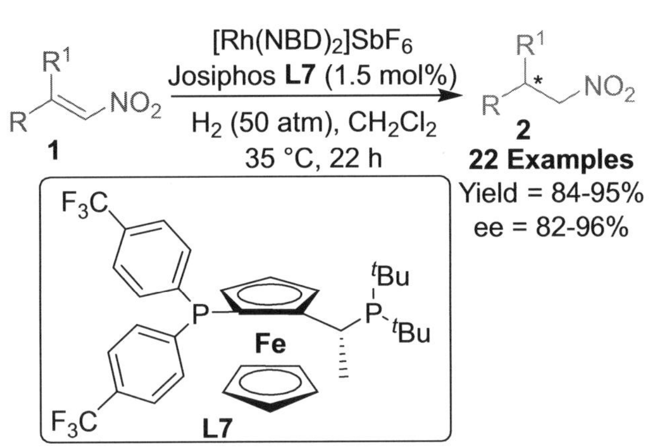

SCHEME 9.10 Rh-Josiphos catalyst system for the asymmetric hydrogenation of β,β′-disubstituted nitroalkenes.

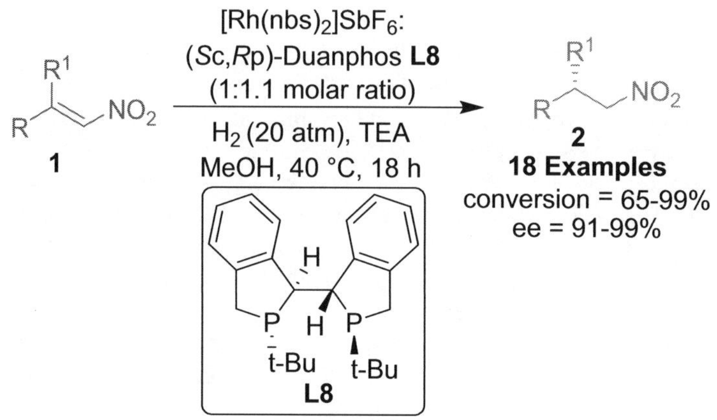

SCHEME 9.11 Rh-Duanphos catalyst system for the asymmetric hydrogenation of β-β′-disubstituted nitroalkenes.

the nitroalkene **1**. Further, F- and Cl-substituted nitroalkenes require additional catalyst loading for complete conversion. The reaction was assumed to proceed through an active species, $Rh^{I}H$. In the proposed pathway, the base converts $Rh^{III}H_2$ to $Rh^{I}H$ and gets oxidized to a neutral complex and thereby facilitate hydrogenation through monohydride transfer, which is responsible for the high enantioselectivity of the reaction under basic conditions.

9.3 ORGANOCATALYZED REDUCTION

Hantzsch esters **3** have been introduced to asymmetric catalysis as a convenient reagent for a number of transfer hydrogenations, and List et al. explored their utility in conjugate reductions of nitroalkenes **1**.[12] As a result of their investigation, a highly enantioselective organocatalytic transfer hydrogenation of β,β-disubstituted nitroalkenes **1** using a Jacobsen-type thiourea catalyst **C2** was demonstrated (Scheme 9.12). The reaction has tolerated a wide range of substrates and gives corresponding products **2** in excellent yields and enantioselectivities.

Fochi group reported a thiourea **C3** catalyzed transfer hydrogenation of β-trifluoromethyl nitroalkenes **1c**, wherein the catalyst binds and stabilizes the negative charge on the nitro group and, at the same time, the oxygen atom of the amide assists in the hydride transfer by coordinating to the –NH group of Hantzsch ester **3**.[13] Here, the transfer hydrogenation occurs in an enantioselective

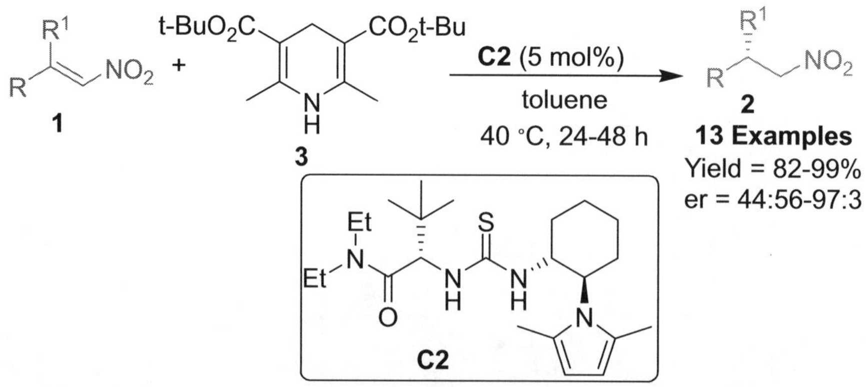

SCHEME 9.12 Hantzsch ester-mediated asymmetric conjugate reduction of nitroalkenes catalyzed by Jacobsen-type thiourea.

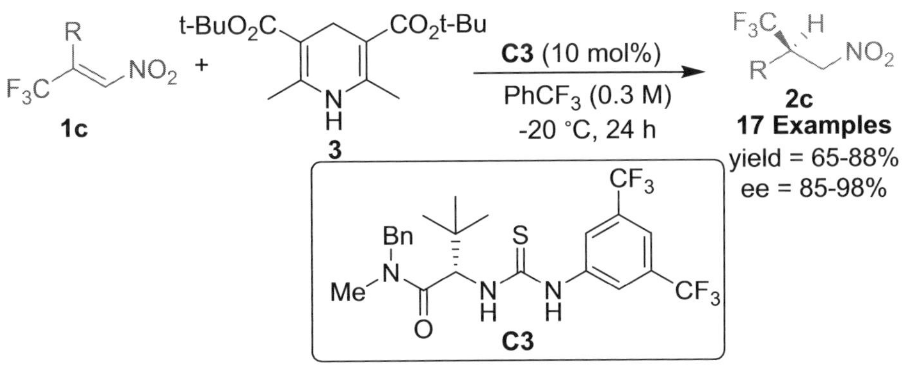

SCHEME 9.13 Hantzsch ester-mediated asymmetric hydrogenation of β-trifluoromethyl nitroalkenes in the presence of thiourea-based catalyst.

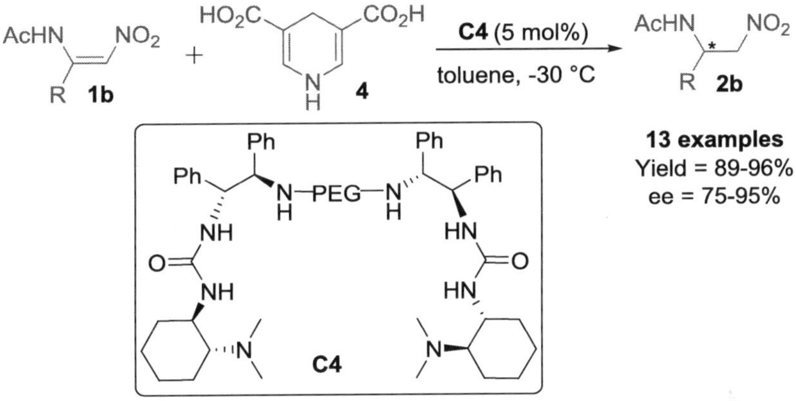

SCHEME 9.14 Asymmetric transfer hydrogenation of β-acylamino nitroalkenes mediated by a novel PEG-supported hydrogen bonding catalyst.

manner and the stereochemistry of the product **2c** directly relies on the stereochemistry of the corresponding nitroalkene **1c** (Scheme 9.13).

Of late, Shi and Zhang prepared a new poly(ethylene glycol)-supported multiple hydrogen bond catalyst **C4** to mediate asymmetric transfer hydrogenation of β-acylamino nitroalkenes **1b** (Scheme 9.14).[14] The authors mentioned that the PEG-bound catalyst **C4** not only provided different β-acylamino nitroalkanes **2b** in excellent yields (up to 96%) and enantioselectivities (up to 95%) but was recyclable up to six times without any loss in catalytic efficiency.

9.4 BIOREDUCTION

Kawai et al. reported the microbial reduction of di and trisubstituted nitroalkenes **1e** using baker's yeast, *Saccharomyces cerevisiae*.[15] This biocatalyst is inexpensive, versatile and readily available. Based on the report of Sakai et al.,[16] the epimerization rate of the α-carbon of *erythro*-3-phenyl-2-nitrobutane was studied and the corresponding half-life was estimated to be 188 h. Authors reasoned that the low selectivity in the earlier report was because of the faster rate of reduction using yeast as biocatalyst. Under these conditions, 25 nitroalkenes were screened and found that stereoselectivity of the reduction was strongly influenced by the substitution pattern of the substrate. The reduction of α,β-disubstituted nitroalkenes occurs with low stereoselectivity (0%–45% *ee*), whereas the trisubstituted nitroalkanes **2g** were formed with satisfactory enantioselectivity (81%–98% *ee*) but with low diastereomeric ratios (Scheme 9.15). The absolute configuration of the product **2g** was confirmed by single-crystal X-ray studies.

Kawai et al. studied the asymmetric reduction of nitroalkenes in the yeast-mediated reduction process, wherein they proposed two stages: (i) a reversible nonstereoselective protonation at the α-carbon and (ii) a stereoselective addition of a hydride from NADPH to α-carbon (Scheme 9.16).[17] However, the detailed mechanism of the nitroalkene reductase is still unclear. These enzymes afforded the product in excellent enantioselectivity and modest diastereoselectivity.

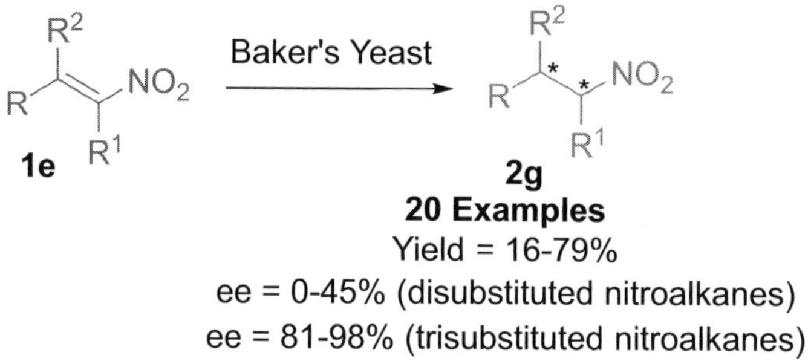

SCHEME 9.15 Microbial reduction of di- and trisubstituted nitroalkenes using baker's yeast.

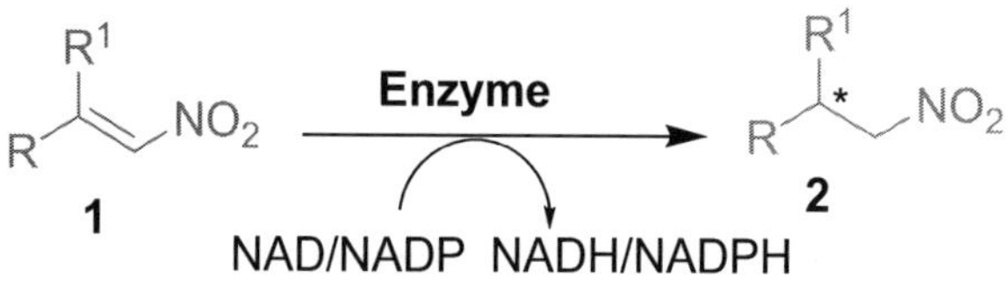

Kwai et al (2001)	**Scrutten et al (2008)**	**Bommarius et al (2011)**	**Stephens et al (2008)**
Reductase	PETN Reductase	KYE1	*Clostridium Spororgenes*
YNAR-1	Yield = 18->99%	Yield = 0.4-100%	Yield = 10-90%
de = 31%	ee = 14->99	ee (*S*) = 47-99%	ee = 30->97%
ee = >98%		ee (*R*) = 32-99%	
YNAR-II		YersER	
ee = >98%		Yield = 42-100%	
		ee (*S*) = 90-99%	
		ee (*R*) = 71-95%	

SCHEME 9.16 Asymmetric bioreduction of nitroalkenes.

The Faber group demonstrated an asymmetric bioreduction of nitroalkenes using an enoate reductase and Old Yellow Enzymes obtained from *Zymomonas mobilis* and yeasts, respectively, to afford the corresponding saturated products in excellent enantioselectivities.[18] The same group in another report described the asymmetric bioreduction of nitroalkene using cloned 12-oxophytodienoate reductase isoenzymes (OPR-1 and OPR-3) from Lycopersiconesculentum (tomato).[19] They employed the isoenzymes OPR1 and OPR3 of 12-oxophytodienoate reductase from *Lycopersicon esculentum* (tomato), and the Old Yellow Enzyme homolog YqjM from *Bacillus subtilis* for the asymmetric bioreduction of nitroalkenes, and the corresponding saturated products were obtained at the expense of NADH or NADPH, which serves as H source.[20] The isoenzymes OPR1 and OPR3 behaved in a stereocomplementary fashion and yielded opposite enantiomeric products.

The Brenna group carried out the asymmetric hydrogenation of β,β-disubstituted nitroalkenes using ene-reductase to afford the reduced products in high yields with satisfactory enantioselectivity.[21] Even the nitroalkenes with long alkyl chains at the β-position of the nitro group as well as different substituents on the aryl ring worked well under this bioreduction process.

The Scrutton group used pentaerythritol tetranitrate reductase obtained from *Enterobacter cloacae* st. PB2 to catalyze the asymmetric reduction of nitroalkenes (Scheme 9.16).[22] The reaction kinetics of this enzyme is comparable to that of Old Yellow Enzyme homologs. However, it shows a preference for *Z*-isomers over *E*-isomers to yield pure enantiomeric reduced products with up to >99% ee.

Bommarius et al. described an asymmetric *trans*-bioreduction of nitroalkenes by employing two ene-reductases, that is, KYE1 and Yers-ER from the Old Yellow Enzyme family (Scheme 9.16).[23] These enzymes offer opposite stereoisomeric products with moderate-to-excellent degree of stereoselectivity.

Stephens and co-workers reported nitroalkene reductase obtained from *Clostridium sporogenes* as an efficient biocatalyst for the enantioselective reduction of nitroalkenes including aromatic and heteroaromatic to yield the corresponding *S*-isomer.[24] This protocol doesn't hold good for the reduction of alkyl derivatives. Here, the absolute configuration of the product was independent, of which geometric isomer of the substrate was used.

9.5 CONCLUSIONS

β,β-Unsymmetrically disubstituted nitroalkenes undergo catalytic asymmetric and bioreductions. Various modes of reduction of the conjugated double bond such as conjugate addition, hydrogenation, metal and organo-catalyzed transfer hydrogenation and biocatalytic reduction have been successfully carried out on conjugated nitroalkenes to afford **β,β**-unsymmetrically disubstituted nitroalkanes in high yield and enantioselectivity.

REFERENCES

1. Czekelius, C.; Carreira, E. M. *Angew. Chem. Int. Ed.* **2003**, *42*, 4793.
2. Czekelius, C.; Carreira, E. M. *Org. Proc. Res. Dev.* **2007**, *11*, 633.
3. Czekelius, C.; Carreira, E. M. *Org. Lett.* **2004**, *6*, 4575.
4. Tang, Y.; Xiang, J.; Cun, L.; Wang, Y.; Zhu, J.; Liao, J.; Deng, J. *Tetrahedron: Asymmetry* **2010**, *21*, 1900.
5. Soltani, O.; Ariger, M. A.; Carreira, E. M. *Org. Lett.* **2009**, *11*, 4196.
6. Zhao, Q.; Li, S.; Huang, K.; Wang, R.; Zhang, X. *Org. Lett.* **2013**, *15*, 4014.
7. Gao, W.; Lv, H.; Zhang, T.; Yang, Y.; Chung, L. W.; Wu, Y.-D.; Zhang, X. *Chem. Sci.* **2017**, *8*, 6419.
8. Chen, L.-A.; Xu, W.; Huang, B.; Ma, J.; Wang, L.; Xi, J.; Harms, K.; Gong, L.; Meggers, E. *J. Am. Chem. Soc.* **2013**, *135*, 10598.
9. Li, P.; Zhou, M.; Zhao, Q.; Wu, W.; Hu, X.; Dong, X.-Q.; Zhang, X. *Org. Lett.* **2016**, *18*, 40.
10. Li, S.; Huang, K.; Cao, B.; Zhang, J.; Wu, W.; Zhang, X. *Angew. Chem. Int. Ed.* **2012**, *51*, 8573.
11. Li, S.; Huang, K.; Zhang, J.; Wu, W.; Zhang, X. *Chem. - Eur. J.* **2013**, *19*, 10840.

12. Martin, N. J. A.; Ozores, L.; List, B. *J. Am. Chem. Soc.* **2007**, *129*, 8976.
13. Martinelli, E.; Vicini, A. C.; Mancinelli, M.; Mazzanti, A.; Zani, P.; Bernardi, L.; Fochi, M. *Chem. Commun.* **2015**, *51*, 658.
14. Zhang, B.; Shi, L. *Catal. Lett.* **2019**, *149*, 2836.
15. Kawai, Y.; Inaba, Y.; Tokitoh, N. *Tetrahedron: Asymmetry* **2001**, *12*, 309.
16. Sakai, K.; Nakazawa, A.; Kondo, K.; Ohta, H. *Agric. Biol. Chem.* **1985**, *49*, 2331.
17. Kawai, Y.; Inaba, Y.; Hayashi, M.; Tokitoh, N. *Tetrahedron Lett.* **2001**, *42*, 3367.
18. Hall, M.; Stueckler, C.; Hauer, B.; Stuermer, R.; Friedrich, T.; Breuer, M.; Kroutil, W.; Faber, K. *Eur. J. Org. Chem.* **2008**, *2008*, 1511.
19. Hall, M.; Stueckler, C.; Kroutil, W.; Macheroux, P.; Faber, K. *Angew. Chem. Int. Ed.* **2007**, *46*, 3934.
20. Hall, M.; Stueckler, C.; Ehammer, H.; Pointner, E.; Oberdorfer, G.; Gruber, K.; Hauer, B.; Stuermer, R.; Kroutil, W.; Macheroux, P.; Faber, K. *Adv. Synth. Catal.* **2008**, *350*, 411.
21. Bertolotti, M.; Brenna, E.; Crotti, M.; Gatti, F. G.; Monti, D.; Parmeggiani, F.; Santangelo, S. *ChemCatChem* **2016**, *8*, 577.
22. Toogood, H. S.; Fryszkowska, A.; Hare, V.; Fisher, K.; Roujeinikova, A.; Leys, D.; Gardiner, J. M.; Stephens, G. M.; Scrutton, N. S. *Adv. Synth. Catal.* **2008**, *350*, 2789.
23. Yanto, Y.; Winkler, C. K.; Lohr, S.; Hall, M.; Faber, K.; Bommarius, A. S. *Org. Lett.* **2011**, *13*, 2540.
24. Fryszkowska, A.; Fisher, K.; Gardiner, J. M.; Stephens, G. M. *J. Org. Chem.* **2008**, *73*, 4295.

10 Catalytic Asymmetric Synthesis of Cycloalkanes via Cascade Reactions of Nitroalkenes

10.1 INTRODUCTION

In recent years, enantioselective cascade reactions have been extensively employed in the total synthesis of various natural products and drug molecules, which allowed a significant reduction in the number of required synthetic steps. In 2012[1,2] and 2014,[3] several authors have reviewed the developments in the enantioselective cascade, as well as multicomponent reactions involving nitroalkenes as the substrates.[4–37] We discuss in Chapters 10–12 the asymmetric versions of cascade and multicomponent reactions of nitroalkenes, which were not covered in those reviews. These chapters are organized based on the bioactive building blocks that are formed, i.e., the formation of optically active (a) cycloalkanes (Chapter 10), (b) aryl-fused heterocycles (Chapter 11) and (c) other five- and six-membered heterocycles (Chapter 12).

This chapter covers the synthesis of different cycloalkane building blocks, especially cyclopentanes and cyclohexanes synthesized via catalytic asymmetric cascade and multicomponent reactions of nitroalkenes with various other components in the presence of mainly proline and cinchona-derived organocatalysts. The highly functionalized monocyclic as well as fused, bridged and spirolinked polycyclic alkanes are potential building blocks for the synthesis of complex molecules, including natural products.

10.2 SYNTHESIS OF CYCLOALKANES

Zhao et al. delineated a tandem Michael–Henry reaction of 1,2-dione **2** with nitroalkenes **1**, which provided cyclopentanone derivatives **3** in high yields and enantioselectivities by employing quinine-derived thiourea catalyst **C1a** (Scheme 10.1).[38]

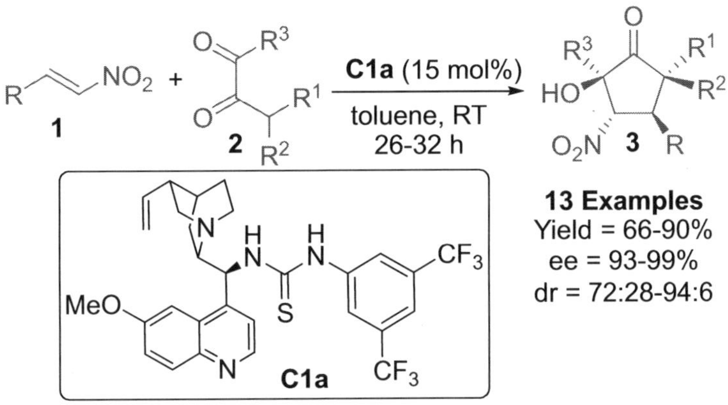

SCHEME 10.1 Quinine-derived thiourea-catalyzed tandem Michael–Henry reaction of 1,2-dione with nitroalkenes.

Zhang and co-workers demonstrated an enantioselective domino double Michael addition of nitroalkenes **1** with conjugated aldehyde esters **4** by employing *trans*-perhydroindolic acid **C2** as an organocatalyst to afford polysubstituted cyclopentanes **5** with excellent enantioselectivities (Scheme 10.2).[39] In this reaction, the conjugated aldehyde ester **4** forms an enamine intermediate with the catalyst **C2**, whereas nitroalkene **1** gets activated by hydrogen bonding with the –NH of catalyst **C2** and will be directed toward the carboxylic group of the catalyst **C2** from above. Now, the enamine adds to nitroalkene **1** (**TS-1**) from behind the plane of the alkene followed by cyclization with the conjugated aldehyde ester **4** (**TS-2**) to yield cyclopentanes **5** with four contiguous stereocenters via the iminium **6** (Scheme 10.2).

The Enders group reported an asymmetric domino Michael addition followed by alkylation of aldehydes **7** with 5-iodo-1-nitropent-1-ene **1a** using proline-derived catalyst **C3** (Scheme 10.3).[40] The reaction follows an enamine–enamine activation pathway to yield cyclopentanes **8** with γ-nitroaldehydes with all-carbon-substituted stereocenter. The absolute configuration of the adduct γ-nitroaldehydes **8** was confirmed by synthesizing its camphynyl derivative **9** (Scheme 10.3).

The reaction begins with activation of the aldehyde by the chiral amine via enamine formation (Scheme 10.4). As the bulky catalyst **C3** containing diphenylsiloxymethyl group shields the *Re*-face at Cα of the enamine intermediate, the Michael addition involving an acyclic synclinal transition state occurs at the *Si*-face to give the intermediate Michael adduct **11**. Subsequently, the intermediate **11** undergoes a proton shift at the expense of the α-carbon chiral center to generate a second enamine intermediate **12**. Further, an intramolecular nucleophilic substitution follows causing in-ring closure to form intermediate **13**. This intramolecular alkylation appears to occur from the *Si*-face at Cα of enamine leading to the *cis*-selective formation of the *S*, *R*-cyclopentane products **8** when R^1 = bulkyl alkyl groups. The reaction was *trans*-selective when R^1 = Me because of the intramolecular alkylation at the diastereotopic *Re*-face at Cα.

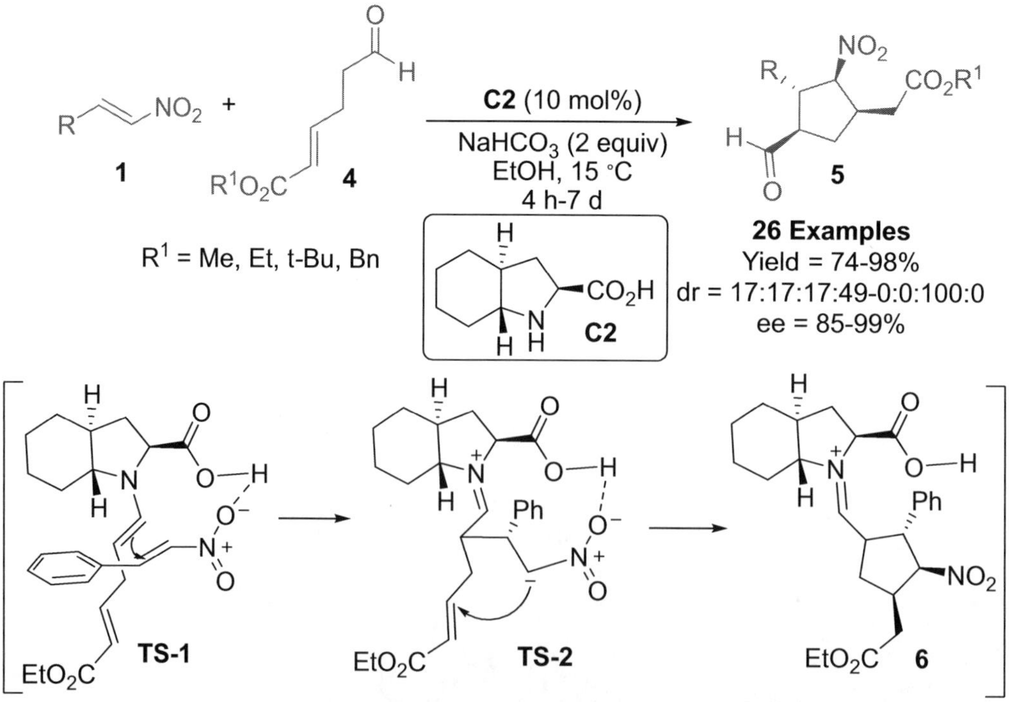

SCHEME 10.2 Enantioselective synthesis of polysubstituted cyclopentanes by domino double Michael addition of nitroalkenes with conjugated aldehyde esters.

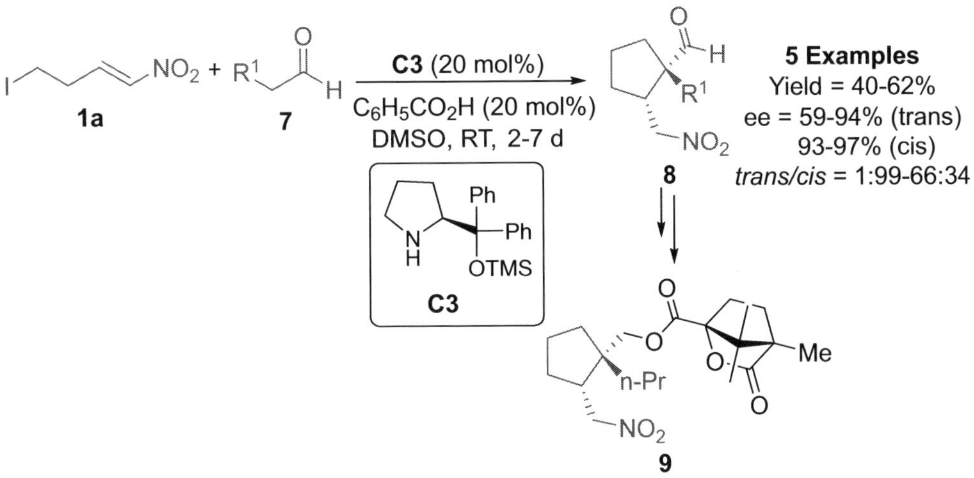

SCHEME 10.3 Enantioselective synthesis of cyclic γ-nitroaldehydes by successive Michael addition-alkylation of aldehydes with 5-iodo-1-nitropent-1-ene using proline-derived catalyst.

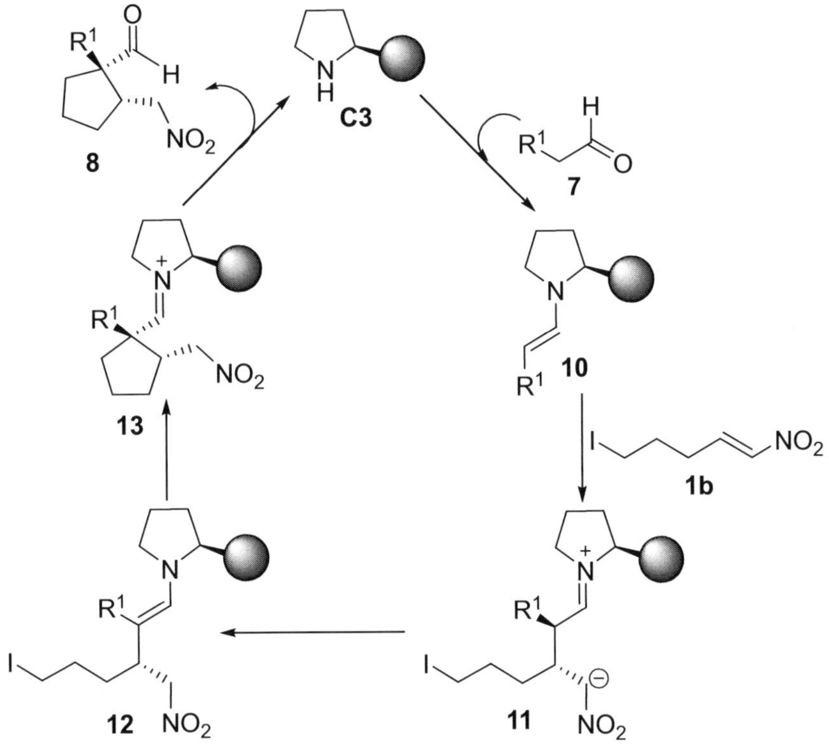

SCHEME 10.4 Proposed mechanism for the enantioselective synthesis of cyclic γ-nitroaldehydes from aldehydes and 5-iodo-1-nitropent-1-ene.

The Enders group reported an organocatalytic three-component quadruple cascade reaction involving a vinylogous Friedel–Crafts-type reaction of electron-rich arenes **15** with nitroalkenes **1**.[41] The reaction afforded the products **16** with three contiguous stereocenters in excellent yields with excellent enantioselectivities (Scheme 10.5). In the proposed mechanism, the catalyst **C3a** reacts

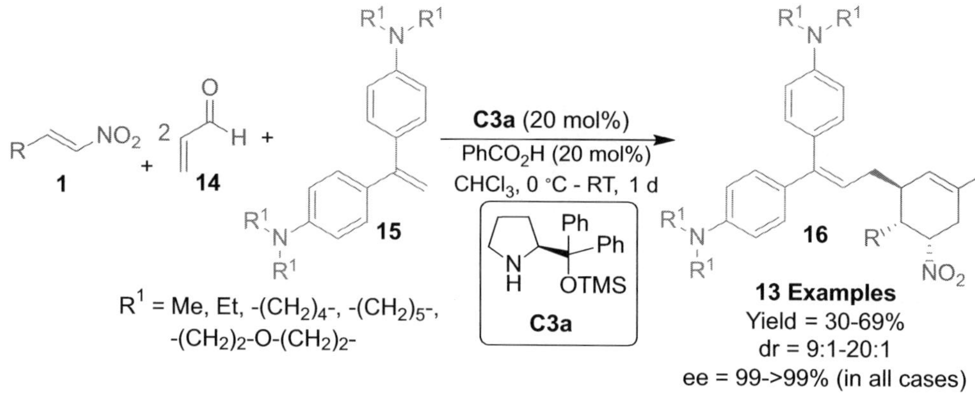

SCHEME 10.5 Synthesis of functionalized cyclohexene-carbaldehydes bearing a 1,1-bis[4-(dialkylamino)phenyl]ethene moiety by employing alkenes, acrolein and β-nitrostyrenes.

with acrolein **14** and forms electrophilic iminium ion. Subsequently, alkene **15** adds to this iminium ion to generate enamine intermediate, which on addition to nitroalkene **1** followed by hydrolysis yields the product **16** with the regeneration of the catalyst **C3a**.

Secondary amine catalyzed domino Michael/retro-oxa-Michael/Michael/aldol condensation reaction was developed by Jiang and co-workers, which afforded optically pure polyfunctionalized cyclohexene-carbaldehydes **18** from propanal derivative **17** and β-nitroalkenes **1** in good yield and excellent stereoselectivity (Scheme 10.6).[42] This domino reaction begins with the formation of enamine intermediate **19** from propanal **17** and catalyst **C3** (Scheme 10.7). The enamine intermediate **19** participates in Michael addition with nitroalkenes **1** to form the Michael adduct **20**. Meanwhile, the enamine intermediate **19** undergoes retro-reaction/elimination to form another iminium intermediate **21** through retro-*oxa*-Michael addition, and the alkoxide deprotonates the Michael adduct to generate anion **22**. Subsequent Michael addition of **22** to **21** followed by intramolecular aldol condensation provides cyclohexene carboxaldehyde **18** (Scheme 10.7).

Zhang et al. developed a one-pot, organocatalytic strategy toward the synthesis of fluorinated cyclohexanols **26** from β-ketoesters **24**, β-nitroalkenes **1** and α,β-unsaturated aldehydes **25** (Scheme 10.8).[43] The cyclization proceeds through quadruple fluorination/Michael/Michael/aldol reaction sequences to generate six contiguous stereocenters under SelectFluor/Cinchona-alkaloid-thiourea catalyst **C4**. The catalyst **C4** controls the formation of the first two stereocenters, whereas enantioenriched substrates control the formation of the remaining four stereocenters. The fluorous catalyst was recovered and reused without losing the reactivity.

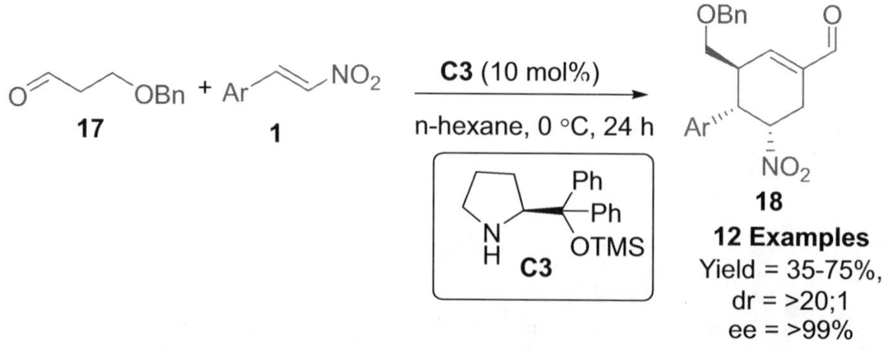

SCHEME 10.6 Diphenyl prolinol TMS-ether catalyzed enantioselective preparation of polyfunctionalized cyclohexene-carbaldehydes from propanal and nitroalkenes.

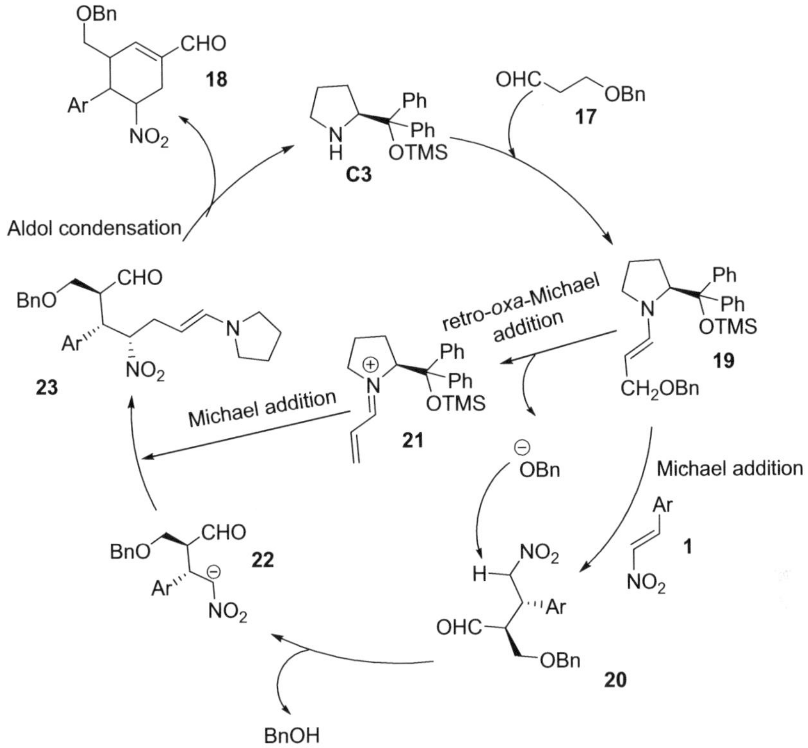

SCHEME 10.7 Proposed mechanism for the formation of cyclohexene-carbaldehydes from proapanal and nitroalkenes.

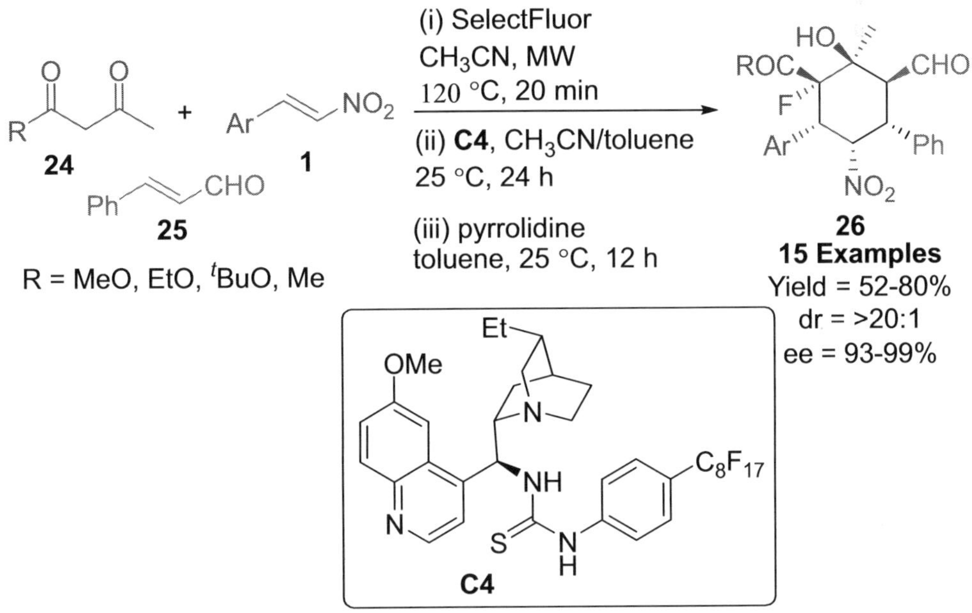

SCHEME 10.8 Fluorous bifunctional cinchona alkaloid/thiourea-catalyzed asymmetric synthesis of 2-fluorocyclohexanols from β-ketoesters, nitroalkenes and α,β-unsaturated aldehydes.

Bonne and co-workers reported an asymmetric synthesis of polysubstituted cyclohexanes **28** by the reaction of 1,2-ketoamides **27** with nitroalkenes **1** employing thiourea-based catalyst **C5** with good yields and enantioselectivity (Scheme 10.9).[44] In the possible pathway, 1,2-ketoamides **27** gets activated with the organocatalyst **C5** and participates in conjugate addition with nitroalkene **1** to form the corresponding Michael adduct. Further addition of another molecule of nitroalkene **1** to the above adduct followed by intramolecular Henry reaction yields the final product **28**.

The Huang group demonstrated an enantioselective **L1**-Cu(II) catalyzed [2+2+2] annulation of α-ketoesters **29a** with nitroalkenes **1a** to afford highly functionalized cyclohexane carboxylates **30a** with six contiguous stereocenters, including one quaternary stereocenter (Scheme 10.10).[45] At first, α-ketoester **29b** forms an enolate **31** on reaction with Lewis acid ($Cu(OAc)_2$), which adds to the nitroalkene **1** in a Michael fashion to give intermediate **32** (Scheme 10.11). The resulting intermediate reacts with another molecule of nitroalkene **1** to afford the intermediate **33**, which further undergoes ring closure via Henry reaction to form six-membered annulation products **30b** with six contiguous stereocenters.

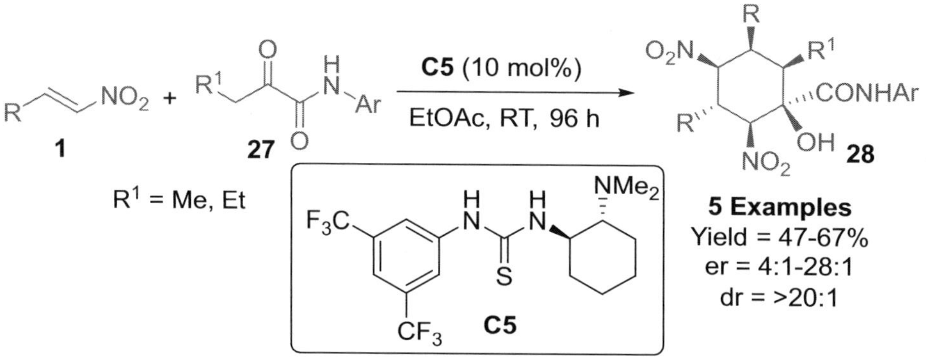

SCHEME 10.9 Thiourea-catalyzed enantioselective domino synthesis of functionalized cyclohexanes from 1,2-ketoamides and nitroalkenes.

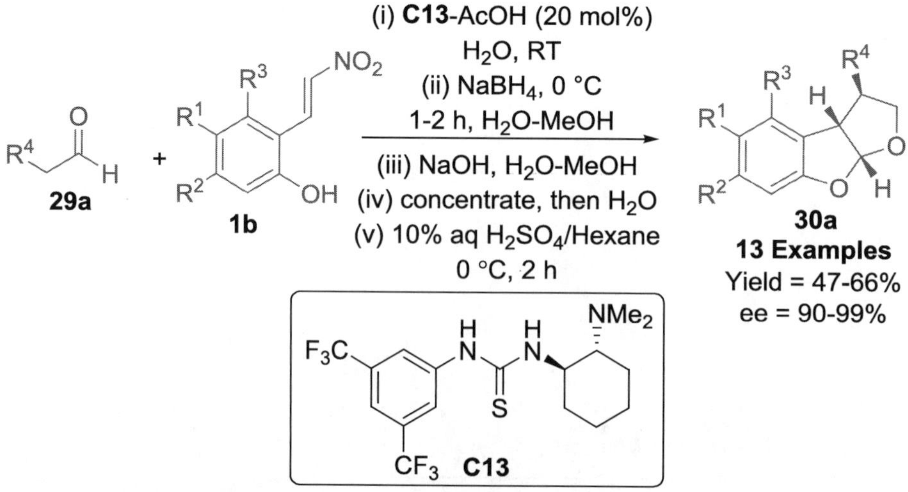

SCHEME 10.10 [2+2+2] Annulation of α-ketoesters with nitroalkenes for the enantioselective construction of functionalized cyclohexanes in the presence of chiral diamine complex of copper catalyst.

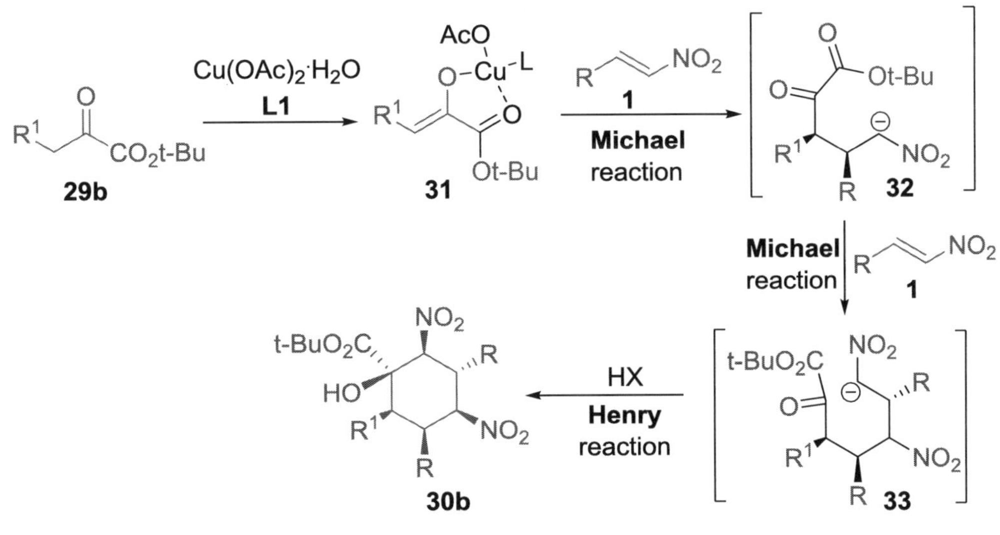

SCHEME 10.11 Proposed mechanism for the formation of functionalized cyclohexanes from α-ketoesters and nitroalkenes.

The Enders group reported a sequential 1,4-/1,6-/vinylogous 1,2-addition cascade reaction of β-dicarbonyl compounds **34**, β-nitroalkenes **1** and 4-nitro-5-styrylisoxazoles **35** by employing cinchona-derived squaramide **C6a** in the presence of an achiral base, resulting in the formation of cyclohexane derivatives **36** with good yields and high enantiomeric excess (Scheme 10.12).[46]

The Enders group developed an asymmetric nitro-Michael/Henry domino reaction of γ-nitro aldehydes **38** or ketones **37** with nitroalkenes **1** employing squaramide-based organocatalyst **C7** to form polysubstituted 1,2,3,4-tetrahydronaphthalen-1-ols **40** or secondary/tertiary cyclohexanols **39** possessing four contiguous stereocenters (Scheme 10.13).[47] Nucleophilic addition of the γ-nitro aldehyde **38** to the nitroalkene **1** followed by an intramolecular Henry reaction yields the products **39/40**.

The Liu group developed an organocatalytic, asymmetric Michael/Michael/Henry cascade reaction of α-ketoesters **29** with three equivalents of nitroalkenes **1** to form *m*-substituted cyclohexane derivatives **41** with six contiguous stereocenters by employing bifunctional guanidine-amide

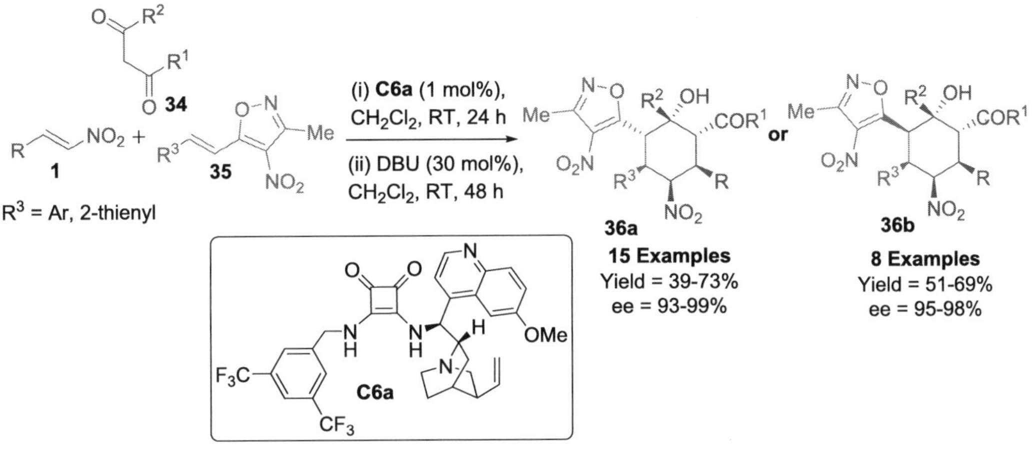

SCHEME 10.12 Cinchona-derived squaramide catalyzed cascade reaction for the synthesis of cyclohexane derivatives.

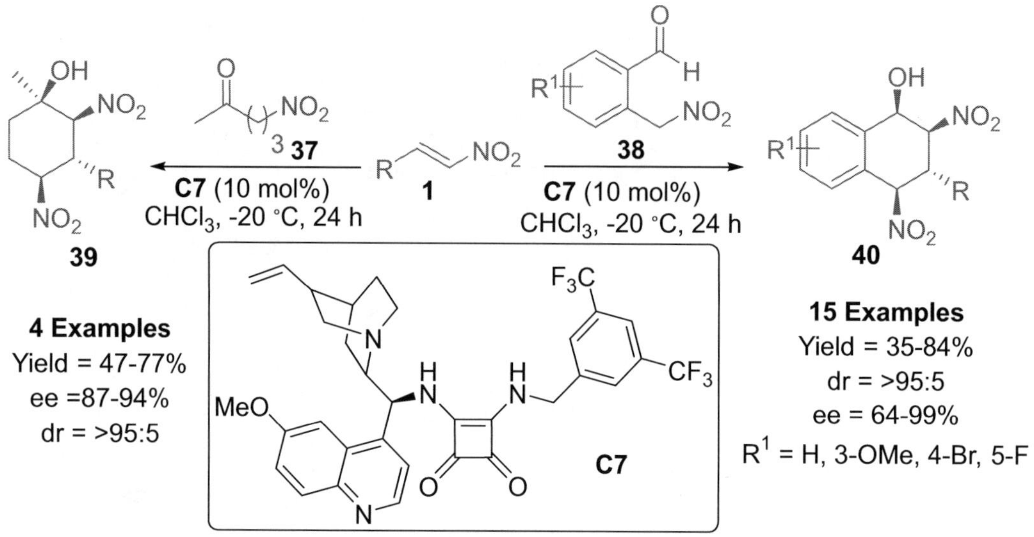

SCHEME 10.13 Nitro-Michael/Henry domino reaction of γ-nitroaldehydes or ketones with nitroalkenes using squaramide-based organocatalyst.

organocatalyst **C8** (Scheme 10.14).[48] In this reaction, the combination of a Brønsted base and hydrogen-bonding catalysis with a guanidine unit allows the reaction to participate in triple-cascade fashion (**TS-3**, Scheme 10.14).

Enders et al. reported an asymmetric synthesis of polyfunctionalized cyclohexane derivatives **43a-b** via a two-component four-step cascade reaction employing aldehydes **42** and nitroalkenes **1**.[49] The reaction involves a Michael addition with parallel oxidation, second Michael addition and an aldol condensation. Interestingly, both the enamine nucleophile and iminium electrophile were formed from the same aldehyde **42** (Scheme 10.15).

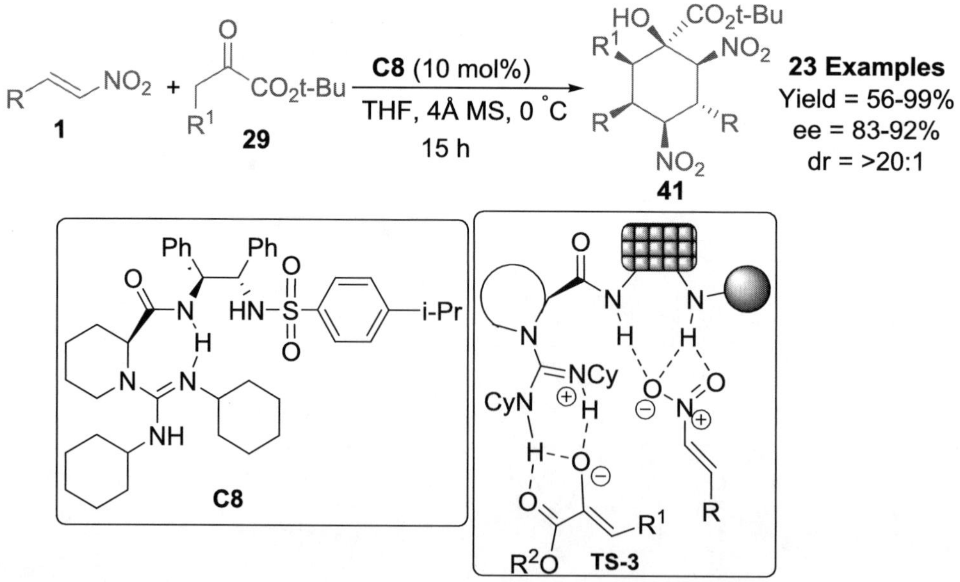

SCHEME 10.14 Asymmetric Michael/Michael/Henry cascade reaction of α-ketoesters with nitroalkenes.

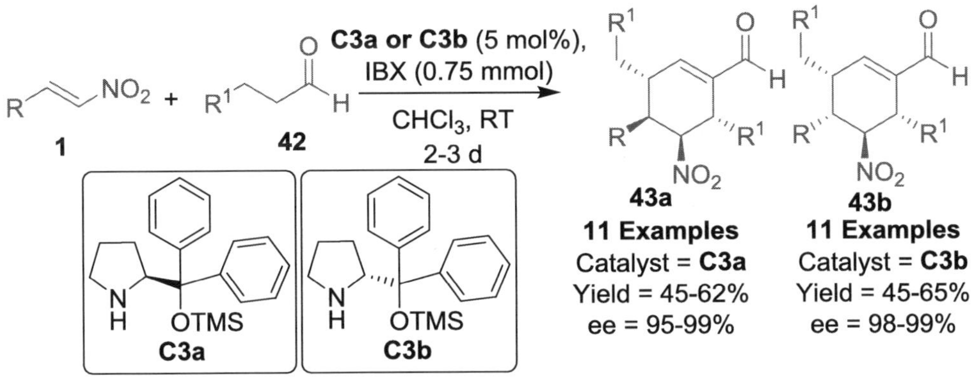

SCHEME 10.15 Asymmetric organocatalytic cascade reaction of aldehydes and nitroalkenes for the preparation of polyfunctionalized cyclohexanes.

In the proposed mechanism, aldehyde **42** reacts with amine catalyst **C3a** to form an enamine intermediate **44**, which adds to nitroalkene **1** to form an adduct **45**. At the same time, enamine gets oxidized to the corresponding iminium ion **46** and the latter reacts as a Michael acceptor with the adduct **45** to form intermediate **47** through aldol condensation. At the end, hydrolysis results in the formation of the cyclohexene product **43a** (Scheme 10.16).

The Chen group reported a formal [3+3] cycloaddition of α,β-unsaturated aldehydes **48** and 2-nitroallylic acetates **1b** by employing a chiral bifunctional secondary amine-thiourea catalyst **C9**.[50] The reaction afforded cyclohexane derivatives **49** with high α,γ-regioselectivity and excellent enantioselectivity (Scheme 10.17). In this reaction, the initial Michael addition generates the second acceptor with the elimination of a molecule of acetic acid, which later participates in another

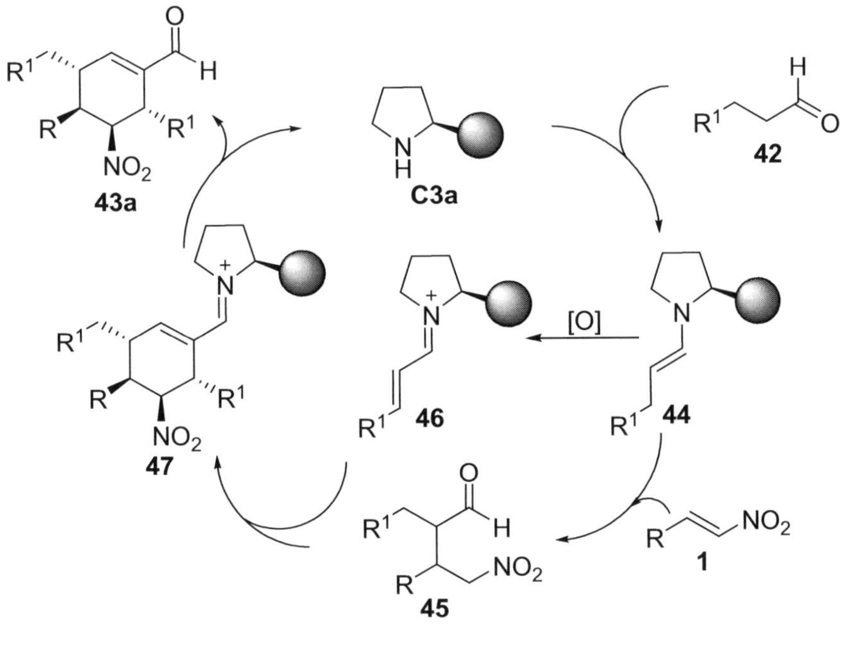

SCHEME 10.16 Proposed mechanism for the formation of polyfunctionalized cyclohexanes from aldehydes and nitroalkenes.

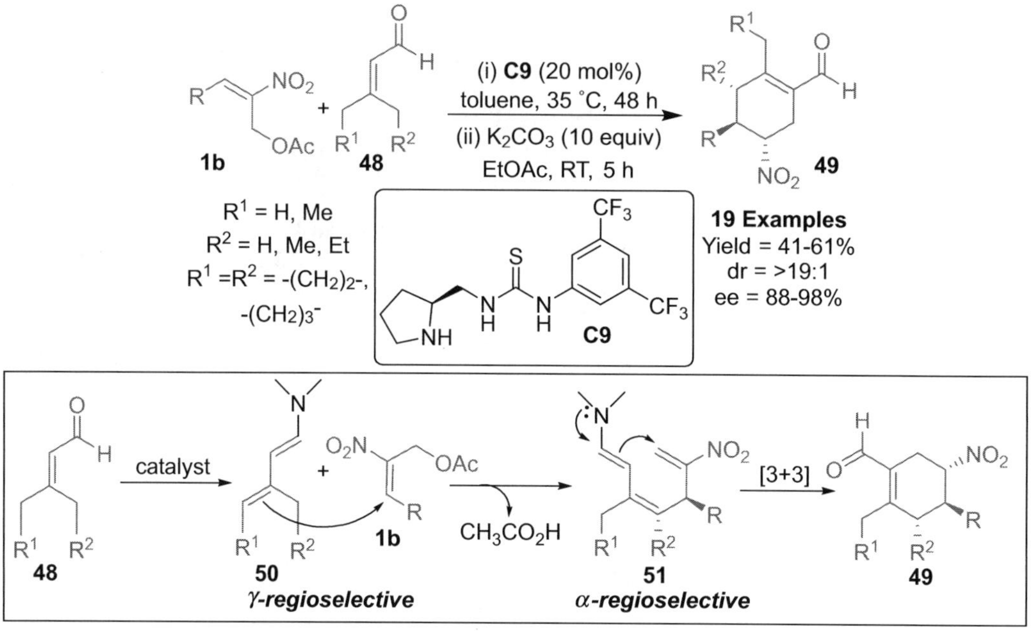

SCHEME 10.17 Chiral bifunctional secondary amine-thiourea-catalyzed [3+3] cycloaddition of α,β-unsaturated aldehydes with 2-nitroallylic acetates.

Michael addition. Likewise, the overall reaction proceeds through a domino double Michael addition via dienamine–dienamine catalysis (Scheme 10.17).

Namboothiri and co-workers reported an asymmetric double Michael addition of curcumins **52** to nitroalkenes **1** for the synthesis of highly functionalized cyclohexanones **53** by employing dihydrocinchonine-thiourea organocatalyst **C1b** and K_2CO_3 (Scheme 10.18).[51] The products **53** were formed in moderate-to-good yields with excellent enantioselectivities.

The Hong group demonstrated an asymmetric synthesis of Hajos–Parrish-type ketones **55** with five to six contiguous stereocenters and two quaternary stereocenters by employing

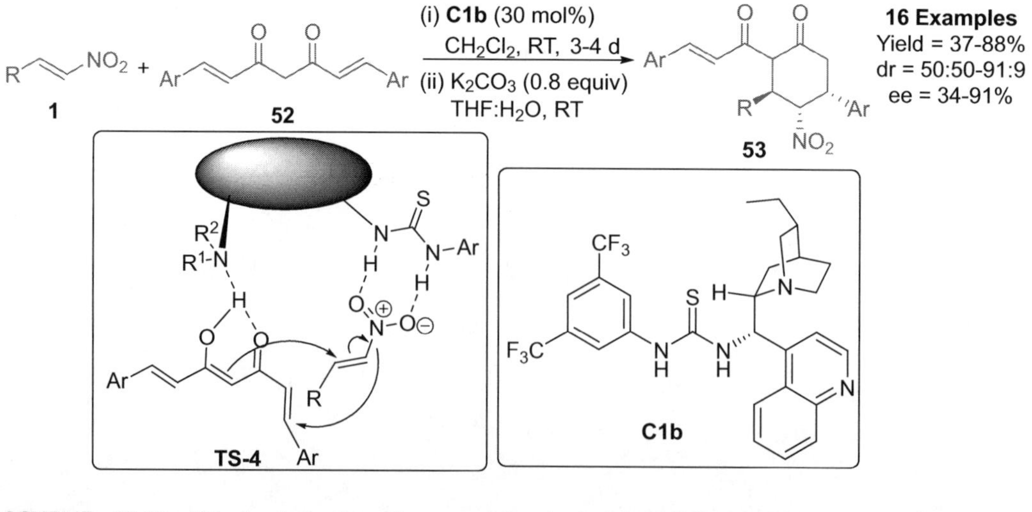

SCHEME 10.18 Dihydrocinchonine-thiourea-catalyzed double Michael addition of curcumins to nitroalkenes.

2-methylcyclopentane-1,3-dione **54**, nitroalkenes **1** and α,β-unsaturated aldehyde **14** in the presence of diphenylprolinol catalyst **C10** (Scheme 10.19).[52]

The Kokotos group described an enantioselective reaction of 1,4-cyclohexanedione **56** with nitroalkenes **1c** involving a domino Michael–Henry reaction employing a proline-based bifunctional organocatalyst **C11** to obtain bicyclo[3.2.1]-octan-2-ones **57** as a single diastereoisomer with four contiguous stereocenters (Scheme 10.20).[53]

Wang et al. reported an asymmetric synthesis of β-naphtholenones **59** through an intermolecular dearomatization of β-naphthols **58** with nitroethylene **1d** using chiral thiourea catalyst **C5** (Scheme 10.21).[54] The β-naphtholenones **59** were obtained in high yields and enantioselectivities with all-carbon quaternary stereocenter. In this reaction, the hydroxyl group of naphthol **58** gets activated by the tertiary amine of the catalyst **C5**, meanwhile, nitroethylene **1d** gets activated by the amide NH of the catalyst **C5** (**TS-5**, Scheme 10.21). Now, the two substrates orient in an ordered conformation, resulting in an *Re*-face attack of naphthol **58** to the nitroethylene **1d** yielding product **59** with *R*-configuration (**TS-5**, Scheme 10.21).

The Zhou group described the asymmetric synthesis of functionalized spiro-pyrazolone derivatives **63** through DBU (an achiral Brønsted base) promoted cascade Michael/aldol reaction

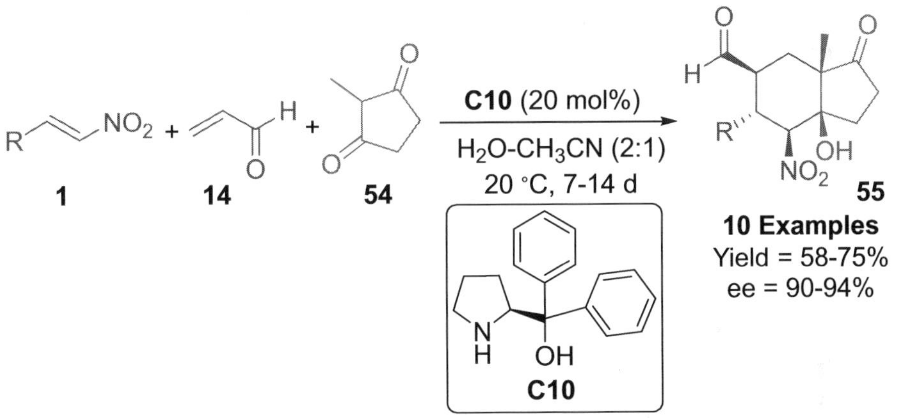

SCHEME 10.19 Diphenylprolinol catalyzed Michael–Michael–Henry cascade reaction of 2-methylcyclopentane-1,3-dione, nitroalkenes and α,β-unsaturated aldehyde.

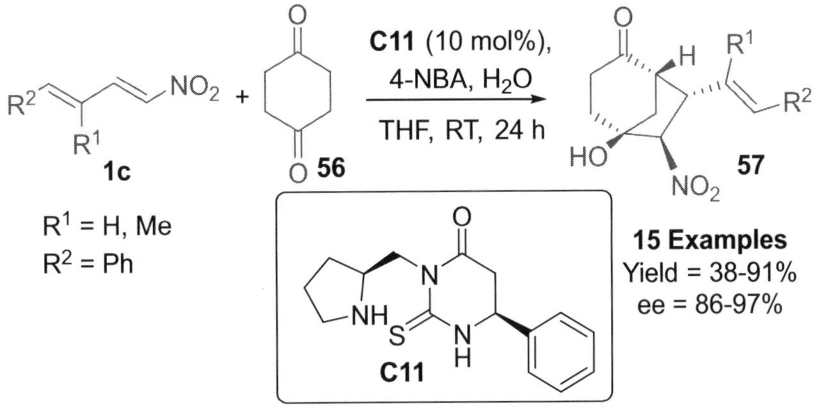

SCHEME 10.20 Domino Michael–Henry reaction mediated by proline-based bifunctional organocatalyst.

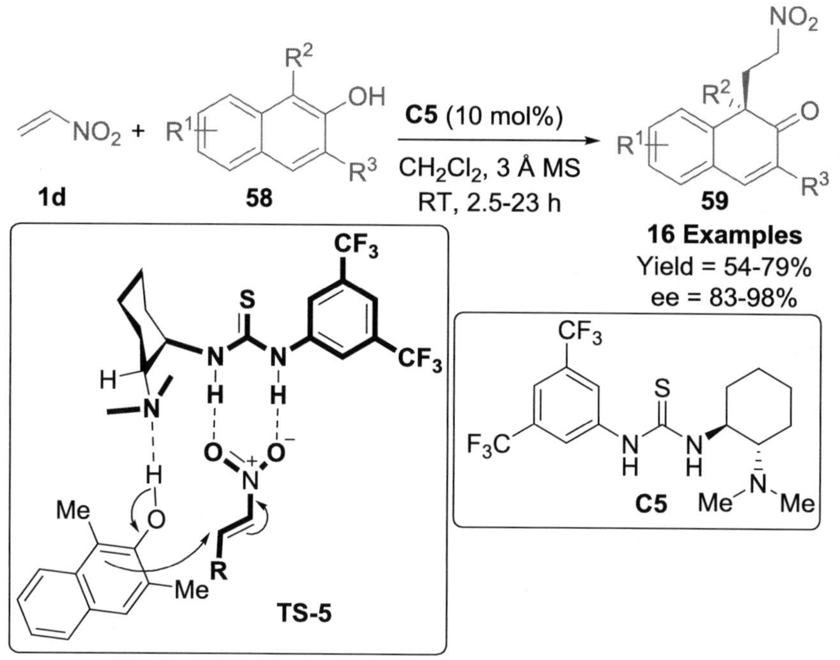

SCHEME 10.21 Chiral thiourea-catalyzed asymmetric synthesis of β-naphtholenones from β-naphthols and nitroethylene.

(Scheme 10.22).[55] Here, the stereochemistry of the product can be deduced from the chair conformation of the transition state **TS-6** (Scheme 10.22), where the phenyl group of the nitroalkene **1** occupies the equatorial position and directs the formation of new chiral centers. Thus, spirocyclohexanepyrazolone derivatives **63** with three tertiary and two quaternary stereocenters were obtained in this reaction.

Lin et al. reported an asymmetric organocatalytic vinylogous Michael/Henry cyclization cascade reaction to synthesize highly substituted tetrahydrofluoren-9-ones **67** as a single diastereomer by the reaction of 1,3-indandione-derived pronucleophiles **66** with nitroalkenes **1** in the

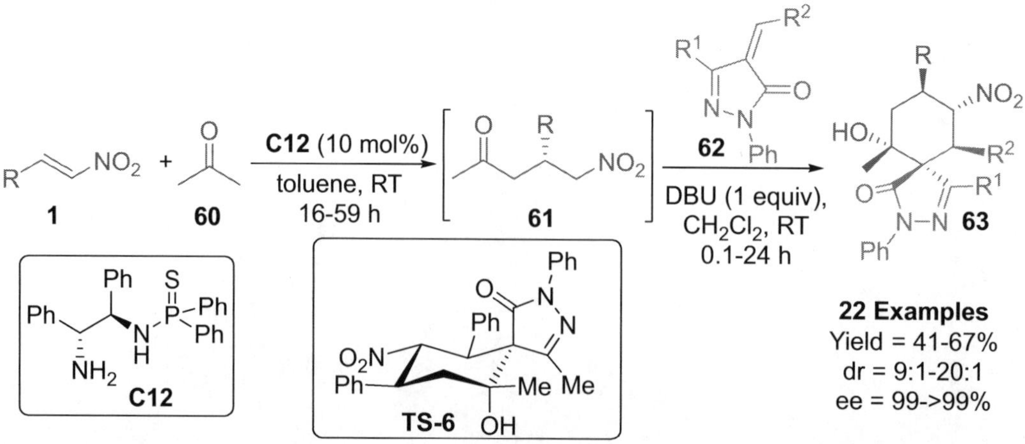

SCHEME 10.22 5-Nitropentan-2-ones as chiral building blocks for the stereoselective synthesis of spiro-pyrazolones.

presence of demethylated quinine-based squaramide catalyst **C6b** with good yields and excellent enantioselectivities (Scheme 10.23).[56] In this reaction, the planar conformation of the incoming nucleophile made it possible for the catalyst to induce a high degree of enantioselectivity. When the catalyst was changed to quinine squaramide **C6a**, only the formation of Michael adducts **65** was observed in good yields and enantioselectivity. The mechanism of the Henry cyclization of Michael adduct **65** to compound **67** is depicted in Scheme 10.24.

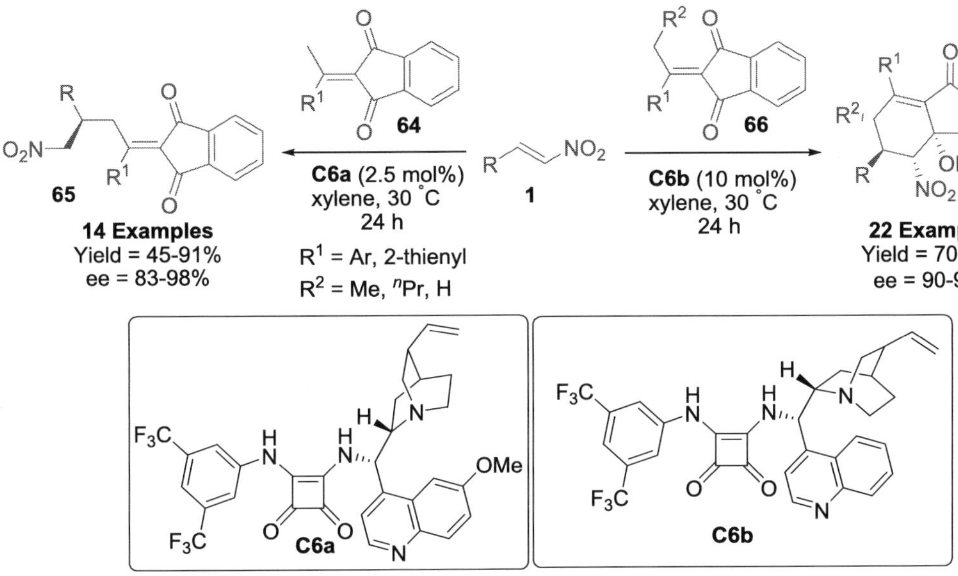

SCHEME 10.23 Asymmetric vinylogous Michael/Henry cyclization of 1,3-indandione-derived pronucleophiles with nitroalkenes.

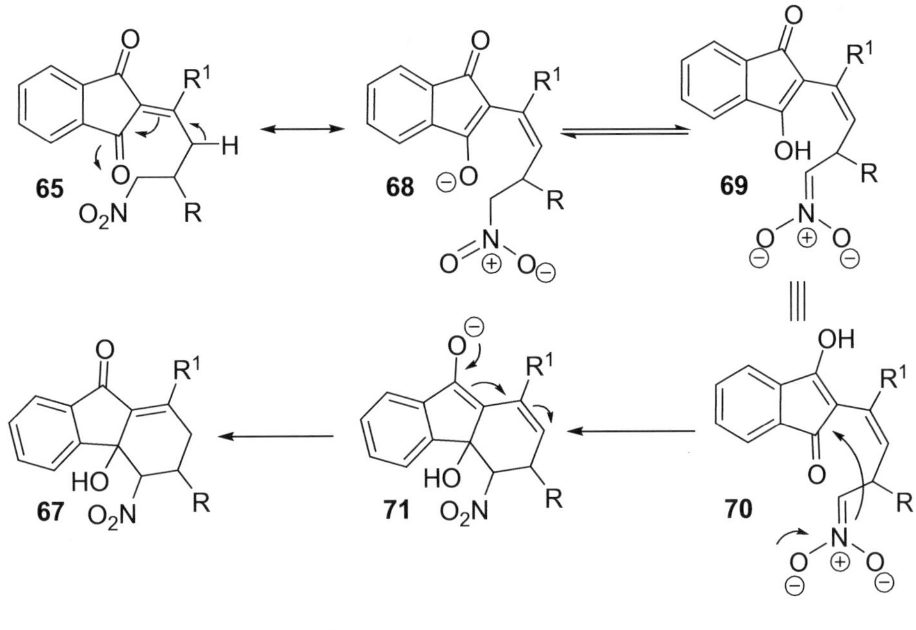

SCHEME 10.24 Proposed mechanism for the formation of tetrahydrofluoren-9-ones.

The Enders group described a one-pot Michael addition/Conia-ene reaction of nitroalkenes **1e** with 1,3-dicarbonyl compounds **34** in a sequential manner by employing a combination of organocatalyst, i.e., squaramide-based catalyst **C6b** and metal catalyst (indium triflate) to synthesize highly functionalized methylene indanes **72** and methyl indenes **73** in good yields and high enantioselectivities (Scheme 10.25).[57] The reaction is assumed to proceed through concerted fragmentation mechanism, where first, enolization of methylene indane **72** R^1 = alkyl) occurs followed by the fragmentation to form methyl indene **73** with the extrusion of ketene, presumably via enolization and a concerted fragmentation, as evidenced by high resolution mass spectrometry (HRMS).

The Enders group described the enantioselective domino Michael/Henry reaction using indoline-3-one **74** and *o*-formyl-β-nitroalkenes **1f** as substrates by employing quinine-derived amine-squaramide catalyst **C6b** to obtain indoline derivatives **75** with good yields and high enantioselectivities (Scheme 10.26).[58] In this reaction, nitroalkene **1f** was activated by the squaramide catalyst **C6b** via H-bonding interaction and, in the meantime, indolinone **74** gets activated by the tertiary N-atom of quinuclidine to form an enolate, which attacks the nitroalkene **1f** from the *Si*-face to form the final product **75** with four contiguous stereocenters (**TS-7** and **TS-8**, Scheme 10.26).

The Enders group developed an enantioselective Michael/Michael/aldol reaction of nitroalkenes **1**, 1,3-dicarbonyls **34a** and arylidene indanediones **77** in one pot to yield substituted spirocyclohexane indan-1,3-diones **78** with five contiguous stereocenters in good yields and enantioselectivities using cinchona-derived squaramide catalyst **C6b** (Scheme 10.27).[59]

Umemiya and co-workers reported a diphenylprolinol silyl ether **C3**-mediated domino Michael/Michael reaction of 2-(2-formylethyl)naphthalene-1,4-dione **79** and nitroalkene **1** (Scheme 10.28).[60] It was observed that the reaction afforded functionalized spirocycles **80** bearing four continuous stereocenters in moderate-to-good yields (59%–75%) along with outstanding diastereo and enantioselectivities (up to 86:7:7 dr and up to >99% ee). It was also mentioned by the authors that the obtained product could be transformed into a spiro[4.5]decan-1,6-dione derivative.

Enders and co-workers reported an organocatalytic, one-pot, four-component, asymmetric quadruple domino reaction of alcohol **81**, acrolein **14** and nitrochromene **1g** as substrates to provide tricyclic chromenes **82** (Scheme 10.29).[61] The reaction involved a sequence of oxa-Michael–Michael–Michael-aldol reactions to generate three contiguous chiral centers. The absolute configuration of the product was assigned as *S, S, S* by single-crystal X-ray analysis.

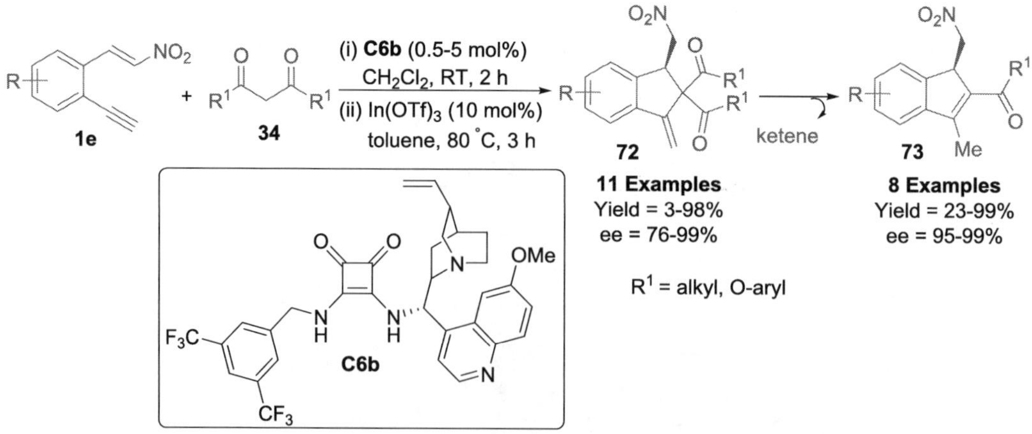

SCHEME 10.25 Asymmetric Michael addition/Conia-ene reaction of nitroalkenes with 1,3-dicarbonyl by employing a squaramide-metal catalyst.

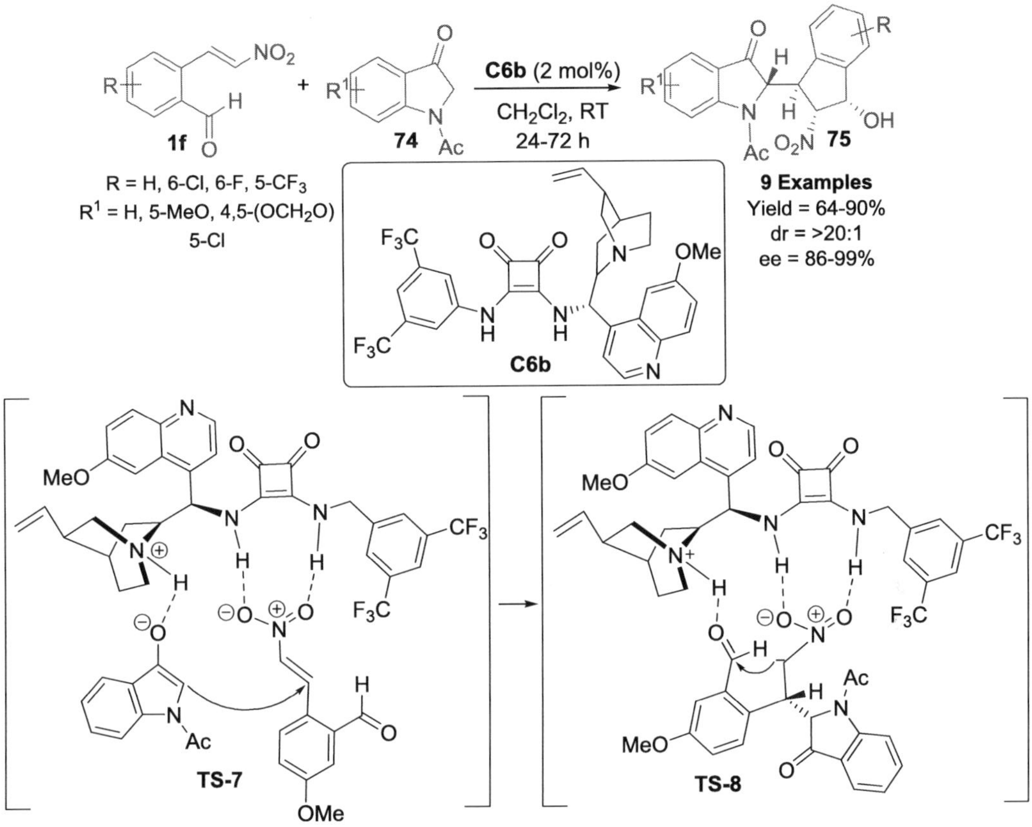

SCHEME 10.26 Asymmetric domino Michael/Henry reaction of indoline-3-ones and *o*-formyl-β-nitrostyrenes using quinine-derived amine-squaramide organocatalyst.

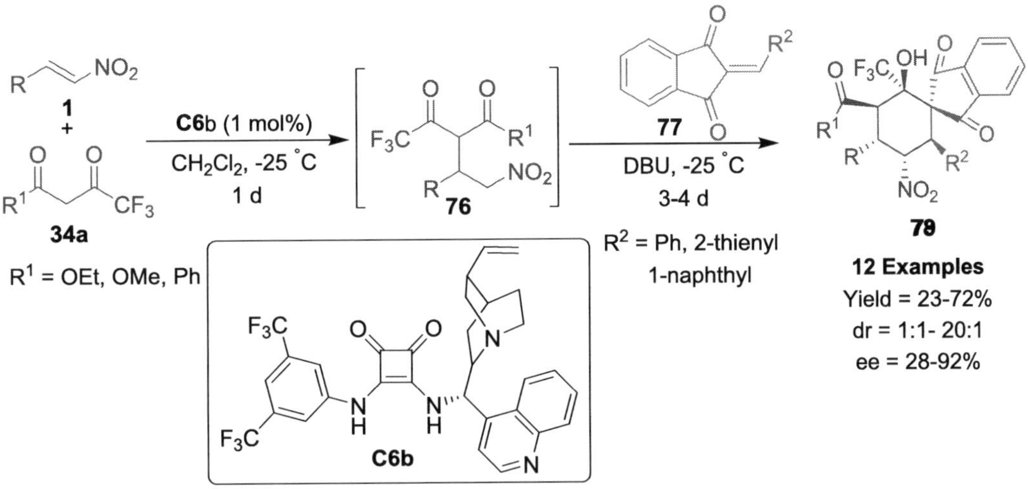

SCHEME 10.27 Cinchona-derived squaramide-catalyzed Michael/Michael/aldol reaction of nitroalkenes, 1,3-dicarbonyls and arylidene indanediones.

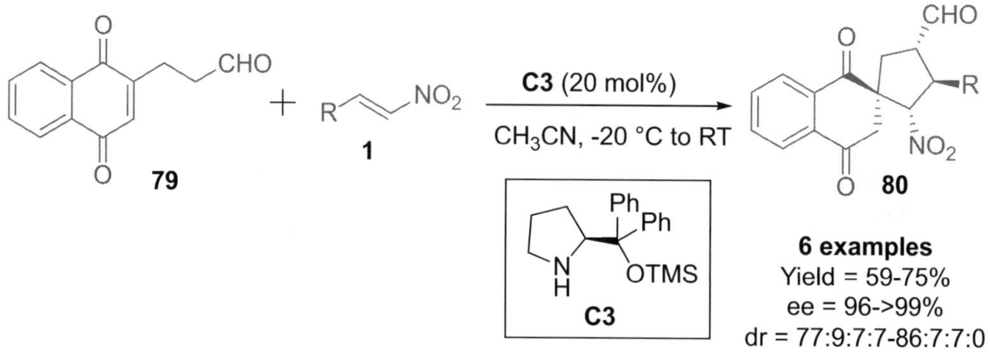

SCHEME 10.28 Domino Michael/Michael reaction for the synthesis of chiral spirocycles mediated by a diphenylprolinol silyl ether.

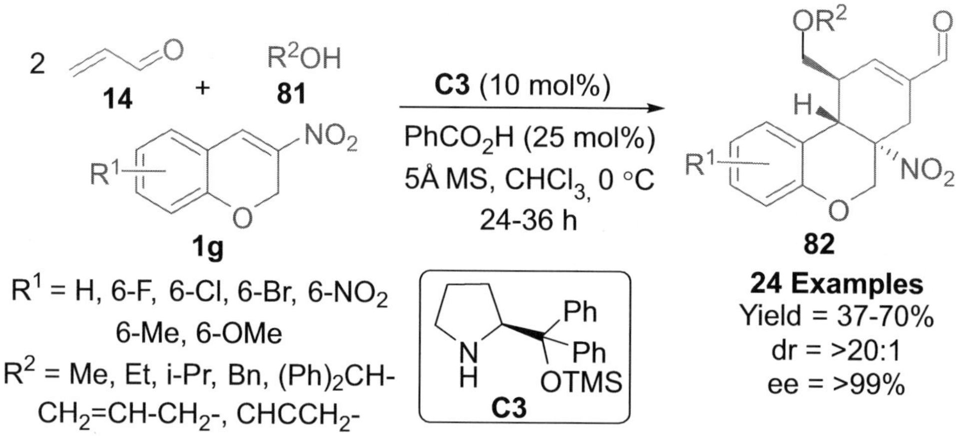

SCHEME 10.29 Asymmetric synthesis of tricyclic chromanes from nitrochromenes, acrylaldehyde and alcohols.

The Du group also reported an enantioselective Michael/Michael cascade reaction of 3-nitro-2H-chromenes **1g-i** with α,β-unsaturated ketones **83** to afford tetrahydro-6H-benzo[*c*]-chromenes **84** in high yields and enantioselectivities using quinine-derived amine catalyst **C12** and benzoic acid as an additive (Scheme 10.30).[62] In the proposed mechanism, the amine motif of the catalyst **C12** reacts with unsaturated ketone **83** to form the dienamine intermediate **85**, which then participates in the intermolecular Michael addition with 3-nitro-2H-chromene **1g** to form intermediate **86**. Subsequently, the intermediate **86** undergoes an intramolecular conjugate addition to form tricyclic chroman derivative **84**. Similar results were also obtained with nitrodihydronapthalenes **1h** and nitrothiochromenes **1i** (Scheme 10.30).

Zhao and co-workers reported a cinchona-derived squaramide catalyst **C13**-mediated three-component reaction of γ-aryl-substituted enals **87** with nitroalkenes **1k** (Scheme 10.31).[63] The reaction occurs via α,γ-dialkylation of **87** followed by intramolecular Henry reaction (**TS-9**). The corresponding products **88** bearing five stereogenic centers were formed in good yields and high enantioselectivities.

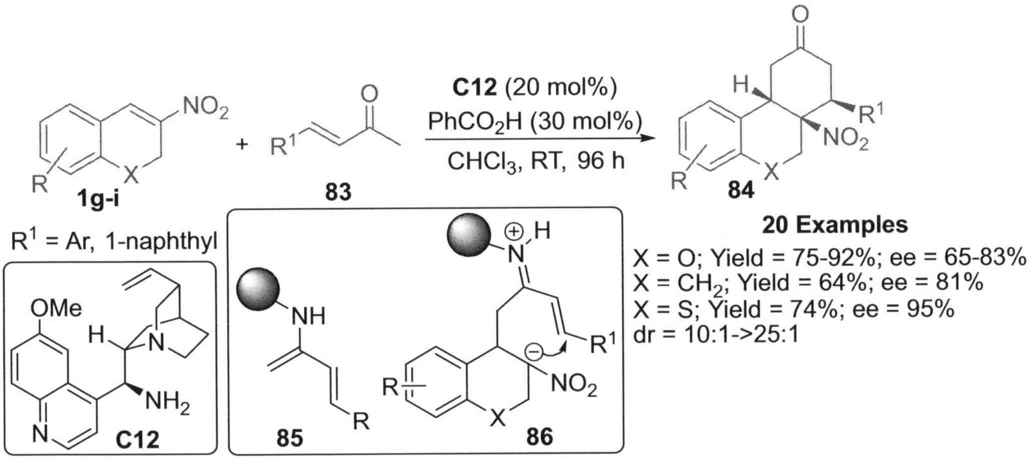

SCHEME 10.30 Quinine-derived amine catalyzed asymmetric Michael/Michael cascade reaction of 3-nitro-2H-chromenes with α,β-unsaturated ketones.

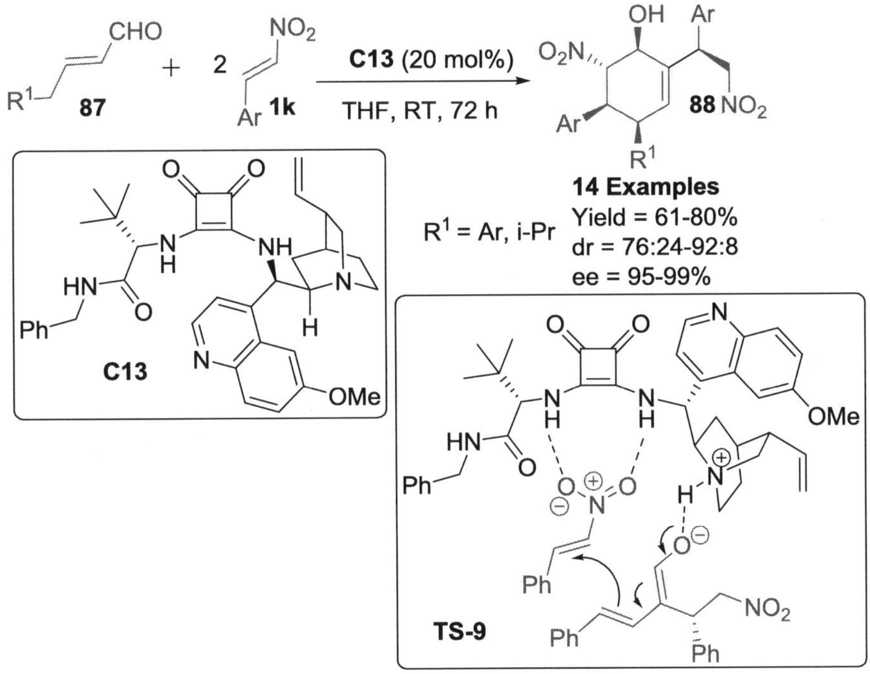

SCHEME 10.31 Three-component reaction mediated by a cinchona-derived squaramide catalyst.

10.3 CONCLUSIONS

Nitroalkenes are efficient substrates in the asymmetric cascade and multicomponent synthesis of functionalized cycloalkanes in the presence of various chiral catalysts. Under suitable conditions, nitroalkenes react as electrophile–nucleophile with diverse nucleophile–electrophile components and provide highly functionalized annulation products with high stereoselectivity. Suitably functionalized nitroalkenes also react as bielectrophiles with various binucleophiles leading to [3+2] and [3+3] annulation products.

REFERENCES

1. Marson, C. M. *Chem. Soc. Rev.* **2012**, *41*, 7712.
2. de Graaff, C.; Ruijter, E.; Orru, R. V. A. *Chem. Soc. Rev.* **2012**, *41*, 3969.
3. Volla, C. M. R.; Atodiresei, I.; Rueping, M. *Chem. Rev.* **2014**, *114*, 2390.
4. Hoashi, Y.; Yabuta, T.; Takemoto, Y. *Tetrahedron Lett.* **2004**, *45*, 9185.
5. Enders, D.; Hüttl, M. R. M.; Grondal, C.; Raabe, G. *Nature* **2006**, *441*, 861.
6. Enders, D.; Hüttl, M. R. M.; Runsink, J.; Raabe, G.; Wendt, B. *Angew. Chem. Int. Ed.* **2007**, *46*, 467.
7. Hayashi, Y.; Okano, T.; Aratake, S.; Hazelard, D. *Angew. Chem. Int. Ed.* **2007**, *46*, 4922.
8. Varela, M. C.; Dixon, S. M.; Lam, K. S.; Schore, N. E. *Tetrahedron* **2008**, *64*, 10087.
9. Cao, C.-L.; Zhou, Y.-Y.; Zhou, J.; Sun, X.-L.; Tang, Y.; Li, Y.-X.; Li, G.-Y.; Sun, J. *Chem. - Eur. J.* **2009**, *15*, 11384.
10. Ishikawa, H.; Suzuki, T.; Hayashi, Y. *Angew. Chem. Int. Ed.* **2009**, *48*, 1304.
11. Kotame, P.; Hong, B.-C.; Liao, J.-H. *Tetrahedron Lett.* **2009**, *50*, 704.
12. Uehara, H.; Barbas, C. F. *Angew. Chem. Int. Ed.* **2009**, *48*, 9848.
13. Wu, L.-Y.; Bencivenni, G.; Mancinelli, M.; Mazzanti, A.; Bartoli, G.; Melchiorre, P. *Angew. Chem. Int. Ed.* **2009**, *48*, 7196.
14. Baslé, O.; Raimondi, W.; Duque, M. M. S.; Bonne, D.; Constantieux, T.; Rodriguez, J. *Org. Lett.* **2010**, *12*, 5246.
15. Enders, D.; Schmid, B.; Erdmann, N.; Raabe, G. *Synthesis* **2010**, *2010*, 2271.
16. Enders, D.; Wang, C.; Mukanova, M.; Greb, A. *Chem. Commun.* **2010**, *46*, 2447.
17. Grondal, C.; Jeanty, M.; Enders, D. *Nat. Chem.* **2010**, *2*, 167.
18. Hong, B.-C.; Dange, N. S.; Hsu, C.-S.; Liao, J.-H. *Org. Lett.* **2010**, *12*, 4812.
19. Hong, B.-C.; Kotame, P.; Tsai, C.-W.; Liao, J.-H. *Org. Lett.* **2010**, *12*, 776.
20. Imashiro, R.; Uehara, H.; Barbas, C. F. *Org. Lett.* **2010**, *12*, 5250.
21. Ishikawa, H.; Suzuki, T.; Orita, H.; Uchimaru, T.; Hayashi, Y. *Chem. - Eur. J.* **2010**, *16*, 12616.
22. Urushima, T.; Sakamoto, D.; Ishikawa, H.; Hayashi, Y. *Org. Lett.* **2010**, *12*, 4588.
23. Chintala, P.; Ghosh, S. K.; Long, E.; Headley, A. D.; Ni, B. *Adv. Synth. Catal.* **2011**, *353*, 2905.
24. Hong, B.-C.; Kotame, P.; Liao, J.-H. *Org. Biomol. Chem.* **2011**, *9*, 382.
25. Hong, B.-C.; Nimje, R. Y.; Lin, C.-W.; Liao, J.-H. *Org. Lett.* **2011**, *13*, 1278.
26. Ishikawa, H.; Honma, M.; Hayashi, Y. *Angew. Chem. Int. Ed.* **2011**, *50*, 2824.
27. Ishikawa, H.; Sawano, S.; Yasui, Y.; Shibata, Y.; Hayashi, Y. *Angew. Chem. Int. Ed.* **2011**, *50*, 3774.
28. Rueping, M.; Haack, K.; Ieawsuwan, W.; Sunden, H.; Blanco, M.; Schoepke, F. R. *Chem. Commun.* **2011**, *47*, 3828.
29. Varga, S.; Jakab, G.; Drahos, L.; Holczbauer, T.; Czugler, M.; Soós, T. *Org. Lett.* **2011**, *13*, 5416.
30. Yu, D.-F.; Wang, Y.; Xu, P.-F. *Adv. Synth. Catal.* **2011**, *353*, 2960.
31. Cagide-Fagín, F.; Nieto-García, O.; Lago-Santomé, H.; Alonso, R. *J. Org. Chem.* **2012**, *77*, 11377.
32. Enders, D.; Urbanietz, G.; Cassens-Sasse, E.; Keeß, S.; Raabe, G. *Adv. Synth. Catal.* **2012**, *354*, 1481.
33. Mao, Z.; Jia, Y.; Xu, Z.; Wang, R. *Adv. Synth. Catal.* **2012**, *354*, 1401.
34. Rajkumar, S.; Shankland, K.; Brown, G. D.; Cobb, A. J. A. *Chem. Sci.* **2012**, *3*, 584.
35. Wong, C. T. *Tetrahedron* **2012**, *68*, 481.
36. Erdmann, N.; Philipps, A. R.; Atodiresei, I.; Enders, D. *Adv. Synth. Catal.* **2013**, *355*, 847.
37. Hong, B.-C.; Lan, D.-J.; Dange, N. S.; Lee, G.-H.; Liao, J.-H. *Eur. J. Org. Chem.* **2013**, *2013*, 2472.
38. Ding, D.; Zhao, C.-G.; Guo, Q.; Arman, H. *Tetrahedron* **2010**, *66*, 4423.
39. An, Q.; Shen, J.; Butt, N.; Liu, D.; Liu, Y.; Zhang, W. *Synthesis* **2013**, *2013*, 1612.
40. Enders, D.; Wang, C.; Bats, J. W. *Angew. Chem. Int. Ed.* **2008**, *47*, 7539.
41. Philipps, A. R.; Fritze, L.; Erdmann, N.; Enders, D. *Synthesis* **2015**, *2015*, 2377.
42. Jiang, H.; Jiang, Q.; Ge, C. *Tetrahedron Lett.* **2017**, *58*, 2355.
43. Huang, X.; Liu, M.; Jasinski, J. P.; Peng, B.; Zhang, W. *Adv. Synth. Catal.* **2017**, *359*, 1919.
44. Raimondi, W.; Sanchez Duque, M. M.; Goudedranche, S.; Quintard, A.; Constantieux, T.; Bugaut, X.; Bonne, D.; Rodriguez, J. *Synthesis* **2013**, *2013*, 1659.
45. Shi, D.; Xie, Y.; Zhou, H.; Xia, C.; Huang, H. *Angew. Chem. Int. Ed.* **2012**, *51*, 1248.
46. Chauhan, P.; Mahajan, S.; Raabe, G.; Enders, D. *Chem. Commun.* **2015**, *51*, 2270.
47. Enders, D.; Hahn, R.; Atodiresei, I. *Adv. Synth. Catal.* **2013**, *355*, 1126.
48. Chen, Y.; Liu, X.; Luo, W.; Lin, L.; Feng, X. *Synlett* **2017**, *2017*, 966.
49. Zeng, X.; Ni, Q.; Raabe, G.; Enders, D. *Angew. Chem. Int. Ed.* **2013**, *52*, 2977.
50. Xiao, W.; Yin, X.; Zhou, Z.; Du, W.; Chen, Y.-C. *Org. Lett.* **2016**, *18*, 116.
51. Ayyagari, N.; Namboothiri, I. N. N. *Tetrahedron: Asymmetry* **2012**, *23*, 605.

52. Raja, A.; Hong, B.-C.; Liao, J.-H.; Lee, G.-H. *Org. Lett.* **2016**, *18*, 1760.
53. Tsakos, M.; Elsegood, M. R. J.; Kokotos, C. G. *Chem. Commun.* **2013**, *49*, 2219.
54. Wang, S.-G.; Liu, X.-J.; Zhao, Q.-C.; Zheng, C.; Wang, S.-B.; You, S.-L. *Angew. Chem. Int. Ed.* **2015**, *54*, 14929.
55. Sun, J.; Jiang, C.; Zhou, Z. *Eur. J. Org. Chem.* **2016**, *2016*, 1165.
56. Möhlmann, L.; Chang, G.-H.; Reddy, G. M.; Lee, C.-J.; Lin, W. *Org. Lett.* **2016**, *18*, 688.
57. Philipps, A. R.; Blümel, M.; Dochain, S.; Hack, D.; Enders, D. *Synthesis* **2017**, *2017*, 1538.
58. Mahajan, S.; Chauhan, P.; Loh, C. C. J.; Uzungelis, S.; Raabe, G.; Enders, D. *Synthesis* **2015**, *2015*, 1024.
59. Blümel, M.; Chauhan, P.; Vermeeren, C.; Dreier, A.; Lehmann, C.; Enders, D. *Synthesis* **2015**, *2015*, 3618.
60. Hayashi, Y.; Nagai, K.; Umemiya, S. *Eur. J. Org. Chem.* **2019**, *2019*, 678.
61. Kumar, M.; Chauhan, P.; Bailey, S. J.; Jafari, E.; von Essen, C.; Rissanen, K.; Enders, D. *Org. Lett.* **2018**, *20*, 1232.
62. Li, J.-H.; Du, D.-M. *Org. Biomol. Chem.* **2015**, *13*, 9600.
63. Majee, D.; Jakkampudi, S.; Arman, H. D.; Zhao, J. C. G. *Org. Lett.* **2019**, *21*, 9166.

11 Catalytic Asymmetric Synthesis of Aryl-Fused Heterocycles via Cascade Reactions of Nitroalkenes

11.1 INTRODUCTION

Heterocycles fused to aromatic rings are attractive targets owing to their presence in numerous natural products and bioactive molecules. Stereoselective synthesis of aryl-fused heterocycles bearing multiple functionalities using nitroalkene as the key substrate in the presence of various chiral catalysts is discussed in this chapter. Diverse chiral organocatalysts have been employed in these cascade reactions of nitroalkenes with other substrates to achieve high yields and selectivity. Five- and six-membered O- and N-heterocycles fused to aromatic rings have been synthesized using various proline, cinchona and other 1,2-diamine and amino alcohol-based chiral catalysts, as described below.

11.2 SYNTHESIS OF ARYL-FUSED HETEROCYCLES

Du et al. reported an asymmetric synthesis of chromans starting from nitroalkenes **1a** and aryl thiols **2** using 1 mol% of squaramide catalyst **C1** (Scheme 11.1).[1] The reaction involves a sequential sulfa-Michael/retro sulfa-Michael/sulfa-Michael/Michael process to yield the product **3** with three contiguous stereocenters, including one quaternary stereocenter.

Inspired by the design of well-known PyBidine-metal systems and chiral NCN-pincer complex, Arai et al. designed halogen bond-mediated chiral bis-imidazolidine ligands **C2** facilitated by the soft

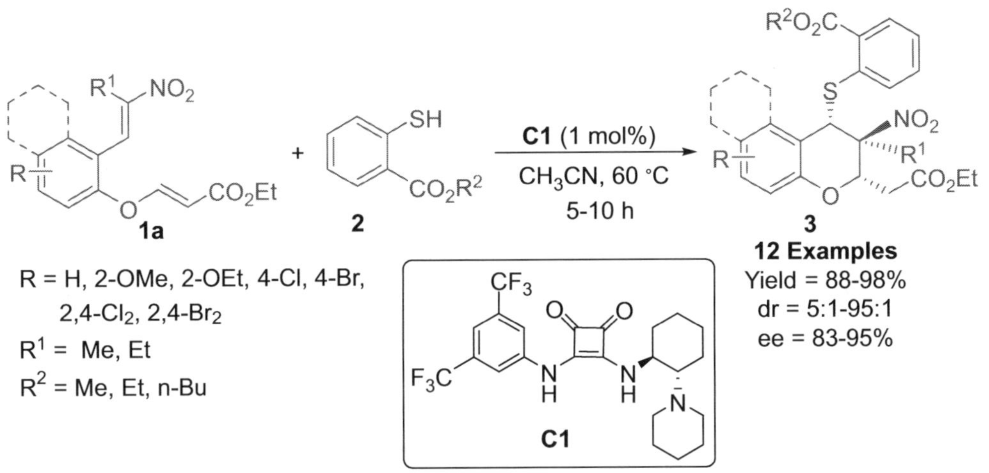

SCHEME 11.1 Squaramide-catalyzed asymmetric synthesis of chromans starting from nitroalkenes and aryl thiols.

Lewis acidity of the iodine. They described an asymmetric Michael/Henry reaction of nitroalkenes **1** with thiosalicyl aldehydes **4** using bis(imidazolidine)iodobenzene **C2** to afford enantiomerically pure thiochromanes **syn-5** with good yields and excellent enantioselectivities (Scheme 11.2).[2]

The Enders group reported an asymmetric synthesis of 3,4-*trans*-disubstituted chromans **7** using nitrovinylphenols **1b** and acrolein **6** as substrates in the presence of diphenylprolinol TMS ether **C3** as the catalyst (Scheme 11.3).[3] The catalyst **C3** activates acrolein **6** by generating iminium ion **8** (Scheme 11.4). Nitrovinyl phenol **1b** participates in oxa-Michel reaction with iminium ion **8** to form enamine **9**. Subsequent intramolecular Michael addition followed by catalyst release affords corresponding chromanes **7**. The reaction followed a domino oxa-Michael/Michael reaction, and the products were transformed to *N*-alkylated *trans*-benzopyrano-[3,4-c]-pyrrolidines by reductive amination followed by *N*-alkylation.

The Alemán group described an asymmetric synthesis of chiral *trans*-arenodihydrofurans **12** employing bromonitroalkenes **1c** and 2-naphthols **11** as substrates through a sequential Michael/Friedel–Crafts reaction followed by nucleophilic substitution on carbon-bearing bromine atom

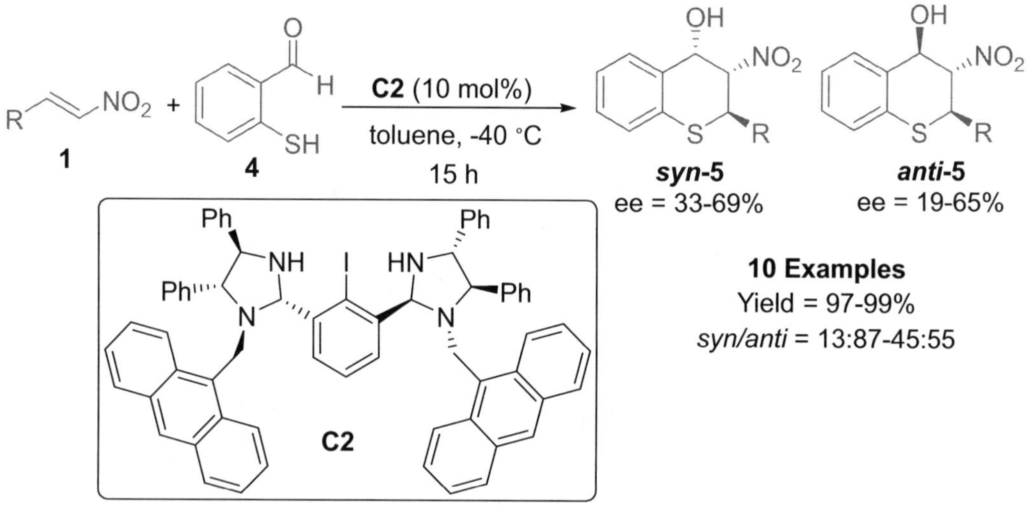

SCHEME 11.2 Asymmetric Michael/Henry reaction of nitroalkenes with thiosalicyl aldehydes using bis(imidazolidine)iodobenzene organocatalyst.

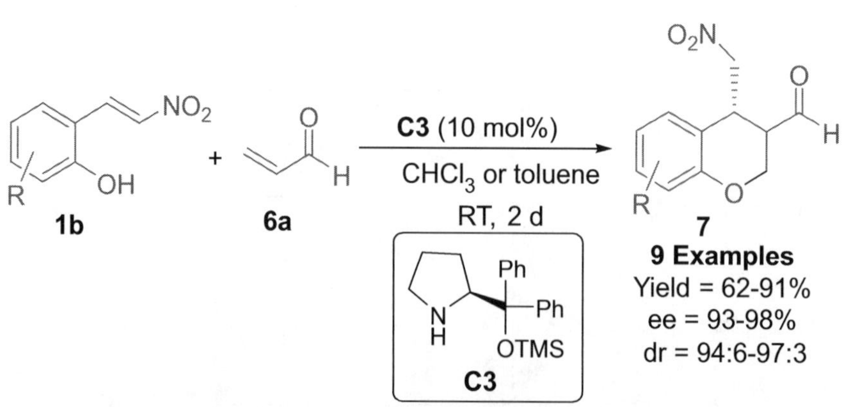

SCHEME 11.3 Diphenylprolinol TMS ether catalyzed asymmetric synthesis of 3,4-*trans*-disubstituted chromans from nitrovinylphenols and acrolein.

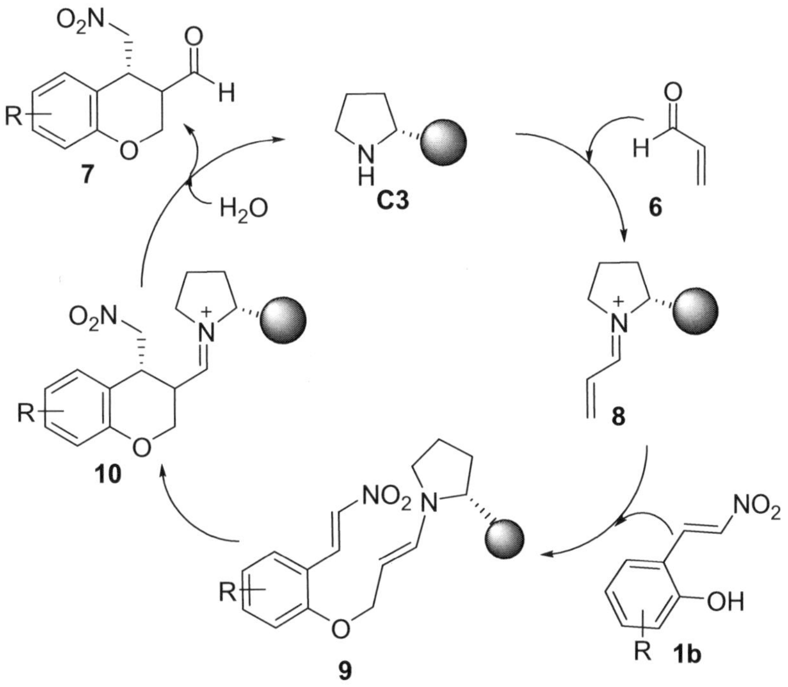

SCHEME 11.4 Proposed mechanism for the formation of 3,4-*trans*-disubstituted chromans from nitrovinylphenols and acrolein.

(Scheme 11.5).[4] The reaction was performed in the presence of squaramide catalyst **C4** and stoichiometric amount of base which was used to neutralize HBr generated during the reaction.

Zhao and co-workers reported a stereoselective domino Michael/hemiacetalization reaction of aliphatic aldehydes **13** and (2-nitrovinyl)phenols **1b** catalyzed by modularly designed organocatalysts (cinchona alkaloid **C5a** and (*L*)-proline (**C6**)), which was subsequently followed by oxidation or dehydroxylation (Scheme 11.6).[5] The corresponding chroman-2-ones **15** and chromanes derivatives **16** were obtained in remarkable yields (up to 97%), as well as excellent diastereo and enantioselectivities (up to 99:1 dr and up to 99% ee). Here, the authors mentioned that the

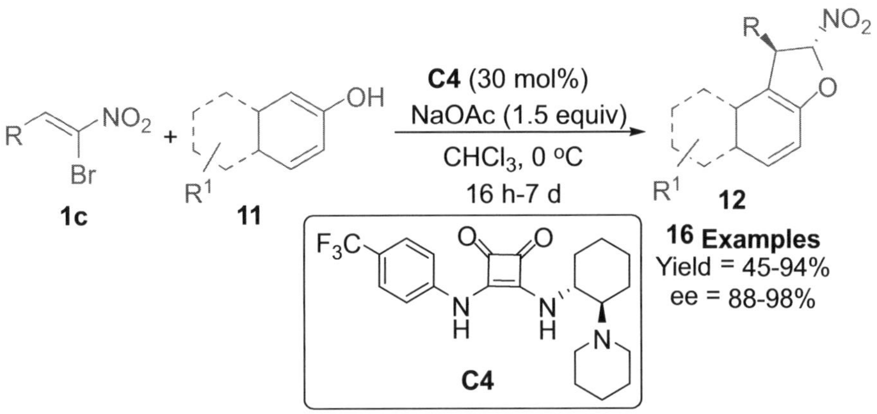

SCHEME 11.5 Squaramide-catalyzed asymmetric synthesis of *trans*-arenodihydrofurans by sequential Michael/Friedel–Crafts reaction of 2-naphthols with bromonitroalkenes.

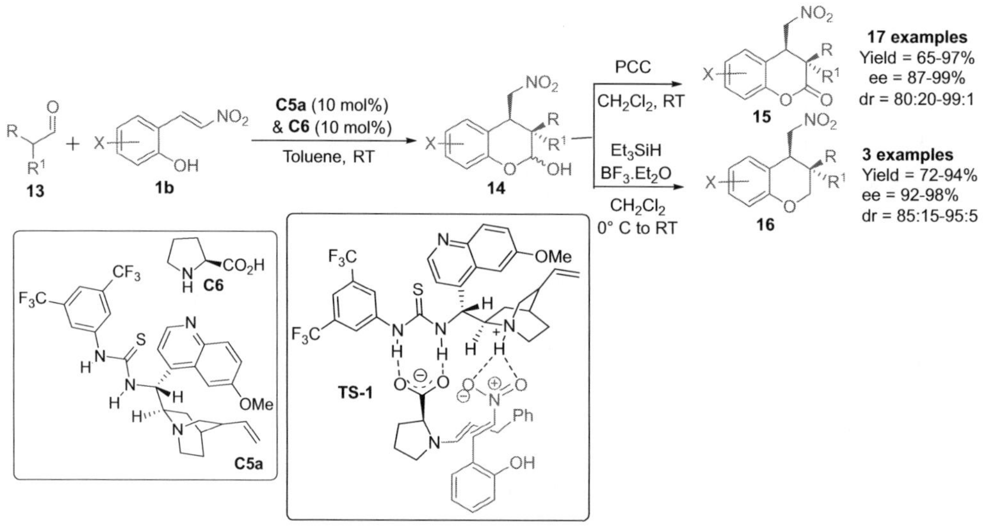

SCHEME 11.6 An asymmetric domino Michael/hemiacetalization reaction of aliphatic aldehydes and (2-nitrovinyl)phenols catalyzed by modularly designed organocatalyst.

(*E*)-enamine formed by the aldehyde **13** preferentially attacks the *Si*-face of the (nitrovinyl)phenol **1b** to form the Michael adduct with the expected stereochemistry, which further undergoes intramolecular hemiacetalization to form compound **14** (**TS-1**, Scheme 11.6). The products so obtained were further oxidized or dehydroxylated to furnish the final compounds.

Xu et al. reported a highly enantioselective oxa-Michael/Henry reaction of nitroalkenes **1** with diversely substituted salicylaldehydes **17** for the synthesis of 3-nitro-2H-chromenes **18** by employing chiral secondary amine-derived organocatalyst **C7** in the presence of salicylic acid as a co-catalyst (Scheme 11.7).[6] This reaction was assumed to proceed via an aromatic iminium

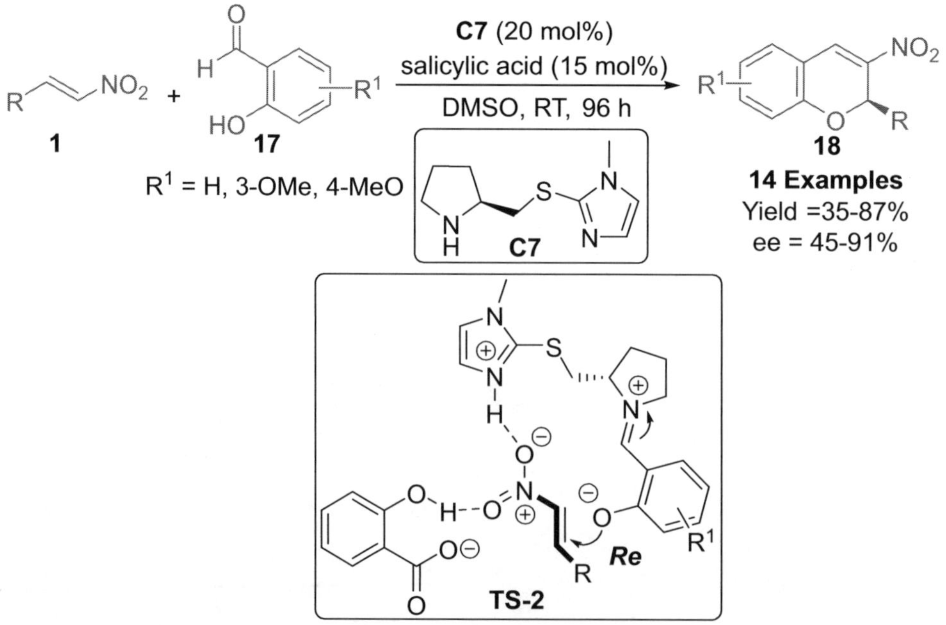

SCHEME 11.7 Asymmetric oxa-Michael/Henry reaction of nitroalkenes with salicylaldehydes.

activation strategy. The unfavorable steric interaction between the aryl group of salicylaldehyde **17** and the catalyst backbone creates the chiral environment needed for the oxa-Michael addition to happen from the *Re*-face of nitroalkene **1** resulting in the expected product **18** with *R*-configuration (**TS-2**, Scheme 11.7). Here, the catalyst **C7** plays a dual role where it assists the asymmetric iminium activation and, at the same time, acts as a Lewis base to participate in the deprotonation of salicylaldehyde **17** to promote Michael addition to nitroalkene **1** (Scheme 11.8). In addition, salicylic acid is presumed to play a dual role by favoring the formation of aromatic iminium and serves for H-bonding interaction with a nitro group with the help of its hydroxyl moiety (**TS-2**, Scheme 11.7).

Xu et al. reported a one-pot, three-component cascade reaction of enyne-tethered nitroalkenes **1d**, phenols **22a-e** and α,β-unsaturated ketones **23** employing bifunctional squaramide catalyst **C8a**.[7] The reaction involves a formal [3+3] annulation and nitro-Michael addition to give rise to highly functionalized 3-nitrochroman derivatives **24** with a quaternary carbon stereocenter (Scheme 11.9). The reaction course involves sequential trapping of in-situ generated nitro-activated allene and the vinylogous nitronate species to furnish the title compounds **24** (**TS-3** and **TS-4**, Scheme 11.9).

Zhao and co-workers reported the diastereodivergent synthesis of hexahydro-6H-benzo[*c*]-chromen-6-one derivatives **26** mediated by modularly designed organocatalysts where proline **C6** and cinchona derivative **C5b** self-assembled to catalyze the domino reaction of (*E*)-2-(2-nitrovinyl) phenols **1b** and substituted *trans*-7-oxo-5-heptenals **25** (Scheme 11.10).[8] It was observed that the corresponding products were obtained with remarkable diastereoselectivities (up to 98:2 dr) and excellent enantioselectivities (up to >99%). Here, the authors explained that the nitrostyrene was hydrogen-bonded to the thiourea functionality of the catalyst while the carbonyl functionality of the enals was hydrogen-bonded to the ammonium moiety of MDO during the transition state which leads to observed stereocontrol.

Xiao et al. reported an asymmetric intramolecular crossed Rauhut–Currier reaction of nitroalkenes **1a** with tethered enoates (Scheme 11.11).[9] The reaction involves a cooperative nucleophilic activation and H-bonding catalytic strategy to yield the corresponding products **27** with good stereoselectivities. The computational studies of this reaction revealed that stereoselectivity of the Rauhut–Currier reaction is determined by the intramolecular Michael addition step and it was found that the retro-aza-Michael addition is the rate-determining step.

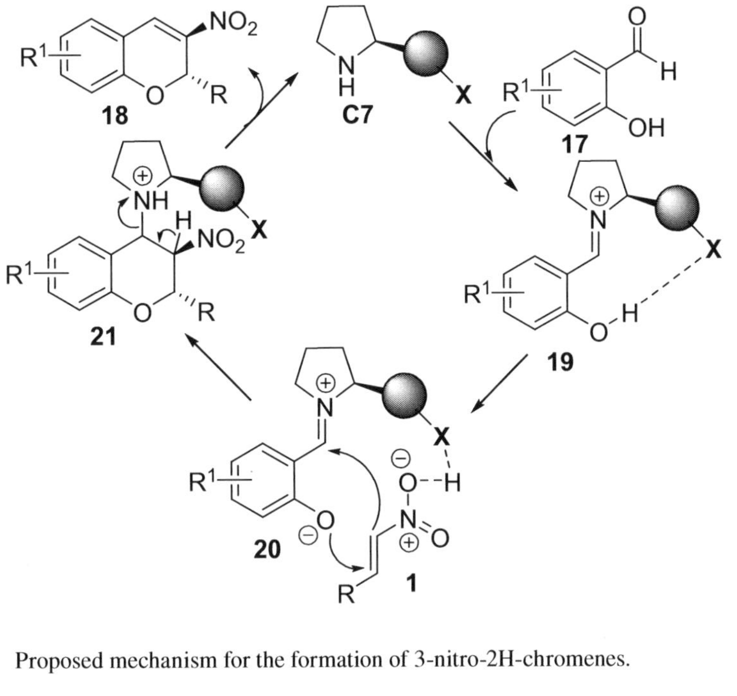

SCHEME 11.8 Proposed mechanism for the formation of 3-nitro-2H-chromenes.

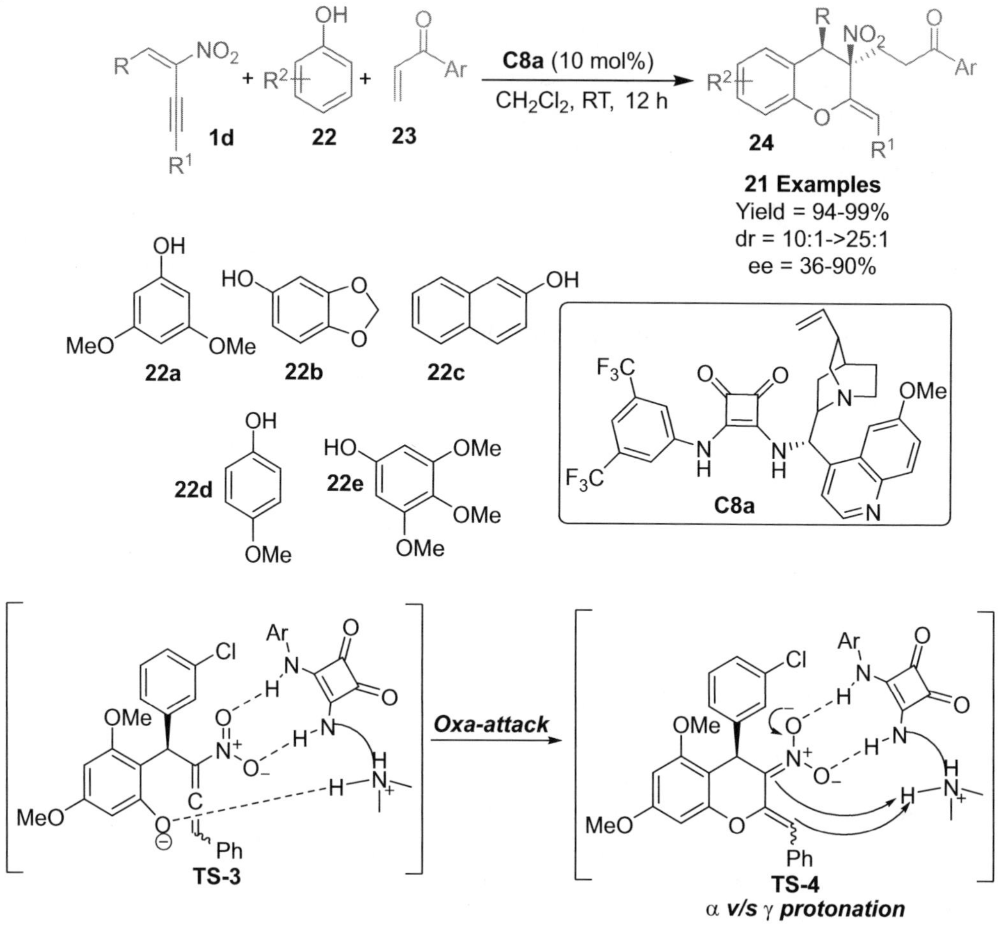

SCHEME 11.9 Bifunctional squaramide-catalyzed, three-component cascade reaction of enyne tethered nitroalkenes, phenols and α,β-unsaturated ketones.

Pedrosa and co-workers synthesized styryl-substituted thiourea catalysts and screened their catalytic activity toward the stereoselective synthesis of 2,3,4-trisubstituted benzopyrans **31** from salicylaldehyde-derived α-amidosulfones **30** and β-nitrostyrenes **1** (Scheme 11.12).[10] Thiourea catalyst **C10** synthesized from *(1R, 2R)*-cyclohexanediamine catalyzed the reaction with a high degree of stereoselection and the reaction proceeds via intramolecular oxa-Michael-aza-Henry involving a ternary complex (**TS-5,** Scheme 11.12). Nitroalkenes **1** were activated through hydrogen bonding with the catalyst **C10**, and the tertiary amine group present in it enhances the nucleophilicity of the α-amidosulfones via deprotonation. The *Re*-face nucleophilic attack on the β-position of the nitroalkene followed by cyclization at the *Re*-face of Boc-imine furnished respective benzopyran products **31**. Moreover, they have synthesized a co-polymer from thiourea **C10**, styrene and divinylbenzene, which acts as an efficient recoverable and reusable catalyst for the cascade reaction without compromising on the enantioselectivity and yield.

Chi and co-workers demonstrated N-heterocyclic carbene (NHC) **C11** and benzenesulfinate co-catalyzed intermolecular Rauhut–Currier reaction of enals **32** with nitrovinyl indoles **1e** to afford the corresponding azepinoindole products **33** with outstanding stereoselectivities (up to >99:1) (Scheme 11.13).[11] Here, the authors explained that the N-heterocyclic carbene catalyst **C11** activated the enal, whereas the nitrovinyl indole **1e** was activated by nucleophilic sulfinate co-catalyst. It was

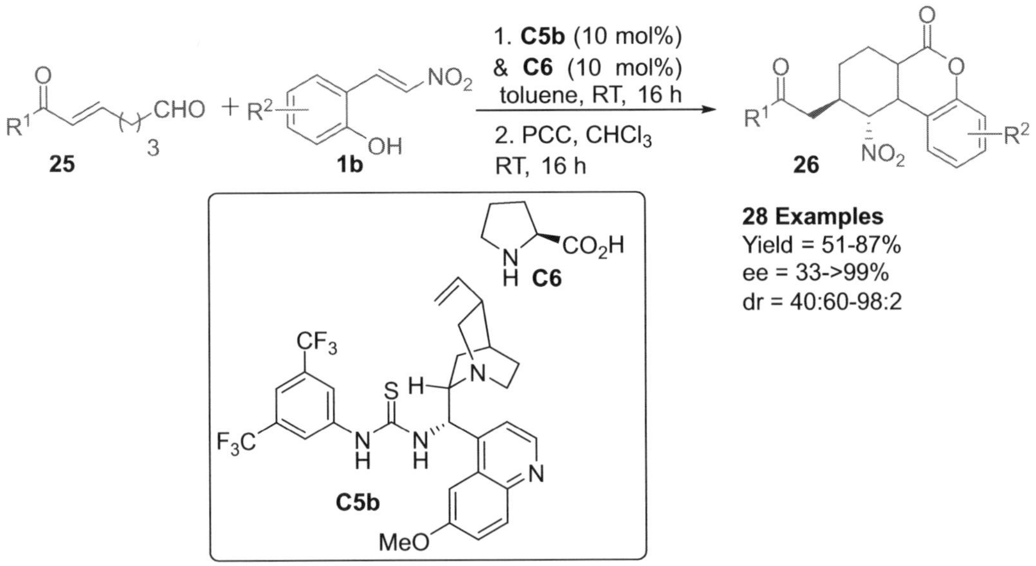

SCHEME 11.10 Stereoselective synthesis of hexahydro-6H-benzo[*c*]chromen-6-one derivatives mediated by modularly designed organocatalyst.

also mentioned that the report was a pioneering research demonstrating the unique involvement of benzenesulfinate as an effective catalyst for the enantioselective Rauhut–Currier reaction.

Kim et al. demonstrated an organocatalytic asymmetric domino *aza*-Michael/Michael reaction using *N*-tosylamino phenyl α,β-unsaturated ketones **34** and nitroalkenes **1** as precursors to synthesize 2,3,4-trisubstituted tetrahydroquinolines **35** in good yields and enantioselectivities (Scheme 11.14).[12] The tertiary amine of the catalyst deprotonates the tosylated amine generating an anion, which is stabilized by hydrogen bonding between carbonyl and thiourea groups. At the same time, nitroalkene **1** gets activated by the protonated amine of the catalyst **C12** (**TS-7**, Scheme 11.14).

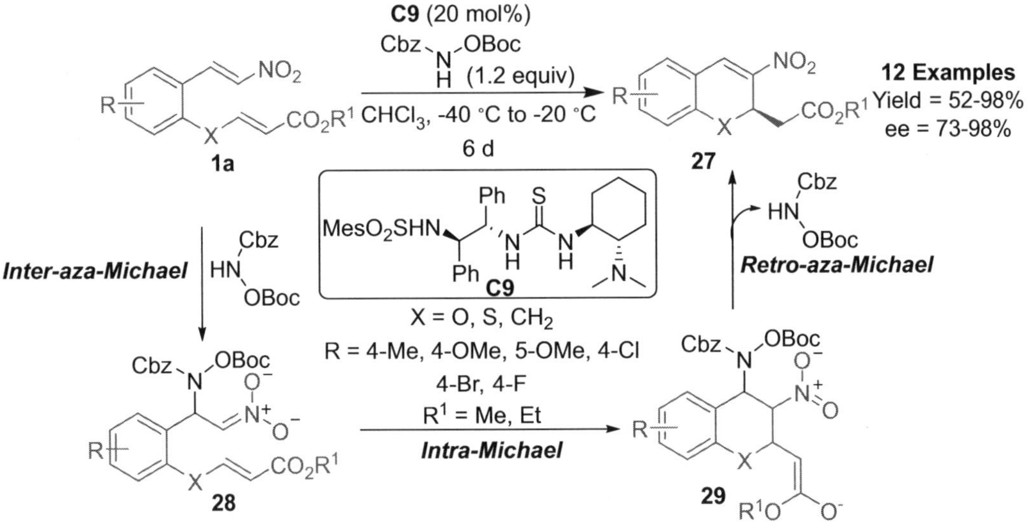

SCHEME 11.11 Enantioselective intramolecular crossed Rauhut–Currier reaction of nitroalkene tethered α,β-unsaturated esters.

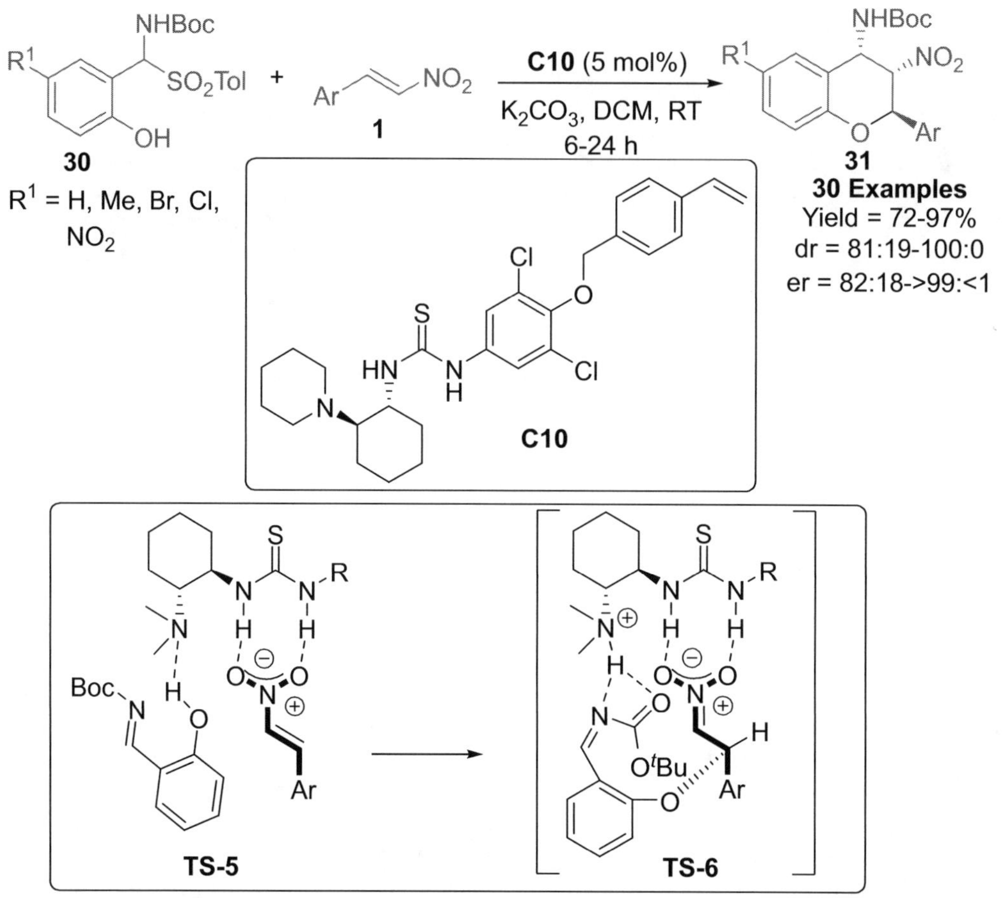

SCHEME 11.12 Enantioselective synthesis of 2,3,4-trisubstituted benzopyrans from salicylaldehyde-derived α-amidosulfones and β-nitrostyrenes.

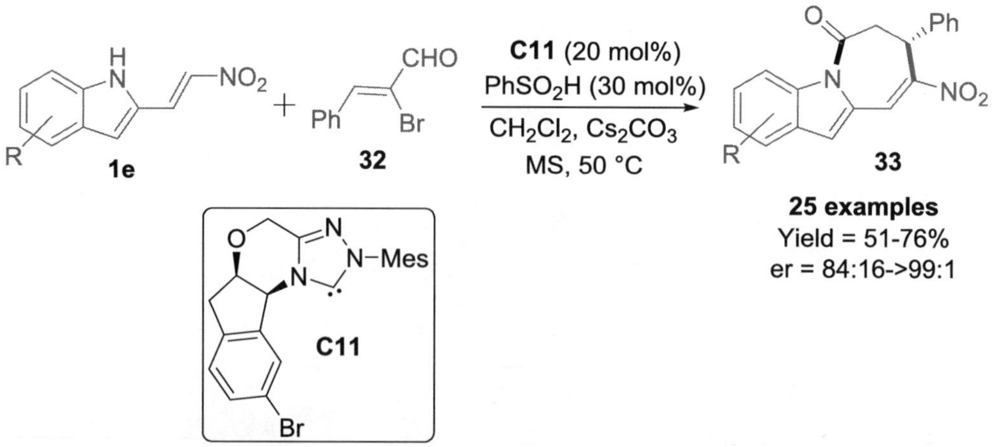

SCHEME 11.13 Rauhut–Currier reaction of enals with nitrovinyl indoles mediated by an N-heterocyclic carbene and sulfinate co-catalyst.

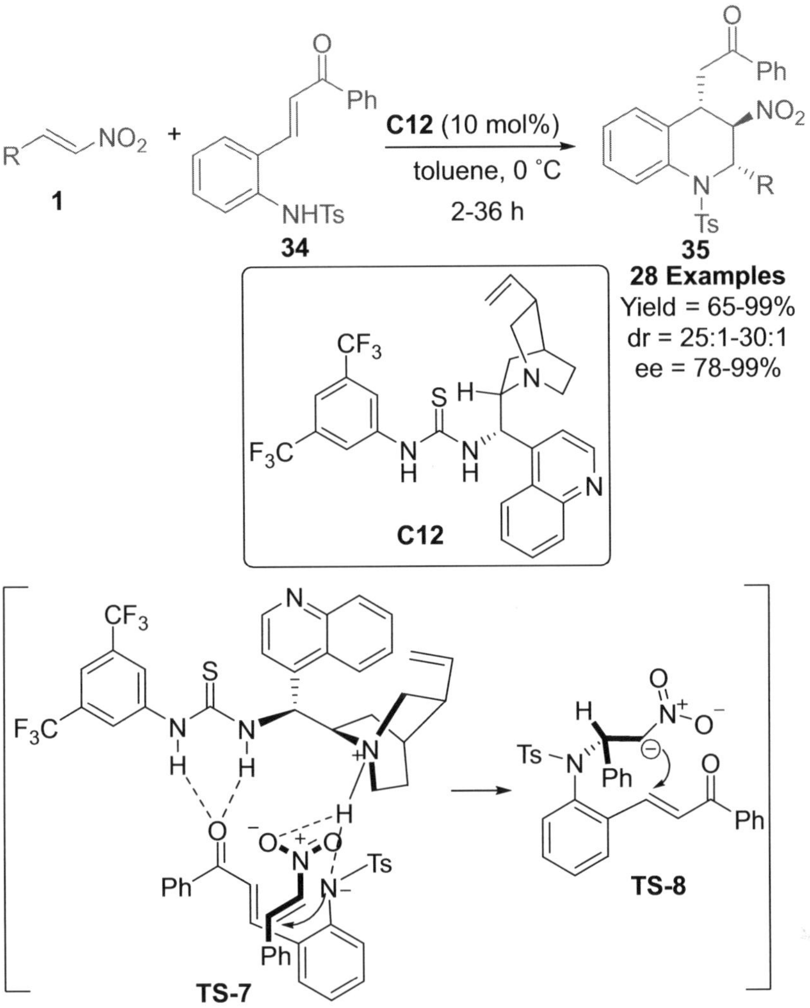

SCHEME 11.14 Enantioselective synthesis of 2,3,4-trisubstituted tetrahydroquinolines by *aza*-Michael/Michael reaction of N-tosylamino phenyl α,β-unsaturated ketones and nitroalkenes.

In the subsequent step, amine anion adds to nitroalkene **1** preferably from the *Si*-face yielding an intermediate with *S*-configuration followed by intramolecular Michael addition to give the final product **35** (**TS-7**, Scheme 11.14).

Wang et al. reported an enantioselective cascade reaction of chalcones **36** with nitroalkenes **1f** in the presence of thiourea catalyst **C5b** to afford optically pure 2-CF_3 tetrahydroquinoline derivatives **37** with three contiguous stereocenters in high yields and excellent stereoselectivities (Scheme 11.15).[13]

Hong et al. described an asymmetric synthesis of 1′, 3-spiro-2′-oxocyclohexan-3,4-dihydrocoumarins **39** with all-carbon stereocenters by employing 2-hydroxynitrostyrenes **1b** and 2-oxocyclohexane carbaldehyde **38** as starting materials.[14] The reaction was carried out using thiourea catalyst **C13** and involves a sequential domino Michael-acetalization reaction followed by oxidation to afford the products **39** in excellent yields and enantioselectivities (Scheme 11.16). The reaction involves a synchronous action of a Brønsted base and H-bonding mode of catalysis. At first, nitroalkene **1b** gets activated by the intermolecular hydrogen bonding of the thiourea moiety of the catalyst **C13**. Meanwhile, the tertiary amine of the catalyst **C13** acts as the Brønsted base, assists

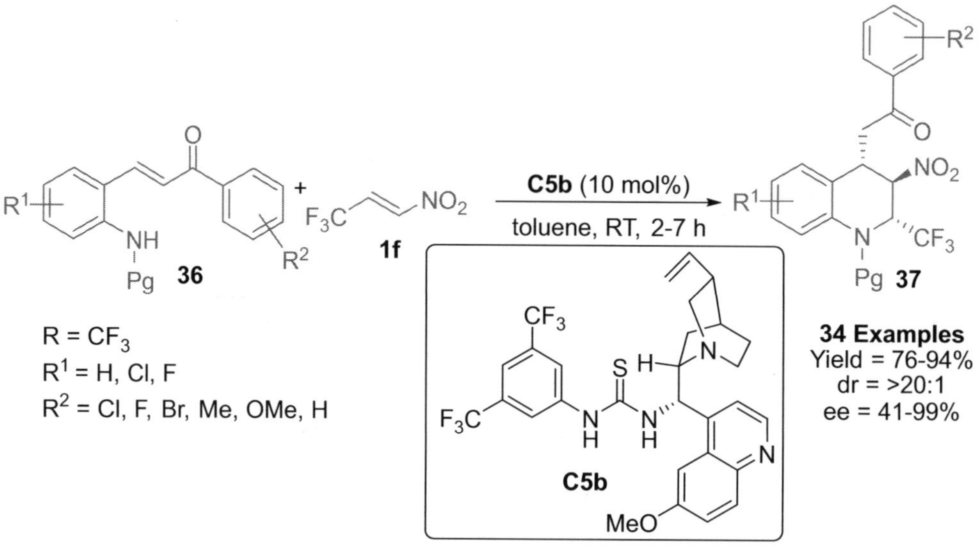

SCHEME 11.15 Enantioselective preparation of 2-CF_3 tetrahydroquinoline derivatives by cascade reaction of *o*-aminochalcones with nitroalkenes.

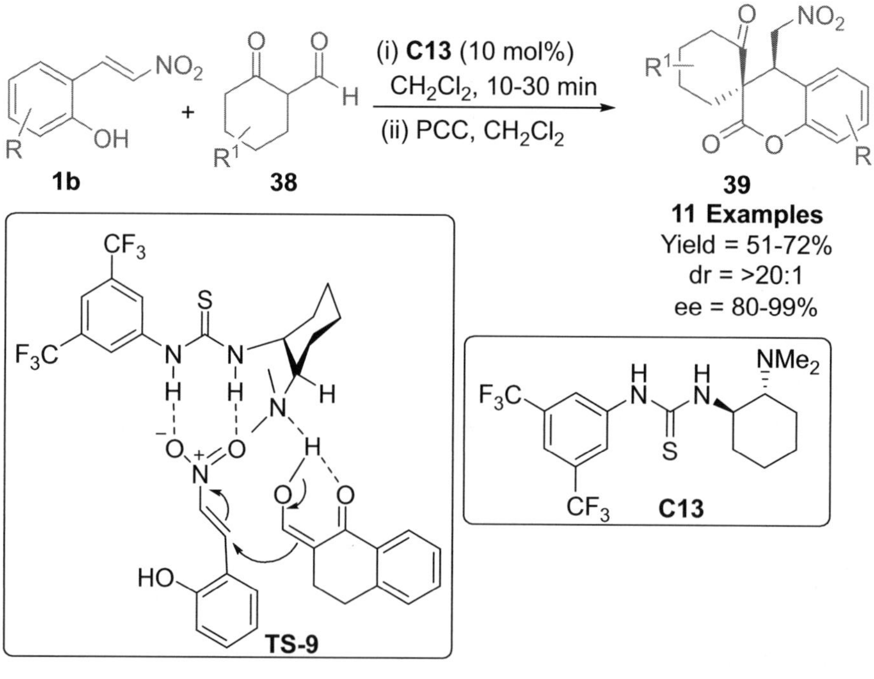

SCHEME 11.16 Bifunctional thiourea *tert*-amine-catalyzed Michael-acetalization reactions of 2-hydroxynitrostyrene and 2-oxocyclohexanecarbaldehyde.

in the enolization of the ketoaldehyde **38** and triggers the Michael addition to the nitroalkene **1b** from the *Re*-face (**TS-6**, Scheme 11.16). The resulting intermediate subsequently underwent acetal formation to give the Michael-acetal adduct.

Later, the same group reported a one-pot organocatalytic synthetic strategy toward biologically potent natural product scaffold aflatoxins **41** (Scheme 11.17).[15] By utilizing the four-step protocol

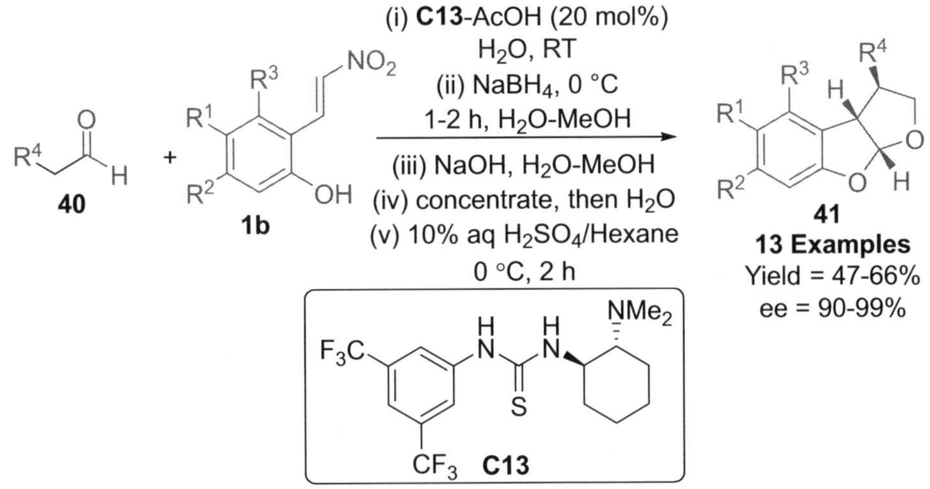

SCHEME 11.17 Asymmetric synthesis of functionalized aflatoxin from 2-hydroxynitrostyrenes and aldehydes.

involving Michael-acetalization of nitroalkenes **1b** and aldehyde **40** followed by reduction and Nef reaction delivered dihydroaflatoxin D_2 and aflatoxin B_2. They have also demonstrated the total synthesis of a natural product (+) and (–)-microminutinin using this protocol.

The Du group reported an enantioselective synthesis of 2-amino-4-(nitromethyl)-4H-chromene-3-carbonitrile derivatives **43** in high yields and enantioselectivities by the reaction of nitroalkenes **1b** with malononitrile **42** in the presence of a bifunctional thiourea catalyst **C5b** (Scheme 11.18).[16]

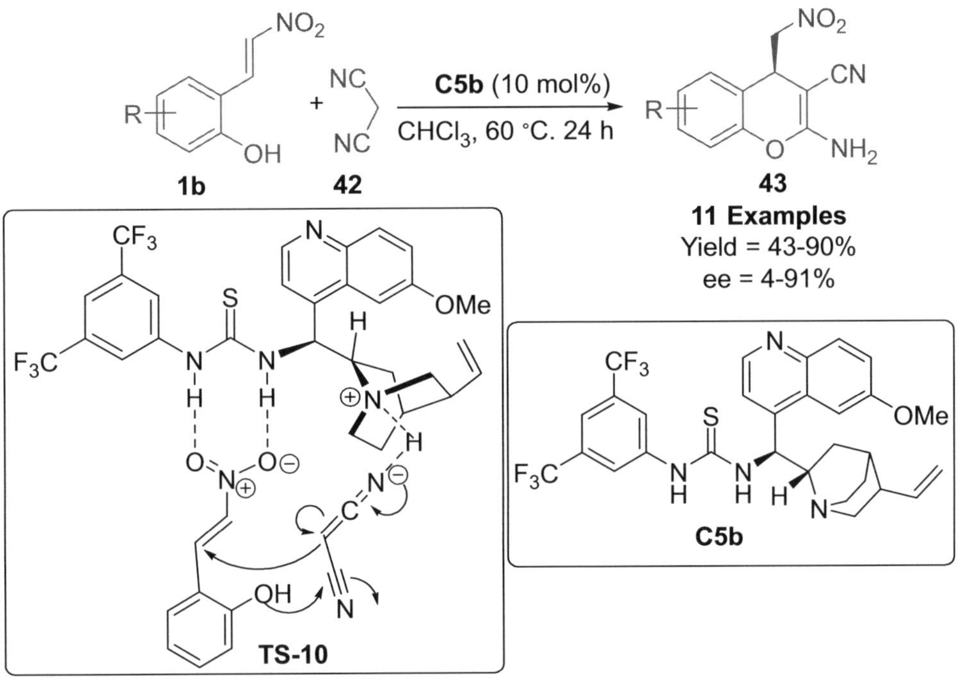

SCHEME 11.18 Enantioselective synthesis of chromene derivatives by bifunctional thiourea-catalyzed reaction of nitroalkenes with malononitrile.

In the proposed mechanism, nitroalkene **1b** gets activated by the catalyst **C5b** through H-bonding interaction and, meanwhile, the tertiary nitrogen atom of the catalyst **C5** deprotonates the malononitrile **42** to form a carbanion, which attacks the activated nitroalkene from the *Si*-face to afford the adduct with *R*-configuration. The resulting intermediate then undergoes cyclization to form the final product **43** (**TS-8**, Scheme 11.18).

Liu and Lu reported a cascade *aza*-Michael/Henry/dehydration reaction for the synthesis of 2-substituted-3-nitro-1,2-dihydroquinolines **45** starting from aldehyde-tethered sulfonamides **44** and nitroalkenes **1** in high yields and excellent enantioselectivities (Scheme 11.19).[17] The electron-withdrawing groups on the amino functionality of 2-aminobenzaldehyde **44** were presumed to increase the acidity of aniline N–H and abstraction of the same by the tertiary amine of the catalyst **C14** resulted in *aza*-Michael addition. Subsequently, Henry reaction occurs between the Michael adduct and the aldehyde followed by dehydration to give 3-nitro-1,2-quinoline **45** (**TS-11**, Scheme 11.19).

The Wang group described a squaramide **C15**-catalyzed cascade reaction of β-CF_3-nitroalkenes **1g** with 2-hydroxychalcones **46** to afford CF_3-containing heterocyclic structures **47/48** in high yields and enantioselectivities (Scheme 11.20).[18]

Wang et al. described an NHC-catalyzed cascade reaction of nitroalkenes **1b** with enals **6** for the synthesis of optically pure 3,4-dihydrocoumarins **49** via an enolate activation strategy (Scheme 11.21).[19] Here, the reaction involves the initial addition of catalyst **C16** to the enal **6** to form a Breslow intermediate **50** in the presence of a base, which then undergoes cycloaddition with nitroalkene **1b** and affords the product **49** (Scheme 11.22).

Xie et al. described an asymmetric tandem Michael–Henry reaction to synthesize chiral 2H-thiopyrano-[2,3-*b*]quinolines **53/54** by H-bonding dependent cooperative organocatalysis (Scheme 11.23).[20] In the possible mechanism, the tertiary amine moiety of the catalyst **C17/C5b** deprotonates the thiol and the resulting thiophenoxide adds to nitroalkene **1** in a Michael fashion. In the second step, the aldehyde is activated by the protonated tertiary amine, which undergoes Henry reaction with the nitronates stabilized by the thiourea moiety (**TS-12**, Scheme 11.23).

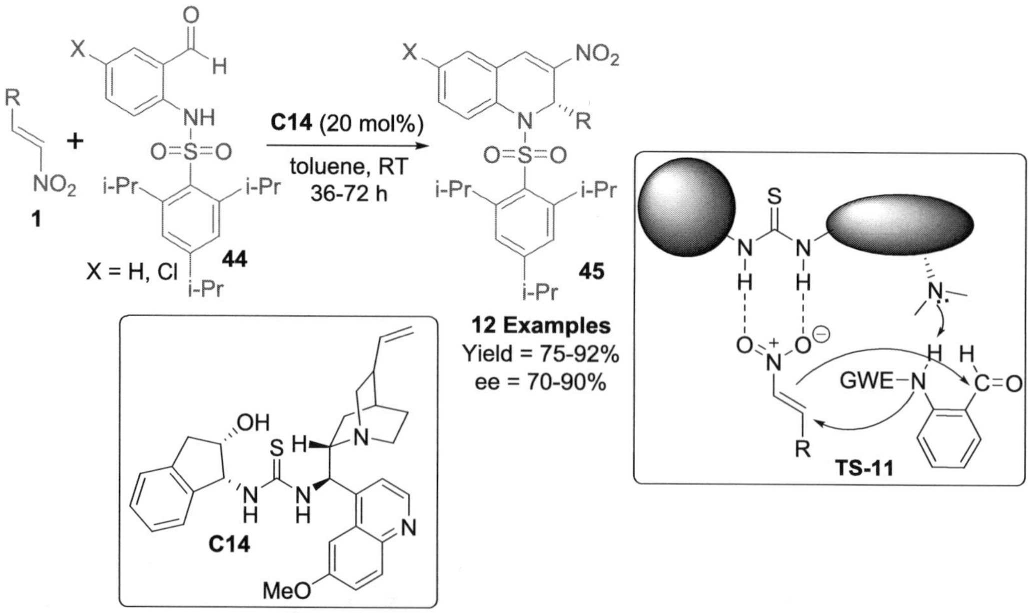

SCHEME 11.19 Asymmetric cascade *aza*-Michael/Henry/dehydration reaction of aldehyde-tethered sulfonamides and nitroalkenes for the preparation of 3-nitro-1,2-quinolines.

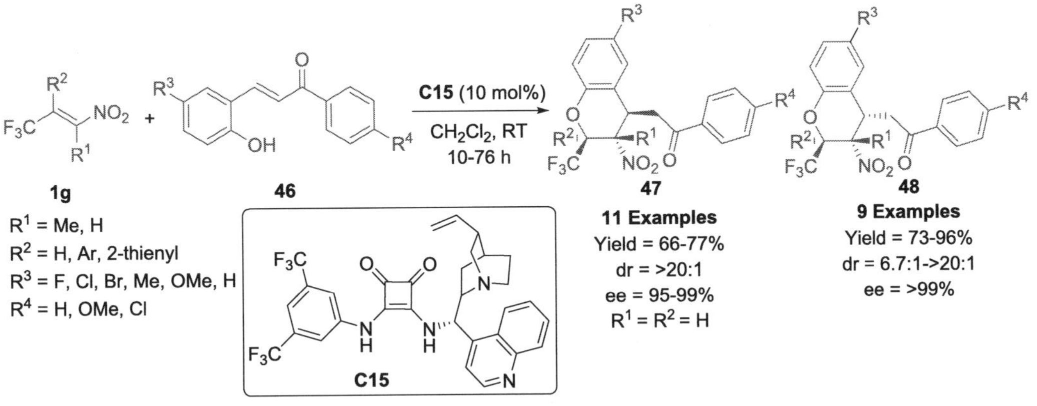

SCHEME 11.20 Synthesis of CF_3-containing chromanes by squaramide-catalyzed cascade reaction of β-CF_3-nitroalkenes with 2-hydroxychalcones.

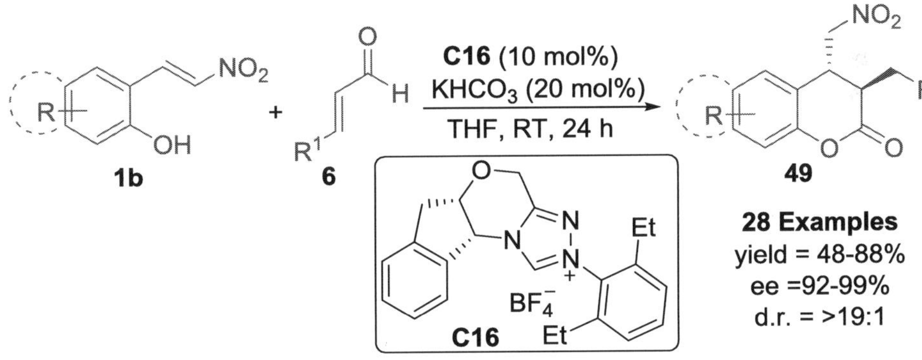

SCHEME 11.21 NHC-catalyzed cascade reaction of nitroalkenes with enals for the enantioselective synthesis of 3,4-dihydrocoumarins.

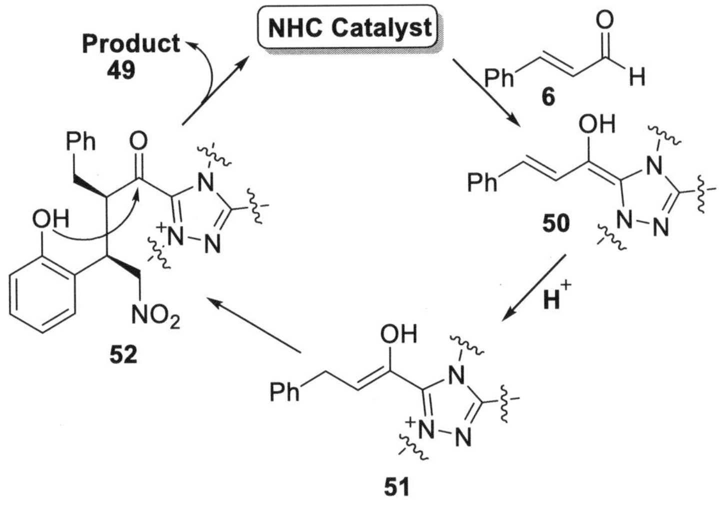

SCHEME 11.22 Proposed mechanism for the NHC-catalyzed enantioselective synthesis of 3,4-dihydrocoumarins.

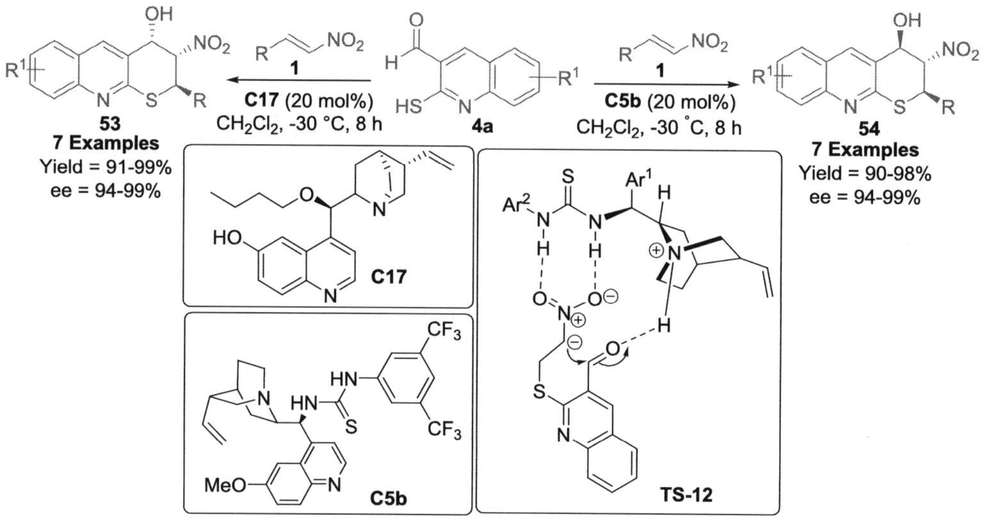

SCHEME 11.23 Asymmetric tandem Michael–Henry reaction of 2-mercaptoquinoline-3-carbaldehydes with nitroalkenes.

The Singh group reported a chiral bifunctional squaramide **C15** catalyzed asymmetric domino Michael-hydroalkoxylation reaction between nitroalkenes **1d** and N-arylpyrazolinones **55** for the synthesis of tetrahydropyrano[2,3-*c*]pyrazoles **56** in good yields and enantioselectivities (Scheme 11.24).[21]

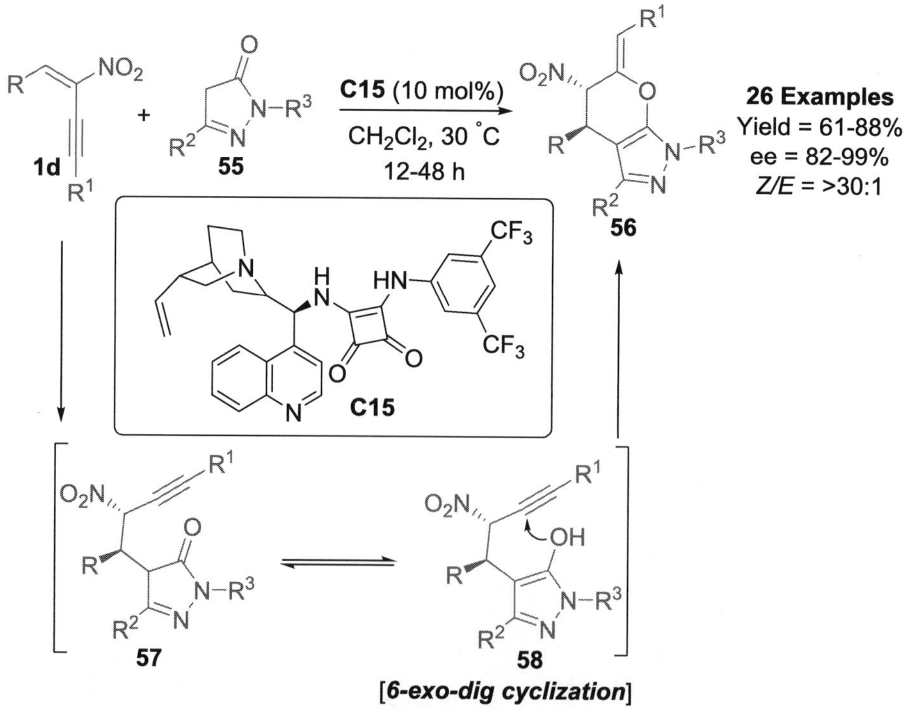

SCHEME 11.24 Asymmetric domino Michael-hydroalkoxylation reaction between trans-α-alkynyl-nitroalkenes and N-arylpyrazolinones.

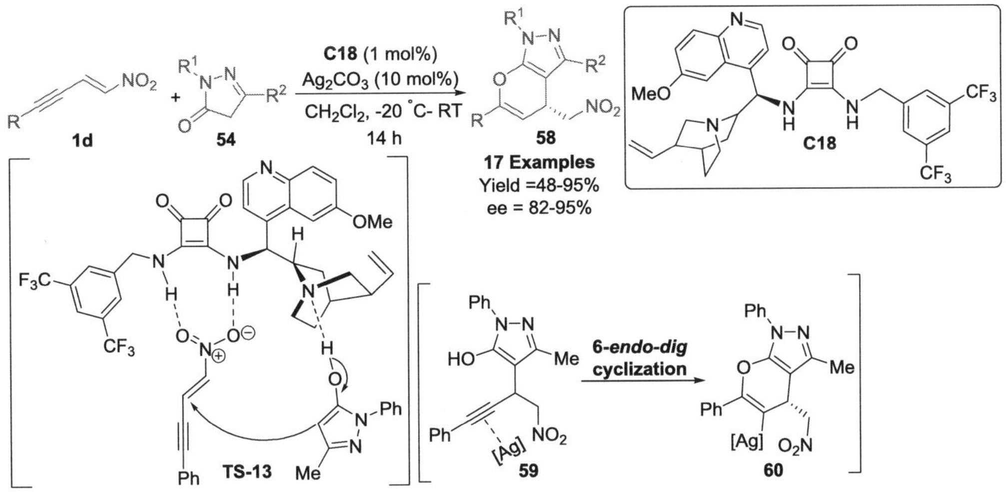

SCHEME 11.25 Asymmetric synthesis of pyrano-annulated pyrazoles by sequential Michael addition and 6-*endo-dig*-hydroalkoxylation of pyrazolinone and enyne-tethered nitroalkene.

The Enders group reported an enantioselective synthesis of chiral pyran-annulated pyrazoles **58** by the sequential Michael addition followed by a stereoselective 6-*endo-dig* hydroalkoxylation of pyrazolinone **55** and enyne-tethered nitroalkenes **1d**.[22] The reaction worked efficiently in the presence of squaramide-based organocatalyst **C18** along with a metal catalyst, Ag_2CO_3 (Scheme 11.25). In this reaction, the catalyst acts in a bifunctional mode, where it activates nitroalkene **1d** by H-bonding interactions between the nitro group and the squaramide. Meanwhile, pyrazolinone **55** gets activated by H-bonding interaction with the quinuclidine moiety of the catalyst **C18**. Simultaneously, the nucleophile will be directed to the *Si*-face of the nitroalkene **1d**. After the Michael addition, the silver-catalyzed electrophilic activation of the internal alkyne allows subsequent hydroalkoxylation. The 6-*endo-dig* cyclization occurs via stereoselective *anti*-addition of the enol to the alkyne to afford the corresponding *Z*-products. Thus formed vinyl silver intermediates **60** because of their instability undergo protodeargentation to give pyrano-annulated pyrazoles **58** (**TS-13**, Scheme 11.25).

Kang and Kim described the synthesis of highly functionalized tetrahydroquinolines **62** in good yields and enantioselectivities by the catalytic aza-Michael/Michael cascade reaction of nitroalkenes **1** with 2-(tosylamino)phenyl-α,β-unsaturated esters **61** by employing cinchona-derived thiourea catalyst **C5b** (Scheme 11.26).[23]

Deng et al. reported the synthesis of diverse polysubstituted chromanes **64** with excellent enantioselectivities in good yields by a stereoselective Michael/hemiacetalization cascade reaction of α-ketoesters **63** with 2-(2-nitrovinyl)phenols **1b**, employing Ni(OAc2)/diamine(**L1**) complex as catalyst (Scheme 11.27).[24] This synthetic strategy showed great potential for the production of chromane and chromanol derivatives with high enantioselectivity.

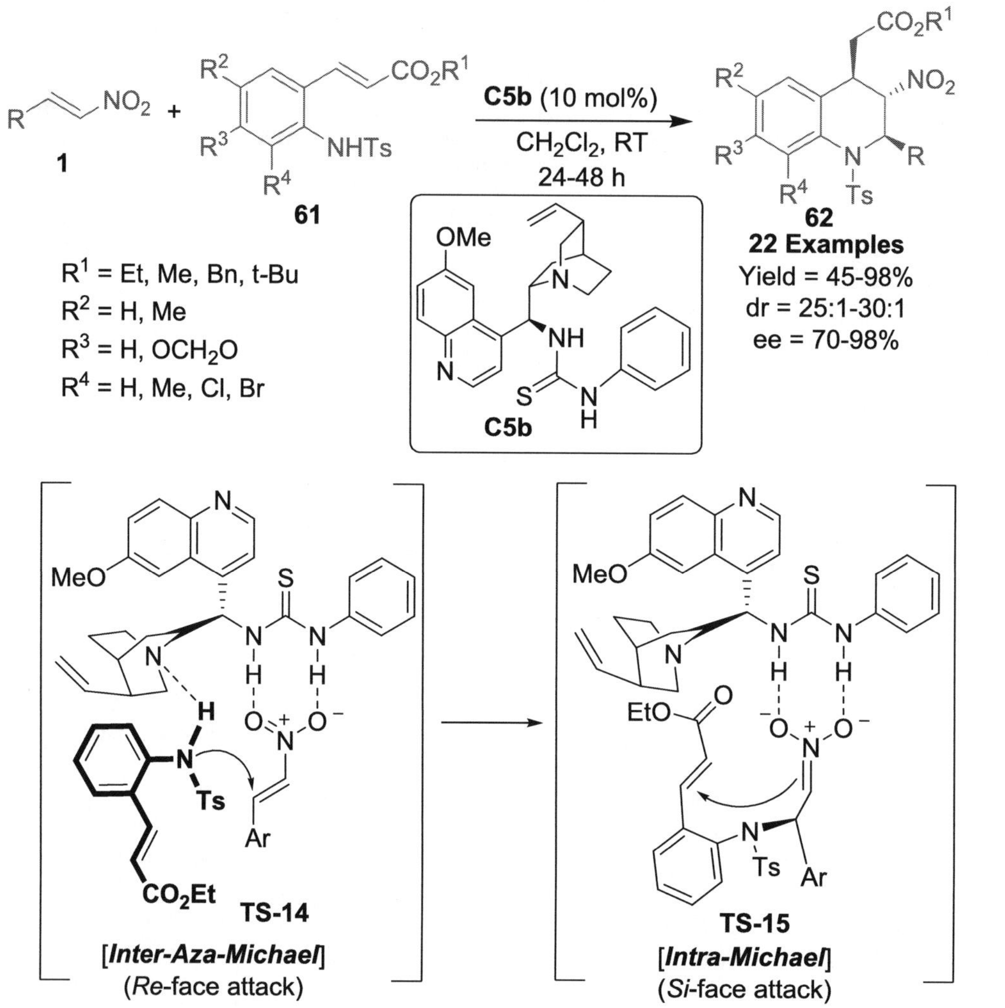

SCHEME 11.26 Asymmetric aza-Michael/Michael cascade reactions of N-protected 2-aminophenyl α,β-unsaturated esters with nitroalkenes.

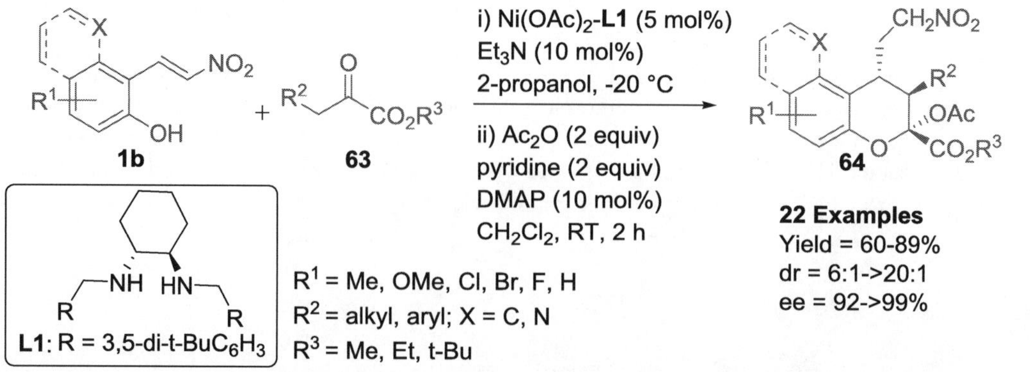

SCHEME 11.27 Stereoselective synthesis of polysubstituted chromanes from α-ketoesters and (2-mitrovinyl) phenols.

11.3 CONCLUSIONS

Synthesis of diverse five- and six-membered O- and N-heterocycles fused to aromatic ring in the presence of various chiral catalysts has been achieved in high yields and stereoselectivity. The synthesis involved multicomponent and cascade reactions of nitroalkenes with other components under mild conditions. The probable reaction pathways have been explained with the help of suitable transition state models.

REFERENCES

1. Yang, W.; Yang, Y.; Du, D.-M. *Org. Lett.* **2013**, *15*, 1190.
2. Arai, T.; Suzuki, T.; Inoue, T.; Kuwano, S. *Synlett* **2017**, *2017*, 122.
3. Wang, C.; Yang, X.; Raabe, G.; Enders, D. *Adv. Synth. Catal.* **2012**, *354*, 2629.
4. Jarava-Barrera, C.; Esteban, F.; Navarro-Ranninger, C.; Parra, A.; Aleman, J. *Chem. Commun.* **2013**, *49*, 2001.
5. Jakkampudi, S.; Parella, R.; Zhao, J. C. G. *Org. Biomol. Chem.* **2019**, *17*, 151.
6. Xu, D.-Q.; Wang, Y.-F.; Luo, S.-P.; Zhang, S.; Zhong, A.-G.; Chen, H.; Xu, Z.-Y. *Adv. Synth. Catal.* **2008**, *350*, 2610.
7. Xiao, Y.; Lin, J.-B.; Zhao, Y.-N.; Liu, J.-Y.; Xu, P.-F. *Org. Lett.* **2016**, *18*, 6276.
8. Jakkampudi, S.; Parella, R.; Arman, H. D.; Zhao, J. C. G. *Chem. - Eur. J.* **2019**, *25*, 7515.
9. Wang, X.-F.; Peng, L.; An, J.; Li, C.; Yang, Q.-Q.; Lu, L.-Q.; Gu, F.-L.; Xiao, W.-J. *Chem. - Eur. J.* **2011**, *17*, 6484.
10. Andres, J. M.; Maestro, A.; Valle, M.; Pedrosa, R. *J. Org. Chem.* **2018**, *83*, 5546.
11. Wu, X.; Zhou, L.; Maiti, R.; Mou, C.; Pan, L.; Chi, Y. R. *Angew. Chem. Int. Ed.* **2019**, *58*, 477.
12. Kim, S.; Kang, K.-T.; Kim, S.-G. *Tetrahedron* **2014**, *70*, 5114.
13. Zhu, Y.; Li, B.; Wang, C.; Dong, Z.; Zhong, X.; Wang, K.; Yan, W.; Wang, R. *Org. Biomol. Chem.* **2017**, *53*, 4544.
14. Hong, B.-C.; Kotame, P.; Lee, G.-H. *Org. Lett.* **2011**, *13*, 5758.
15. Huang, W.-L.; Raja, A.; Hong, B.-C.; Lee, G.-H. *Org. Lett.* **2017**, *19*, 3494.
16. Gao, Y.; Yang, W.; Du, D.-M. *Tetrahedron: Asymmetry* **2012**, *23*, 339.
17. Liu, X.; Lu, Y. *Org. Biomol. Chem.* **2010**, *8*, 4063.
18. Zhu, Y.; Li, X.; Chen, Q.; Su, J.; Jia, F.; Qiu, S.; Ma, M.; Sun, Q.; Yan, W.; Wang, K.; Wang, R. *Org. Lett.* **2015**, *17*, 3826.
19. Wu, Z.; Wang, X.; Li, F.; Wu, J.; Wang, J. *Org. Lett.* **2015**, *17*, 3588.
20. Ping, X.-N.; Wei, P.-S.; Zhu, X.-Q.; Xie, J.-W. *J. Org. Chem.* **2017**, *82*, 2205.
21. Rana, N. K.; Jha, R. K.; Joshi, H.; Singh, V. K. *Tetrahedron Lett.* **2017**, *58*, 2135.
22. Hack, D.; Chauhan, P.; Deckers, K.; Mizutani, Y.; Raabe, G.; Enders, D. *Chem. Commun.* **2015**, *51*, 2266.
23. Kang, K.-T.; Kim, S.-G. *Synthesis* **2014**, *2014*, 3365.
24. Chen, L.; Yang, W.-L.; Shen, J.-H.; Deng, W.-P. *Adv. Synth. Catal.* **2019**, *361*, 4611.

12 Catalytic Asymmetric Synthesis of Five- and Six-Membered Heterocycles via Cascade Reactions of Nitroalkenes

12.1 INTRODUCTION

Heterocycles bearing multiple functionalities and fused and spiro-linked to other rings are attractive synthetic targets owing to their presence in drugs and pharmaceuticals, as well as bioactive natural products. In this chapter, we focus on the synthesis of chiral five- and six-membered heterocycles, including nitrogen heterocycles, oxygen heterocycles, spirocycles and some miscellaneous reactions.

12.2 SYNTHESIS OF FIVE AND SIX-MEMBERED NITROGEN HETEROCYCLES

Kanger et al. developed an efficient synthetic route to enantiomerically enriched 2,3,4-trisubstituted piperidines **3a-b** by reacting nitroalkenes **1** and Bn-protected 5-aminopentanoate **2** in the presence of quinine-derived thiourea catalyst **C1a** (Scheme 12.1).[1] The method is based on organocatalytic *aza*-Michael–Michael cascade reaction. The activation of the first Michael acceptor via amino catalysis or H-bond catalysis triggers the following cascade process. The electronic effect of the substituents at the aromatic rings has little influence on this cascade reaction.

The Enders group reported an enantioselective Michael/*aza*-Henry [3+2] cycloaddition of trifluoromethyl-substituted iminomalonates **4** and nitroalkenes **1** in the presence of quinine-derived squaramide **C2a** as an organocatalyst for the preparation of pyrrolidine derivatives **5** (Scheme 12.2).[2] The absolute configuration of the product was assigned as *3R, 4S, 5R* based on

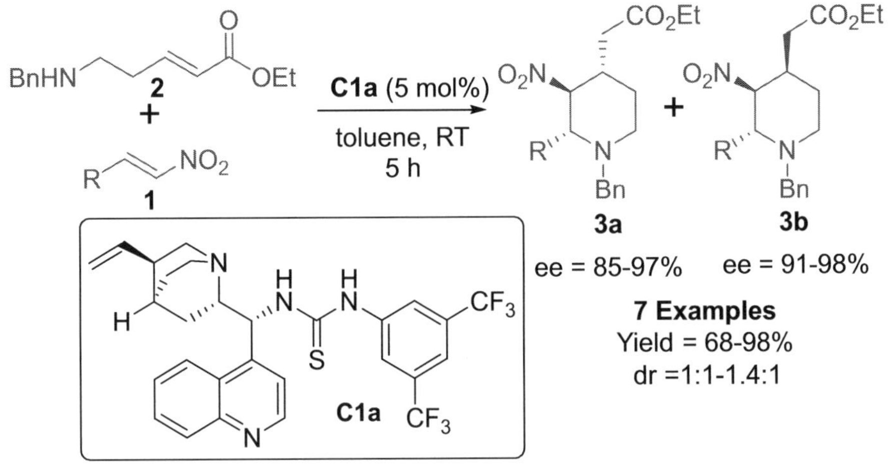

SCHEME 12.1 Asymmetric synthesis of 2,3,4-trisubstituted piperidines mediated by quinine-derived thiourea.

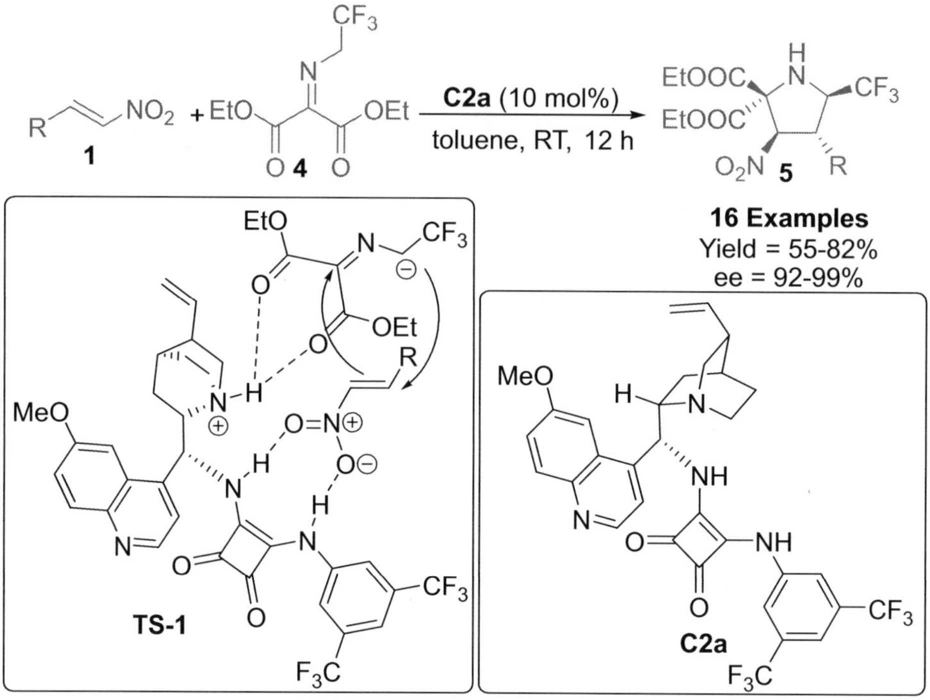

SCHEME 12.2 Domino Michael/aza-Henry cyclization of trifluoromethyl substituted iminomalonate with nitroalkene.

X-ray crystallographic analysis. During the reaction, the iminomalonate **4** gets activated by the catalyst **C2a** and adds to the activated nitroalkene from the *Re*-face followed by an *aza*-Henry reaction of nitronate with the imino group leading to the formation of highly functionalized 5-trifluoromethyl and 3-nitro-substituted pyrrolidine derivatives **5** with three contiguous stereocenters (**TS-1**, Scheme 12.2).

Zhang et al. described a cascade Michael/*aza*-Henry/lactamization reaction of diester **6**, nitroalkenes **1**, aldehydes and NH_4OAc in one pot using recyclable fluorous organocatalyst **C3** to afford fluorinated 2-piperidinones **7** in high yields and enantioselectivities (Scheme 12.3).[3] In this reaction, the catalyst induces Michael reaction of nitroalkene **1** with diester **6** to form an adduct, which then undergoes aza-Henry reaction with aldehyde and NH_4OAc, where in-situ generated imine acts as an aza-Henry donor to give an intermediate which on further cyclization, that is, lactamization, yields the final product **7** with four stereocenters.

Cossío et al. presented a three-component reaction with nitroalkene **1a**, ketone **9** and carboxylic acid **10** substrates in the presence of catalyst **C4** to afford bicyclic octahydro-2H-indol-2-one skeleton **11** possessing three contiguous chiral centers (Scheme 12.4).[4] The reaction proceeds through thermal rearrangement of the nitro group and mechanism (Scheme 12.5) of the reaction was proposed based on density functional theory calculations and isotopic labeling experiment. According to the mechanistic proposal, the ketone **9** forms an enamine intermediate **12** with the catalyst **C4**, which reacts with nitroalkene **1a** to form protonated nitronate **13**. The carboxylic acid **10** adds to intermediate **13**, which upon dehydration and tautomerization gives **14**. Cyclization of the amide moiety formed from the zwitterionic structure leads to **16** and, finally, the product is formed by hydration of **17**, which is formed after removal of the catalyst from **16**. The synthetic potential of the reaction was proven by the concise total synthesis of (+) enantiomer of a polycyclic natural compound pancracine.

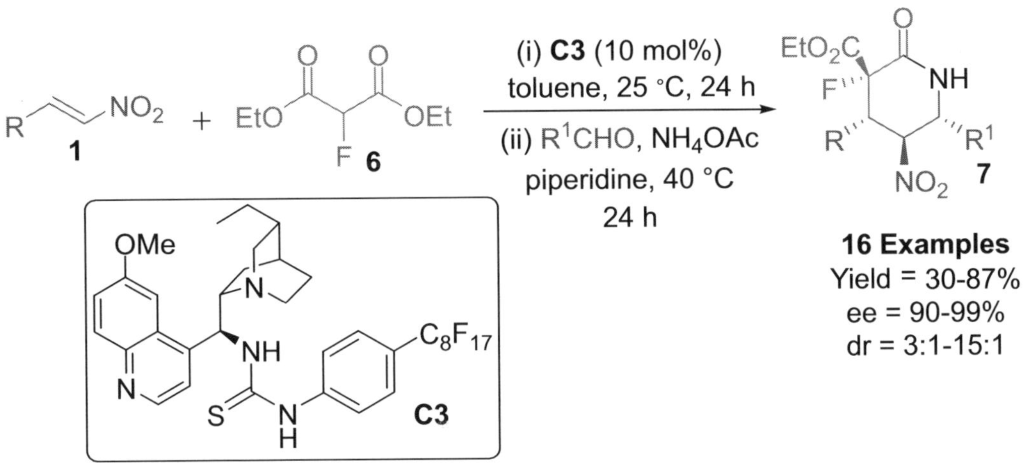

SCHEME 12.3 Asymmetric cascade Michael/*aza*-Henry/lactamization reaction of diester, nitroalkenes, aldehydes and NH_4OAc.

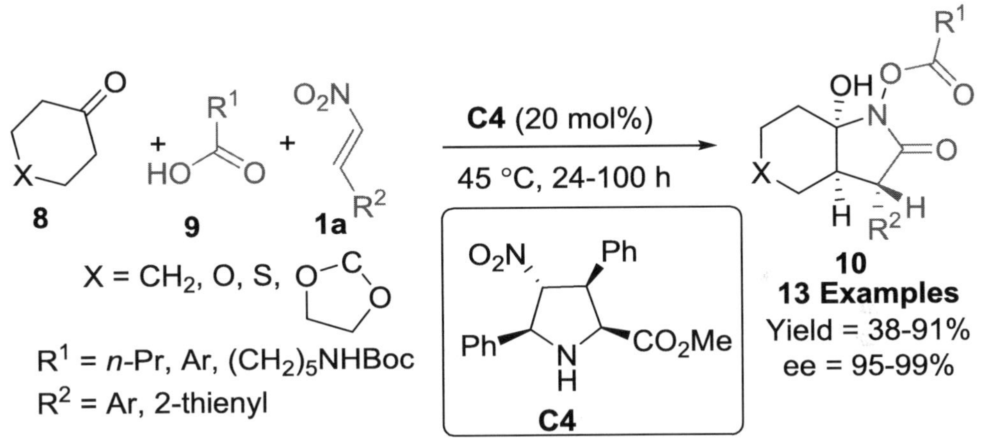

SCHEME 12.4 Enantioselective synthesis of bicyclic octahydro-2H-indol-2-ones by unnatural amino acid-catalyzed three-component reaction with nitroalkenes, ketones and carboxylic acids.

Chen and co-workers reported the use of Barton–Zard reaction for efficient synthesis of axially chiral 3-(hetero)arylpyrrole skeletons **20** from substituted nitroalkenes **1b** and α-isocyano substrates **18**.[5] Here, the authors employed Ag_2O along with a cinchona-derived phosphine ligand **L1** as a catalytic assembly, which helped in the efficient generation of enantioenriched atropoisomers via a central-to-axial chirality transfer approach. In the plausible mechanism, it was mentioned that Ag will bind with the phosphine moiety present in the ligand, as well as it will coordinate with the enolate of α-isocyano substrate **18** and nitro-functionality of nitroalkene **1b**. Moreover, the nitro-functionality will also exhibit hydrogen bonding with the protonated form of quinuclidine (**TS-2**, Scheme 12.6). This spatial orientation of the transition state will facilitate the preferential *Re*-face attack during the Michael addition, subsequently followed by the intramolecular cyclization (**TS-3**, Scheme 12.6), leading to the generation of rotation-restricted cyclized products **19**. It was treated with DBU to provide axially chiral 3-(hetero) arylpyrrole skeletons **20**.

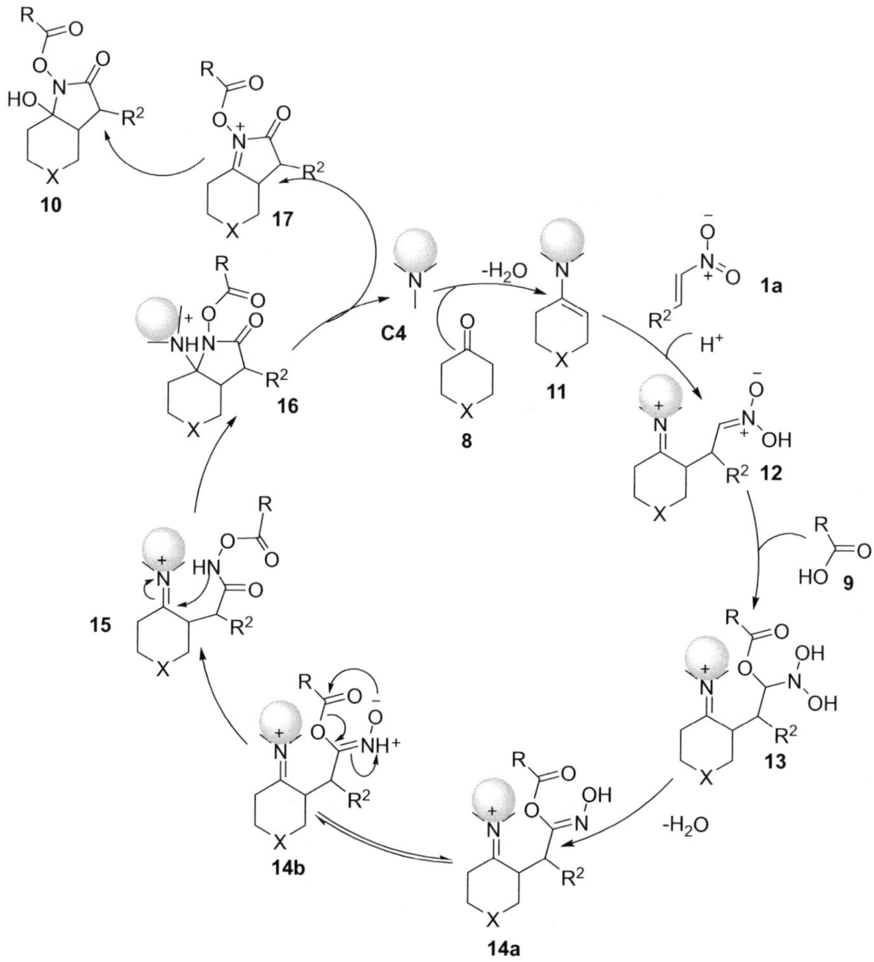

SCHEME 12.5 Proposed mechanism for the stereoselective synthesis of bicyclic octahydro-2H-indol-2-ones.

12.3 SYNTHESIS OF FIVE- AND SIX-MEMBERED OXYGEN HETEROCYCLES

The Feist–Benary reaction is a base-catalyzed reaction between α-halogenated ketones and 1,3-dicarbonyl compounds for the synthesis of substituted furans. It can also be considered as a domino aldol-alkylation reaction, where the α-halogenated ketones are used as bielectrophiles and 1,3-dicarbonyl compounds act as binucleophiles.

The Lu group designed a Feist–Benary reaction for the enantioselective preparation of 3(2H)-furanones **22** from bromo-substituted β-ketoester **21** and nitroalkenes **1** by employing a domino Michael-alkylation reaction sequence (Scheme 12.7).[6] In this method, ketone electrophile plays a significant part in the proton transfer, and the presence of the halogen atom is critical for the cyclization. Mechanistically, it was envisaged that the tertiary amine group of the catalyst **C5** would deprotonate the β-ketoester **21** to form an enolate, which, in turn, forms hydrogen bonding with the thiourea functionality of the catalyst. The ionic interaction between the positively charged ammonium with the nitroalkene **1** is assumed to be significant for the substrate-binding, as indicated by theoretical studies (**TS-4**, Scheme 12.7).

A modified Feist–Benary reaction for an enantioselective synthesis of highly functionalized 2,3-dihydrofurans **24** was described by the same group by employing acyclic β-ketoesters **23** and α,β-disubstituted nitroalkenes **1c** in the presence of l-threonine-derived tertiary amine/thiourea

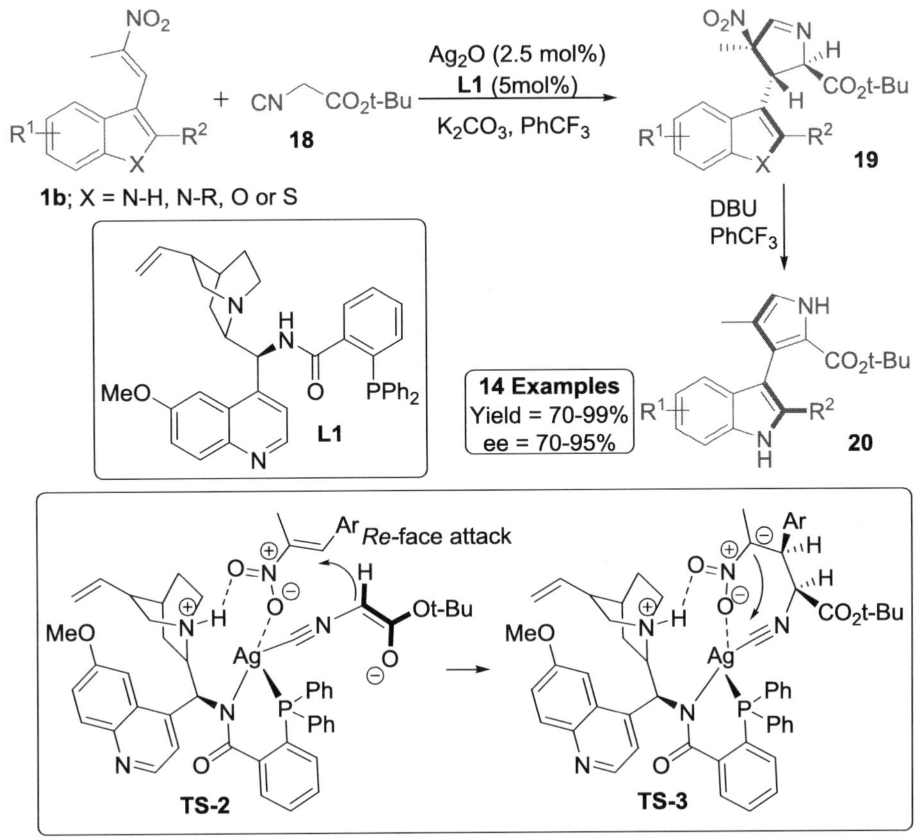

SCHEME 12.6 Synthesis of axially chiral 3-(hetero)arylpyrrole skeletons from substituted nitroalkenes.

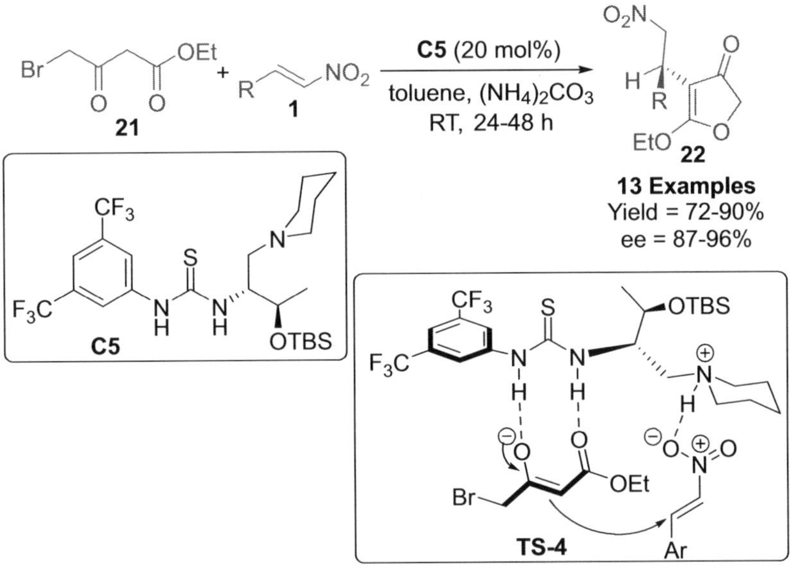

SCHEME 12.7 Chiral *tert*-amine-thiourea catalyzed enantioselective synthesis of furanones from bromo-substituted β-ketoester and nitroalkenes.

bifunctional catalyst **C5** (Scheme 12.8).[7] Complete *trans*-diastereoselectivity was observed for all the examples examined in this study, which could be attributed to the stereospecific protonation of the nitronate anion, following the initial Michael addition of keto enolate to nitroalkene.

Feng et al. demonstrated an organocatalytic asymmetric domino Michael alkylation reaction of bromonitrostyrenes **1c** with cyclohexane-1,3-dione derivatives **8a**.[8] Desymmetrization of 5-monosubstituted cyclohexane-1,3-diones **8a** (R^1 = H) was achieved leading to the formation of bicyclic 2,3-dihydrofurans **25** with three chiral stereogenic centers (Scheme 12.9). In this reaction, the N-oxide moiety of the catalyst **C6** activates the enolate form of diketone **8a** and the amide –NH group of the catalyst **C6** activates bromonitrostyrene **1c** at the same time to promote domino Michael-alkylation reaction to afford the products **25** (**TS-5**, Scheme 12.9).

Recently, Li and co-workers reported a cascade involving the Michael addition- alkylation process of gem-benzoylnitrostyrenes **1d** and 1,3-dicarbonyl compounds **8b** using a bifunctional

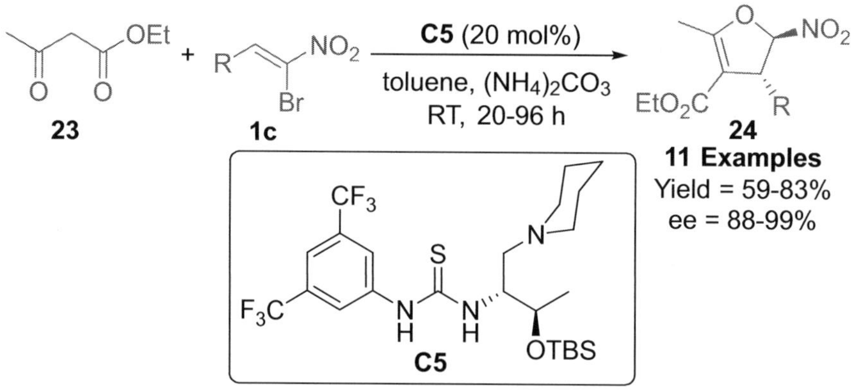

SCHEME 12.8 Chiral *tert*-amine-thiourea catalyzed enantioselective synthesis of 2,3-dihydrofurans from acyclic β-ketoesters and α,β-disubstituted nitroalkenes.

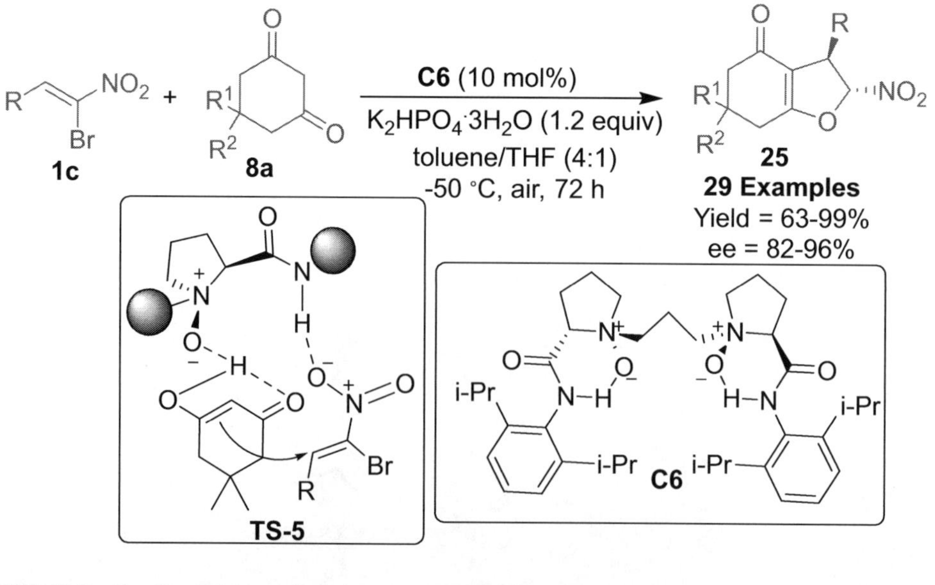

SCHEME 12.9 Domino Michael-alkylation reaction of bromonitrostyrenes with cyclohexane-1,3-diones for the enantioselective synthesis of dihydrofurans.

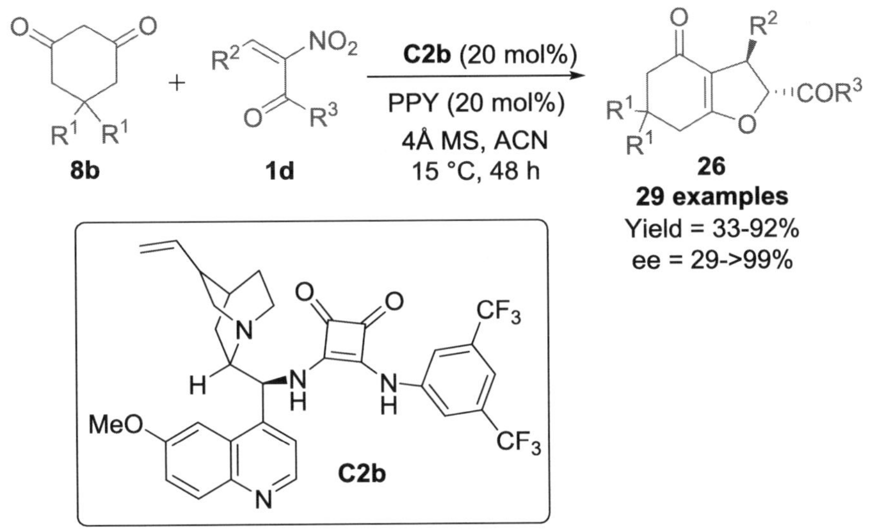

SCHEME 12.10 Stereoselective synthesis of 2,3-dihydrofurans using a Michael addition-alkylation cascade.

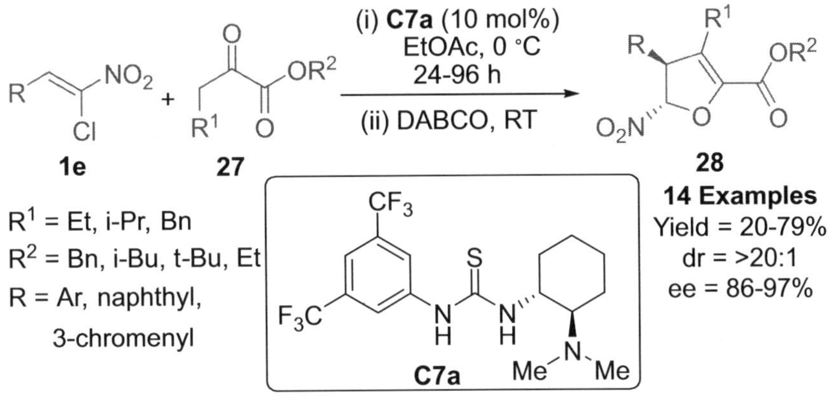

SCHEME 12.11 Synthesis of alkyl 4,5-dihydrofuran-2-carboxylates from α-ketoesters and (Z)-β-chloro-β-nitrostyrene.

squaramide **C2b** as the catalyst (Scheme 12.10).[9] The reactions afforded *trans*-diastereomers of 2,3-dihydrofurans **26** exclusively in moderate-to-good yields (up to 92%) along with excellent enantioselectivities (up to >99% ee). The authors also mentioned here that the report was the pioneering example showing replacement of the nitro group in an asymmetric process where the nitro group was well employed as a leaving group in the enantioselective cascade reaction.

The Rodriguez group reported an asymmetric synthesis of alkyl *trans*-4,5-dihydrofuran-2-carboxylates **28** by the reaction of α-ketoesters **27** and β-chloro-β-nitroalkenes **1e** via Michael/*O*-alkylation reaction (Scheme 12.11).[10] The reaction was performed in the presence of Takemoto thiourea catalyst **C7a**. In addition, DABCO has been used to assist in the intramolecular *O*-alkylation to afford the product **28** in excellent yields and good enantioselectivities.

The Shen group described an enantioselective Michael/cyclization tandem reaction of trifluoromethylated nitroalkenes **1f** with cyclohexanediones **8c** for the synthesis of trifluoromethylated hydroxyimino tetrahydrobenzofuranones **29** employing bifunctional squaramide catalyst **C8** (Scheme 12.12).[11] In this reaction, the enol attacks at the β-position of the nitroalkene **1f** from

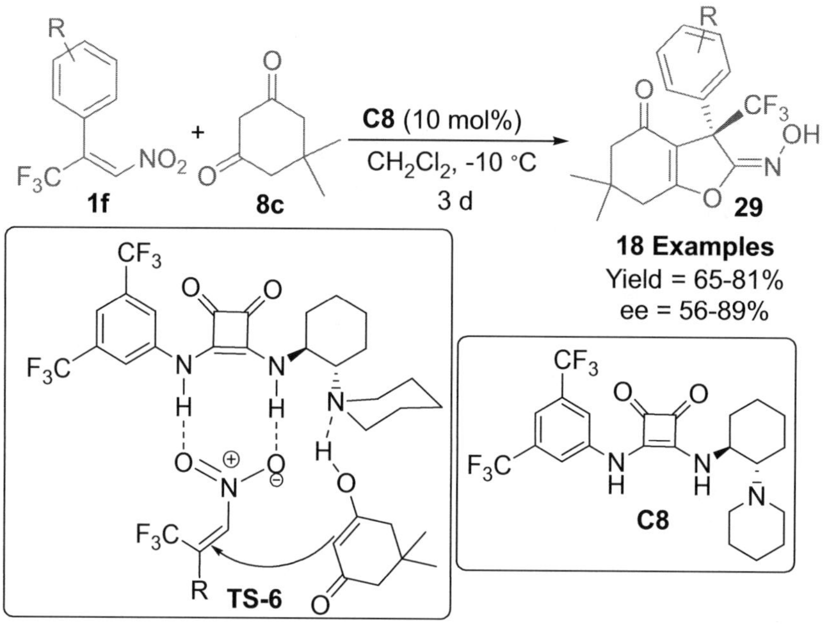

SCHEME 12.12 Asymmetric tandem Michael/cyclization of trifluoromethylated nitroalkenes with cyclohexanediones in the presence of bifunctional squaramide catalyst.

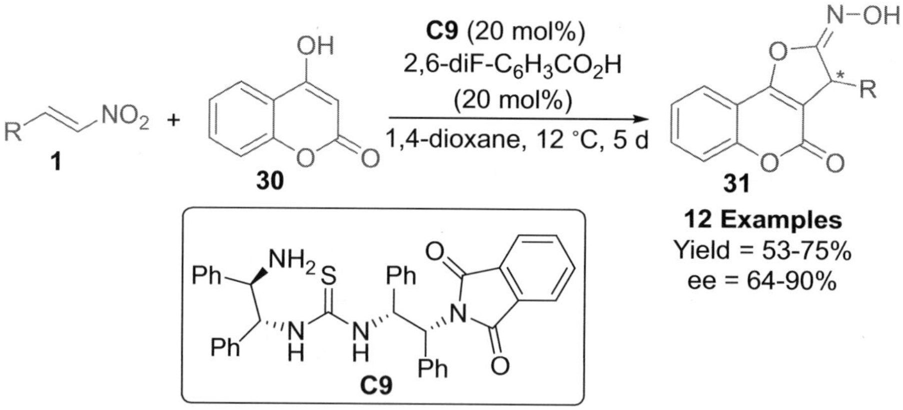

SCHEME 12.13 Bifunctional thiourea-catalyzed asymmetric Michael addition-cyclization of 4-hydroxycoumarin with nitroalkenes for the synthesis of dihydrofuran-fused coumarins.

the *Re*-face (**TS-6**, Scheme 12.12) and delivered tetrahydrobenzofuranones **29** in excellent enantioselectivities.

The Wang group reported an asymmetric tandem Michael addition-cyclization of 4-hydroxycoumarin **30** to nitroalkenes **1** using bifunctional thiourea catalyst **C9** to afford 2,3-dihydrofuro[3,2-c]-coumarin-type adducts **31** (Scheme 12.13).[12] Irrespective of the nature of the electronic properties of the substituents on the aryl ring of nitroalkene **1**, the corresponding adducts **31** were obtained in high yields and excellent enantioselectivities.

The proposed mechanism for the formation of dihydrofuran-fused coumarin is shown in Scheme 12.14. At first, hydroxycoumarin **30** participates in the Michael addition with nitroalkene **1a** to form an adduct **32** in the presence of chiral catalyst **C10** through enamine intermediate.

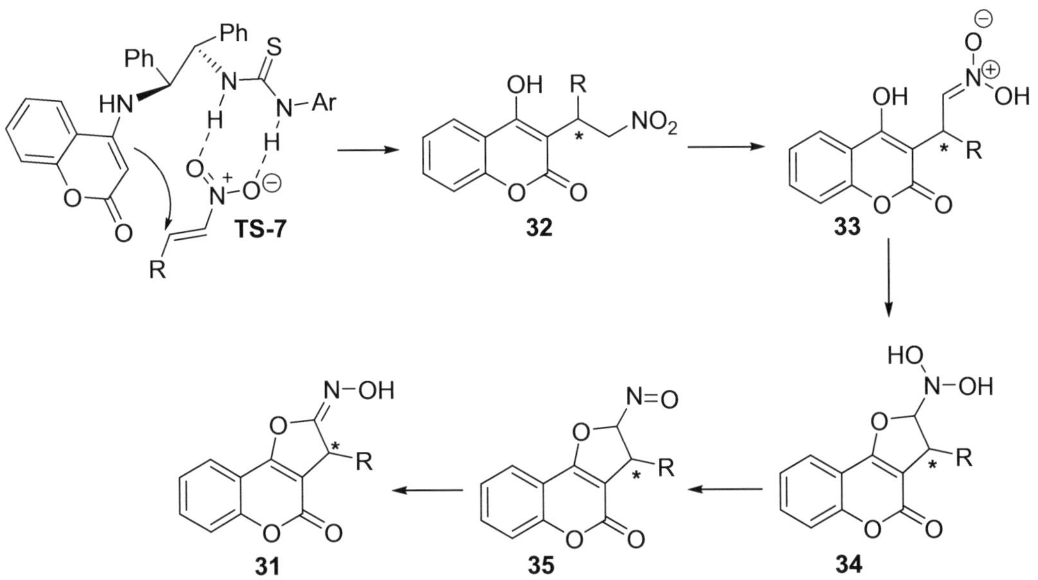

SCHEME 12.14 Mechanism of formation of dihydrofuran-fused coumarin from 4-hydroxycoumarin and nitroalkene.

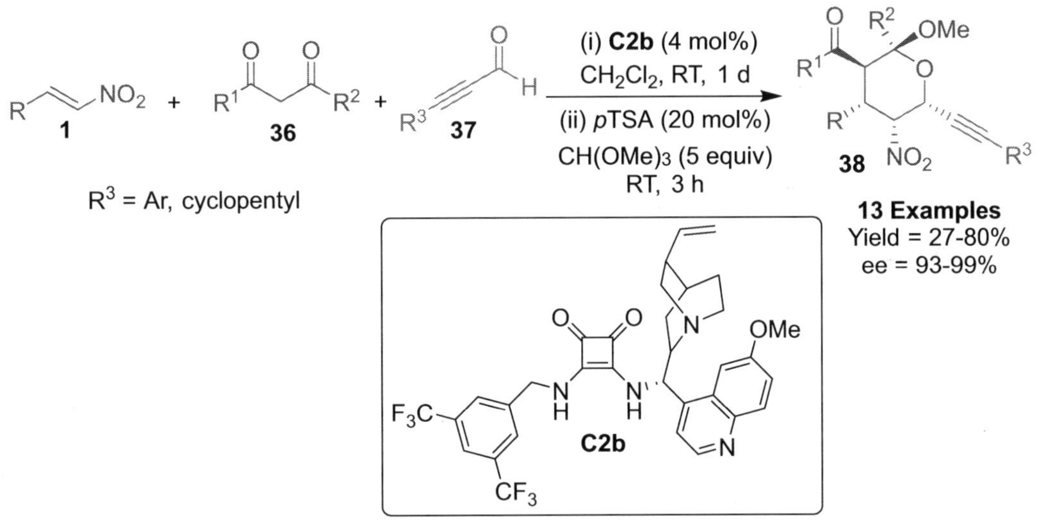

SCHEME 12.15 Quinine-based squaramide-catalyzed sequential Michael/Henry/ketalization reaction of β-ketoesters, β-nitrostyrenes and alkynyl aldehydes.

The adduct thus formed then undergoes a sequence of reactions, i.e., enolization, cyclization, protonation, dehydration and nitroso-oxime tautomerization to offer the final product **31**.

The Enders group described an asymmetric synthesis of tetrahydropyrans **38** with five contiguous stereocenters via a sequential Michael/Henry/ketalization reaction employing β-ketoesters **36**, β-nitrostyrenes **1**, and alkynyl aldehydes **37** as reactants (Scheme 12.15).[13] A bifunctional quinine-based squaramide **C2b** was used as an organocatalyst along with an additional base to afford the tetrahydropyrans **38** in excellent yields and enantioselectivities. The products were isolated in the protected form with improved yields.

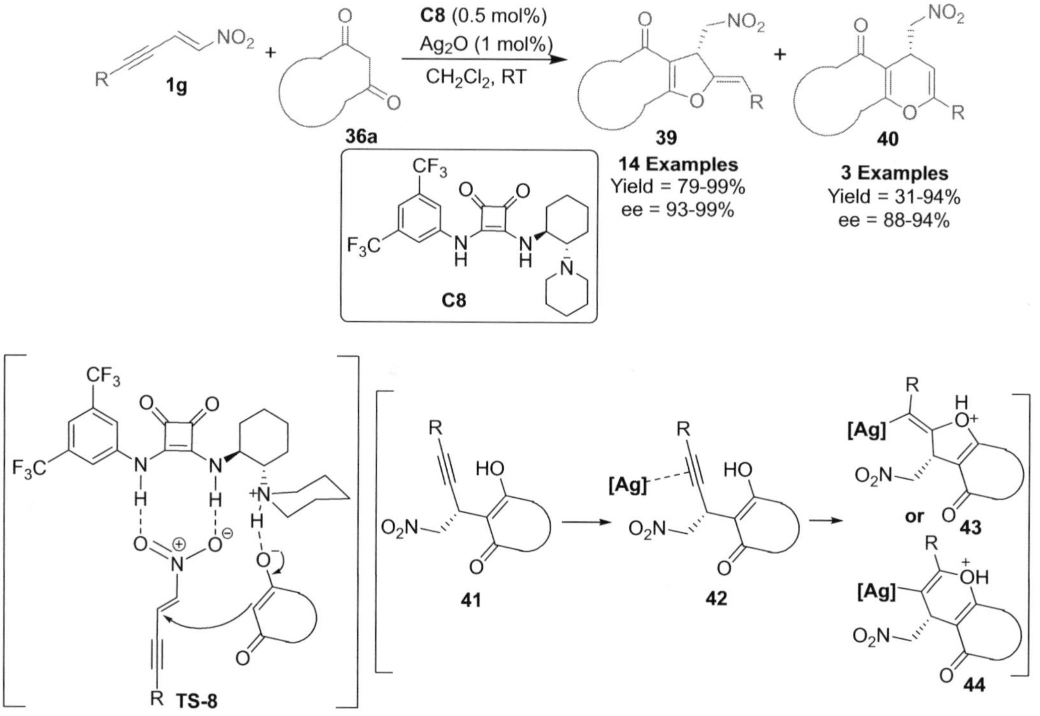

SCHEME 12.16 Asymmetric Michael addition/hydroalkoxylation of 1,3-diketones with alkyne-tethered nitroalkenes under cooperative organo- and silver-catalysis.

The Enders group described an enantioselective Michael addition/hydroalkoxylation reaction of 1,3-diketones **36a** with alkyne-tethered nitroalkenes **1g** by employing a combination of bifunctional squaramide catalyst **C8** and a silver (I) salt to obtain tetrahydrobenzofurans **39** and pyrans **40** in good yields and excellent enantioselectivities in one pot (Scheme 12.16).[14] In the possible reaction pathway, both ketone **36a** and nitroalkene **1g** get activated synergistically by the tertiary amine and squaramide **C8** through the deprotonation and hydrogen bonding of 1,3-diketone. The enolate then attacks the nitroalkene **1g** from the *Re*-face to form an adduct **41**. In the next step, silver salt forms complex **42** with internal alkyne and thereby facilitate 5-*exo-dig* or 6-*endo-dig* annulation reaction resulting in vinyl silver intermediate **43/44**. The formed intermediate then undergoes protodeargentation to yield the final product **39/40** (**TS-8**, Scheme 12.16).

Huang et al. reported an asymmetric Michael/Henry/acetalization reaction of aldehydes **45**, nitroalkenes **1** and ninhydrin **46** to form six-membered chiral oxa-spirocyclic indanones **47** by employing silylprolinol catalyst **C10** in good yields and enantioselectivities (Scheme 12.17).[15]

Zhao and co-workers also reported the stereoselective synthesis of 3-oxabicyclo[3.3.1]nonan-2-ones **49** through a domino reaction of aryl-2-nitroprop-2-enols **1h** and aryl-7-oxo-hept-5-enals **48** mediated by MDOs (quinidine squaramide **C11** and octahydroindolecarboxylic acid **C12**) and followed by oxidation (Scheme 12.18).[16] Here, it was mentioned by the authors that the obtained stereoselectivity was substrate-controlled and the reaction was not catalyzed by Hayashi–Jørgensen catalyst.

The Zhang group reported an asymmetric synthesis of highly functionalized oxazolidine-fused tetrahydronaphthalenes **50** via an organocatalytic tandem Michael/nitrone formation followed by intramolecular [3+2] nitrone-olefin cycloaddition reaction. The reaction was performed in one pot using an aqueous medium (Scheme 12.19).[17] The reaction proceeds through the formation of enamine **52** from aldehyde **45** and catalyst **C10** (Scheme 12.20).

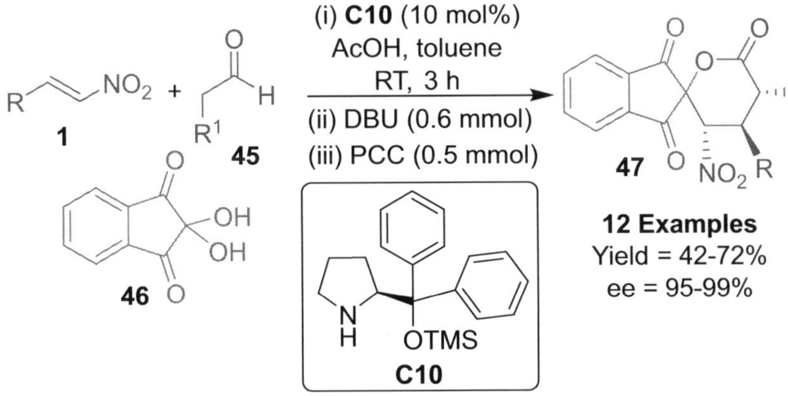

SCHEME 12.17 Asymmetric Michael/Henry/acetalization reaction of aldehydes, nitroalkenes and ninhydrin.

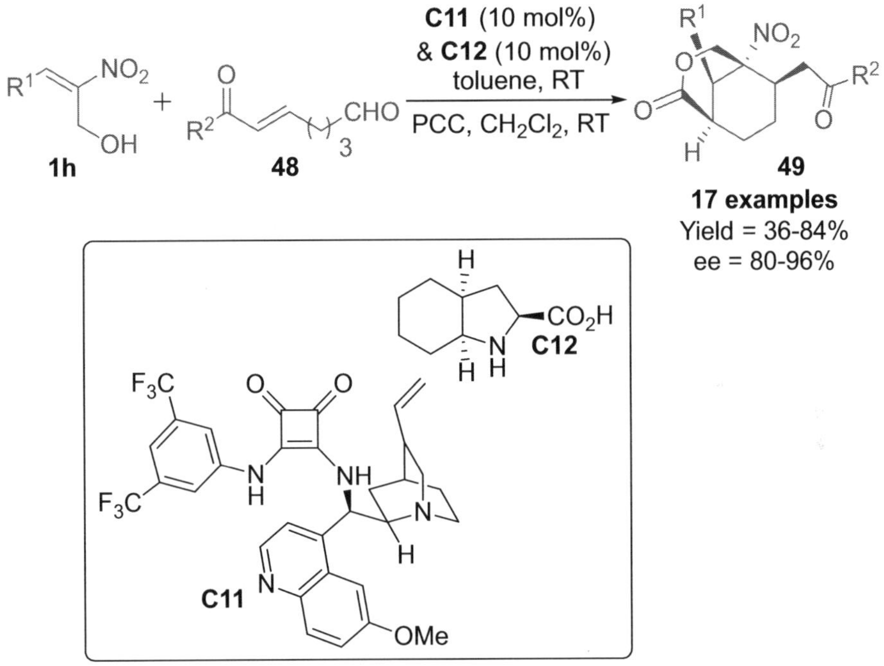

SCHEME 12.18 Asymmetric synthesis of 3-oxabicyclo[3.3.1]nonan-2-ones through a domino reaction mediated by modularly designed organocatalyst.

The biologically relevant tetrahydro-1,2-oxazine core was constructed via the cascade aza-Michael–Michael addition reaction of nitroalkenes **1** and amino-oxylating reagents **56** using cinchonidine-derived bifunctional organocatalyst **C1b** (Scheme 12.21).[18] The reaction proceeded with a broad scope, high yields (up to 97%) and excellent enantioselectivity (ee up to 99.5:0.5). The reaction is activated through the H-bonding interaction of thiourea moiety with nitroalkene **1** and generation of nucleophile via deprotonation of the nitrogen in **56** by the quinuiclidine moiety present in the catalyst **C1b** (**TS-9**, Scheme 12.21). Configuration of the product **57** is assigned as *3R, 4S, 5S* by single-crystal X-ray analysis and configuration of the product was the consequence of the chair conformation of the transition state **TS-11** (Scheme 12.21), in which all the bulky groups occupy an equatorial position.

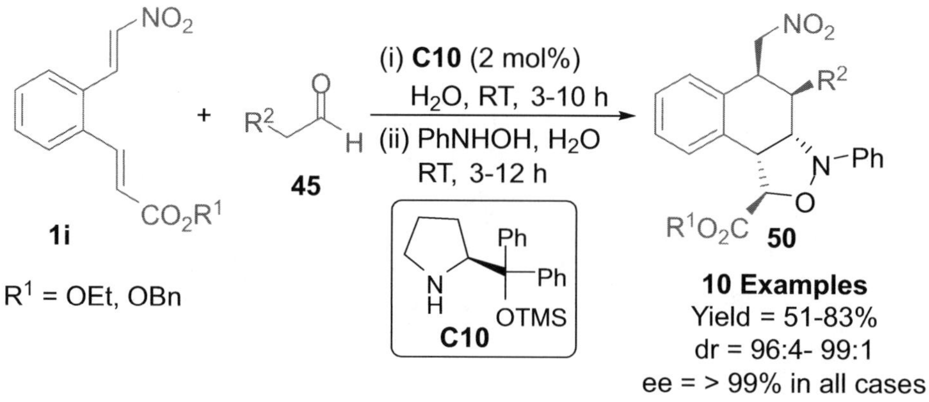

SCHEME 12.19 Asymmetric synthesis of functionalized oxazolidine-fused tetrahydronaphthalenes from nitroalkene acrylates and aldehydes.

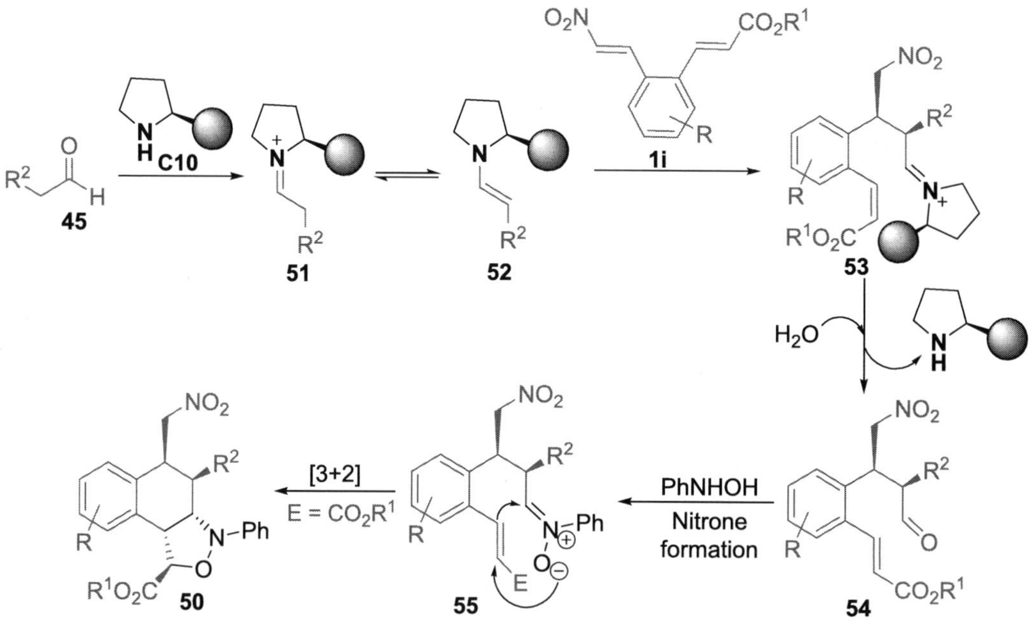

SCHEME 12.20 Proposed mechanism for the formation of oxazolidine fused tetrahydronaphthalenes from nitroalkene acrylates and aldehydes.

The Chen group employed bifunctional secondary amine-thiourea catalyst **C12** for the domino Michael–Michael addition reaction of barbiturates **58** with MBH acetates of nitroalkenes **1j** to afford barbiturate-fused teterahydropyrans **59** in high yields and excellent stereoselectivities (Scheme 12.22).[19] This protocol was also successfully employed for the construction of pharmaceutically relevant pyranocoumarin scaffolds **61** (Scheme 12.22). According to the proposed mechanism, H-bonding interaction of the catalyst **C12** activates both the substrates, and the intermediate **TS-12** thus formed (Scheme 12.22) controls the enantioselectivity. Further, intramolecular double Michael addition leads to the product **61**. Formation of the stabilized transition state **TS-14** (Scheme 12.22) during the second Michael addition is responsible for the formation of the *trans*-isomer of the product **61**.

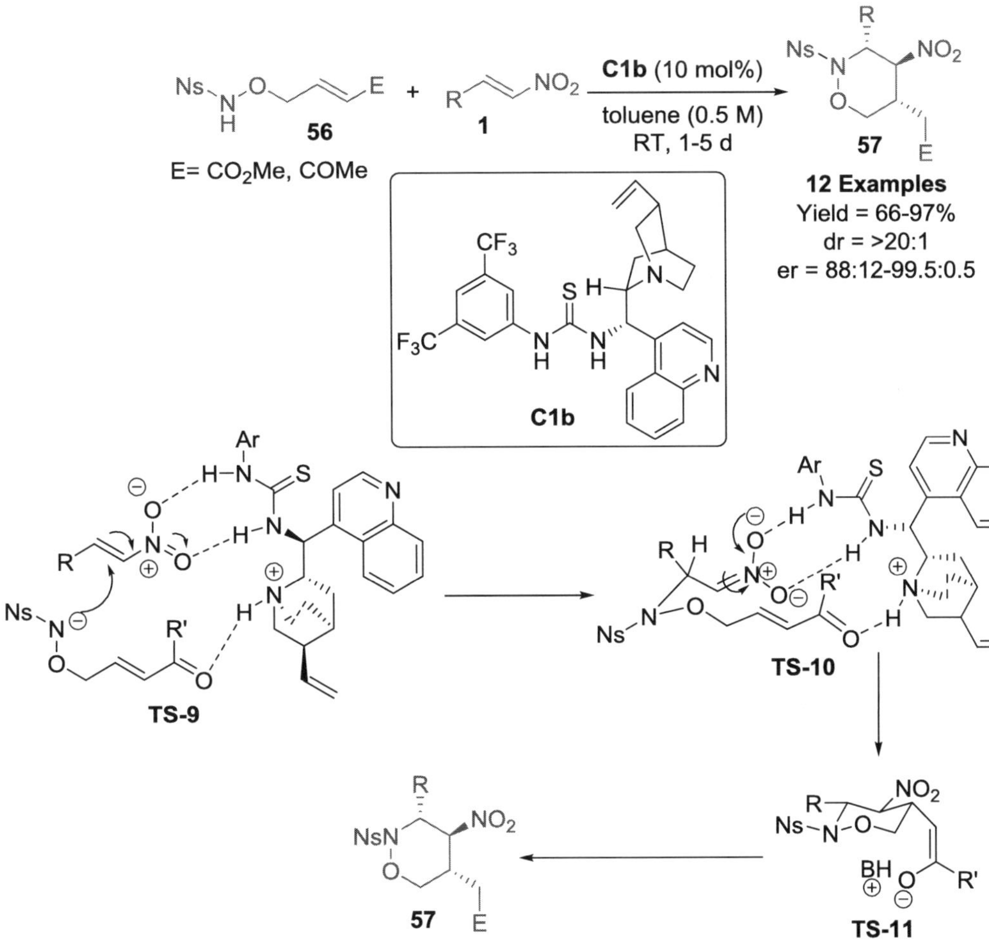

SCHEME 12.21 Synthesis of tetrahydro-1,2-oxazines from nitroalkenes and amino-oxylating reagents in the presence of cinchonidine-derived bifunctional organocatalyst.

12.4 SYNTHESIS OF SPIRO-OXINDOLES

Chen and co-workers reported an asymmetric synthesis of spirooxindoles **63** by an α-regioselective asymmetric [3+2] annulation of MBH carbonates **62** of isatin with enyne-tethered nitroalkene **1k**.[20] The reaction employs cinchona-derived amine **C13** as a chiral catalyst to afford the products **63** in high yields and enantioselectivities (Scheme 12.23).

The Shao group described an asymmetric synthesis of spirocyclopentaneoxindoles **65** with four contiguous stereocenters including one spiro-quaternary stereocenter through a combined Ru-catalyzed cross-metathesis followed by an organocatalyzed asymmetric double Michael addition reaction (Scheme 12.24).[21] Nitroalkenes **1** and oxindole-derived weak Michael acceptor **64** were employed as substrates to obtain the products **65** in good yields with excellent enantioselectivities.

Zhou and co-workers synthesized chromanone-based spirocyclohexaneoxindoles **67** using a Takemoto's organocatalyst **C7b**-based domino formal double Michael cycloaddition reaction of bifunctional chromone-oxindole moieties with nitroalkenes with good yields in a highly stereoselective fashion (up to >99% ee and >20:1 dr) (Scheme 12.25).[22] The products formed using this strategy possessed five contiguous stereocenters including a spiro-quaternary carbon, as well as possessed good biological activity. During the reaction, the tertiary amine moiety of the catalyst

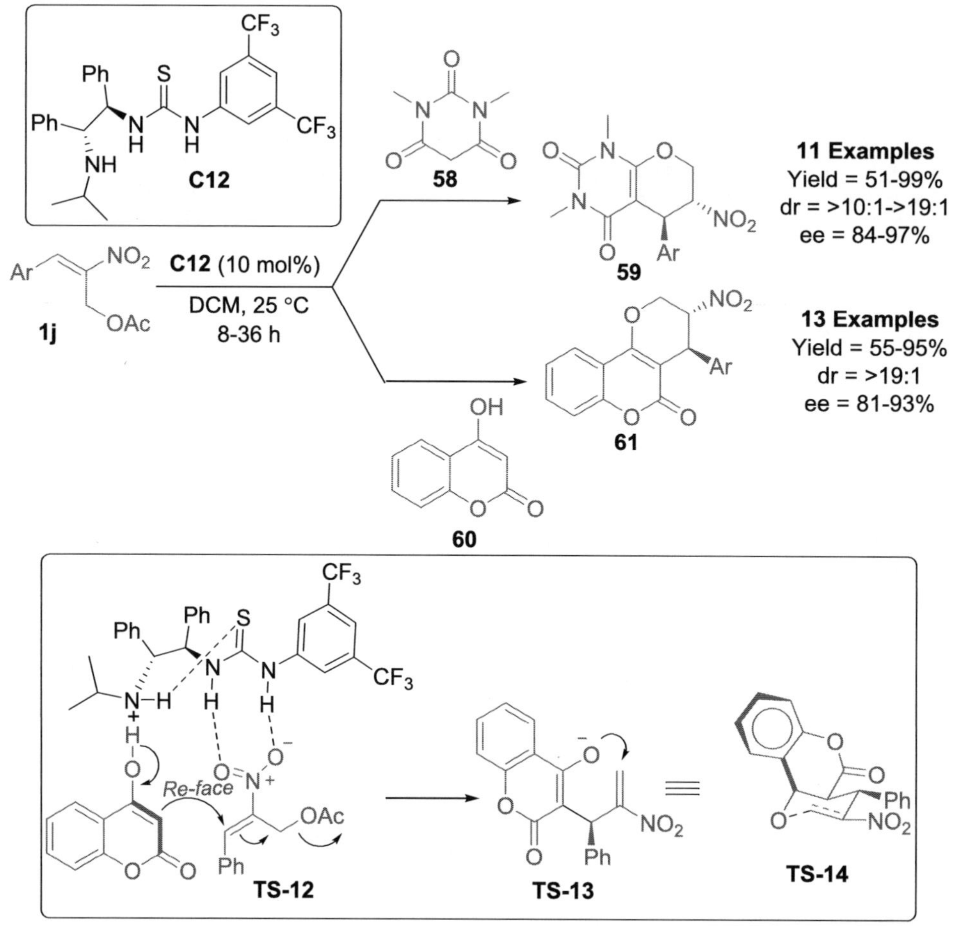

SCHEME 12.22 Stereoselective synthesis of barbiturate-fused teterahydropyrans by bifunctional secondary amine-thirourea catalyzed domino Michael–Michael addition of barbiturates with MBH acetates of nitroalkenes.

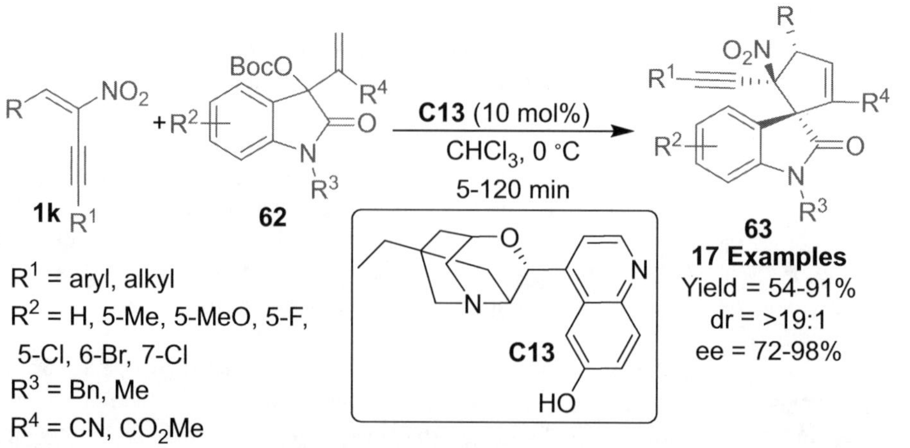

SCHEME 12.23 Asymmetric synthesis of spirooxindoles by [3+2] annulation of MBH carbonates of isatin and enyne-tethered nitroalkenes.

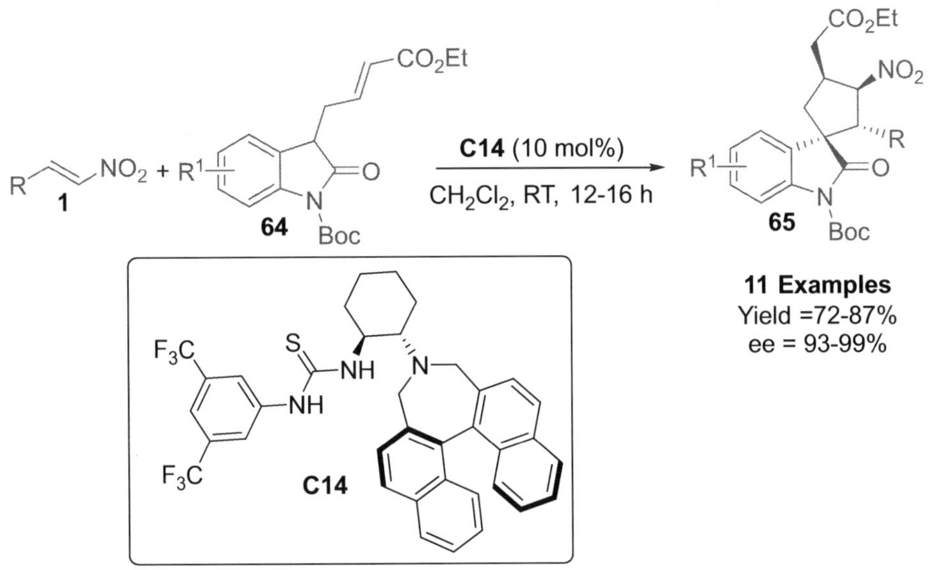

SCHEME 12.24 Combined Ru-catalyzed sequential cross-metathesis/double Michael addition of nitroalkenes and donor-acceptor oxindoles.

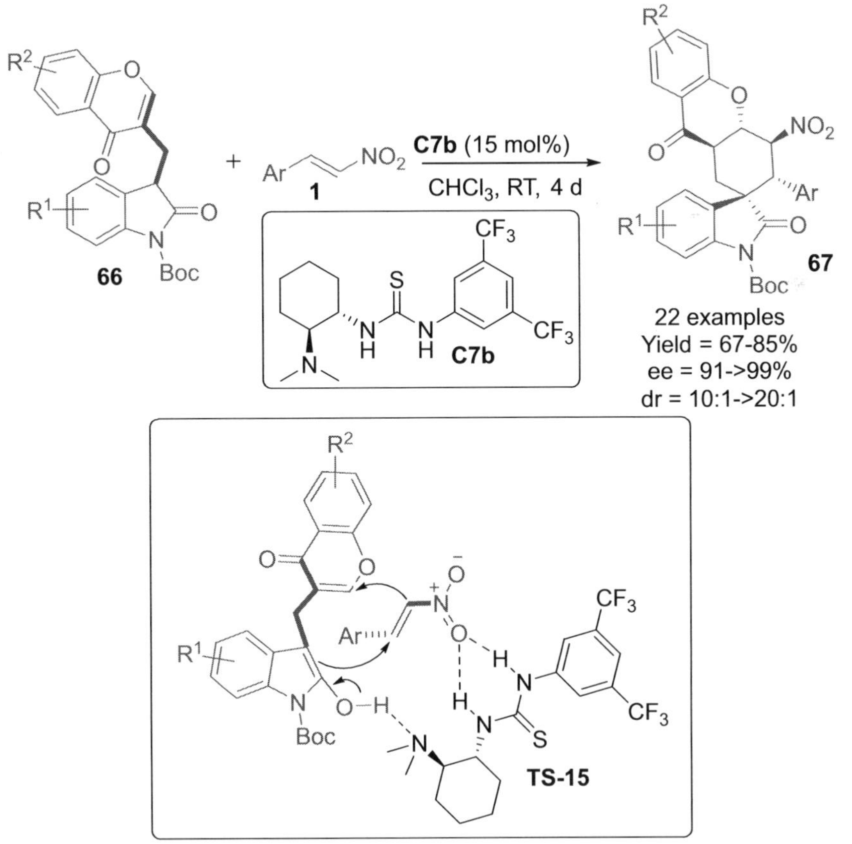

SCHEME 12.25 Domino formal double Michael cycloaddition of chromanone-oxindole moiety and nitroalkenes mediated by a bifunctional tertiary amine-thiourea-based catalyst.

C7b activated the nucleophile by protonation, while the nitroalkene was activated by hydrogen-bonding interactions with the thiourea group (**TS-15**, Scheme 12.25). In the first step, Michael addition of the oxindole **66** to nitroalkene **1** occurs under the stereochemical influence of the catalyst **C7b**, followed by annulation of the resulting nitronate with the electron-deficient chromone unit in the second step.

Lee and co-workers reported a chiral squaramide-amine catalyst **C15**-mediated enantioselective domino Michael–Michael reaction of nitroolefins **1l** and 2-nitro-3-arylacrylates **68** (Scheme 12.26).[23] The reaction afforded a series of spirocyclopentane oxindoles **69a-b** bearing four consecutive stereocenters including quaternary α-nitroesters with good yields (up to 73%) and remarkable enantioselectivities (up to 97% ee). The stereoselective construction of cyclopentane-fused spirooxindoles **69a-b** using cascade reactions was considered to be of significance in asymmetric catalysis.

In general, diazo compounds in the presence of noble metals form an active zwitterionic intermediate, which can be trapped in situ by various Michael acceptors to form an acyclic product. With this idea, Tan et al. reported an asymmetric multicomponent reaction employing nitrosoarenes **70**, diazooxindole **71** and nitroalkenes **1** in the presence of bis-thiourea catalyst **C16** to form spirooxindole derivatives **72** with three contiguous stereocenters in excellent yields and enantioselectivities (Scheme 12.27).[24]

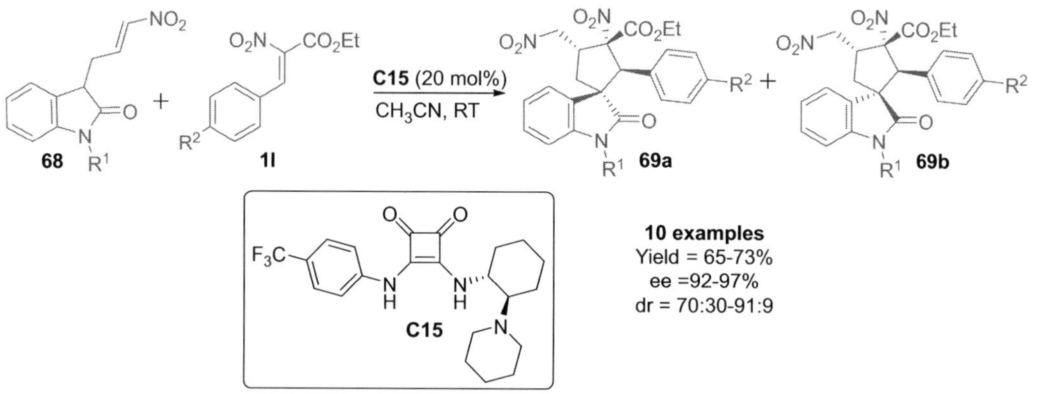

SCHEME 12.26 Asymmetric synthesis of spirocyclopentane oxindoles using a chiral squaramide-amine catalyst.

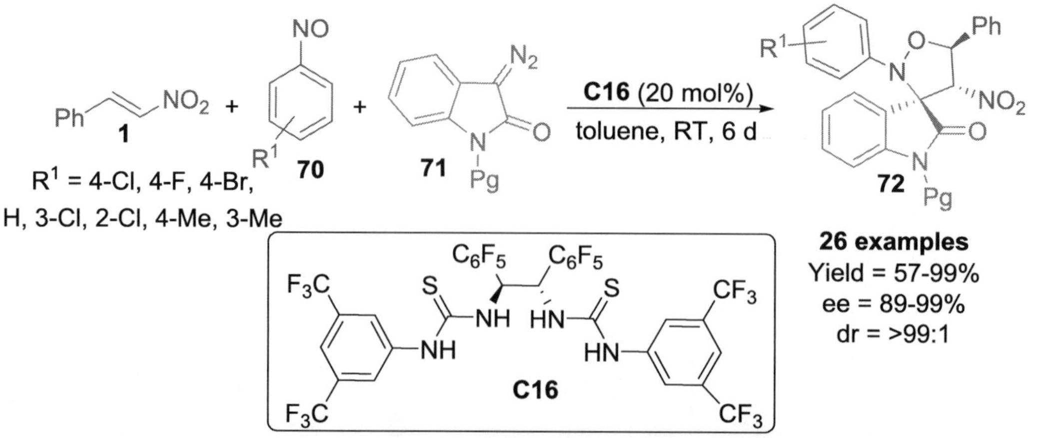

SCHEME 12.27 Bis-thiourea-catalyzed enantioselective synthesis of spirooxindoles by asymmetric multicomponent reaction of diazooxindole, nitroalkenes and nitrosoarenes.

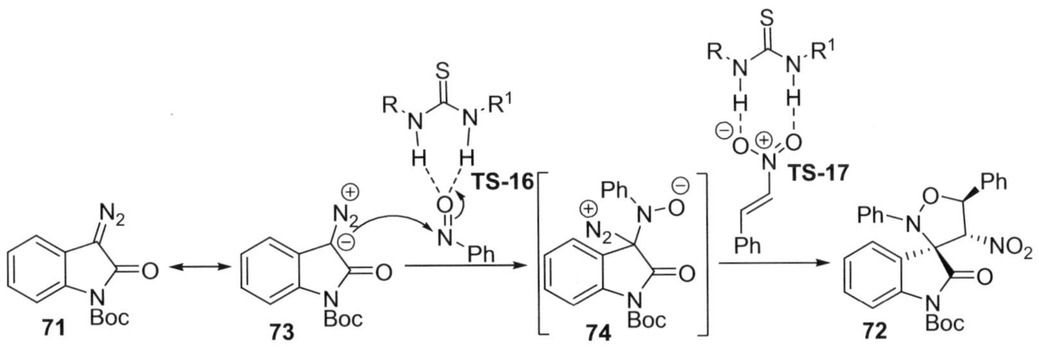

SCHEME 12.28 Mechanism of formation of spirooxindoles by the multicomponent reaction of diazooxindole, nitroalkenes and nitrosoarenes.

The mechanism of formation of spirooxindoles **72** through multicomponent reaction is outlined in Scheme 12.27. Initially, diazooxindole **71** and nitrosoarene **70** forms a chiral intermediate through the H-bonding interaction by the chiral catalyst **C16**, and it was trapped by the nitroalkene **1** to form another intermediate which undergoes cyclization to yield spirooxindole **72** (Scheme 12.28).

The Xiao group reported **L2**-Zn(OTf)$_2$-catalyzed enantioselective Michael addition/cyclization cascade reaction of 3-nitro-2H-chromenes **1m** with 3-isothiocyanato oxindoles **75** to deliver highly functionalized spirooxindoles **76** in high yields and excellent enantioselectivities (Scheme 12.29).[25] In this reaction, the zinc complex acts as Lewis acid and activates the olefin via coordination between Zn and a nitro group (**TS-18**, Scheme 12.29). At the same time, the nitrogen of oxindole –NH group acts as a Lewis base and directs the incoming nucleophile to attack from the *Re*-face through H-bonding.

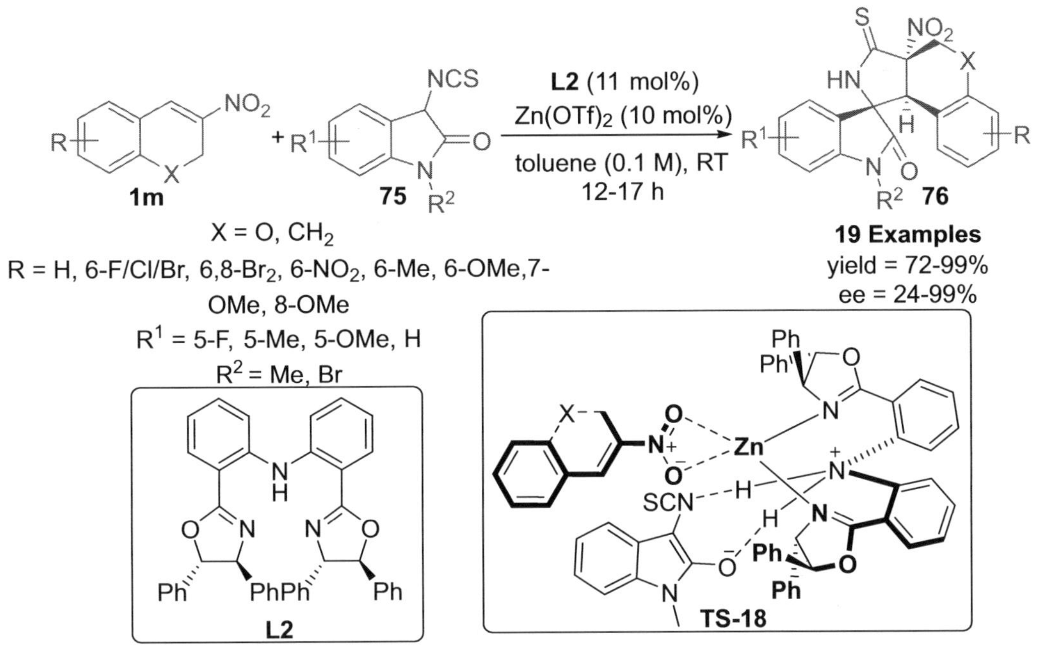

SCHEME 12.29 Asymmetric Michael addition/cyclization cascade of 3-nitro-2H-chromenes with 3-isothiocyanato oxindoles in the presence of Zn(OTf)$_2$/bis(oxazoline) complex.

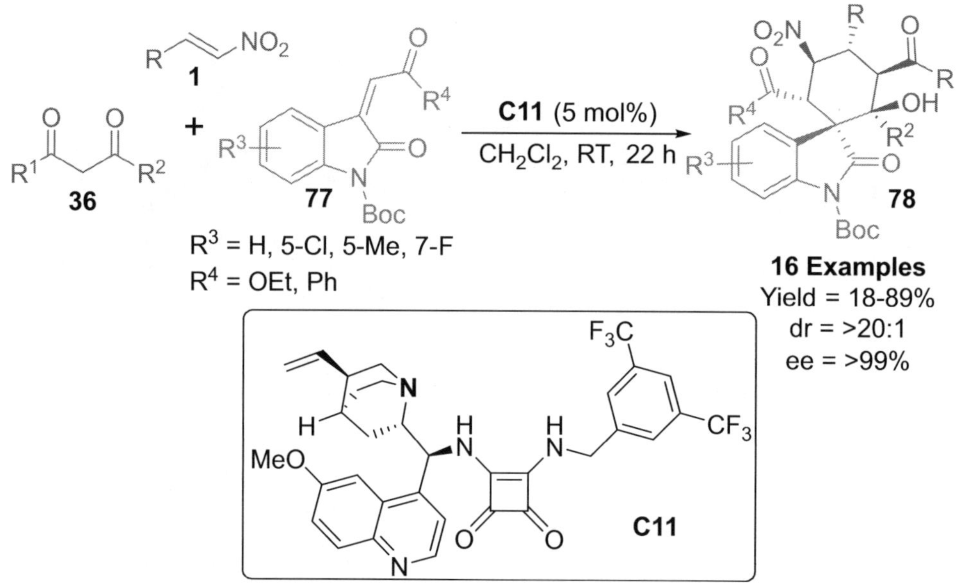

SCHEME 12.30 Bifunctional squaramide-catalyzed asymmetric Michael–Michael-aldol reaction of 1,3-dicarbonyl compounds, nitroalkenes and methyleneindolinone for the synthesis of polysubstituted spirocyclohexane oxindoles.

Sun et al. described an asymmetric Michael–Michael-aldol reaction of 1,3-dicarbonyl compounds **36**, nitroalkenes **1** and methyleneindolinone **77** in the presence of bifunctional squaramide catalyst **C11** for the enantioselective preparation of polysubstituted spirocyclohexane oxindoles **78** (Scheme 12.30).[26] Here, initially, 1,3-dicarbonyl compound **36** participates in Michael addition reaction with the nitroalkene **1** to form corresponding adducts, which later attacks another Michael acceptor, that is, methyleneindolinone **77** to undergo aldol reaction under the same reaction conditions to form spirocyclohexane derivatives **78**.

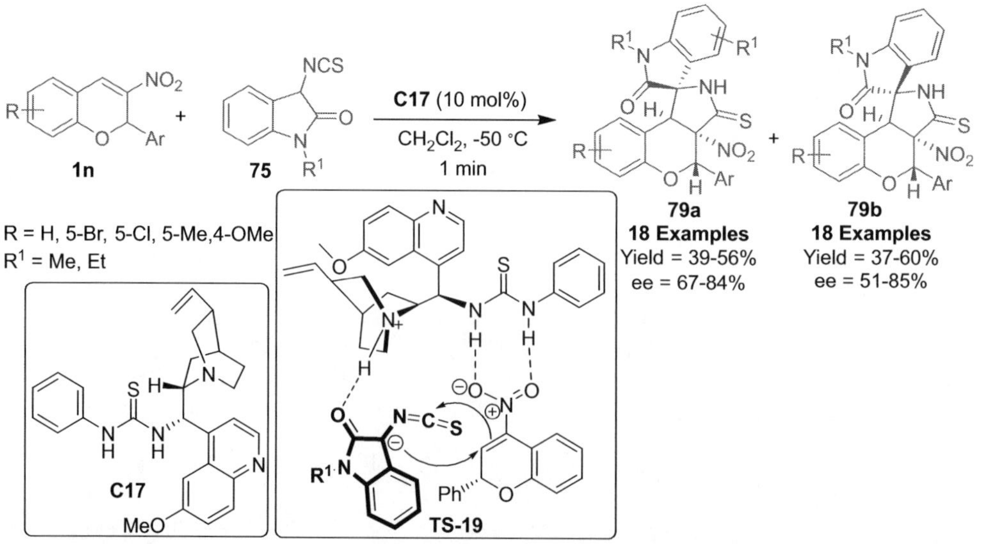

SCHEME 12.31 Bifunctional thiourea-catalyzed asymmetric synthesis of tetracyclic spiro[chromeno[3,4-*c*] pyrrole-1,3′-indoline] derivatives from 3-nitro-2H-chromene derivatives and 3-isothiocyanato oxindole.

Xie et al. demonstrated an asymmetric synthesis of highly functionalized tetracyclic spiro [chromeno[3,4-*c*] pyrrole-1,3′-indoline] derivatives **79a-b** with four vicinal chiral carbon centers, including two quaternary stereocenters *via* the domino reaction of 3-nitro-2H-chromene derivatives **1n** with 3-isothiocyanato oxindoles **75** in high yields with moderate-to-good enantioselectivities (Scheme 12.31).[27] Bifunctional thiourea **C1a** was employed as an organocatalyst for this reaction, wherein it activates the nitroalkene **1n** by H-bonding interactions and tertiary amine of the catalyst **C17** activates 3-isothiocyanato oxindole **75** by increasing its nucleophilicity. These interactions led to a well-defined orientation facilitating 3-isothiocyanato oxindoles **75** to attack the *Re*-face of the nitroalkene **1n** to form adduct. The adduct formed on subsequent intramolecular cyclization through isothiocyanato group results in the final product **79a-b** (**TS-19**, Scheme 12.31).

12.5 CONCLUSIONS

Synthesis of functionalized five- and six-membered heterocycles and oxindole-derived spiro-compounds via asymmetric cascade reactions of nitroalkenes in the presence of a wide variety of organocatalysts has been reported in recent years. Such cascade processes, primarily involving Michael addition and cyclization, are an attractive means for the highly efficient synthesis of complex molecular architectures, including natural products.

REFERENCES

1. Kriis, K.; Melnik, T.; Lips, K.; Juhanson, I.; Kaabel, S.; Järving, I.; Kanger, T. *Synthesis* **2017**, *2017*, 604.
2. Liu, Q.; Zhao, K.; Zhi, Y.; Raabe, G.; Enders, D. *Org. Chem. Front.* **2017**, *4*, 1416.
3. Huang, X.; Pham, K.; Zhang, X.; Yi, W.-B.; Hyatt, J. H.; Tran, A. P.; Jasinski, J. P.; Zhang, W. *RSC Adv.* **2015**, *5*, 71071.
4. de, G. R. M.; Ruiz-Olalla, A.; Bello, T.; de, C. A.; Cossio, F. P.; de, C. A. *Angew. Chem. Int. Ed.* **2018**, *57*, 668.
5. He, X.-L.; Zhao, H.-R.; Song, X.; Jiang, W.; Chen, Y.-C. *ACS Catal.* **2019**, *9*, 4374.
6. Dou, X.; Han, X.; Lu, Y. *Chem. - Eur. J.* **2012**, *18*, 85.
7. Dou, X.; Zhong, F.; Lu, Y. *Chem. - Eur. J.* **2012**, *18*, 13945.
8. Feng, J.; Lin, L.; Yu, K.; Liu, X.; Feng, X. *Adv. Synth. Catal.* **2015**, *357*, 1305.
9. Mo, Y.; Liu, S.; Liu, Y.; Ye, L.; Shi, Z.; Zhao, Z.; Li, X. *Chem. Commun.* **2019**, *55*, 6285.
10. Becerra, D.; Raimondi, W.; Dauzonne, D.; Constantieux, T.; Bonne, D.; Rodriguez, J. *Synthesis* **2017**, *2017*, 195.
11. Liu, W.; Lai, X.; Zha, G.; Xu, Y.; Sun, P.; Xia, T.; Shen, Y. *Org. Biomol. Chem.* **2016**, *14*, 3603.
12. Mei, R.-Q.; Xu, X.-Y.; Peng, L.; Wang, F.; Tian, F.; Wang, L.-X. *Org. Biomol. Chem.* **2013**, *11*, 1286.
13. Hahn, R.; Raabe, G.; Enders, D. *Org. Lett.* **2014**, *16*, 3636.
14. Kaya, U.; Chauhan, P.; Deckers, K.; Puttreddy, R.; Rissanen, K.; Raabe, G.; Enders, D. *Synthesis* **2016**, *2016*, 3207.
15. Xie, X.; Peng, C.; Leng, H.-J.; Wang, B.; Tang, Z.-W.; Huang, W. *Synlett* **2014**, *2014*, 143.
16. Parella, R.; Jakkampudi, S.; Arman, H.; Zhao, J. C. G. *Adv. Synth. Catal.* **2019**, *361*, 208.
17. Tan, B.; Zhu, D.; Zhang, L.; Chua, P. J.; Zeng, X.; Zhong, G. *Chem. - Eur. J.* **2010**, *16*, 3842.
18. Moczulski, M.; Drelich, P.; Albrecht, L. *Org. Biomol. Chem.* **2018**, *16*, 376.
19. Zhang, J.; Yin, G.; Du, Y.; Yang, Z.; Li, Y.; Chen, L. *J. Org. Chem.* **2017**, *82*, 13594.
20. Ran, G.-Y.; Wang, P.; Du, W.; Chen, Y.-C. *Org. Chem. Front.* **2016**, *3*, 861.
21. Li, Y.-M.; Li, X.; Peng, F.-Z.; Li, Z.-Q.; Wu, S.-T.; Sun, Z.-W.; Zhang, H.-B.; Shao, Z.-H. *Org. Lett.* **2011**, *13*, 6200.
22. Zuo, X.; Liu, X.-L.; Wang, J.-X.; Yao, Y.-M.; Zhou, Y.-Y.; Wei, Q.-D.; Gong, Y.; Zhou, Y. *J. Org. Chem.* **2019**, *84*, 6679.
23. Chaudhari, P. D.; Hong, B.-C.; Wen, C.-L.; Lee, G.-H. *ACS Omega* **2019**, *4*, 655.
24. Wu, M.-Y.; He, W.-W.; Liu, X.-Y.; Tan, B. *Angew. Chem. Int. Ed.* **2015**, *54*, 9409.
25. Tan, F.; Lu, L.-Q.; Yang, Q.-Q.; Guo, W.; Bian, Q.; Chen, J.-R.; Xiao, W.-J. *Chem. - Eur. J.* **2014**, *20*, 3415.
26. Sun, Q.-S.; Chen, X.-Y.; Zhu, H.; Lin, H.; Sun, X.-W.; Lin, G.-Q. *Org. Chem. Front.* **2015**, *2*, 110.
27. Fu, Z.-K.; Pan, J.-Y.; Xu, D.-C.; Xie, J.-W. *RSC Adv.* **2014**, *4*, 51548.

Index

abietic-acid 145
acetylacetone 7, 8
acetylene 178
p-acetyl-α-methyl styrene 227
aci-nitro compound 141
acrolein 240, 250, 258
acyclic α-branched enone 101
β-acylamino nitroalkene 229
Adrio, J. 224
aflatoxin 266
AgOAc/ThioClickFerrophos 115
Akiyama, T. 158
Aleman, J. 75, 114, 134. 224, 273
Alexakis, A. 4, 73–75
alginate gel 60
alkenylpyrazolone adducts 132
alkoxymethyl ynones 102
alkylation of aldehyde 238
alkylidenecyclopropane 65
N-alkylpyridinium 142
β-alkynyl ester 24
all-carbon quaternary stereocenter 94, 129, 148
Alonso, D. A. 4, 5, 40, 73
Alonso, R. 254
α-amination 62
amino-acetal 176
amino alcohol 29, 137, 178, 208, 257
β-amino alcohol 176
amino amide organocatalyst 29
2-aminobenzimidazole catalyst 17
5-amino-3-(2-pyridyl)-1H-pyrazole 230
2-amino-DMAP-squaramide 10
Andrés, J. M. 41, 134, 273
anthrone 104
anti-enamine 45, 53, 55, 56, 60, 85
anti-α-imino-γ-nitrobutyrate 115
Antonchick, A. P. 224
Arai, M. A. 158
Arai, T. 158, 159, 224, 273
aromatic thiol 191
arylacetyl phosphonate 109
arylaminophosphonium barfate 195
aryl boronic 169, 170, 172
aryl-fused heterocycles 237, 257
arylidene indanediones 250
aryl ketones 77, 91, 107, 112
aryl methyl ketone 78
aryl-2-nitroprop-2-enols 284
2-aryl-3-nitropropionamides 169
aryl-7-oxo-hept-5-enals 284
aryloxyacetaldehyde 61
N-arylpyrazolinone 270
aryl thiol 257
axially chiral 17, 102, 119, 185, 192, 277
aza-allyl anion 205
azaarylacetates 115
aza-Henry reaction 276
azanorbornane-based amino alcohol 29
azepine 60
azepinoindole 262
Azerad, R. 4
azetidinyl alkyl halide 124

Bacillus subtilis 235
Baker's yeast 227, 233
Bakó, P. 40
Ballini, R. 4
Ban, S. 113, 134, 135
Barbas, C. F. 73, 74, 254
barbiturate 286
barbituric acid 37
Barnes, D. M. 40
Barrett, A. G. M. 4
Barros, M. T. 74
Bartlett, M. J. 2
Bartoli, G. 5
Barton-Zard reaction 277
Basceken, S. 75
Benaglia, M. 73
benzenesulfinate 262
benzisoxazoles 117
benzofuranone 281
benzophenone imine 192
benzothiazoles 117
benzoxazole ligand 230
benzylic ynones 102
Berkessel, A. 3
Bernardi, L. 75
bicyclic γ-amino acids 24
bicyclic skeleton 138
bicyclo[3.3.0]diene 170
bicyclo[2.2.1]heptane-1-carboxylate 217
bicyclo[3.2.1]-octan-2-ones 247
bidentate activation 192
Billault, I. 40
BINAP (2,2′-bis(diphenylphosphino)-1,1′-binaphthyl) 170, 227
binaphthylsulfonimide 90
BINOL (1,1′-Binaphthalene-2,2′-diol) 176
bioreduction 227, 233, 235
biphenanthrylbis (oxazoline) 145
bipyridine-Zn $(OTf)_2$ 139
bis-alkaloid 19
bis-imidazolidine 257
bis(imidazolidine) iodobenzene 258
bis-(imidazolinyl) phenyl (Phebim) 144
Bis-phosphine-thiourea 229
bis(phenylthio) propan-2-one 97
bis(imidazolidine) pyridine (PyBidine) 148
bis(imidazolidine) pyridine-Cu complex 204
bis(phosphine)-thiourea 231
Bis-thiourea 156
N-Boc-1-amino acid 176
N-Boc-3-fluorooxindole 124
Boc-L-prolinol 80
N-Boc-protected oxindole 123

Bolm, C. 39, 74
Bommarius, A. S. 236
Bonne, D. 254
BOPA (bis(oxazolinylphenyl)amide) 220
Bornscheuer, U. T. 4
Brenna, E. 236
Breslow intermediate 268
Brière, J.-F. 41
bromonitroalkenes 258
Brønsted acid 137, 158, 183, 223
Brønsted base activation 122
Broughton, H. B. 74
Bugaut, X. 254
γ-butenolide 119
γ-butyrolactam 130

calcium enolate 20
calcium-pybox 20
camphor scaffold 54
camphynyl derivative 238
Carmona, D. 159
Carreira, E. M. 235
Carretero, J. C. 224
Carrillo, L. 73, 224, 225
chalcone 265
Chandrasekhar, S. 113
Chen, J.-R. 75, 113, 199, 224, 293
Chen, K. 74
Chen, L. 113
Chen, W. 135
Chen, W.-Y. 40
Chen, Y.-C. 135
Chen, Y.-J. 158
Cheng, H. 223, 224, 273
Cheng, J.-P. 73
Chi, Y. R. 273
chiral calcium enolate 20
chiral isoxazol-5(4H)-one 117
chiral nickel(II)-diamine catalyst 24
chiral phosphoric acid 141, 149, 150
chiral P,N-ligand 180
chiral pocket 11
chiral squaramide-amine 11, 290
chiral thiophosphinamide-1,2-diphenylethane-1,2-diamine 102
chromane 22, 205, 258, 271
chroman-2-one 259
Chung, L. 235
Clarke, M. L. 74
Clostridium sporogenes 235
Cobb, A. J. A. 73
Connon, S. 52
Constantieux, T. 5, 154, 254, 293
contiguous stereocenters 213, 218, 221, 238, 240, 243, 247, 250, 287
copper-amidophosphane 178, 180
copper-bis-(1-phenylethyl)amine 174
copper(I) perchlorate 203
copper-phosphane complex 227
Cordova, A. 74
Corma, A. 73
Co_2-Schiff base complex 26
Cossio, F. P. 223
Couty, F. 41
Cozzi, F. 73
Cp/Rh(*R*)-prophos 156
crown ether 21
cryptophycins 176
C_2 symmetric 108
C_3-symmetric 10, 111
$Cu(OTf)_2$/bis(sulfonamide)-diamine 139
Cu(II)-bound enolate 29
Cu/Eu/Cu (Trimetallic complex of copper/Europium/copper) 141
Cu-hydroxo complex 108
CuI/Si-FOXAP 201
Cu/Mn (Copper-manganese heterobimetals) 124
curcumins 246
cyanosilylation 167
cyclohexene-carbaldehydes 240
cyclopentanecarbaldehyde 213
Czekelius, C. 40, 235

Dalko, P. I. 3
Dalpozzo, R. 5, 158
dearomative 213
dehydroabietic amine 79, 105
dehydroxylation 259
demethylated quinine salt 17
Demir, A. S. 74, 75
dendritic catalyst 93
Deng, J. 235
Deng, J.-G. 181
Deng, W.-P. 224
Deng, X. 223
Denmark, S. E. 5, 223
Deschamp, J. 74
α,γ-dialkylation 252
diastereodivergent synthesis 219
α,α-dicyanovinyl group 173
dienamine intermediate 252
diethyl malonate 14, 19, 21
Diez, D. 74
2,3-dihydrofuran 278, 280, 281
3,3-dimethylbutyraldehyde 46
Ding, C.-H. 134
dinuclear nickel-Schiff base 109
dinuclear zinc bis-prophenol 152
dioxindole 126
di(methylimidazole)prolinol silyl ether 47
α,α-disubstituted aldehydes 44, 46, 47, 53, 55, 62
3,3-disubstituted 3,4-dihydro-2-quinolones 30
β,β-disubstituted nitroalkanes 170
α,β-disubstituted nitroalkene 66
divinylbenzene 262
Dixon, D. J. 40, 41, 224
α-D-mannopyranoside 21
Dong, C. 3, 39, 41, 114
Dong, X.-Q. 40, 180, 235
Duanphos 231
Duan, W.-L. 199
Du, D.-M. 3, 41, 113, 134, 158, 159, 180, 181, 199, 255, 273
Du, W. 293

electron-rich arenes 158, 239
Ellman, J. A. 41, 135, 199
Enders, D. 2, 4, 5, 73, 114, 134, 199, 223, 224, 254, 255, 273, 293

ene-reductase 235
enoate reductase 235
enolizable ketones 112
enyne tethered nitroalkenes 262
ephedrine 7, 27
epi-quinine 60, 119
Evans, D. A. 40

Faber, K. 236
Fang, X. 114
Feist-Benary reaction 278
Feng, X. 293
Fe_3O_4/PVP@SiO_2/ProTMS 50
Fe_3O_4@SiO_2 48
Feringa, B. L. 3
Fernandez, R. 180, 199
Ferreira, A. M. 4
fluorinated amines 165
fluorinated cyclohexanols 240
fluorinated pyrrolidines 124
fluorinated quaternary center 13
fluoroalkylation 141
2-fluoro-1,3-diketone 13
fluoromalonate 16, 17, 20
2-fluoro-2-(2-nitro-1-arylethyl) malonate 16
β-fluoro-β-proline 124
Fochi, M. 75, 236
formal [3+3] annulation 261
formal [3+3] cycloaddition 245
formaldehyde N-tert-butyl hydrazone 192
o-formyl-β-nitroalkene 250
γ-formyl nitro compound 43, 64
FOXAP 201
Fraile, A. 134, 224
Franz, A. K. 74
Fréchet-type dendrimer 137
Frost, C. G. 74
Fu, B. 114
Fujioka, H. 41
Fukuzawa, S. 134
functionalized γ-butenolides 119
functionalized cyclohexane carboxylates 242
functionalized 2,3-dihydrofurans 278
2-(*5H*)-furanone 121
2-(3*H*)-furanones 121
3(2*H*)-furanones 278

GABA (γ-aminobutyric acid) 25
Gandelman, M. 40
Gao, L. 113
Gao, L.-X. 74
Garcia, H. 73
Ge, C. 254
gem-benzoylnitrostyrenes 280
Genc, H. N. 75, 114
Ghosh, S. K. 40, 74, 254
Giacalone, F. 2, 4
Glorius, F. 2
α-D-glucopyranoside 21
glycine ketimine 115, 208
Goldfuss, B. J. 41
Gong, J.-F. 114, 159
Gong, L. 158, 159, 235
Gong, L.-Z. 74, 223, 224
Gong, Y. 74, 293
Gong, Y.-F. 224
Gonzalez, J. 5
Govender, T. 40
Graboski, G. G. 4
Greck, C. 75
Groerger, H. 3
Gruttadauria, M. 2, 4
guanidine-amide 243
guanine unit 17
Gu, F.-L. 273
Guiry, P. J. 158
Guo, H. 181
Guo, Q.-X. 69
Gutnov, A. 180
Gutteridge, C. E. 5

Hajra, A. 3
α-halogenated ketones 278
Hammett substituent constant 173
Hantzsch ester 227
Hashimoto, T. 5, 223
Hatanaka, Y. 134
Hayashi, T. 3, 4, 74, 181
Hayashi, Y. 5, 74, 224, 254, 255
Headley, A. D. 74, 113, 254
Henry cyclization 248
Heravi, M. M. 39
Herrera, R. P. 86
β-heteroaryl-nitro adduct 148
α-heteroatom ketone 99
heterocycle-bearing ketones 79
heterotrimetallic Pd-Sm-Pd catalyst 152
Hexahydropyrrolo[2,3-b]indole 60
Hirose, T. 75
homodiphenylprolinol methyl ether 81
Hong, B.-C. 224
Hou, C.-J. 181
Hou, X.-L. 223
Houk, K. N. 225
Hoveyda, A. H. 181
H-Pro-Pro-D-Gln-OH 66
Hu, X.-P. 181
Huang, H. 40, 73, 254
Huang, W. 158, 293
Huang, X. 254, 293
Huang, Y. 135
Huang, Z.-Z. 74
hydrazide 183, 192
α-hydrazino aldehyde 62
hydrazones 183
hydroarsination 197
hydrocarbazole 221
hydrogen cyanide 166
hydrogen/hydride sources 227
hydroperoxides 183
hydrophobic hydration 8, 25
2-hydroxychalcones 268
hydroxyindole 152
α-hydroxyketone 101
hydroxylation 196
α-hydroxymethylstyrene 222
2-Hydroxy-1,4-naphthoquinone 33

Iklariya, T. 3
Imamoto, T. 199
imidazoline-oxazolone 94
imidazolium-aryloxide betaine 28
imides 183
iminoesters 203
α-imino γ-lactone 201
immobilized squaramide 11
indanol 142
indole-3-carboxaldehyde 67
indoline-3-one 250
indolyl-nitroalkene 172
Islam, M. S. 159
isobutyraldehyde 45
isocyano substrate 277
isomannide 141
isoquinolines 117
isoxazol-5(4H)-one 117
isoxazolones 115
Itsuno, S. 41

Jacobsen, E. N. 74
Jew, S.-s. 40
JH-CPP (Jorgensen-Hayashi chiral porous polymer) 50
Ji, J. 40
Ji, Y. 18
Jia, Y.-X. 159
Johnson, J. S. 113
Jørgensen, K. A. 134
Josiphos 227, 228, 231
Juaristi, E. 113

Kabalka, G. W. 4
Kanger, T. 134
Kang, J. 181
Kang, Q. 114
Kass, S. R. 159
Kawai, Y. 236
Kazlauskas, R. 4
Kelleher, F. 75
Kesavan, V. 41
ketimineylide 115
Khan, N. H. 199
Kim, D. Y. 39
Kim, S.-G. 273
kinetic studies 52, 156
Kiliç, H. 159
Kleczkowska, E. 41
Kobayashi, S. 40
Kocovsky, P. 3
Kokotos, C. G. 41, 112
Kokotos, G. 112
Koskinen, A. M. P. 41
Kotsuki, H. 73
Kowalczyk, R. 199
Kundu, M. 113
Kwon, E. 41
KYE1 (Enoate reductase 1 from *Kluyveromyces lactis*) 234, 235

lactams 115, 122, 134, 167
γ-lactams 167
Lam, H. W. 113, 134
L-amino acid 77
Lassaletta, J. M. 180, 199
Lattanzi, A. 40, 199
Lecouvey, M. 74
Leung, P.-H. 199
Ley, S. V. 5
Li, G.-Y. 254
Li, Q. 113
Li, R. 113, 134, 135, 224
Li, X. 39, 40, 75, 159, 180, 224, 273, 293
Li, X.-S. 40, 224
Li, Y. 39, 40, 73, 114, 134, 199, 293
Li, Z. 114
Liang, Y.-M. 113
Liao, J. 181, 235
Lin, G.-Q. 40, 181, 224, 293
Lin, W. 49, 255
linear aldehydes 44
liquid carbon dioxide 14
List, B. 3, 73, 74, 236
Liu, D. 75, 254
Liu, G. 40
Liu, J.-T. 41
Liu, L. 158, 224
Liu, Q. Z. 40
Liu, X. 40, 41, 134, 180, 254, 273, 293
Liu, X.-W. 75
Liu, X.-Y. 73, 224, 293
Loh, T.-P. 75
Lombardo, M. 75
Lou, B. 74
Lu, G. 180
Lu, Y. 40, 41, 113, 135, 180, 223, 224, 273, 293
Luo, R.-S. 74
Luo, S. 73
Luo, X. 224
Lycopersicon esculentum 235

Ma, B. C. 74
Ma, D. 75
Ma, J.-A. 40, 113, 114, 134, 135
Ma, J.-T. 113
Ma, X. 74
magnetic nanoparticles 48
Maji, B. 75
Al Majid, A. M. A. 159
Malkov, A. V. 3
Mancuso, R. 5
Marques-Lopez, E. 158, 198, 199
Marson, C. M. 254
Maruoka, K. 3, 5, 180, 199, 223
Matsunaga, S. 41, 114, 134, 135
Meggers, E. 158, 159, 235
Melchiorre, P. 254
Merino, P. 75
2-methoxyfuran 153, 155
o-methylbenzoic acid 81
2-methylcyclopentane-1, 3-dione 247
methylene indane 250
methyleneindolinone 292
methyl indene 250
α-methyl prolinamide 61
p-methyl styrene 170
Minnaard, A. J. 3, 180, 181
Miura, T. 39, 75, 113

monoamine oxidase (MAO) inhibitor 149
Moorthy, J. N. 113
Moreau, X. 75
Moyano, A. 223
Mukherjee, S. 113
Müller, C. 159

Nagasawa, K. 159
Naicker, T. 4, 40, 41
Najera, C. 3, 5, 40, 73, 224
Nakamura, S. 180
Nakano, H. 41
Namboothiri, I. N. N. 4, 41, 198, 254
nanoporous polymer 50
2-(2-formylethyl)naphthalene-1,4-dione 250
β-naphtholenones 247
naphthylprolinol silyl ether 64
NCN pincer 142
Nef reaction 177
Ni, B. 254
nitro-activated allene 261
nitroalkadienes 56
nitroalkylated indole 141
nitroallylic acetate 245
2-nitroallylic acetates 245
3-nitro-2-arylpropanamide 152
1-nitrocyclohexene 170
nitroenynes 24, 73, 195
γ-nitro-α-fluorocarbonyl compound 27, 98
γ-nitro heteroaromatic ketone 79
γ-nitro-imidazolyl ketone 99
δ-nitroketones 77
nitro-Michael addition 77, 261
β-nitronitrile 166, 192
β-nitrophosphonates 183
nitroso-oxime tautomerization 283
β-nitro sulfides 188
North, M. 180
Nugent, T. C. 75

octahydro-2H-indol-2-one skeleton 276
octahydroindolecarboxylic acid 284
off-cycle resting state 47
Oh, K. 159, 224
Ohta, H. 236
Oiarbide, M. 113
Ojima, I. 4, 181
Old Yellow Enzymes 235
Ooi, T. 180, 199
Orru, R. V. A. 254
oxazolines 64
oxazolones 115
oxidative cleavage 152

Paixao, M. W. 41, 74
Palladium–diphosphine 169
Palmieri, A. 4
Palomo, C. 74, 113, 114
Pan, L. 273
Pan, S.C. 39, 41, 114
Pan, Y. 158
Pansare, S. V. 73
Papai, I. 40
Park, H.-g. 40
Park, Y. 40
Parra, A. 273
Patel, R. N. 4
Pd/Mn 124
Pd-Sm-Pd 152
Pedro, J. R. 135, 159
Pedrosa, R. 41, 134, 273
Pellissier, H. 3, 4, 73
Peng, C. 293
Peng, F. 39, 40, 293
Peng, Y. 75, 113, 199
Peng, Y.-G. 113
pentamethyl cyclopentadiene 215
perhydroindole 50, 60
perhydroindolic acid 60
Pericàs, M. A. 39, 41, 74
periselective Diels–Alder reaction 215
β-peroxidation 195
peroxides 183, 195
Peters, R. 41
L-phenylalanine lithium salt 55
N-phenylethylated azanorbornane 29
3-(aminomethyl)-5-phenylpentanoic acid 105
2-(tosylamino)phenyl-α,β-unsaturated esters 271
polyurethanes 104
porous polymer 84
Portnoy, M. 112
Pramanik, A. 113
L-prolinamide 55
L-proline-derived diamine 44
protodeargentation 271, 284
prototropic path 141
PyBidine-metal 257
pyranocoumarin 286
pyranonaphthoquinones 35
pyrazines 117
pyrazolinone 270, 271
pyrazol-5-one 128
pyrazolone acetate 131
Pyrazolones 15, 122, 128, 129, 132
pyrrolidinyl-camphor 87
pyrrolidinyl-sulfamide 87

Qin, Z. 114
quadruple cascade reaction 239
α-quaternary amino acid 117
quaternary ammonium bifluoride 161
quaternary ammonium ionic liquid 80
quinidine 19, 23, 94, 130, 183, 188, 284
quinine-squaramide 35
quinuclidine 16, 17, 120, 191, 250, 271, 277

Rachwalski, M. 75
Ramapanicker, R. 75
Ramaraj Ramanathan, C. 159
Rawal, V. H. 39
R-(-)-baclofen 21
Reddy, B. V. S. 41
Rh-aqua complex 141
Rh complex 170
Rh/Duanphos 231
rhodium/olefin-sulfoxide 169
Ricci, A. 4, 159
Rios, R. 5, 223

Rodriguez, J. 5, 75, 114, 254, 293
Rovis, T. 2, 75, 225
Rueping, M. 5, 224, 254

salicylaldehyde 207, 260–262
salicylic acid 92
Salunkhe, M. M. 73
Saluzzo, C. 159
Sansano, J. M. 5, 224
Schmatz, S. 74
Schmitzer, A. R. 113
Schore, N. E. 254
Schreiner, P. R. 2, 223
Scrutton, N. S. 236
Sebesta, R. 40, 41, 75
Seebach, D. 74
Seebach's model 93
Seidel, D. 158
Sekikawa, T. 134
SelectFluor 240
self-assembled organocatalyst 56, 57, 77
Sewald, N. 181
Shao, Z. 39, 40
Shao, Z.-H. 39, 293
Shen, Y. 135, 293
Shibasaki, M. 3, 41, 114, 134, 135
Shi, L. 236
Shi, M. 135, 158
silyl prolinol ether 47
Singh, K. N. 113
Singh, V. K. 37, 273
Sirit, A. 74
Sodeoka, M. 114
Sohtome, Y. 159
solvent-free reaction 47, 87
Song, C. E. 39
Song, M.-P. 114, 159
Song, Z. 39
Soós, T. 40
de Souza, R. O. M. A. 4
spacer length 52
spirocycles 250, 275
spirocyclohexane 287
spiro-nitroprolinate 201
spirooxindole-chromane 205
spiro-oxindoles 287
spiro-pyrazolone 247
Stephens, G. M. 236
stereodivergent synthetic route 117
Stetter reaction 43, 71, 73
o-substituted boronic acid 172
γ-substituted butenolides 121
α-substituted carboxylic acid 101
N-substituted diaminocyclohexane 108
α-substituted malonates 21
α-substituted β-nitroacrylate 97
α-substituted β-phosphine oxide 186
succinaldehyde 210
N-sulfinylurea 128
Sun, B.-F. 224
Sun, J. 254
Sun, X.-W. 293
superparamagnetic nanoparticle 48
syn-2-alkyl-3-(1H-indol-3-yl)-4-nitrobutanal 66
syn and anti diastereomer 61
synclinal transition state 45, 56, 83, 85, 238
synergistic activation pathway 231
syn-selectivity 56

Takabe, K. 73
Takemoto's catalyst 14, 37, 108, 115
Takemoto, Y. 3, 4, 39, 40, 223, 254
Takenaka, N. 2, 159, 224
Takeshi, O. 73
Tanaka, F. 73, 74
Tan, B. 39, 40, 113, 223, 224, 293
Tan, C.-H. 199
Tanchoux, N. 75
Tang, Y. 73, 113, 235, 254
Tanyeli, C. 39
Terada, M. 134
terephthalic acid 78
ternary complex 7, 262
α-tert-butylthio substituted furanone 119
L-tert-leucine 122
tetrahydronaphthalenes 286
tetrahydro-1,2-oxazine 285
tetrahydropyrans 283
tetrahydropyrimidine 165
tetrahydroquinolines 263
β-tetralone 104
Tetralone 104
thermodynamic (Z)-enolate 109
ThioClickFerrophos ligand 117
thiophenoxide 268
2H-thiopyrano-[2,3-b] quinolines 268
thioureido acid 92
thioxazolines 117
Thorarensen, A. 5, 223
threonine 59, 60, 127, 278
Tian, P. 40
Toma, S. 73
Tomioka, K. 181
Torres, R. R. 3
3,4-trans-disubstituted chromans 258
transfer hydrogenation 228, 232, 233, 235
trans-perhydroindolic acid 238
triazines 117
triazole linker 47
tricyclic chroman derivative 252
tridentate bis(oxazoline) 141
trimethylenemethane 209
trimethylsilyl cyanide 166
tripeptidic catalysts 51
triple-cascade fashion 244
2,3,4-trisubstituted piperidines 275
Trivedi, R. 39
Trost, B. M. 2, 5, 73, 134, 159, 224
Tsogoeva, S. B. 73

umpolung 1, 43, 71, 73, 166, 205
α-unsubstituted nitroalkene 104

Varma, R. S. 4
Vetticatt, M. J. 40
Vicario, J. L. 73, 224, 225
Vila, C. 135, 159
vinylogous Michael addition 119

vinylogous Michael/Henry cyclization 248
vinyl silver intermediate 271, 284

Waldmann, H. 224
Walphos 203
Wan, B. 159, 180
Wang, B. 224
Wang, C.-J. 40
Wang, D.-C. 224
Wang, J. 39, 40, 74, 113, 114, 134, 198, 199, 223, 273, 293
Wang, K. 114
Wang, L. 75, 113, 114, 159, 199, 235
Wang, L.-X. 74, 112, 293
Wang, R. 112, 114, 134, 224, 235, 273
Wang, S. 75, 113, 159, 225, 255
Wang, W. 39–41, 74, 75, 113, 198, 199, 223
Wang, X.-W. 75, 114
Wang, Y. 40, 41, 75, 112–114, 159, 199, 224, 225, 235, 254, 273
Wang, Z. 75, 114, 135, 159, 181
Waser, M. 3
Wei, Y. 75, 113, 224
Wennemers, H. 74, 75
Wolf, C. 134
Wong, C. T. 254
Wu, F. 40, 114, 159
Wu, W. 40, 224, 225, 235
Wu, X.-Y. 75, 114
Wulff, W. D. 180

Xia, C. 40, 254
Xiao, J.-C. 159
Xiao, W. 254
Xiao, W.-J. 73, 75, 113, 199, 224, 225, 273, 293
Xie, B. 74
Xie, J.-W. 40, 224, 273, 293
Xu, D.-Q. 113, 223, 224, 273
Xu, L.-W. 41
Xu, M.-H. 40, 181
Xu, P.-F. 224, 225, 254, 273
Xu, X.-Y. 74, 112, 114, 224, 225, 293
Xu, Z.-H. 4
Xu, Z.-Y. 113, 223, 273

Yamanaka, M. 158
Yan, M. 40, 75, 112, 113
Yan, W. 114, 224, 273
Yan, X. 113, 223
Yang, D. 114
Yang, W.-L. 224, 273
ynones 102
Yoshida, M. 74, 75
You, S.-L. 2, 4, 159, 255
You, T. 159
Yuan, W.-C. 114, 135, 181, 199, 224, 225
Yu, C. 199
Yu, X. 74, 114

Zeitler, K. 225
Z-enamine 64, 77
Zeng, M. 2
Zeng, X. 40, 113, 159, 224, 254, 293
Zhang, G. 40, 75, 159
Zhang, S.-Q. 113
Zhang, W. 75, 199, 224, 254, 293
Zhang, X. 39, 75, 235, 293
Zhang, X.-M. 114, 135, 181, 199, 224, 225
Zhang, Y. 9, 40, 74, 112–114, 199
Zhao, C.-G. 112, 198, 199, 224, 254
Zhao, G. 73, 114
Zhao, J. 181, 255
Zhao, J. C. G. 255, 273, 293
Zheng, L.-Y. 113
Zhong, G. 39, 40, 113, 223, 224, 293
Zhou, E. 41
Zhou, J. 134, 254
Zhou, Q.-L. 159
Zhou, Y. 59
Zhou, Z.-M. 41
zwitterion intermediate 47

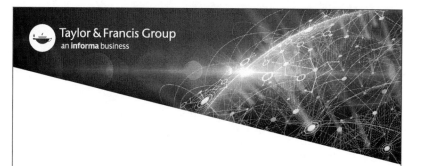

9780367535629